# Plant Secondary Metabolites

Anil Kumar Sharma • Ajay Sharma
Editors

# Plant Secondary Metabolites

## Physico-Chemical Properties and Therapeutic Applications

Springer

*Editors*
Anil Kumar Sharma
Department of Biotechnology
Maharishi Markandeshwar
(Deemed to be University)
Mullana, Ambala, Haryana, India

Ajay Sharma
Department of Chemistry
Career Point University
Hamirpur, Himachal Pradesh, India

ISBN 978-981-16-4778-9 ISBN 978-981-16-4779-6 (eBook)
https://doi.org/10.1007/978-981-16-4779-6



This Springer imprint is published by the registered company Springer Nature Singapore Pte Ltd.
The registered company address is: 152 Beach Road, #21-01/04 Gateway East, Singapore 189721, Singapore

# Preface

Plant secondary metabolites (PSM) have been the mainstay of broad-spectrum therapeutic implications as plants as they are the rich reservoirs of candidate compounds which could be further explored for drug discovery. Extensive efforts have been devoted during the past decade or so to explore the potential of such compounds for therapeutic use. Research and development of chemotherapeutic drugs derived from plants have led to the identification of a variety of candidate molecules that have therapeutic roles facilitated through a variety of mechanisms. These plant-based natural product metabolites have been known to display a variety of pharmacological effects which have been discussed in the context of their invaluable bioactivity and multifaceted potential. The book has included a diversity of chapters covering the physico-chemical and biochemical aspects of PSM along with the chemistry, therapeutic mechanisms, and future relevance of such compounds. Moreover, the book also covers various sources of PSM and the metabolite determination through various analytical techniques. Further description of the potential applications of PSM as anticancer and chemopreventive agents, their role as cosmetic ingredients, and activity in skin cancer therapy have also been the highlights of the book. Some of the chapters have focused on the plethora of roles of PSM, including those as antivirals, antibacterial, anti-inflammatory drugs, and cardioprotective agents as well. The book culminates with chapters on the impact of certain PSM on plant defence and human healthcare.

With the advent of newer technologies such as combinatorial chemistry, high-throughput screening, and next generation sequencing, it is now possible to consider that PSM could sound the death knell for a variety of dreadful diseases including cancer. Moreover, PSM could provide novel lead molecules which would be used as templates for restructuring them for potential therapeutic drug candidates with enhanced bioactive properties. Despite the increasing interest in natural products research, to our knowledge this area still requires the attention of the scientific community to explore the wide-scale therapeutic mechanisms with PSM being the lead compound further redressing the growing problem of drug resistance as well. In order to fill these gaps and what kind of therapeutic roles PSM play in the treatment and management of diseases, this book titled *Plant Secondary Metabolites: Physico-Chemical Properties and Therapeutic Applications* has been able to successfully address the remarkable therapeutic potential of bioactive secondary metabolites.

Once again, my sincere thanks to all the contributing authors who worked as a team to let us complete this book. Special thanks and appreciation to Dr. Madhurima Kahali, publishing editor, who was available all the time to impart her valuable inputs and assistance. Words of appreciation also go to Mr. Lenold Esithor and the whole production team.

Mullana, Ambala, Haryana, India Anil Kumar Sharma
Hamirpur, Himachal Pradesh, India Ajay Sharma

# Contents

# About the Editors

**Anil Kumar Sharma** is a Professor and Head of the Department of Biotechnology at M.M. (Deemed to be University), Mullana (Ambala), Haryana, India, with a vast research experience of working as a Post-Doctoral Scientist (2003–2010) in the Micro and Immunology Department at UIC, Chicago, IL, USA. He worked in diverse scientific fields in the area of microbial, medical, and environmental biotechnology. He has published 165 articles in peer-reviewed, high-impact journals of international repute. He has more than 4920 citations [H-Index ~33; i10 index: 71; Cumulative IF ~336]. He has published 06 books and 30 chapters with Springer and other well-reputed publishers. He has been felicitated with many awards for scientific excellence during his career such as International Adroit Scientist Award-2021, Abdul Kalam Azad Award-2018, BRICPL Eminent Scientist Award (2017 and 2018), Achiever Award-2017, Appreciation Award-2008 from UIC Illinois at Chicago, USA, MGIMS Young Scientist Award-2000, VCS Memorial Young Scientist Award-1999, etc. He has been enlisted in top 2% scientists working in the life sciences field as per the list recently released by Stanford University, California, during 2019 and 2020 and has been on the editorial board of many international and national journals.

**Ajay Sharma** is currently an Assistant Professor in the Department of Chemistry, Career Point University, Tikker-kharwarian, Hamirpur, HP, India, while previously at Maharishi Markandeshwar University, Mullana, Ambala (2014–2015). He also served as a teaching associate in the School of Studies in Chemicals Sales and Marketing Management, Jiwaji University, Gwalior, MP, India (2008–2010). His research focused on the development in the synthesis of heterocyclic organic compounds of different classes and natural metabolites through greener and conventional methods having pharmacological activity. He has also published more than 26 papers in peer-reviewed international and national journals and also authored/coauthored book chapters from reputed publishers. He is a member of many international scientific societies and organizations as well.

# Plant Secondary Metabolites: An Introduction of Their Chemistry and Biological Significance with Physicochemical Aspect

1

Ajay Sharma, Saurabh Sharma, Anil Kumar, Vinod Kumar, and Anil Kumar Sharma

**Abstract**

Nature is the immense source of chemical compounds or substances usually acknowledged as natural products. Plants are the considerable source of an inestimable diversity of natural products with manifold chemical structures and functionalities. These natural compounds have various healthcare benefits and are commonly termed as the primary metabolites (PPMs) and secondary metabolites (PSMs). Biogenically, there are basically four main classes of secondary metabolites: phenolics and polyphenolics, terpenes, nitrogen-containing alkaloids, and sulfur-containing compounds. The functional groups and other structural features present in a molecule are accountable for delivering a variety of characteristics including therapeutic applications and the solubility and stability of the molecules. The structural features of compounds useful to understand the relations between their molecular structures and biological or pharmacological activities are attributed to being responsible for arousing a target biological effect in the organism. This chapter will delineate the chemistry, classification, physicochemical properties, and numerous biological activities of plant secondary metabolites. In order to develop the new derivatives of therapeutic value, the

A. Sharma · S. Sharma
Department of Chemistry, Career Point University, Tikker-Kharwarian, Hamirpur, Himachal Pradesh, India

A. Kumar
Department of Microbiology, DAV University, Jalandhar, Punjab, India

V. Kumar
Department of Chemistry, Central University of Haryana, Mahendragarh, Haryana, India

A. K. Sharma (✉)
Department of Biotechnology, Maharishi Markandeshwar (Deemed to be University) Mullana, Ambala, Haryana, India





A. K. Sharma, A. Sharma (eds.), *Plant Secondary Metabolites*,
https://doi.org/10.1007/978-981-16-4779-6_1

present chapter would be useful for medicinal chemists to design and introduce new chemical groups into these biomedical compounds and to explore their biological effects.

**Keywords**

Natural products · Plant secondary metabolites · Therapeutic applications · Structure-function relationship · Biological effects · Pharmacological properties

## 1.1 Introduction

Nature is an essential and the best source for life that contributes toward the physical well-being along with the better emotional feelings of an individual. It provides enormous health benefits due to the availability of foods such as maize, wheat, rice, pulses, spices, vegetables, fruits, beverages, herbs, and other crops from plants. Plants are able to use simple, inorganic precursors to produce a larger chemically diverse classes of organic compounds. These organic compounds are generally divided into plant primary metabolites (PPMs) and plant secondary metabolites (PSMs) on the basis of their perspective functions as well as commonly nutritive and nonnutritive components of plant foods. These have been widely used in human nutrition and medicine from ancient times. PPMs such as acyl lipids phytosterols, proteins, nucleotides, organic acids, sugar, etc. are necessary for plant survival along with cellular functions and play a pivotal role in the plant life cycle such as photosynthesis, respiration, development, and reproduction processes. While the secondary metabolites were earlier regarded as waste products because of their adaptive therapeutic significance not known earlier, nowadays, their numerous functions in plants and human health along with therapeutic roles have been evaluated and explored as a source of natural drugs, allelopathic agents, anticancer agents, antibiotics, flavoring agent, perfuming agent, pigments, signal molecules, UV protectants, etc. (Takshak and Agrawal 2019; Crozier et al. 2006; Anwar et al. 2019; Erb and Kliebenstein 2020). Besides, PSMs bestow the various functions and benefits to the plants such as to attract the pollinators and seed dispersers, help in the recovery from injury, protect from ultraviolet radiation, and enhance the defense against pathogens, diseases, and herbivores. PSMs have been the source of lead compounds for the development of drugs and alternative therapeutics to treat/prevent the broad spectrum of human health problems and diseases such as neurodegenerative diseases, cancer, inflammation, infectious and blood vessel diseases, etc. Over the last few decades, plant secondary metabolites attracted the attention of researchers, scientists, academicians, and industrialists, owing to their multifaceted role and contributions in plants (specific odors, color, and taste as well defense system) and in human nutrition along with the commercial importance as cosmetics, drugs, dyes, polymers, herbicides, insecticides, etc. (Korkina et al. 2018; Alamgir 2017; Sato 2014; Rattan 2010; Ibanez et al. 2012; Ntalli et al. 2019; Gorlenko et al. 2020; Crozier et al. 2006; Anwar et al. 2019). The plant secondary metabolites are

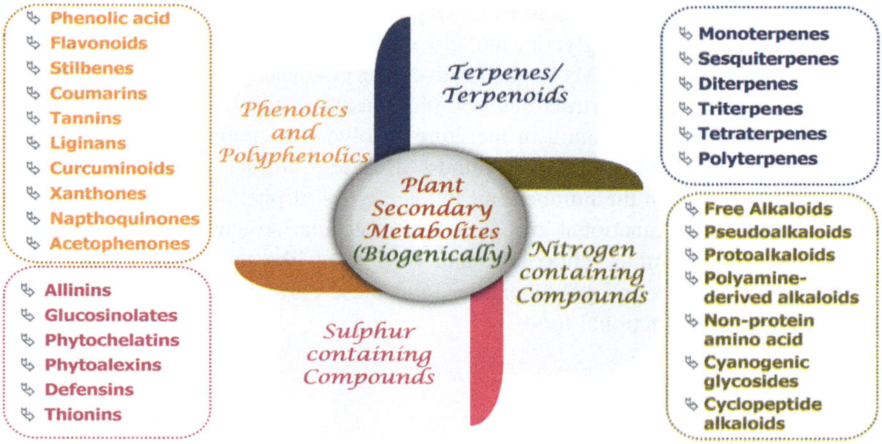


**Fig. 1.1** The general classification of plant secondary metabolites into four major categories and further subclasses on the basis of their biosynthetic routes and common structural features

biogenically divided into main four classes (phenolics and polyphenolics, terpenes, and nitrogen-containing (alkaloids) and sulfur-containing compounds (Fig. 1.1)) and further subdivided into subclasses on the basis of their molecular structure, substituent groups, functional groups, and linkage types with substitution pattern. These compounds are widely distributed in the plant realm and ubiquitous in plant-based foods. The infinite combination of structural features of PSMs such as complex and specifically arranged aromatic rings, chiral centers, and the number and the ratio of heteroatoms might be essential and responsible for their interaction with one or more compounds of living tissues and showing an effect on the human health as the starting point for drug discovery and novel synthetic drug developments. PSMs belong to the different subclasses of compounds on the basis of chemical structures, structural features, and compositions (Fig. 1.1) (Crozier et al. 2006; Wink 2016; Bottger et al. 2018; Wang et al. 2019; Anwar et al. 2019).

## 1.2 Phenolics and Polyphenolics

Phenolics and polyphenolics are omnipresent natural compounds in plant kingdom and plant-based foods. These have received humongous attention of consumers, biologists, nutritionists, chemists, and food scientists because of their role in the prevention of various diseases such as cardiovascular, neurodegenerative, degenerative diseases, and particularly cancers. This class of plant secondary metabolites (PSMs) is a highly diverse class of natural products and chemically typified with phenolic structural features. In general, these possess an aromatic ring with one hydroxyl group (OH) at least, and their structures diversified from simple to high-molecular-weight complex molecules as well as infinite combination of functional groups, namely, hydroxyl, alcohols, alkoxy, acetyl, aldehydes, alkyls, and benzyl rings, existing as either in aglycone and glucoside forms. The varied distribution and diversity of this class of PSMs in plants have been characterized on the basis of their source of origin, chemical structures, and biological functions. The most commonly used and adopted classification in literature involves two main groups: flavonoids and non-flavonoid phenolic compounds. Besides, further categorization into subgroups is based on the number and linkage types of phenolic units, the number of substituents, and functional groups within molecular structure (Fig. 1.2). Fruits, vegetables, whole grains, chocolate, and beverages like tea and wine are good sources of phenolics and polyphenolics, and these enriched foods are considered as being potential functional foods.

### 1.2.1 Flavonoids

Flavonoids constitute the prevalent (up to 60%) and meticulously investigated class of polyphenolics compounds on account of their impressive biological activities and therapeutic significance such as antioxidant, anticancer, antimicrobial, anti-inflammatory, immunomodulatory, vasodilating, prebiotic-like properties, etc. (González-Vallinas et al. 2013, 2012; Sharma et al. 2018, 2019; Batra and Sharma 2013). Flavonoid-rich foods are considered as being potential drugs and food supplements for the prevention and treatment of various diseases and human ailments (Hernandez-Rodríguez et al. 2019; Biharee et al. 2020; Rengasamy et al. 2019; Uddin et al. 2020; Maaliki et al. 2019; Boniface and Ferreira 2019). Various therapeutic roles and biological functions associated with this class of compounds have been owing to their structural features such as carbon skeleton, substitution pattern, oxidation level, etc. Flavonoids consist of the basic diphenylpropane ($C_6$-$C_3$-$C_6$) carbon skeleton in which two benzene rings (A and B) and a ring/chain of carbon atom system either as a heterocyclic ring C (pyran-4-one fused with ring A)/ pyrylium ion/$\alpha\beta$-unsaturated carbonyl system/furan-3-one ring (fused with ring A) with exo-double bond system. The chalcones also belong to the flavonoid class of compounds because of two benzene rings are attached by open $\alpha\beta$-unsaturated carbonyl system even if they have no ring C as pyran ring system. Besides, aurones family of flavonoids consist of the benzofuran ring connected with a benzylidene at position 2. Although the Pyridium ion as ring C is present in the class of anthocyanidin compounds. Flavonoids are categorized into subclasses, namely, anthocyanidins, aurones, catechins (flavanols), chalcone, flavanones (dihydroflavones), flavones, flavonols, flavanonol (dihydroflavonols), isoflavones, and neoflavone, on the basis of the oxidation level and substitution pattern of the ring C. However, the individual compounds of subclasses differ in the substitution pattern of rings A and B, that is, the position of hydroxyl and methoxy substituents (Yoshida et al. 2012; Corcoran et al. 2012; Sharma et al. 2019; Apetrei et al. 2016). The structural features of each subclass and few representative compounds have been mentioned in Table 1.1.

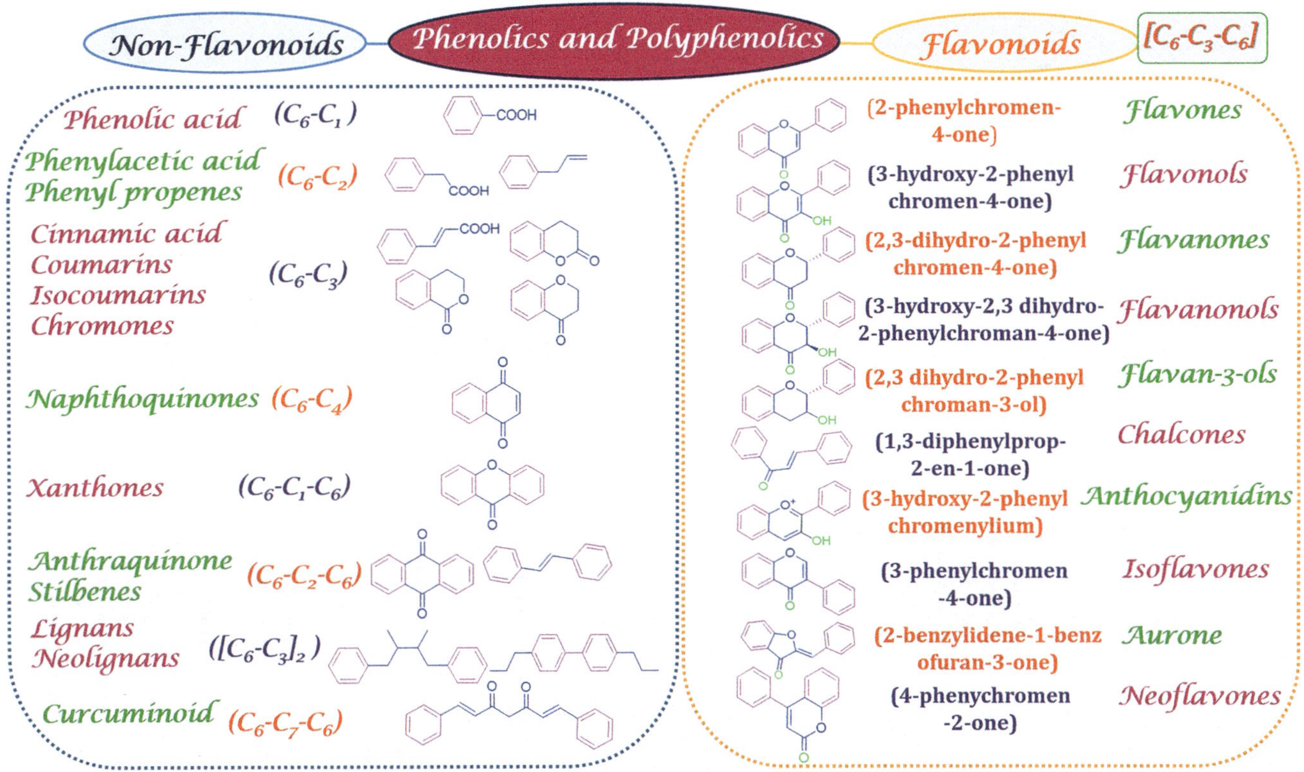


**Fig. 1.2** The most adopted classification of phenolics and polyphenolics: flavonoids and non-flavonoids. The different subclasses of flavonoids and non-flavonoids with their common nucleus carbon skeleton

**Table 1.1** Different subclasses of flavonoids with few representative compounds belonging to these subclasses with common structural features

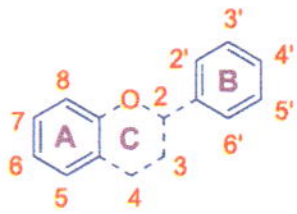


($C_6$-$C_3$-$C_6$) Diphenylpropane carbon skeleton

| Subclass name with structural features | Compound name and structure |
|---|---|
| **Flavones**<br>• $C_2 = C_3$ double bond and 4-oxo group on ring C<br>• Ring B linked at 2 position of ring C<br>• Ring B is either monohydroxylated (4′-OH), pyrocatechol (3′-, 4′-OH), resorcinol (3′-, 5′-OH), pyrogallol (3′-, 4′-, 5′-OH)<br>• Ring A is either monohydroxylated and *meta*-dihydroxylated (5- and 7-OH groups) | 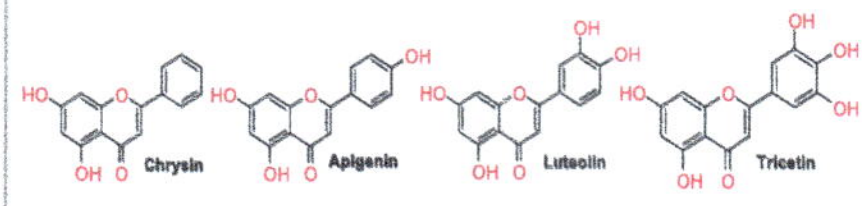  |
| **Flavonols**<br>• $C_2 = C_3$ double bond, 3-OH and 4-oxo group in ring C<br>• Ring B linked at 2 position of ring C<br>• Ring B is either monohydroxylated (4′-OH), pyrocatechol (3′-, 4′-OH), resorcinol (3′-, 5′-OH), pyrogallol (3′-, 4′-, 5′-OH)<br>• Ring A is either monohydroxylated and *meta*-dihydroxylated (5- and 7-OH groups) | 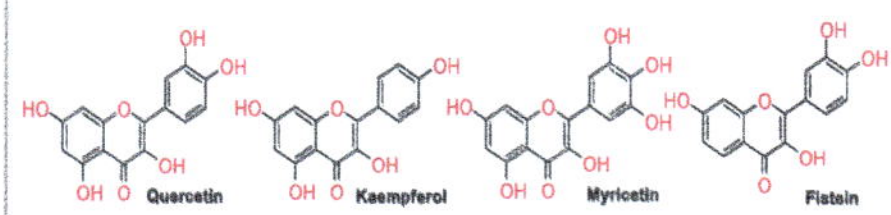  |
| **Flavanones (dihydroflavones)**<br>• $C_2$-$C_3$ single bond, $C_2$ is a center of asymmetry (epimeric center) and 4-oxo group in ring C<br>• Ring B linked at 2 position of ring C<br>• Ring B is either monohydroxylated (4′-OH), pyrocatechol (3′-, 4′-OH), resorcinol (3′-, 5′-OH), pyrogallol (3′-, 4′-, 5′-OH)<br>• Ring A is either monohydroxylated and *meta*-dihydroxylated (5- and 7-OH groups) | 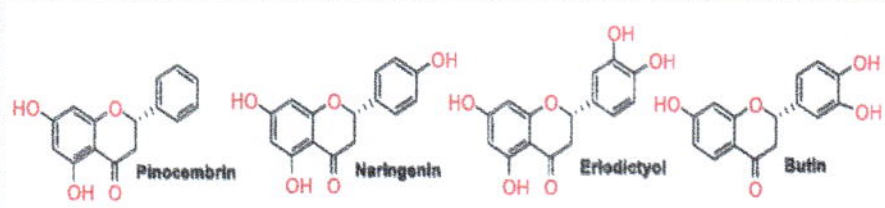  |
| **Flavanonol (dihydroflavonols)**<br>• $C_2$-$C_3$ single bond, chiral center at $C_2$ and $C_3$, 3-OH, and 4-oxo group on ring C<br>• Ring B linked at 2 position of ring C<br>• Ring B is either monohydroxylated (4′-OH), pyrocatechol (3′-, 4′-OH), resorcinol (3′-, 5′-OH), pyrogallol (3′-, 4′-, 5′-OH) | 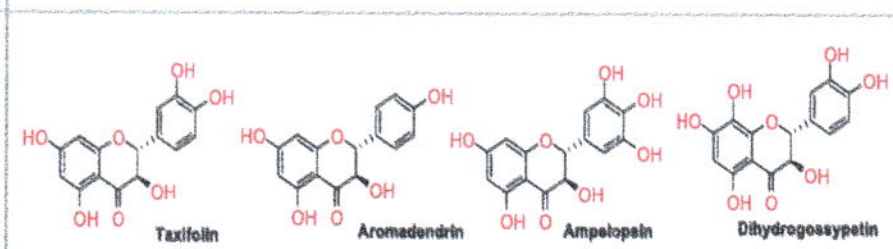  |

(continued)

**Table 1.1** (continued)

| | |
|---|---|
| • Ring A is either monohydroxylated, *meta*-dihydroxylated (5- and 7-OH groups), or trihydroxlated (5-, 7-, and 8-OH groups) | |
| **Flavanols (catechins)**<br>• $C_2$-$C_3$ single bond, 3-OH or 3-gallate ester, chiral center at $C_2$ and $C_3$, 4-oxo group on ring C<br>• Ring B linked at 2 position of ring C<br>• Ring B is either monohydroxylated (4′-OH), pyrocatechol (3′-, 4′-OH), resorcinol (3′-, 5′-OH), pyrogallol (3′-, 4′-, 5′-OH)<br>• Ring A is either monohydroxylated, *meta*-dihydroxylated (5- and 7-OH groups) | 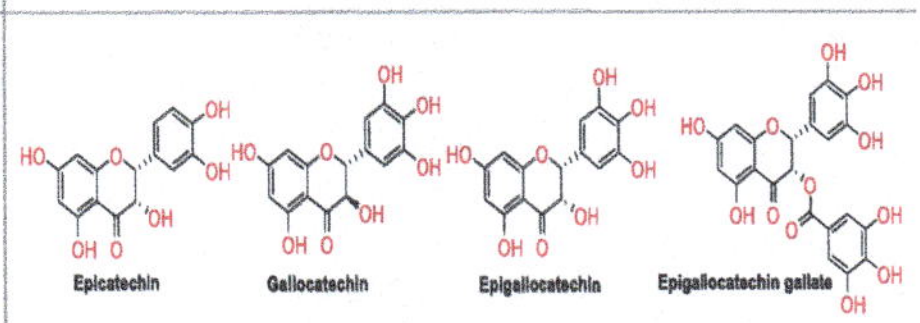 |
| **Isoflavones**<br>• $C_2 = C_3$ double bond and 4-oxo group on ring C<br>• Ring B linked at 3 position of ring C<br>• Ring B is either *mono*hydroxylated (4′-OH group), methoxy substituted (4′-$OCH_3$ group)<br>• Ring A is *mono*hydroxylated (7-OH groups), *meta*-dihydroxylated (5- and 7-OH groups). | 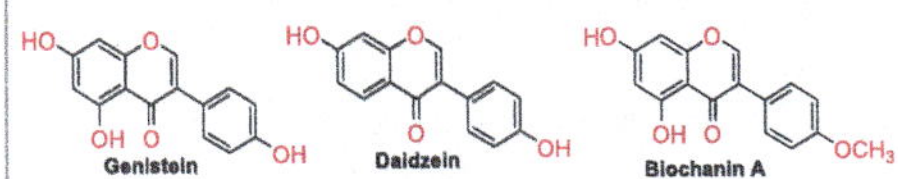 |
| **Anthocyanidins**<br>• Ring C as pyrylium ion with 3-OH group<br>• Ring B linked at 2 position of ring C<br>• Ring B is either monohydroxylated (4′-OH), pyrocatechol (3′-, 4′-OH), resorcinol (3′-, 5′-OH), pyrogallol (3′-, 4′-, 5′-OH)<br>• Ring A is either monohydroxylated or *meta*-dihydroxylated (5- and 7-OH groups) | 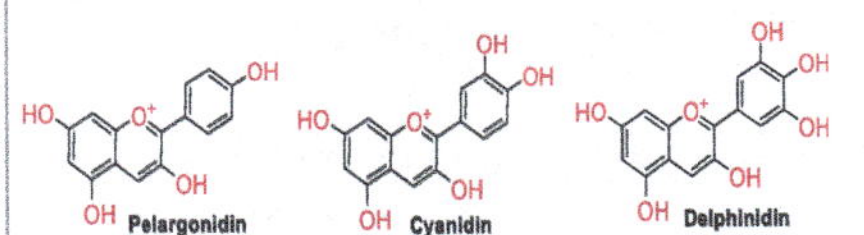 |
| **Aurones**<br>• Ring C is furan-3-one with methylene group at $C_2$<br>• Ring B is attached with methylene group.<br>• Ring B is either monohydroxylated (4′-OH), pyrocatechol (3′ -, 4′ -OH) or chloro-substituted (4′-Cl)<br>• Ring A is either monohydroxylated or methoxy substituted<br>• (*E*)- and (*Z*)-isomeric configurations | 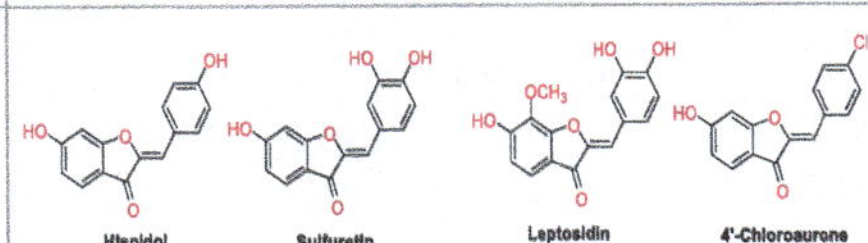 |
| **Neoflavonoids**<br>**Dalbergin type**<br>• 2-oxo group on ring C and either presence or absence of $C_3 = C_4$ double bond<br>• Ring B linked at 4-position of ring C<br>• Ring B is either *mono*hydroxylated (4′-OH group), pyrocatechol (3′-, 4′-OH), | |

(continued)

**Table 1.1** (continued)

| | |
|---|---|
| resorcinol (3′-, 5′-OH), pyrogallol (3′-, 4′-, 5′-OH), or methoxy substituted (4′-$OCH_3$ group)<br>• Ring A is *mono*hydroxylated (7-OH groups), *meta*-dihydroxylated (5- and 7-OH groups), methoxy substituted (7-$OCH_3$ groups)<br>**Latifolin type**<br>• Diphenyl allyl compounds – Open-chain neoflavonoids<br>• Rings A and B are either (i) both hydroxyl-/methoxy-substituted aromatic in nature; (ii) one is diketo substituted cyclohexene, and second is hydroxyl-/ methoxy-substituted aromatic | 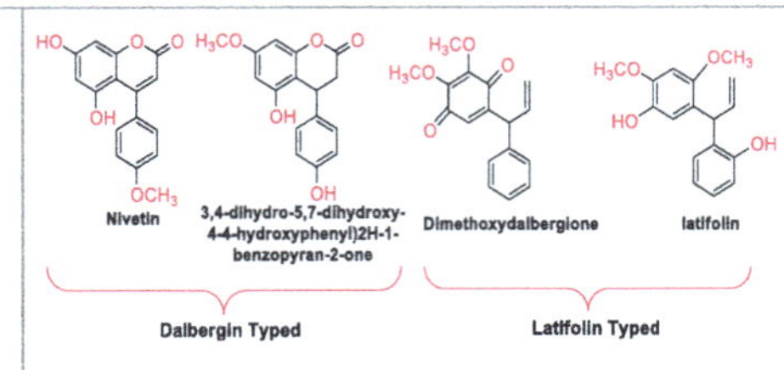 |
| **Chalcones**<br>• Three carbon atom system with $\alpha\beta$-unsaturated ketone functionality<br>• Both aromatic rings A and B are either unsubstituted, monohydroxylated, dihydroxylated, trihydroxylated, or methoxy substituted | 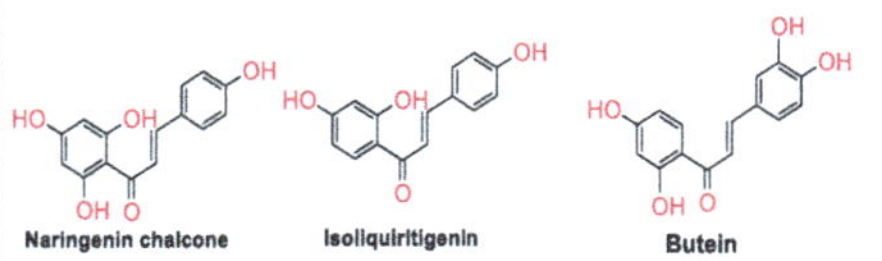 |

## 1.2.2 Non-flavonoids

The non-flavonoid compounds have slightly more variable structures and received considerable attention due to their intake related to healthcare benefits in order to lower the incidence of chronic degenerative diseases such as Alzheimer's disease, cancer, cardiovascular diseases, and diabetes. Phenolic compounds are responsible for the nutritional benefits and sensory properties of foods due to the hydroxylated aromatic rings structural features. The hydroxylated aromatic rings consists of the direct attachment of hydroxy group (-OH) to either the phenyl, substituted phenyl, or other aryl group, for example, phenolic acids, phenylacetic acids, phenylpropenes, and cinnamic acids (Van sumere 1989; Harborne 1989; Apetrei et al. 2016; Rentzsch et al. 2009; Pereira et al. 2009; Cheynier 2012; Kumar and Goel 2019). Phenolic acids such as hydroxybenzoic acids ($C_6$-$C_1$), hydroxyphenyl acetic acids ($C_6$-$C_2$), and hydroxycinnamic acids ($C_6$-$C_3$) are components of complex structures (like lignin and hydrolysable tannins) and are commonly present in bound form in foods (fruits, vegetables, grains, nuts, seeds, etc.). They are connected to cell wall structures such as cellulose, protein, sugar, and lignin through ester bonds. Hydroxybenzoic acids are the metabolites of hydroxycinnamic acid and flavonoids (Apetrei et al. 2016; Rentzsch et al. 2009; Pereira et al. 2009; Kumar and Goel 2019; Velderrain-Rodríguez et al. 2014; Bodoira and Maestri 2020). Phenylpropenes are a diverse class of free volatiles and sequestered glycosides of plant secondary metabolites. The structural features such as the variation of substituents on phenyl ring bearing a propenyl side chain with a different position of double bond are responsible for the flavor and aroma of various herbs, spices, and fruits (Atkinson 2016; Koeduka 2014). Coumarins, isocoumarins, and chromones belong to the non-flavonoid compound class of $C_6$-$C_3$ carbon skeleton (benzopyrone nucleus) and associated with antioxidant, antimicrobial, anti-inflammatory, and anticancer activities. Coumarins are widely distributed in plants in a free state as well as in the form of glycosides and consist of 1-benzopyran-2-one nucleus as the unsaturated lactones functionality (-C(=O)-O-) (Shahidi and Ambigaipalan 2015). However, isocoumarins are the isomers of coumarins and consist of 2-benzopyran-1-one nucleus as the inverted lactone functionality (-O-C(=O)-). Chromones are also the isomers of coumarins and consist of 1-benzopyran-4-one nucleus, that is, the keto group present at 4 position of the pyran ring (Saeed 2016; Nazhand et al. 2020; Semwal et al. 2020; Matos et al. 2015; Yerer et al. 2020). Naphthoquinones are composed of the $C_6$-$C_2$ carbon skeleton which are effective quinones specifically because of two carbonyl groups present at 1 and 4 position of one aromatic ring fused with another aromatic ring (Wellington 2015). Xanthones constitute an important class of oxygenated heterocycles as a distinct ($C_6$-$C_1$-$C_6$) tricyclic ring system (dibenzo-$\gamma$-pyrone) and often comprise multiple hydroxyl, methoxy, and methyl groups (Negi et al. 2013; Klein-Júnior et al. 2020). Anthraquinones are composed of the $C_6$-$C_2$-$C_6$ carbon skeleton (9,10-anthracenedione core) which are effective specifically because of two carbonyl groups present at 9 and 10 position of one aromatic ring fused with another two aromatic rings. These compounds impart the color to plants and are generally used as natural dyes (Malik and Müller 2016, Chien et al. 2015). Stilbenes also belong to the class of $C_6$-$C_2$-$C_6$ carbon skeleton non-flavonoids, characterized by 1,2-diphenylethylene structural features and often substituted with hydroxyl, methoxy, and methyl groups as in free state and glycosides (Chong et al. 2009; Reinisalo et al. 2015). Lignans and neolignans are also the classes of phenolic secondary metabolites which consist of dimeric $C_6$-$C_3$ carbon nucleus. In the lignan class of compounds, both phenyl propane units are attached with each other through $\beta$-$\beta'$ linkage having different substitution pattern in the aryl groups and a different degree of oxidation state in the side chain as well. There is an absence of $\beta$-$\beta'$ linkage between two phenyl propane dimeric rings, and both the aromatic rings are substituted with different groups in the case of neolignans (Ward 2000; Teponno et al. 2016). Curcuminoids is an important class of dietary heptanoid compounds of $C_6$-$C_7$-$C_6$ nucleus which are responsible for therapeutic significance of *Curcuma longa* (Li et al. 2019; Hewlings and Kalman 2017). The different classes of non-flavonoid compounds and their structural features are enlisted in Table 1.2 and Fig. 1.2 based on literature sources.

## 1.3 Nitrogen-Containing Compounds

Nitrogen-containing compounds constitute another important class of plant secondary metabolites including mainly alkaloids, cyanogenic glycosides, glucosinolates, and nonprotein amino acids. Most of the nitrogen-containing PSMs are derived commonly from amino acids. Alkaloids are a well-known and structurally diverse

**Table 1.2** Different subclasses of non-flavonoids with few representative compounds belonging to these subclass and having common structural features

| Name of non-flavonoid compound classes and their structural features | Name and chemical structure of few common representative compounds |
|---|---|
| **Phenolic acids**<br>• Phenolic acid are aromatic acid and composed of phenyl and carboxylic acid group<br>• The aromatic ring substituted either by hydroxyl, alkoxy, or alkyl groups. | 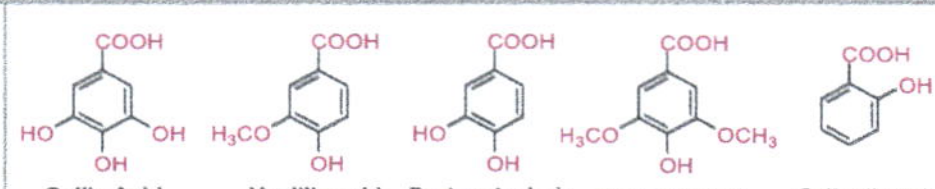<br>Gallic Acid, Vanillic acid, Protocatechuic acid, Syringic acid, Salicylic acid |
| **Phenyl acetic acids and alcohol**<br>• Phenyl acetic acids/alcohols are aromatic acid and composed of phenyl and carboxylic acid/alcoholic group<br>• The aromatic ring substituted either by hydroxyl, alkoxy, or alkyl groups | 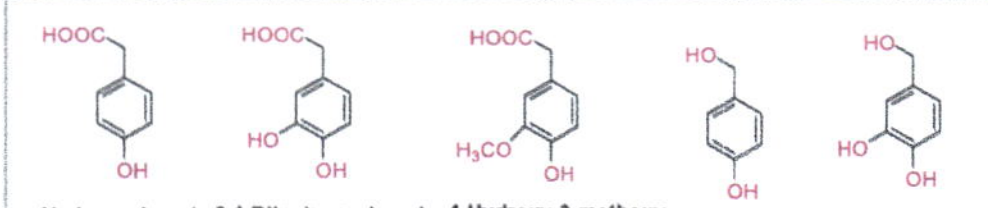<br>p-Hydroxyphenyl acetic acid, 3,4-Dihydroxyphenyl acetic acid, 4-Hydroxy-3-methoxy phenylacetic acid, Tyrosol, Hydroxytyrosol |
| **Phenylpropenes**<br>• Consists of a benzene ring (phenyl group) attached either to (prop-1-en-1-yl) or (prop-2-en-1-yl) group<br>• Benzene ring substituted either by hydroxyl, methoxy, or acetate groups<br>• Exist in both cis and trans configuration | 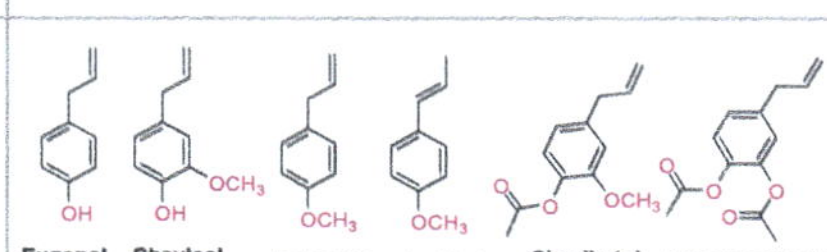<br>Eugenol, Chavicol, Estragole, Anethole, Chavibetol acetate, Allylpyrocatechol diacetate |
| **Cinnamic acid**<br>• Organic aromatic compounds containing an aromatic ring and a carboxylic acid group forming 3-phenylprop-2-enoic acid. The aromatic ring substituted either by hydroxyl or methoxy<br>• Ester of these acids with quinic acid is also included in this class | 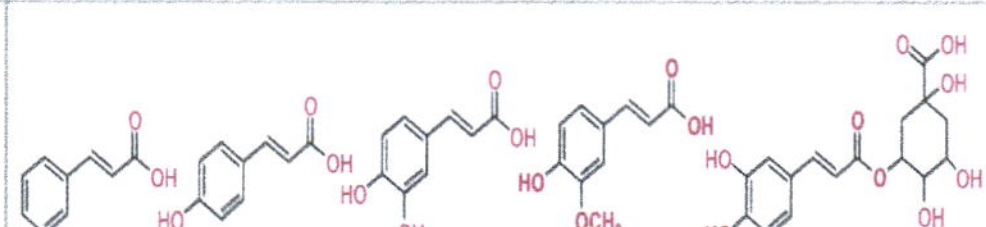<br>Cinnamic acid, p-coumaric acid, Caffeic acid, Ferulic acid, Chlorgenic acid |
| **Coumarins**<br>• 1-benzopyran-2-one chemical class and considered as a lactone<br>• Different substituents such as hydroxyl, alkoxy, alkyl, prenyl, etc. are present and various types, viz., simple, Furano, pyrano, phenyl, and biscoumarins | 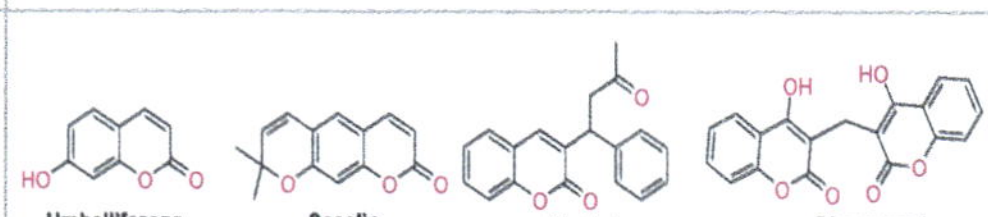<br>Umbelliferone, Seselin, Warfarin, Dicoumarol |
| **Isocoumarins**<br>• 2-benzopyran-1-one chemicals class and inverted lactone functionality<br>• Alkyl-, alkoxy-, alkenyl-, hydroxyl-, and alkenyl ester-substituted derivatives | 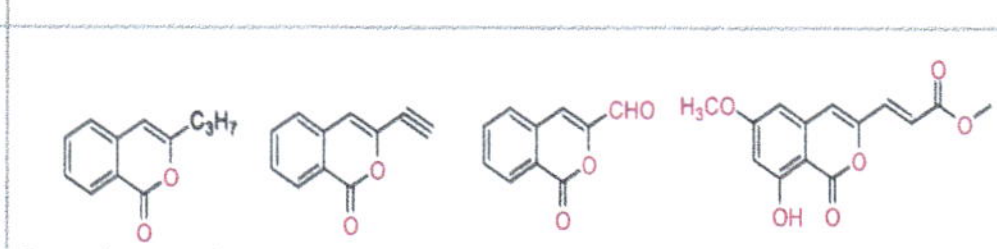<br>3-propylisocoumarin, Capillarin, Artemidiol, Aspergisocoumrin B |

(continued)

**Table 1.2** (continued)

| Name of non-flavonoid compound classes and their structural features | Name and chemical structure of few common representative compounds |
|---|---|
| **Chromones**<br>• 1-benzopyran-4-one nucleus that is the keto group present at 4 position of pyran ring<br>• Various derivatizations due to substitution of aromatic ring, 2 and 3 position of pyran ring, heteroaryl ring, flavone and isoflavone types, etc. | 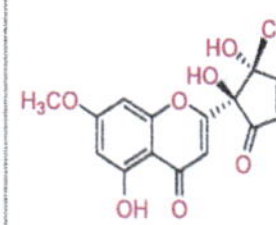 Chrysograyanon 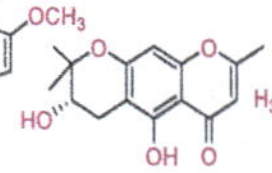 Hamaudol 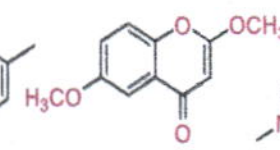 Dennettine 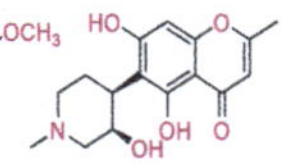 Dysoline |
| **Naphthoquinones**<br>• One aromatic ring fused to quinone subunit that is two keto group present in carbocyclic ring<br>• The different substitution of either hydroxyl, alkoxy, alkenyl, alkenols, etc. on the aromatic and/or quinone ring<br>• 1,4 isomer widely present naturally | 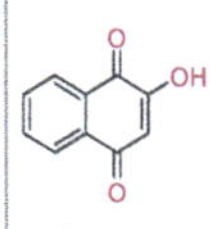 Lawsone 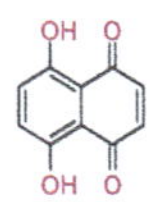 Naphthazarin 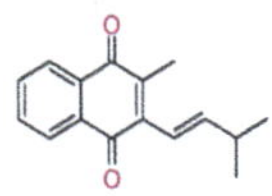 Lapachole 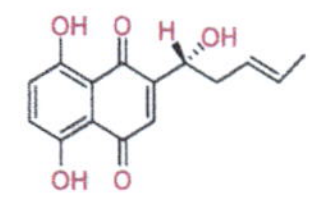 Shikone |
| **Xanthones**<br>• Three carbocyclic ring system that is two benzene rings fused with γ-pyrone ring<br>• di-/tri-/tetrahyroxy, alkoxylated, prenylated, and glycosides derivatives of xanthones | 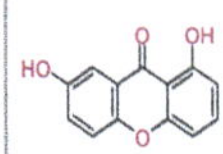 Euxanthone 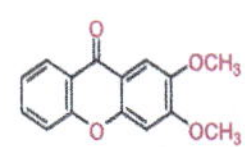 2,3-Dimethoxyxanthone 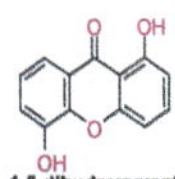 1,5-dihydroxyxanthone 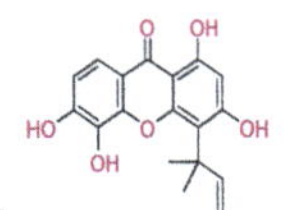 Isocudraniaxanthone A |
| **Anthraquinones**<br>• Three fused carbocyclic ring system wherein two keto groups are present in the central ring<br>• The different substitution pattern of either hydroxy, methoxy, hydroxymethyl, etc. groups on outsider aromatic rings | 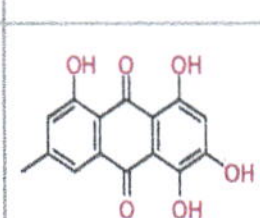 Emodin 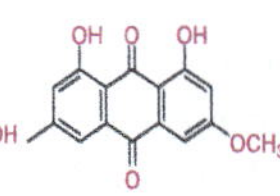 Physcion 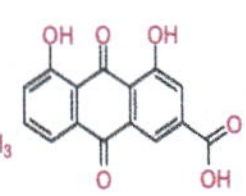 Rhein 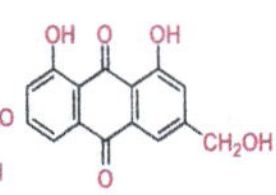 Aloe-emodin |
| **Stilbenes**<br>• Two aromatic rings connected through ethylene moiety<br>• The different pattern of substituents such as hydroxyl, methoxy, isoamylene, *O*-glycosides, etc. on both aromatic rings<br>• Exist as *trans*- and *cis*-stilbenes | 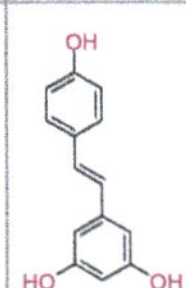 Resveratrol 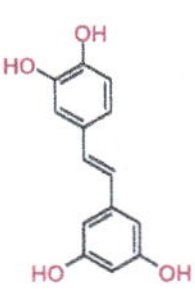 Piceatannol 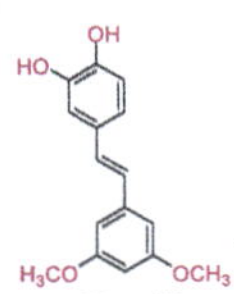 Pterostilbene 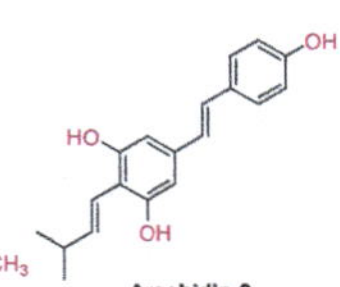 Arachidin-3 |
| **Lignans**<br>• Two phenyl propane units attached through *β*-*β*′ linkage<br>• Different degrees of oxidation state on side chains and functional groups<br>• Rings substituted by either OH, $OCH_3$,or $CH_3$ or fused with another | 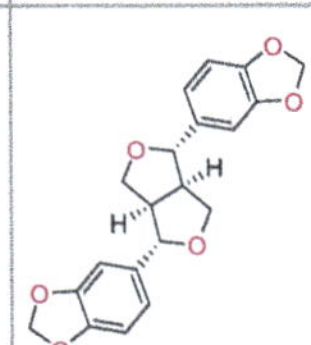 Sesamin 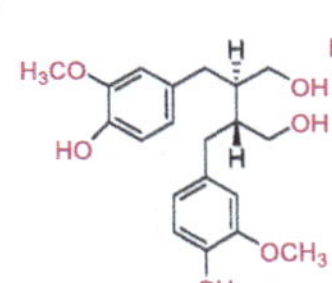 Secoisolariciresinol 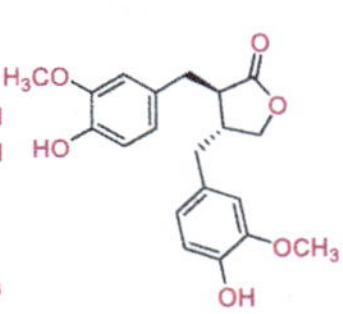 Matairesinol |

(continued)

**Table 1.2** (continued)

| Name of non-flavonoid compound classes and their structural features | Name and chemical structure of few common representative compounds |
|---|---|
| heterocyclic or aromatic ring system<br>**Neolignans**<br>• Two phenylpropanes units are connected through a carbon-carbon bond except $\beta$-$\beta'$ linkage<br>• Substituents on aromatic ring including hydroxy, methoxy, and methylenedioxy groups.<br>• Different degrees of oxidation in the side chain | 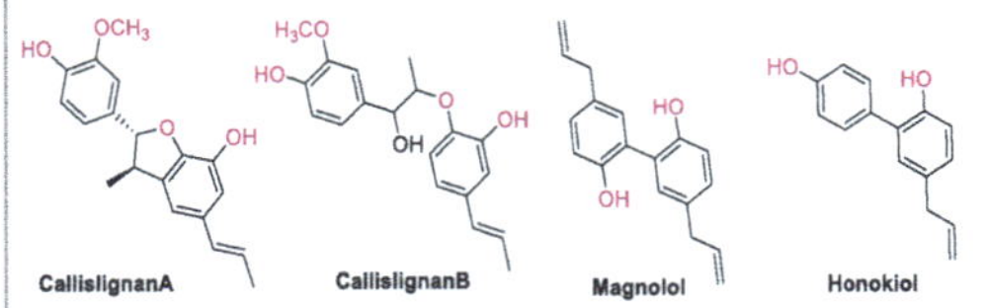 |
| **Curcuminoids**<br>• Two aromatic ring system allied by seven carbon linker comprising of an $\alpha,\beta$- unsaturated $\beta$-diketones motif<br>• Para-hydroxyl groups and meta-methoxy groups either on one or both aromatic rings | 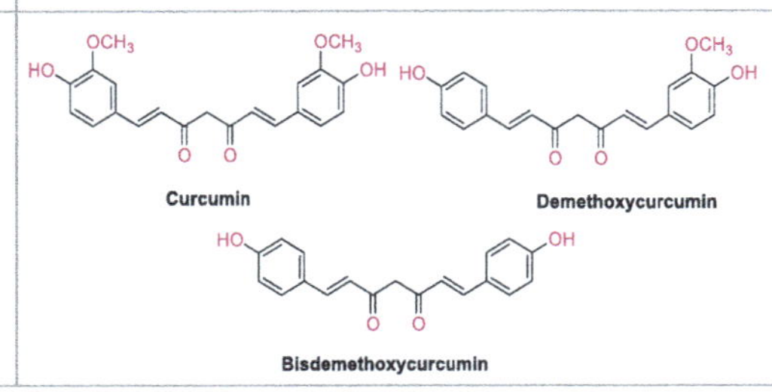 |

class of nitrogen-containing PSMs. Among nitrogen-containing PSMs, there is no certain boundary line between the alkaloids and non-alkaloids. Moreover, cyanogenic glycosides also belong to the nitrogen-containing secondary metabolites which are present in a number of food plants and seeds as well. Upon hydrolysis, these are able to release hydrogen cyanide. Besides, glucosinolates are natural compounds generally obtained from pungent plants and consist of sulfur and nitrogen. These are derived from glucose and amino acid and on degradation produce the characteristic isothiocyanate compounds which are responsible for plant defense against microbes due to the pungent or irritating taste and odor. Also, nonprotein amino acids are widely distributed nitrogen-containing secondary metabolites in plant kingdom and are very similar to the protein amino acid known as their functionally substituted and alkylated derivatives (Figs. 1.3 and 1.4). The diverse structural features of these compounds are responsible for their biological effects as well as roles in plant defense system (Guggisberg and Hesse 2003).

### 1.3.1 Alkaloids

Alkaloids are one of the largest classes of secondary metabolites and preliminary consist of at least one nitrogen atom and termed as the class of biological active compounds. These are widely present in plant kingdom, especially medicinal plant families in particular plant parts such as leaves, bark, or roots at higher concentration levels. Biosynthetically, they are derived from one of the few common amino acids

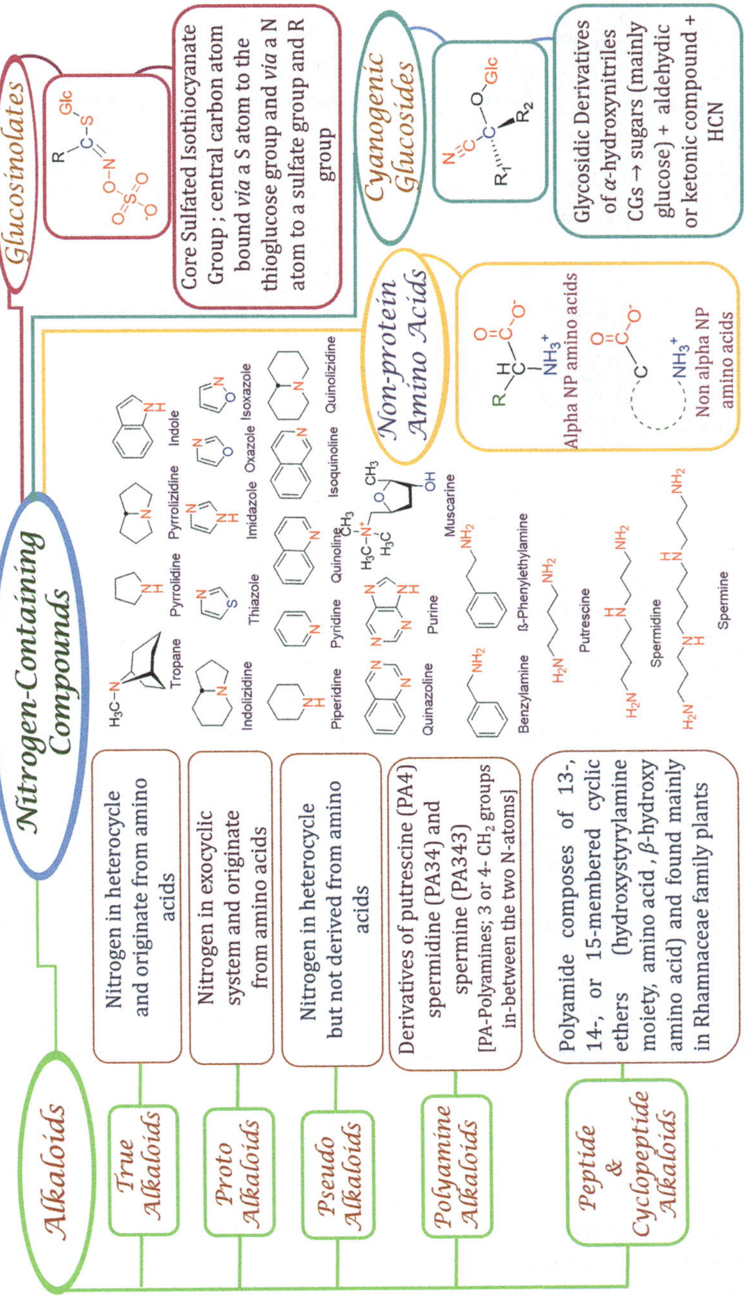


**Fig. 1.3** Different classes of nitrogen-containing plant secondary metabolites: alkaloids, cyanogenic glycosides, glucosinolates, and nonprotein amino acids and their common structural features

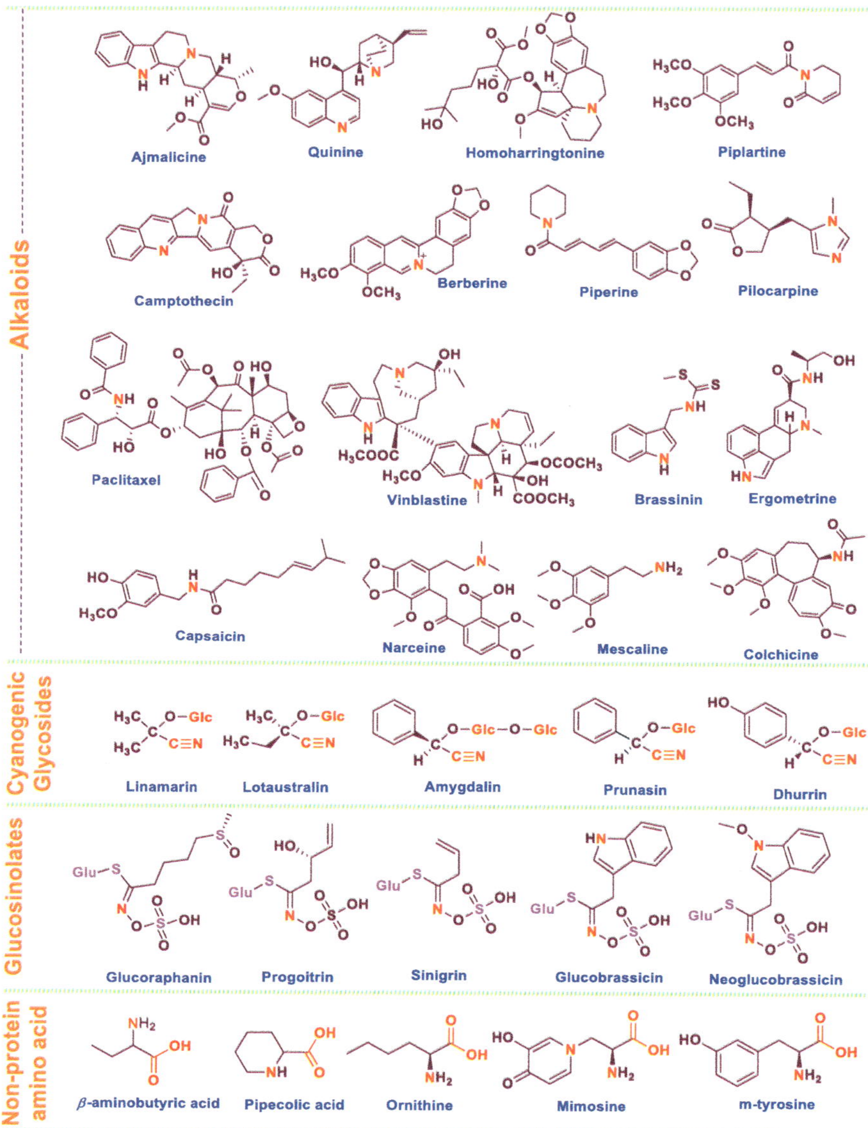


**Fig. 1.4** Chemical structures of some of the compounds of different nitrogen-containing plant secondary metabolite classes

such as lysine, tryptophan, and tyrosine. More than 12,000 alkaloids have been identified in plants and widely distributed in plant tissues as water-soluble salts of organic acids (acetic, citric, lactic, oxalic, malic, and tartaric acids), esters, and glycosides of sugar (glucose, galactose, and rhamnose) and in combined form with tannins. About 20% of the flowering plant species contain alkaloids. These are isolated as crystalline, amorphous, nonvolatile, and non-odorous compounds from the different plant matrices. The widespread biological values (therapeutic, addictive, poisonous, and protector against pathogens and predators) have been found to be associated with alkaloids and also assist the survival of plants in the ecosystem. The dose, duration of exposure, and biologic factors (age, sites of action, and sensitivity) are responsible for their beneficial or toxic effects. There has been enormous structural diversity reported to be involved among the class of alkaloid compounds as compared to other classes of natural products. These compounds have been generally classified on the basis of either the carbon skeleton similarity characteristics or biochemical precursors. Generally, most alkaloids comprise of heterocyclic groups, while polyamine subclass is the smallest subclass of alkaloids. Alkaloids are commonly divided into major groups with respect to the position/location of nitrogen atom, and this classification is mostly used irrespective of whether the nitrogen is the part of ring or not, that is, heterocyclic and non-heterocyclic alkaloids (Fig. 1.3) (Eguchi et al. 2019; Aniszewski 2015; O'Connor 2010).

#### 1.3.1.1 True Alkaloids

This class of alkaloids also known as heterocyclic alkaloids have been derived from amino acids and have the nitrogen atom in heterocyclic ring. These also consist of one or more heterocyclic rings with either mono- or polycyclic compounds as well as either with oxygen, sulfur, or more than one nitrogen atom, for example, pyridine alkaloids, pyrrolidine alkaloids, piperidine alkaloids, tropane alkaloids, quinoline alkaloids, isoquinoline alkaloids, quinolizidine alkaloids, indole alkaloids, pyrrolizidine alkaloids, and imidazole alkaloids (Aniszewski 2015).

#### 1.3.1.2 Protoalkaloids

Protoalkaloids are also known as non-heterocyclic alkaloids which also originated from amino acids, while nitrogen atom is not part of the heterocyclic ring but present in the exocyclic system which means that it is a part of a side chain (Aniszewski 2015).

#### 1.3.1.3 Pseudoalkaoids

Pseudoalkaoids are the compounds of elementary carbon skeleton which are derived from the precursors or post cursors of amino acid as well as nonamino precursors. Terpenoidal alkaloids (sterol alkaloids), phenanthrene alkaloids. Tropolone alkaloids are the subcategories of pseudoalkaoids (Aniszewski 2015).

#### 1.3.1.4 Polyamines Alkaloids

Polyamines alkaloids are the derivatives of polyamines consisting of at least two or more amino groups as part of an aliphatic chain,and these amino groups are separated by three or four methylene group unit linkages as the characteristic feature. These alkaloids consist of basically three basic skeletons of polyamines functionalities such as putrescine (PA4), spermidine (PA34), and spermine (PA343) (Bienz et al. 2005).

#### 1.3.1.5 Peptide and Cyclopeptide Alkaloids

Peptide and cyclopeptide alkaloids contain two or more amino acids linked by peptide bonds. These are derived from the peptide-bound and subsequently cyclized amino acids. Cyclic peptides generally possess 13-, 14-, and 15-membered cycle containing an aromatic ring. These macrocycles consist a peptide unit which is connected either through 1,4- or 1,3- orientation with benzene ring. Linear peptide alkaloids formally obtained from cyclopeptide alkaloids through elimination reaction by the scission of the bridge (Joullié and Richard 2004; Schmidt et al. 1985).

### 1.3.2 Cyanogenic Glycosides

Cyanogenic glycosides are the O-$\beta$-glycosides of $\alpha$-hydroxynitriles (cyanohydrins) in which glucose is directly attached to the hydroxy of the cyanohydrin kept in plant tissues. These are derived from protein amino acids as well as nonprotein amino acids and serve as the storage compounds for reduced nitrogen and sugar. When the plant tissues get disrupted, these are degraded into $\alpha$-hydroxynitriles. Being unstable, hydroxynitriles further undergo hydrolysis and produce hydrogen cyanide by the action of plant enzymes (Vetter 2000; Moller et al. 2016).

### 1.3.3 Glucosinolates

Glucosinolates is another class of nitrogen-containing secondary metabolites that contain sulfur and nitrogen atoms, chemically known as $\beta$-thioglucosides of *N*-hydroximinosulfate esters. These are derived from aliphatic amino acids and glucose. The molecular structure consists of a central atom which is bonded via a sulfur atom to glycone group and via a nitrogen atom to a sulfonated oxime group. Besides, the central carbon is also bound with side group. Different glucosinolates have different side groups, and variations in their biological activities are also reported due to the variation in side groups structural features (Blazevic et al. 2020; Agerbirk and Olsen 2012).

### 1.3.4 Nonprotein Amino Acid

Nonprotein amino acid are the functionally substituted derivatives and C-alkylated analogs of protein amino acids. The diverse structural features are associated with them due to the presence of various functional groups substituents such as α/δ-amino, α-/β-/γ-carboxylic group, thiol, hydroxyl, imidazole, guanidine groups, etc. and a variety of alkyl, aryl, and heterocyclic substituents. Generally, these are classified into two types, i.e., alpha and non-alpha amino acids (Saghyan and Langer 2016).

## 1.4 Terpenes/Terpenoids

Terpenes constitute the most abundant and structurally diverse group of plant secondary metabolites. Terpenes are mostly examined in plants and structurally comprise of the carbon backbones made up of isoprene (2-methyl-1,3-butadiene) units which are connected to each other by thousands of combinations. The terms "terpenes" and "terpenoids" are often used interchangeably. Terpenes are simple organic hydrocarbons that contain skeletons which are exact multiples of the isoprene building units, while the terpenoids are the modified terpenes that include some oxygen functionality or some rearrangement. In other words, terpenoids contain the additional functional groups and oxidized methyl groups replaced or removed at various positions. Terpenoids possess varied structures (acyclic, monocyclic, and polycyclic) and functional groups, usually oxygen-containing groups. The word terpenoids has been broadly used to cover ionones' natural degradation products as well as the terpene's natural and synthetic derivatives with oxygen-containing functional groups such as alcohol, aldehydes, ketones, carboxylic acids, epoxides, esters, and nitriles as well as hydrogenated and dehydrogenated compounds. From many authors' point of view, the term "terpene" is broadly used and includes the terpenoids. Terpenes have been classified by the number of isoprene units in a molecule. Like terpenes, terpenoids also have been classified on the basis of isoprene unit number involved in the parent terpenes. Further each class of terpenoids has also been classified on the basis of rings present in the molecules that are acyclic, having an open-chain structure; monocyclic, having one ring in their structure; bicyclic, having two rings in their structure; tricyclic, having three rings in their structure; tetracyclic, having four rings in their structures; and pentacyclic, having five rings in their structures. Most of the terpenoids are optically active compounds and can be isolated as in isomeric form because the isomerism such as structural as well as stereoisomerism is common in this class of compounds due to the presence of unsaturation and/or functional groups (LaLonde 2005; Harborne 1984; Dev 1989; Sell 2007). The different subclasses with common structural features have been mentioned (Figs. 1.5 and 1.6):

### 1.4.1 Hemiterpenoids

Hemiterpenoids are the simplest class of terpenoids. In fact, isoprene is a well-known member and has great importance as building block of isoprenoids and for polymers production. Other members of this subclass are angelic, isovaleric, senecioic, tiglic acids, and isoamyl alcohol.

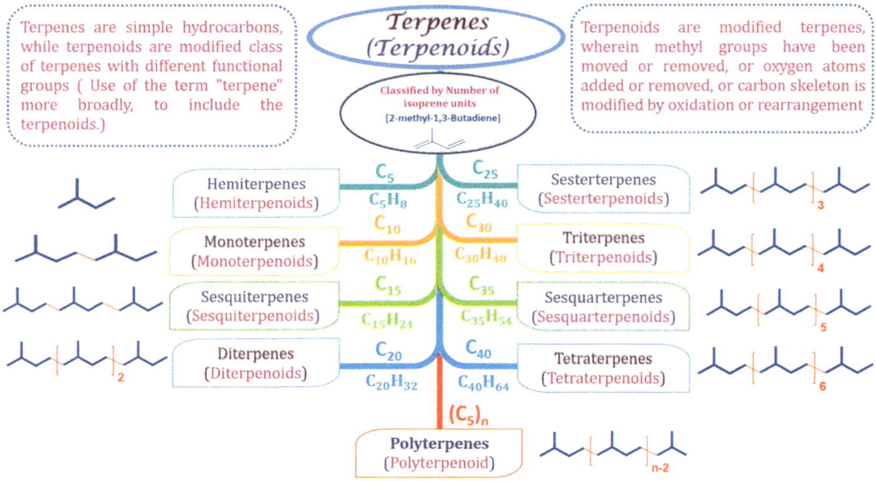


**Fig. 1.5** The different classes of terpene/terpenoid plant secondary metabolites on the basis of isoprene units and their common structural features based upon the number of carbon atoms

### 1.4.2 Monoterpenoids

These are naturally occurring compounds containing two isoprene units and derived from monoterpenes by rearrangement or modified skeleton. Due to their pleasant odor, these are used in perfumery industries as the main constituents of essential oils obtained from the leaves, bark, and roots of various plants. These are subdivided as acyclic (citral, geraniol, lavandulol, linalool, and myrcene), monocyclic (carvone, eucalyptol, limonene, menthol, α-terpineol, perillaldehyde, and thymol), and bicyclic (carene, sabinene, camphene, thujene, camphor, borneol, eucalyptol, and ascaridole).

### 1.4.3 Sesquiterpenoids

Sesquiterpenoids consist of three isoprene units and generally derived from sesquiterpenes through rearrangement and modification in skeleton, and they form the higher boiling fraction of essential oils. Like monoterpenoids, these also exist in a variety of forms such as acyclic (farnesol, β-nerolidol), monocyclic (β-bisabolene, α-zingiberene, α-humulene), bicyclic (β-santalol, β-caryophyllene, δ-cadinene), and tricyclic (khushimol, thujopsene, patchoulol) sesquiterpenoids. Another subclass is sesquiterpene lactones which is chemically distinct from other sesquiterpenoids and consists of a γ-lactone system (artemisinin, alantolactone, costunolide, alantolactone, thapsigargin) (Chadwick et al. 2013; Andrade and de Sousa 2015).

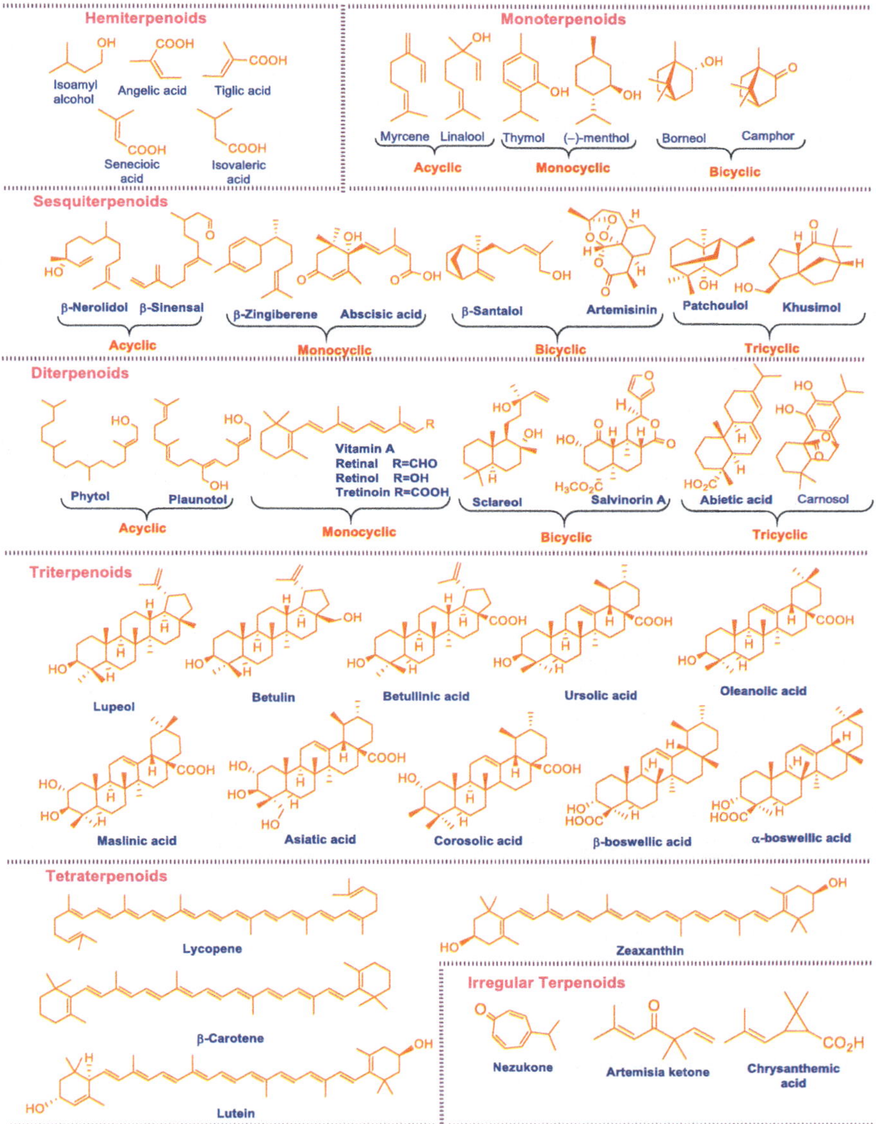


**Fig. 1.6** Chemical structures of few compounds of different terpenoid plant secondary metabolites classes

### 1.4.4 Diterpenoids

Diterpenoid compounds are highly distributed in plant kingdom and are composed of four isoprene units (20 carbon atoms) as the diterpenes skeleton has been rearranged or modified by transfer or replacing one or more skeleton atoms. These are subdivided into acyclic (phytol, retinol), bicyclic (9-geranyl-α-terpineol), tricyclic (sclareol, marrubiin, salvinorin A), tetracyclic (abietic acid, carnosic acid, tanshinone I), pentacyclic (gibberellin A1, steviol), or macrocyclic diterpenes (taxol) on the basis of their skeletal core. These commonly exist in polyoxygenated form with esterified keto and hydroxyl group by small aliphatic and aromatic acids.

### 1.4.5 Sesterterpenoids

Sesterterpenoids is a rare class of terpenes which are comprised of five isoprene units (25 backbone carbon) and derived by rearrangement or amended sesterterpene skeleton atoms. This subclass of terpenoids is rarely found in higher plants, while other widespread sources include fungi, bacteria, marine sponge, etc. The examples of some compounds belonging to sesterterpenoid structural frameworks are linear (hippolide E), monocyclic (manoalide), bicyclic (leucosceptrine), tricyclic (heliocide H1), tetracyclic (sesterstatin 7), and macrocyclic (nitinol) as well.

### 1.4.6 Triterpenoids

Triterpenoids are a large group of natural products which contain six isoprene units as modified and rearranged derivatives of triterpenes with oxygen-containing functional groups. The different types of triterpenoid skeleton are derived from C30 (squalene or oxidosqualene) precursors through the different ring closure ways of squalene or oxidosqualene. They have relatively complex cyclic skeletal structures either with alcohols, aldehydes, or carboxylic acids. The triterpenic skeletons such as tetracyclic and pentacyclic are mostly abundant, while the acyclic, mono-, di-, tri-, and hexacyclic triterpenoids have also been studied and exist in natural sources, for example, α-amyrin, β-amyrin, ursolic acid, oleanolic acid, betulinic acid, cucurbitacin B, sitosterol, stigmasterol, campesterol, α-spinasterol, asiatic acid, celastrol, etc. (Xu et al. 2004).

### 1.4.7 Sesquarterpenoids

Sesquarterpenoids consist of seven isoprene units and are a rare group of terpenoids typically found in microbial origin.

### 1.4.8 Tetraterpenoids

Tetraterpenoids (including many carotenoids) are chemically modified tetraterpenes through rearrangement or modification of skeleton atoms, as indicated by the presence of oxygen-containing functional groups. These consist of eight isoprene units, that is, the C40 skeleton of the parent tetraterpene, and produce the particular pigmentation found in plants especially in flowers and fruits such as red, yellow, and orange colors as well as useful in photosynthesis as accessory pigments.

### 1.4.9 Polyterpenoids

This group of compounds is polymeric hydrocarbon (natural rubber) and consist of more than eight isoperene units. The isoperene units exits either in *cis* (natural rubber) or *trans* configuration (gutta-percha, balata rubber).

### 1.4.10 Irregular Terpenoids

Another class of terpenoids is also identified and known as irregular terpenoids. The compounds such as tropones (due to unknown ring expansion of cyclohexane skeleton), artemisane chrysanthemane, lavandulane, santolinane, sesquilavandulane (formed by condensation of isoprene units), and norterpenoids (ionones: α- and β-ionone; damascones: α- and β-damascone) (Baser and Demirci 2007; Liu et al. 2007; Sell 2007).

Terpenes/terpenoids are natural compounds synthesized exclusively by plants with chemical features such as molecular structure, substituent groups, functional groups, and/or the linkage type. The structural diversity among this class of compounds is responsible for their function roles (food flavor additives, perfume fragrances, in aromatherapy as traditional or alternate medicines, etc.) (Lyu et al. 2019; Harman-Ware 2020; Leavell et al. 2016) and therapeutic properties (antimicrobial, antioxidants, antifungal, antiviral, antiparasitic, antihyperglycemic, anti-inflammatory, immunomodulatory, skin permeation enhancer, and drugs in cancer chemotherapy) (Bergman et al. 2019; Boncan et al. 2020; Zeng et al. 2019; Kim et al. 2020; Yang et al. 2020)

## 1.5 Sulfur-Containing Compounds

Sulfur-containing compounds is another important class of small group of low-molecular-weight plant secondary metabolites (PSMs). Sulfur-containing PSMs often have a characteristic smell and play various role such as signaling molecules (for fundamental cellular functions) as well the defense compounds against herbivores and pathogenic organism. The structural diversity in sulfur PSMs such as different sulfur bonds is the main and an important feature together with other structural characteristics responsible for their interesting biological activities with medicinal effect and therapeutic significance. The sulfur-containing PSMs are sub-divide into different subclasses (Figs. 1.7 and 1.8).

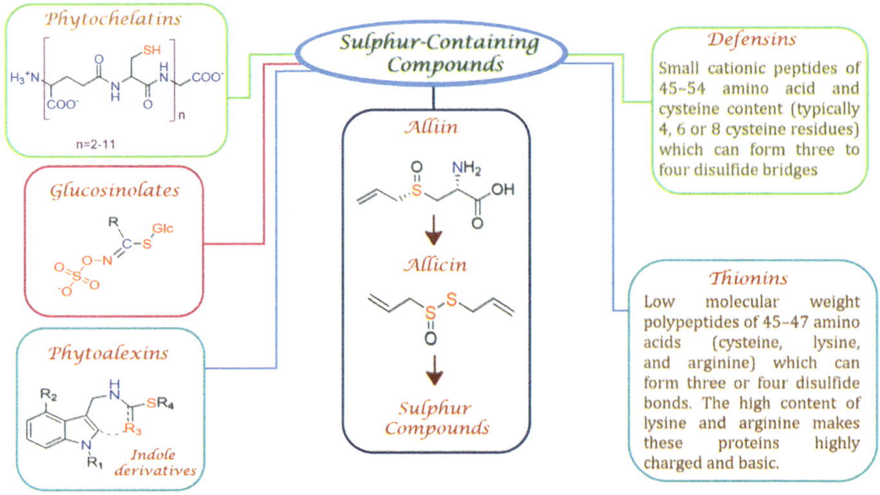


**Fig. 1.7** The different classes of sulfur-containing plant secondary metabolites and their common structural features

### 1.5.1 Phytochelatins

Phytochelatins are ubiquitous in plants as sulfur-containing small metal-binding peptides with a universal structure ($\gamma$-glutamyl-cysteine) *n*-glycine, where $n = 2$–11 known as oligomers of glutathione. They play a vital role in heavy metal management in plants and act as sequester of heavy metals through metal chelation reaction with these metals. They are also regarded as heavy metal hyperaccumulators and have the ability to prevent the binding and reaction of heavy metals with sulphydryl groups of vital enzymes and proteins of plants in order to control and detoxify the effect of heavy metals (Inouhe 2005).

### 1.5.2 Glucosinolate

This diverse class of compounds is also termed as sulfur-containing as well as nitrogen-containing PSMs because the core structure consists of $\beta$-*D*-thioglucose group linked to an *N*-hydroximinosulfate ester along with a variable side chain (aliphatic, aromatic, or indolic) that is derived from an amino acid. The enzymatic hydrolysis of thioglucose causes the conversion of these compounds into respective isothiocyanates, for example, allylglucosinolate (sinigrin)-allyl isothiocyanate; benzylglucosinolate (glucotropaeolin)-benzyl isothiocyanate; phenethylglucosinolate (gluconasturtiin)-phenethyl isothiocyanate; and glucoraphanin-sulforaphane (Agerbirk and Olsen 2012).

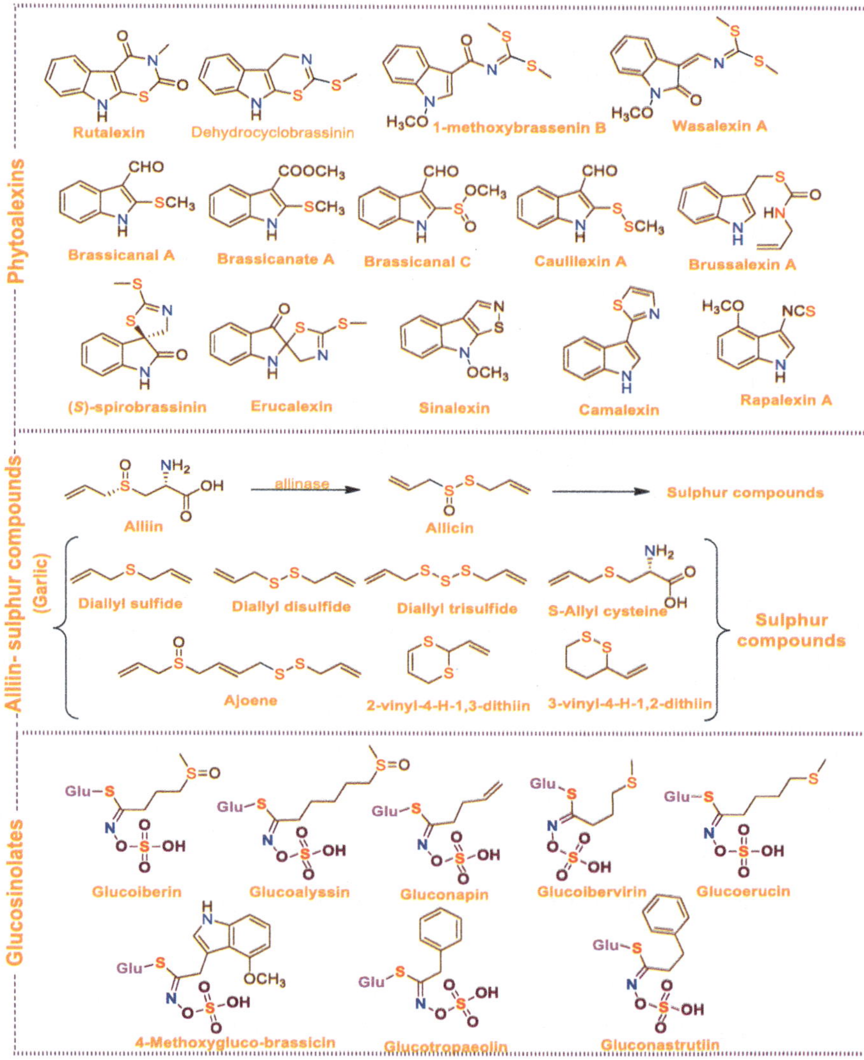


**Fig. 1.8** Chemical structure of some few compounds of different plant secondary metabolites of sulfur-containing compound classes

### 1.5.3 Phytoalexins

Phytoalexins are a diverse and important class of compounds including alkaloids, phenolic, and also sulfur-containing compounds. These are low-molecular-weight antibiotics as inducible PSMs produced by a plant in response to pathogens attack caused the environmental stresses. They are secreted and accumulated temporarily by plants in response to biotic and abiotic stresses. The names of a few compounds of sulfur-containing phytoalexins comprise at least one sulfur atom in the side chain or ring: camalexins, brassinin, brassitin, brassicanal A and B, spirobrassinin and rutalexin, wasalexin A, rapalexin A and B, etc. (Ruszkowska and Wróbel 2003; Baur et al. 1998; Ahuja et al. 2012).

### 1.5.4 Defensins

Defensins are cationic antimicrobial peptides and exhibit direct antimicrobial activity as well as immune signaling activities. They are host defense cysteine-rich proteins with characteristic $\beta$-sheet-rich fold and a framework of six disulfide-linked cysteines. Defensins are generally categorized into three types, namely, α-, β-, θ-defensins, on the basis of disulfide pairing of cysteines among the members of the class (Machado and Ottolini 2015; Ganz 2003).

### 1.5.5 Thionins

Thionins are cysteine-rich amino acids containing peptides and present in the leaves, stems, roots, and seeds of a number of plant species. These folded plant peptides are generally 4 types, consisting of 45–47 amino acids with 3 or 4 disulfide bonds displaying toxic and antimicrobial effects against pathogenic bacteria, yeast, and fungi. Thionins (crambin viscotoxins, phoratoxin, etc.) are important components of plant defense system and protect the plants from various pathogens by acting directly on the membranes of microorganisms (Hayes and Bleakley 2018; Odintsova et al. 2018).

## 1.6 Physiochemical Aspect of Plant Secondary Metabolites (PSMs)

Plants are the rich source of drug molecules; these are the various phytochemicals found to be active against various diseases. The demand of the drugs for their economical use is the main point of attention; for this plants can be a very good option for drug development and discovery. Plant phytochemicals already are well-known for their biological and clinical applications. Research clearly indicates that phytochemicals have a significant approach in various diseases and antitumor treatments. Nearly 50% of approved anticancer drugs are derived from natural sources. The approach of the natural products can be seen as 63% of the anticancer molecules are from natural sources that either is the natural molecules or are derived from natural sources. The data is from 1940 to 2014 by the US Food and Drug Administration (FDA) against approved 175 anticancer molecules. With this, the natural products and their semisynthetic derivatives have high therapeutic rate than the chemically synthesized compounds (Boufridi and Quinn 2018). Some of the classified phytochemicals like terpenes, phenolics, alkaloids, etc. belonging to PSMs are very important structural drug molecules. The chemical relations along with some physiochemical properties like poor aqueous solubility, surface area, etc. are the major concern toward the drugs discovery and their activity. The different groups of PSMs have been examined for their potential as various biologically active and drug molecules; however, the step for the molecules to be studied under the detailed clinical trial is very tedious and lengthy. In order to resolve these, the study of preliminary properties of the molecules in the clinical action is very important. The acid-base dissociation constant (pKa), lipophilicity/hydrophobicity, solubility, etc. are the most important and significant physical parameters in drug discovery and development (Young 2014).

The properties of rapidly growing phytochemical drug molecules rely on the absorption, distribution, metabolism, and excretion normally abbreviated under the term ADME. The ADME of the drug molecule can be correlated with their action in the biological system. All these and some other physiological aspects like molecular weight (MW), partition coefficient (XlogP), hydrogen bond acceptor (HBA), hydrogen bond donor (HBD), and total polar surface area are the properties that govern the applicability of these molecules as a drug compound. The value of the different pharmacokinetic/physiochemical properties runs further the efficiency of the various drug molecules (Jablonsky et al. 2019). The structural features of the PSM are correlating the application of the naturally available drug molecules. The availability and presence of various functional groups, stereocenters, and bond type in the active molecules are sources of applicability of these in various clinical actions.

Choudhari et al. have reported nearly 50 compounds; some of them are under clinical trial or used as an anticancer agent (Choudhari et al. 2020). Various research reports revealed the significant potential of plant phytochemicals; however, some of the limitations associated are dose or maximum/minimum concentration with their effective and efficient biological and therapeutic role after the clinical studies (Cheuka et al. 2017; Newman and Cragg 2016). The study or the classification of the nondrug and high-quality drug molecule is a notable aspect to defining which molecules can be carried for the detailed clinical trials. To work on drugs, the simple properties of the molecule under investigation for the drug likeness such as molecular mass, partition coefficient, hydrogen bond acceptance, donation ability, etc. are very important (Mignani et al. 2018). Toward this, the Lipinski criteria are very important and widely used to filter the molecules. Lipinski criteria or rules can be seen to represent with the "rule of five" (Ro5) and based on the chemo-informatic filters, also known as the drug-likeness filter. These filters reduce the entry of the drug and satisfy 90% of the orally absorbed compounds as drugs. The rule governs the absorption/permeation of a drug in the system is more valuable relation and can be correlated with the following criteria or filters:

1. Molecular weight (MW) $\leq$500
2. Octanol-water partition coefficient (LogP) $\leq$5
3. Number of hydrogen bond donner (HBD) $\leq$5
4. Number of hydrogen bond acceptor (HBA) $\leq$10

Lipinski's set has been derived by taking the 90th percentile of phase II clinical trial drug. Ro5 based on the algorithm consists of four main rules in which the maximum of the numbers are five or multiples of five factor; from this nature the name of the rule has assigned. The histograms expressed as the compounds count and highlight the number of compounds for the exact physiochemical assessment. The algorithm has governed the above values given for the four physiological properties. To be drug-like, a candidate should have to follow and submit the analysis for the phyto-molecule. The aim of the "Lipinski's" rule is to pre-highlight the bioavailability of the drug molecule or possible limitation if two or more properties are violated (Lipinski et al. 1997). However, this rule do not calculate the pharmacological activity of drug molecules (Kumar et al. 2010). Lipinski's rule does not absolutely group all finally and poorly absorbed molecules/compounds, though it provides a fast, simple, and reasonable degree of classification. However, the well-absorbed compounds can be less active or have lower bioavailability as a result of the high presystemic clearance (Turner and Agatonovic-Kustrin 2007). In a study of 814 natural drug compounds, the fulfillment of Lipinski's rule criteria have been conducted. Among these drug molecules, about 85% of the molecules have no violations; however, 95% of the compounds have less than two violations (Quinn et al. 2008).

In addition to the Lipinski's rule, the other parameter that has been used to predict the drug-like or DMPK (drug metabolism and pharmacokinetics) behavior. The log distribution coefficient (logD), polar surface area (PSA), rotatable bonds, and counts of nitrogen and oxygen atoms are important aspects. However, the synthetic drug molecules has been found to have less complex structural compounds, while the natural phytochemicals generally show a wide range of structural complexity. Therefore, the compounds with the smaller molecular weight, small number of bonds, and less rotatable bond lead to more active molecules. Therefore, the rule of "lead-likeness" gives some little rigid criteria than Lipinski such as molecular weight $<450$ and $\log P < 4$. An even more restrictive set of rules have been defined for fragment screening (Congreve et al. 2003). Both these rules collectively can be designed for finding the compound activity as the proficient drug-like molecules through medicinal chemistry optimization.

The extension to Lipinski's rule has been given by various researchers like Veber and co-workers, suggesting an important role of rotatable bond counts. They reported that a maximum of seven rotatable bond counts is optimum for the drug-likeness or bioavailability (Veber et al. 2002). The molecules that have the polar surface area (PSA) of 110–140 $Å^2$ are supposed to have low oral bioavailability (Nagren 2003). Therefore, these factor values' can be counted in the discovery of drug molecules. With the discussion, the area to discover natural molecules as the new drug molecules that can be correlating the physiochemical and biological aspects in nature requires further research and investigation. Physiochemical properties of the molecules are helpful in the examination of their therapeutic applications and pharmacological action of the drug molecules. Exploring the physiochemical as well as the biochemical properties of PSMs is very important in order to develop and discover drug molecules with ease and effectively in the biosystem. Table 1.3 represents the active PSMs of

**Table 1.3** Some of the PSMs of different classes as the drug molecules under various clinical trials with different physiochemical properties

| PSMs class | PSMS subclass | Individual PSM compound | M.W. (g/mol)[a] | HBD[a] | HBA[a] | RC[a] | Xlog P[a] | TPSA[a] | Biological significance on the basis of clinical trial [b, c] |
|---|---|---|---|---|---|---|---|---|---|
| Phenolics and polyphenolics | Flavonoid (flavone) | Apigenin | 270.24 | 3 | 5 | 1 | 1.7 | 87 | In dietary supplement, placebo tablet, treat insomnia |
| | | Baicalein | 270.24 | 3 | 5 | 1 | 1.7 | 87 | As drug for influenza |
| | | Diosmetin | 300.26 | 3 | 6 | 2 | 1.7 | 96.2 | Weak TrkB receptor agonist |
| | | Hispidulin | 300.26 | 3 | 6 | 2 | 1.7 | 96.2 | Potent benzodiazepine receptor ligand with positive allosteric properties |
| | | Luteolin | 286.24 | 4 | 6 | 1 | 1.4 | 107 | In memory health, on tongue squamous cell carcinoma cell line, and as dietary supplement |
| | Flavonoid (flavonol) | Fisetin | 286.24 | 4 | 6 | 1 | 2 | 107 | In COVID-19, kidney diseases, diabetes |
| | | Quercetin | 302.23 | 5 | 7 | 1 | 1.5 | 127 | Chronic obstructive pulmonary disease, stroke, aging problem, chronic hepatitis C, prevention of adenocarcinoma of the prostate/recurrent prostate cancer/stage I prostate cancer/stage IIA prostate cancer/stage IIB prostate cancer/stage III |

(continued)

**Table 1.3** (continued)

| PSMs class | PSMS subclass | Individual PSM compound | M.W. (g/mol)[a] | HBD[a] | HBA[a] | RC[a] | Xlog P[a] | TPSA[a] | Biological significance on the basis of clinical trial [b, c] |
|---|---|---|---|---|---|---|---|---|---|
| | | | | | | | | | prostate cancer/stage IV prostate cancer |
| | | Kaempferol | 286.24 | 4 | 6 | 1 | 1.9 | 107 | In diabetes, intrauterine growth restriction |
| | | Morin | 302.23 | 5 | 7 | 1 | 1.5 | 127 | In pulmonary hypertension, insomnia, schizophrenia, schizoaffective disorder, urinary incontinence |
| | | Myricetin | 318.23 | 6 | 8 | 1 | 1.2 | 148 | In urinary tract infections |
| | | Isorhamnetin | 316.26 | 4 | 7 | 2 | 1.9 | 116 | In intrauterine growth restriction |
| | | Rutin | 610.5 | 10 | 16 | 6 | −1.3 | 266 | In hyperuricemia, diabetes, colorectal cancer |
| | Flavonoid (flavanone) | Pinocembrin | 256.25 | 2 | 4 | 1 | 2.7 | 66.8 | In ischemic stroke |
| | | Eriodictyol | 288.25 | 4 | 6 | 1 | 2 | 107 | In breast cancer, diabetes |
| | | Butin | 272.25 | 3 | 5 | 1 | 1.8 | 87 | In decontamination as antiseptic molecule |
| | | Hesperetin | 302.28 | 3 | 6 | 2 | 2.4 | 96.2 | In cardiovascular disease and as dietary supplement |
| | Flavonoid (flavanonols) | Dihydromyricetin | 320.25 | 6 | 8 | 1 | 1.1 | 148 | In diabetes |
| | Flavonoid (isoflavones) | Genistein | 270.24 | 3 | 5 | 1 | 2.7 | 87 | In cancer, Alzheimer's disease, lymphoma |

| | | | | | | | | | |
|---|---|---|---|---|---|---|---|---|---|
| | | Alpinumisoflavone (pyranoisoflavone) | 336.3 | 2 | 5 | 1 | 3.9 | 76 | In erectile dysfunction, sexual function |
| | | Daidzein | 254.24 | 2 | 4 | 1 | 2.5 | 66.8 | In diabetes, postmenopause, cardiovascular diseases |
| | | Puerarin | 416.4 | 6 | 9 | 3 | 0 | 157 | In arthritis, cardiovascular risk, alcohol abuse |
| | Flavonoid (flavan-3-ol) | Catechin | 290.27 | 5 | 6 | 1 | 0.4 | 110 | In influenza, diabetes, overweight, cancer |
| | | Epicatechin | 290.27 | 5 | 6 | 1 | 0.4 | 110 | In Becker muscular dystrophy, diabetes |
| | | Epigallocatechin (flavonoids) | 306.27 | 6 | 7 | 1 | 0 | 131 | In diabetic nephropathy, hypertension, Alzheimer's disease |
| | | Epigallocatechin-3-gallate | 442.4 | 7 | 10 | 4 | 1.5 | 177 | In cardiac treatment, fetal alcohol syndrome, prostate adenocarcinoma |
| | Flavonoids (chalcone) | Xanthohumol | 354.4 | 3 | 5 | 6 | 5.1 | 87 | In metabolic syndrome, Crohn disease, oxidative stress |
| | | Licochalcone A (chalcone) | 338.4 | 2 | 4 | 6 | 4.9 | 66.8 | In relapsing of acne, carcinoma |
| | Phenolic acid | Gallic acid | 170.12 | 4 | 5 | 1 | 0.7 | 98 | In cardiovascular and as dietary supplement |
| | | Vanillic acid | 168.15 | 2 | 4 | 2 | 1.4 | 66.8 | In renal disease, dietary modification |
| | | Apocynin | 166.17 | 1 | 3 | 2 | 0.5 | 46.5 | In asthma, osteoarthritis, cardiovascular diseases |

(continued)

**Table 1.3** (continued)

| PSMs class | PSMS subclass | Individual PSM compound | M.W. (g/mol)[a] | HBD[a] | HBA[a] | RC[a] | Xlog P[a] | TPSA[a] | Biological significance on the basis of clinical trial[b, c] |
|---|---|---|---|---|---|---|---|---|---|
| | Phenylacetic acid | Ferulic acid | 194.18 | 2 | 4 | 3 | 1.5 | 66.8 | In pediatric acute kidney injury, COVID-19, dietary modification |
| | | P-coumaric acid | 164.16 | 2 | 3 | 2 | 1.5 | 57.5 | In COVID-19 and as dietary supplement |
| | Stilbenoid | Resveratrol | 228.24 | 3 | 3 | 2 | 3.1 | 60.7 | In proinflammatory cytokines, Friedreich ataxia, systemic inflammation |
| | | Piceid | 390.4 | 6 | 8 | 5 | 1.7 | 140 | In chronic pelvic pain, as polydatin injectable |
| | | Combretastatin A4 | 316.3 | 1 | 5 | 6 | 3.7 | 57.2 | In choroidal neovascularization, myopia, ovarian cancer |
| | | Pterostilbene | 256.3 | 1 | 3 | 4 | 3.8 | 38.7 | Hyperlipidemia, blood pressure, muscle injury |
| | Naphthoquinone | Shikonin | 288.29 | 3 | 5 | 3 | 3 | 94.8 | In breast cancer, bladder urothelial carcinoma |
| | | Lawsone | 174.15 | 1 | 3 | 0 | 0.9 | | In chemotherapy, HIV/AIDS, kidney injury, and as henna drug |
| | Curcuminoids | Curcumin | 368.4 | 2 | 6 | 8 | 3.2 | 93.1 | In oral lichen planus, psoriasis, and as drug/dietary supplement |
| | | Demethoxycurcumin | 338.4 | 2 | 5 | 7 | 3.3 | 83.8 | In breast cancer and as drug/dietary supplement |
| | | Bisdemethoxycurcumin | 308.3 | 2 | 4 | 6 | 3.3 | 74.6 | In pharmacokinetics, dietary supplement, and as drug |

| | | | | | | | | | |
|---|---|---|---|---|---|---|---|---|---|
| | Coumarins | Decursin | 328.4 | 0 | 5 | 3 | 3.9 | 61.8 | In pharmacokinetics and as dietary supplement |
| | | Decursinol | 246.26 | 1 | 4 | 0 | 1.9 | 55.8 | As dietary supplement |
| | Anthraquinone | Emodin | 270.24 | 3 | 5 | 0 | 2.7 | 94.8 | In polycystic kidney, breast cancer, and as drug |
| | Lignans | Podophyllotoxin | 414.4 | 1 | 8 | 4 | 2 | 92.7 | In lung cancer, lymphoma, neuroendocrine carcinoma, and as podofilox drug |
| Nitrogen containing compounds | Alkaloid (isoquinoline) | Berberine | 336.4 | 0 | 4 | 2 | 3.6 | 40.8 | In metabolic syndrome, COVID-19, nonalcoholic fatty liver disease, and as berberine drug |
| | Alkaloid (quinolizidine) | Matrine | 248.36 | 0 | 2 | 0 | 1.6 | 23.6 | In antitumor, as dietary supplement, and as drug |
| | Alkaloid (pyridine) | Piperine | 285.34 | 0 | 3 | 3 | 3.5 | 38.8 | In deglutition disorders, chronic kidney, and as piperine drug |
| | Alkaloid (isoquinoline) | Sinomenine | 329.4 | 1 | 5 | 2 | 2.2 | 59 | In breast cancer, and as cocculine drug |
| | Alkaloid (bisbenzylisoquinoline) | Tetrandrine | 622.7 | 0 | 8 | 4 | 6.4 | 61.9 | In COVID-19 and as tetrandrine drug |
| | Alkaloids (bisindole) | Vinblastine | 81 | 3 | 12 | 10 | 3.7 | 154 | Solid or central nervous system tumors, lymphoma, and as vinblastine drug |

(continued)

**Table 1.3** (continued)

| PSMs class | PSMS subclass | Individual PSM compound | M.W. (g/mol)[a] | HBD[a] | HBA[a] | RC[a] | Xlog P[a] | TPSA[a] | Biological significance on the basis of clinical trial [b, c] |
|---|---|---|---|---|---|---|---|---|---|
| | Alkaloids (bisindole) | Vincristine | 825 | 3 | 12 | 10 | 2.8 | 171 | In myeloid leukemia, peripheral neuropathy, and as vincristine drug |
| Terpenes/ Terpenoids | Monoterpene | Limonene | 136.23 | 0 | 0 | 1 | 3.4 | 0 | In breast cancer, xerostomia, diarrhea, and as placebo drug |
| | | Thymol | 150.22 | 1 | 1 | 1 | 3.3 | 2.2 | In celiac disease, periodontitis, diabetes, and as chlorhexidine/ thymol varnish drug |
| | | Thymoquinone | 164.2 | 0 | 2 | 1 | 2 | 34.1 | In premalignant lesion, diabetes, COVID-19, and as *Nigella sativa* buccal drug |
| | Diterpenoid | Triptolide | 360.4 | 1 | 6 | 1 | 0.2 | 84.1 | In HIV/AIDS, polycystic kidney, inflammatory, and as triptolide drug |
| | | Andrographolide | 350.4 | 3 | 5 | 3 | 2.2 | 87 | In carcinoma, increased insulin, migraine, and as andrographolides/ placebo drug |
| | | Ingenol mebutate | 430.5 | 3 | 6 | 4 | 2 | 104 | In actinic keratosis, verruca vulgaris, and as ingenol mebutate drug and gel |
| | Triterpenoid | Asiaticoside | 959.1 | 12 | 19 | 10 | 0.1 | 315 | In diabetes and as WH-1 ointment drug |

| | | | | | | | | | |
|---|---|---|---|---|---|---|---|---|---|
| | | Ursolic acid | 456.7 | 2 | 3 | 1 | 7.3 | 57.5 | In prostate cancer, sarcopenia, sclerosing cholangitis, and as ursolic acid drug |
| | | Glycyrrhizin | 822.9 | 8 | 16 | 7 | 3.7 | 267 | In chronic hepatitis B/C, liver inflammation, hypercortisolism, and as diammonium glycyrrhizinate drug |
| | | Lupeol | 426.7 | 1 | 1 | 1 | 9.9 | 20.2 | In acne |
| | | Oleanolic acid | 456.7 | 2 | 3 | 1 | 7.5 | 57.5 | In echolocation and as dietary supplement |
| | Carotenoids | Lycopene | 536.9 | 0 | 0 | 16 | 15.6 | 0 | In prostate cancer patients, inflammation, and as lycopene drug/softgel |
| | | Carotene | 536.9 | 0 | 0 | 10 | 13.6 | | In digestion, nutrient deficiency, and as dietary supplement and drug |
| | | Cryptoxanthin | 552.9 | 1 | 1 | 10 | 12.3 | 20.2 | In hypercholesterolemia, osteoporosis, and as dietary supplement and drug |
| Sulfur containing compounds | Glucosinolate | Allicin | 162.3 | 0 | 3 | 5 | 1.3 | 61.6 | In atherosclerosis, HIV infections, hypercholesterolemia, and as dietary supplement and drug |

(continued)

**Table 1.3** (continued)

| PSMs class | PSMS subclass | Individual PSM compound | M.W. (g/mol)[a] | HBD[a] | HBA[a] | RC[a] | Xlog P[a] | TPSA[a] | Biological significance on the basis of clinical trial [b, c] |
|---|---|---|---|---|---|---|---|---|---|
| | | Glucobrassicin | 448.5 | 6 | 11 | 7 | −0.1 | 216 | In metabolism and as deuterated phenanthrene drug |
| | | Gluconasturtiin | 423.5 | 5 | 11 | 8 | 0.1 | 200 | Detoxification and as dietary supplement |
| | | Glucoraphanin | 437.5 | 5 | 13 | 10 | −2.1 | 236 | Reducing core symptoms of autism spectrum disorder (neurodevelopmental disorder) |
| | | Sinigrin | 397.5 | 4 | 11 | 7 | – | 203 | In glucose regulation and as dietary supplement |
| | Isothiocyanate (from glucosinolate) | Phenethyl isothiocyanate | 163.24 | 0 | 2 | 3 | 3.5 | 44.4 | In lung cancer, skin aging, and as dietary supplement and drug |
| | | Sulforaphane (isothiocyanate) | 177.3 | 0 | 4 | 5 | 1.4 | 80.7 | In breast cancer, schizophrenia, frontal lobe, and as dietary supplement and sulforaphane drug |

*M.W.* molecular weight, *HBD* hydrogen bond donor, *HBA* hydrogen bond acceptor, *RC* rotatable bond count, *Xlog P* partition coefficient, *TPSA* topological polar surface area ($Å^2$)

The data retrieved and collected: [a]PubChem, ChemSpider, ChEMBL (Chemical Database of Bioactive Molecules with Drug-Like Properties) and DNP (Dictionary of Natural Products); [b] clinicaltrials.gov; [c]drugbank different classes as the drug molecules under various clinical trial with different physiochemical properties.

## 1.7 Biological Activities of Plant Secondary Metabolites (PSMs)

PSMs have gained increasing attention because of their key role in plants such as protecting plants from microbial infection and herbivores, as allelopathic agents, signal molecules, UV protectants, and pollinator's attractants. PSMs are also of interest due to their use as flavoring agents, dyes, waxes, fibers, oils, glues, drugs, and perfumes. PSMs has been reported for their anticancer, antioxidant, antibacterial, anti-inflammatory, immune stimulatory, and cardioprotective activities, which make them a potential source of novel natural drugs for therapeutic applications.

### 1.7.1 Antioxidant Effects

Antioxidant compounds protect the cells from the oxidative stress-causing factors which include free radicals like hydroxyl radicals, superoxide, and singlet oxygen. These factors produce reactive oxygen species (ROS) and damage the nucleus of the healthy cells (Velu et al. 2018). The ROS are highly reactive oxidative molecules that are constantly produced by cellular metabolic processes. The free radicals are chemically unstable and acquire electrons from the adjacent nuclear levels like the mitochondria, DNA, lipid membrane, and other proteins in the cell nucleus to become stable which cause the destruction of cellular metabolic pathway and ultimately cell death (Uddin et al. 2008; Lea et al. 2015). This destruction leads to the development of several chronic disorders including diabetes, cardiovascular diseases, tumor, and metabolic syndrome and effects aging (Lin et al. 2016). To scavenge free radicals, the human body synthesizes few antioxidants naturally as its own defense mechanism. However the remaining untreated free radicals are scavenged with the help of antioxidants procured from plant sources like ascorbic acid, vitamin E, β-carotene, and other phytochemicals. The phenolic compounds (phenolic acids, flavonoids, and other phenolics) in plants are widely and effectively recognized as direct antioxidants and indirect antioxidants (Kumar et al. 2016; Stevenson and Hurst 2007). There are several mechanisms contributing to the antioxidant potential of phenolic acid, e.g., reactivity of phenol moiety and hydroxyl substituent on aromatic ring. However, the radical scavenging via hydrogen atom donation is assumed as the key method. Other than phenolic compounds, terpenes can act as potential antioxidants by modulating the endogenous enzyme system and through directly scavenging ROS.

### 1.7.2 Antimicrobial Effects

Since the ancient times, plants were traditionally used as medicine. The increase in antibiotic-resistant microorganisms has directed the attention of researchers onto plants to develop new antimicrobial compounds. A variety of plants produce secondary metabolites to combat against microbial and insect/pest attack. For example, *Salvia officinalis* (sage) a plant native to the Middle East contains a number of plant secondary metabolites. The aqueous and alcoholic extracts of *S. officinalis* contain elevated concentrations of flavonoids and phenolic acids and showed bacteriostatic and bactericidal activities against both Gram-negative and Gram-positive species (Al-Juraifani 2011; Ghorbani and Esmaeilizadeh 2017). *Salvadora persica* L. (miswak) chewing sticks are used as a source of oral hygiene. Aqueous extracts of miswak or the arak tree exhibited antimicrobial activities against seven microbial species, i.e., *Streptococcus faecalis*, *Streptococcus mutans*, *Streptococcus pyogenes*, *Staphylococcus aureus*, *P. aeruginosa*, *Lactobacillus acidophilus*, and *Candida albicans*, tested using the micro-well dilution and disc diffusion methods (Al-Bayati and Sulaiman 2008). Shehadi et al. (2014) tested the extracts of four plants, i.e., mint (*Mentha piperita*), cloves (*Eugenia caryophyllata*), cherry (*Prunus avium*), and rosemary (*R. officinalis*), for their antimicrobial activities against *B. subtilis*. The results showed that cloves exhibited highest antibacterial activity. The phytochemical analysis showed the presence of phenolics, flavonoids, and terpenoids in cloves. A variety of mechanisms including possible reaction with proteins through SH group, enzyme inhibition by oxidized compounds, and other nonspecific interactions are attributed with antimicrobial activities of these polyphenolic compounds (Mason 1987). Some phenolics, e.g., quinones, are the source of stable free radicals and bind irreversibly with proteins, leading to its loss of function. Other targets include inactivating enzymes, binding to cell wall proteins, binding to adhesins on the microbial cell surface, interacting with substrates, and complexing with metal ions (Cowan 1999).

Alkaloids are another class of plant secondary metabolites known for their pharmacological activities. Darabpour et al. (2011) reported antibacterial activities of methanolic extracts from the roots and seeds of *Peganum harmala* against 13 MDR (multidrug-resistant) Gram-positive and Gram-negative bacterial clinical isolates. The roots and seeds of *P. harmala* are rich in alkaloids. Khalil (2012) investigated the antimicrobial activity of the ethanolic leaf extract of *Catharanthus roseus* (100 mg/mL). The results of the disc diffusion method showed inhibition zone of 11, 12, and 15 mm against *E. coli*, *C. albicans*, and *S. aureus*, respectively. Further, they reported that *C. roseus* contains $>$130 different alkaloids. Alkaloids exhibit antimicrobial activities through inhibiting enzyme activity, affecting cell division, inhibiting respiration, and bacterial membrane affecting virulence genes (Othman et al. 2019).

### 1.7.3 Anticancer Effects

Globally, cancer is a most daunting challenge medical science is facing today. The treatment includes surgical removal of solid tumor, radiations, and chemotherapy which reduces the quality of life or have severe side effects. There is a pressing need to invent potential anticancer drug molecules with reduced side effects. In this context, plants are considered as a promising source, and over 60% of anticancer drugs are derived from the kingdom Plantae (Gordaliza 2007). These plant bioactive molecules belong to numerous groups of chemicals such as flavonoids, phenolic acids, carotenoids, tannins, and saponins. These bioactive compounds inactivate enzymes or chemicals and prevent the formation of active species by scavenging. A few of these compounds have become successful for the pharmaceutical industry. For example, Vincristine is a plant-derived anticancer drug and is the first to be approved by the Food and Drug Administration (FDA). It is extracted from the leaves of *Catharanthus roseus* and is an alkaloid, occurring naturally. The drug has been used in chemotherapy in adults against acute lymphoblastic leukemia. It is also used in the treatment of lymphomas, neuroblastoma, rhabdomyosarcoma, and nephroblastoma (Seca and Pinto 2018). Paclitaxel is another novel molecule sold as Taxol® since 1993 and has become the most active cancer chemotherapeutic drug (Bernabeu et al. 2017). It is isolated from the bark of *Taxus brevifolia* Nutt. (Pacific yew) and is a pseudoalkaloid. Numerous studies both in vitro and in vivo have been published on curcumin (diferuloylmethane), a yellow-orange turmeric powder (a polyphenol accumulates in rhizome of *Curcuma longa*). The probable mechanism may be the suppression of transcription factor NF-κ-B, a key protein in many forms of cancer. The inhibition reduces the expression of NF-κ-B target genes such as *COX-2* and *cyclin D1*, leading to apoptosis (Fridlender et al. 2015; Thangapazham et al. 2013). The apoptosis induction by curcumin was also shown when this was jointly given with the estrogen receptor antagonist Tamoxifen for melanoma treatment. *Cannabis sativa* is a plant that contains >60 terpenophenolic compounds, called phytocannabinoids (Chatterjee & Pandey 2011). In the last two centuries, phytocannabinoids are used as supporting drugs for the patients that receive either radiation or chemotherapies (Fridlender et al. 2015).

### 1.7.4 Antidiabetic Effects

Diabetes is another complex metabolic disorder which causes adversities in sexual, cerebrovascular, renal, neurological, and pathophysiological functioning, leading to sublethal to lethal conditions in an individual (El-Abhar and Schaalan 2014). Therefore, arouses need to find antidiabetic agents to restore normal metabolic functioning of a diabetic individual. In this context, plant-based drugs in the form of secondary metabolites have the potential of regulating β-cell functioning, glucose metabolism, GLP-1 homeostasis, and insulin restoration and providing cost-economic and safer treatment for diabetes. Kumar et al. (2017) reported significant antihyperglycemic and antioxidative activity along with anti-inflammatory properties from methonolic extracts of *Gymnema sylvestre* and *Andrographis paniculata* leaves, as shown from the treatment in streptozotocin-induced diabetes in Sprague Dawley rats. Aloe vera gel contains a variety of active ingredients like amino acids, minerals, polyphenols, and polysaccharides. The gel reduces triglycerides and blood glucose level when continuously administrated twice a day (Jerine Peter and Sabina 2016). Similarly, the whole plant extract of *Ocimum sanctum* (Indian traditional therapeutic plants) efficiently improves diabetes mellitus by its antioxidant and insulin-potentiating activities (Kumar and Kumar 2016).

### 1.7.5 Anti-inflammatory Effect

Since antiquity, plant extract has been used as anti-inflammatory compounds. Inflammation occurs in the case of pathogen attack, cell death, tissue injury, degeneration of a part of body, and cancer (Artis and Spits 2015). The natural phytochemicals contain some active ingredients, having soothing and healing effects to wounds and inflammation (Isailovic et al. 2015; Pedraza-Alva et al. 2015). For example, the extract of *Ceratonia siliqua L.* (ripe carob) has been used for curing mouth infections and inflammation (Khalifa 2004). Phytochemical analysis of species *Schinus terebinthifolius* Raddi, *Myracroduo nurundeuva* Allemão, and *Spondias tuberosa* Arruda revealed the presence of numerous secondary metabolites. The most prominent are cinnamic acid, flavonoids, triterpenes, and phenols and are responsible for their anti-inflammatory activities. Similarly, *Ruellia asperula* (Mart. Ex Ness) Lindau, *Solanum paniculatum* L., and *Zingiber officinale* Roscoe also show the presence of biomolecules having anti-inflammatory activity. There are plenty of other compounds that exhibit anti-inflammatory activities and have their origin from plants. Further, the continuous use of these natural phytochemicals may become successful, cost-effective, and safer practice to treat chronic inflammatory ailments.

### 1.7.6 Antidepressant

Depression is another serious health issue faced throughout the world. There are synthetic drugs but have negative impacts, hence, urges researchers to find novel stress-relieving natural molecules. Plant secondary metabolites like terpenes can be used to design natural antidepressant drugs (Bahramsoltani et al. 2015). According to Saki et al. (2014), 25% of drugs used to treat depression are procured from herbs through various extracts. In terpenes, beta-pienes and linalool are commonly active and are extracted from flowers of lavender and medicinal plants: *Tagetes lucida* and *Litsea glaucescens*. In the serotonergic pathway, the monoterpenes interact with the 5HT1A receptors. Serotonins are vital as their release and reuptake levels can be altered to overcome stress (Chaouloff 2000; Guzman-Gutierrez et al. 2012). They also interact with the body's adrenergic receptors that play a key role in stress-induced behavioral changes (Guzman-Gutierrez et al. 2015). Another key finding is the interaction of dopaminergic receptors (D1 receptors) with beta-pinene and is the mechanisms followed by majority of antidepressant drugs accessible in the market (Guzman-Gutierrez et al. 2015). Similarly, hyperforin present in the extracts of *Hypericum perforatum* is another effective antidepressant terpenes and works by inhibiting the neuronal uptake of mood regulators including norepinephrine, serotonin, and dopamine. Additionally, it inhibits the neurotransmitters L-glutamate and GABA uptake, thereby, controls depression (Muller et al. 2001; Cox-Georgian et al. 2019).

## 1.8 Conclusion

PSMs are widely distributed in plant kingdom with infinite combinations of functional groups, viz., hydroxyls, amino, alcohols, aldehydes, carboxyl, alkyls, benzyl, heterocyclic, fused rings, etc., leading to a great diversity of plant compounds. The structural diversity arises due to the presence of various structural features, resulting in a wide range of compounds among the classes of PSMs. The most adopted classification of PSMs depends on the basis of structural nucleus/core structure and functionalities such as molecular structure, substituent groups with their substitution pattern, the linkage type present within molecular structures, etc. The structural features of PSMs are important for the determination of their physicochemical properties which are accountable for their targeted action on living tissues and biological significance. On account of the structural features, these compounds exhibit various biological and therapeutic effects on human health due to their ability to interact with the living tissues.

**Conflict of Interest** None.

## References

Agerbirk N, Olsen CE (2012) Glucosinolate structures in evolution. Phytochemistry 77:16–45

Ahuja I, Kissen R, Bones AM (2012) Phytoalexins in defense against pathogens. Trends Plant Sci 17(2):73–90

Alamgir ANM (2017) Medicinal, non-medicinal, biopesticides, color-and dye-yielding plants; secondary metabolites and drug principles; significance of medicinal plants; use of medicinal plants in the systems of traditional and complementary and alternative medicines (CAMs). In: Therapeutic use of medicinal plants and their extracts, vol 1. Springer, Cham, pp 61–104

Al-Bayati FA, Sulaiman KD (2008) In vitro antimicrobial activity of Salvadora persica L. extracts against some isolated oral pathogens in Iraq. Turk J Biol 32(1):57–62

Al-Juraifani AA (2011) Antimicrobial activity of some medicinal plants used in Saudi Arabia. Can J Pure Appl Sci 5(2):1509–1512

Andrade LN, de Sousa DP (2015) Sesquiterpenes from essential oils and anti-inflammatory activity. Nat Prod Commun 10(10):1767–1774

Aniszewski T (2015) Alkaloid chemistry. In: Alkaloids chemistry, biology, ecology, and applications. Elsevier, pp 99–193

Anwar N, Teo YK, Tan JBL (2019) The role of plant metabolites in drug discovery: current challenges and future perspectives. In: Natural bio-active compounds. Springer, Singapore, pp 25–51

Apetrei C, Ghasemi-Varnamkhasti M, Apetrei IM (2016) Olive oil and combined electronic nose and tongue. In: Electronic noses and tongues in food science. Academic Press, pp 277–289

Artis D, Spits H (2015) The biology of innate lymphoid cells. Nature 517(7534):293–301

Atkinson RG (2016) Phenylpropenes: occurrence, distribution, and biosynthesis in fruit. J Agric Food Chem 66(10):2259–2272

Bahramsoltani R, Farzaei MH, Farahani MS, Rahimi R (2015) Phytochemical constituents as future antidepressants: a comprehensive review. Rev Neurosci 26(6):699–719

Baser KHC, Demirci F (2007) Chemistry of essential oils. In: Berger RG (ed) Flavours and fragrances: chemistry, bioprocessing and sustainability. Springer, New York, pp 43–86

Batra P, Sharma AK (2013) Anti-cancer potential of flavonoids: recent trends and future perspectives. 3Biotech 3(6):439–459

Baur R, Städler E, Monde K, Takasugi M (1998) Phytoalexins from Brassica (Cruciferae) as oviposition stimulants for the cabbage root fly, Delia radicum. Chemoecology 8(4):163–168

Bergman ME, Davis B, Phillips MA (2019) Medically useful plant terpenoids: biosynthesis, occurrence, and mechanism of action. Molecules 24(21):3961

Bernabeu E, Cagel M, Lagomarsino E, Moretton M, Chiappetta DA (2017) Paclitaxel: what has been done and the challenges remain ahead. Int J Pharm 526(1–2):474–495

Bienz S, Bisegger P, Guggisberg A, Hesse M (2005) Polyamine alkaloids. Nat Prod Rep 22(5):647–658

Biharee A, Sharma A, Kumar A, Jaitak V (2020) Antimicrobial flavonoids as a potential substitute for overcoming antimicrobial resistance. Fitoterapia, p 104720

Blazevic I, Montaut S, Burčul F, Olsen CE, Burow M, Rollin P, Agerbirk N (2020) Glucosinolate structural diversity, identification, chemical synthesis and metabolism in plants. Phytochemistry 169:112100

Bodoira R, Maestri D (2020) Phenolic compounds from nuts: extraction, chemical profiles, and bioactivity. J Agric Food Chem 68(4):927–942

Boncan DAT, Tsang SS, Li C, Lee IH, Lam HM, Chan TF, Hui JH (2020) Terpenes and terpenoids in plants: interactions with environment and insects. Int J Mol Sci 21(19):7382

Boniface PK, Ferreira EI (2019) Flavonoids as efficient scaffolds: recent trends for malaria, leishmaniasis, Chagas disease, and dengue. Phytother Res 33(10):2473–2517

Bottger A, Vothknecht U, Bolle C, Wolf A (2018) Plant secondary metabolites and their general function in plants. In: Lessons on caffeine, Cannabis & Co. Springer, Cham, pp 3–17

Boufridi A, Quinn RJ (2018) Harnessing the properties of natural products. Annu Rev Pharmacol Toxicol 58:451–470

Chadwick M, Trewin H, Gawthrop F, Wagstaff C (2013) Sesquiterpenoids lactones: benefits to plants and people. Int J Mol Sci 14(6):12780–12805

Chaouloff F (2000) Serotonin, stress and corticoids. J Psychopharmacol 14(2):139–151

Chatterjee SJ, Pandey S (2011) Chemo-resistant melanoma sensitized by tamoxifen to low dose curcumin treatment through induction of apoptosis and autophagy. Cancer Biol Ther 11(2):216–228

Cheuka PM, Mayoka G, Mutai P, Chibale K (2017) The role of natural products in drug discovery and development against neglected tropical diseases. Molecules 22(1):58

Cheynier V (2012) Phenolic compounds: from plants to foods. Phytochem Rev 11(2–3):153–177

Chien SC, Wu YC, Chen ZW, Yang WC (2015) Naturally occurring anthraquinones: chemistry and therapeutic potential in autoimmune diabetes. Evid Based Complement Alternat Med 2015:357357. https://doi.org/10.1155/2015/357357

Chong J, Poutaraud A, Hugueney P (2009) Metabolism and roles of stilbenes in plants. Plant Sci 177(3):143–155

Choudhari AS, Mandave PC, Deshpande M, Ranjekar P, Prakash O (2020) Phytochemicals in cancer treatment: from preclinical studies to clinical practice. Front Pharmacol 10:1614

Congreve M, Carr R, Murray C, Jhoti H (2003) A 'rule of three' for fragment-based lead discovery? Drug Discov Today 19(8):876–877
Corcoran MP, McKay DL, Blumberg JB (2012) Flavonoid basics: chemistry, sources, mechanisms of action, and safety. J Nutr Gerontol Geriatrics 31(3):176–189
Cowan MM (1999) Plant products as antimicrobial agents. Clin Microbiol Rev 12(4):564–582
Cox-Georgian D, Ramadoss N, Dona C, Basu C (2019) Therapeutic and medicinal uses of terpenes. In: Medicinal plants. Springer, Cham, pp 333–359
Crozier A, Jaganath IB, Clifford MN (2006) Phenols, polyphenols and tannins: an overview. In: Plant secondary metabolites: occurrence, structure and role in the human diet, pp 1–24
Darabpour E, Bavi AP, Motamedi H, Nejad SMS (2011) Antibacterial activity of different parts of Peganum harmala L. growing in Iran against multi-drug resistant bacteria. EXCLI J 10:252
Dev S (1989) Terpenoids. In: Natural products of woody plants. Springer, Berlin, pp 691–807
Eguchi R, Ono N, Morita AH, Katsuragi T, Nakamura S, Huang M, Altaf-Ul-Amin M, Kanaya S (2019) Classification of alkaloids according to the starting substances of their biosynthetic pathways using graph convolutional neural networks. BMC Bioinformatics 20(1):380
El-Abhar HS, Schaalan MF (2014) Phytotherapy in diabetes: review on potential mechanistic perspectives. World J Diabetes 5(2):176
Erb M, Kliebenstein DJ (2020) Plant secondary metabolites as defenses, regulators, and primary metabolites: the blurred functional trichotomy. Plant Physiol 184(1):39–52
Fridlender M, Kapulnik Y, Koltai H (2015) Plant derived substances with anti-cancer activity: from folklore to practice. Front Plant Sci 6:799
Ganz T (2003) Defensins: antimicrobial peptides of innate immunity. Nat Rev Immunol 3(9):710–720
Ghorbani A, Esmaeilizadeh M (2017) Pharmacological properties of Salvia officinalis and its components. J Tradit Complement Med 7(4):433–440
Gonzalez-Vallinas M, González-Castejón M, Rodríguez-Casado A, Ramírez de Molina A (2013) Dietary phytochemicals in cancer prevention and therapy: a complementary approach with promising perspectives. Nutr Rev 71(9):585–599
Gordaliza M (2007) Natural products as leads to anticancer drugs. Clin Transl Oncol 9(12):767–776
Gorlenko CL, Kiselev HY, Budanova EV, Zamyatnin AA, Ikryannikova LN (2020) Plant secondary metabolites in the battle of drugs and drug-resistant bacteria: new heroes or worse clones of antibiotics? Antibiotics 9(4):170
Guggisberg A, Hesse M (2003) Alkaloids. In: Meyers RA (ed) Encyclopedia of physical science and technology, 3rd edn. Academic Press, pp 477–493., ISBN 9780122274107. https://doi.org/10.1016/B0-12-227410-5/00021-1
Guzman-Gutierrez SL, Gómez-Cansino R, García-Zebadúa JC, Jiménez-Pérez NC, Reyes-Chilpa R (2012) Antidepressant activity of Litsea glaucescens essential oil: identification of β-pinene and linalool as active principles. J Ethnopharmacol 143(2):673–679
Guzman-Gutierrez SL, Bonilla-Jaime H, Gómez-Cansino R, Reyes-Chilpa R (2015) Linalool and β-pinene exert their antidepressant-like activity through the monoaminergic pathway. Life Sci 128:24–29
Harborne JB (1984) The terpenoids. In: Phytochemical methods. Springer, Dordrecht, pp 100–141. https://doi.org/10.1007/978-94-009-5570-7_3
Harborne JB (1989) General procedures and measurement of total phenolics. Methods Plant Biochem 1:1–28
Harman-Ware AE (2020) Conversion of terpenes to chemicals and related products. In: Chemical catalysts for biomass upgrading, pp 529–568
Hayes M, Bleakley S (2018) Peptides from plants and their applications. In: Peptide applications in biomedicine, biotechnology and bioengineering. Woodhead Publishing, pp 603–622
Hernandez-Rodríguez P, Baquero LP, Larrota HR (2019) Flavonoids: potential therapeutic agents by their antioxidant capacity. In: Bioactive compounds. Woodhead Publishing, pp 265–288
Hewlings SJ, Kalman DS (2017) Curcumin: a review of its' effects on human health. Foods 6(10):92

Ibanez S, Gallet C, Després L (2012) Plant insecticidal toxins in ecological networks. Toxins 4(4):228–243
Inouhe M (2005) Phytochelatins. Braz J Plant Physiol 17(1):65–78
Isailovic N, Daigo K, Mantovani A, Selmi C (2015) Interleukin-17 and innate immunity in infections and chronic inflammation. J Autoimmun 60:1–11
Jablonsky M, Haz A, Sladkova A, Strizincova P, Skulcova A, Majova V, Jablonsky J (2019) Nutraceuticals as phenolic bioactive compounds analysis of softwood bark and their possibilities of industry applications. J Hygienic Eng Design 26:93–99
Jerine Peter S, Sabina EP (2016) Global current trends in natural products for diabetes management: a review. Int J Pharm Sci 8:20–28
Joullié MM, Richard DJ (2004) Cyclopeptide alkaloids: chemistry and biology. Chem Commun 18: 2011–2015
Khalifa AB (2004) Herbs: nature's pharmacy, 1st edn. Arab Cultural Center, Casablanca, pp 286–288
Khalil A (2012) Antimicrobial activity of ethanol leaf extracts of Catharanthus roseus from Saudi Arabia. In: 2nd International Conference on Environment Science and Biotechnology 48 (2):6–11
Kim T, Song B, Cho KS, Lee IS (2020) Therapeutic potential of volatile terpenes and terpenoids from forests for inflammatory diseases. Int J Mol Sci 21(6):2187
Klein-Júnior LC, Campos A, Niero R, Corrêa R, Vander Heyden Y, Filho VC (2020) Xanthones and cancer: from natural sources to mechanisms of action. Chem Biodiv 17(2):e1900499
Koeduka T (2014) The phenylpropene synthase pathway and its applications in the engineering of volatile phenylpropanoids in plants. Plant Biotechnol 31(5):401–407
Korkina L, Kostyuk V, Potapovich A, Mayer W, Talib N, De Luca C (2018) Secondary plant metabolites for sun protective cosmetics: from pre-selection to product formulation. Cosmetics 5(2):32
Kumar N, Goel N (2019) Phenolic acids: natural versatile molecules with promising therapeutic applications. Biotechnol Rep 24:e00370
Kumar B, Kumar A (2016) Diabetes mellitus and its control by Ocimum sanctum extract in mice diabetic model. Int J Curr Microbiol App Sci 5:795–810
Kumar BP, Soni M, Bhikhalal UB, Kakkot IR, Jagadeesh M, Bommu P, Nanjan MJ (2010) Analysis of physicochemical properties for drugs from nature. Med Chem Res 19(8):984–992
Kumar N, Kumar S, Abbat S, Nikhil K, Sondhi SM, Bharatam PV, Roy P, Pruthi V (2016) Ferulic acid amide derivatives as anticancer and antioxidant agents: synthesis, thermal, biological and computational studies. Med Chem Res 25(6):1175–1192
Kumar P, Rani S, Arunjyothi B, Chakrapani P, Rojarani A (2017) Evaluation of antidiabetic activity of Gymnema sylvestre and Andrographis paniculata in streptozotocin induced diabetic rats. Int J Pharmacogn Phytochem Res 9:22–25
LaLonde RT (2005) Terpenes and terpenoids. In: Considine GD (ed) Van Nostrand's encyclopedia of chemistry. https://doi.org/10.1002/0471740039.vec2473
Lea AJ, Tung J, Zhou X (2015) A flexible, efficient binomial mixed model for identifying differential DNA methylation in bisulfite sequencing data. PLoS Genet 11(11):e1005650
Leavell MD, McPhee DJ, Paddon CJ (2016) Developing fermentative terpenoid production for commercial usage. Curr Opin Biotechnol 37:114–119
Li C, Miao X, Li F, Adhikari BK, Liu Y, Sun J, Zhang R, Cai L, Liu Q, Wang Y (2019) Curcuminoids: implication for inflammation and oxidative stress in cardiovascular diseases. Phytother Res 33(5):1302–1317
Lin D, Xiao M, Zhao J, Li Z, Xing B, Li X, Kong M, Li L, Zhang Q, Liu Y, Chen H (2016) An overview of plant phenolic compounds and their importance in human nutrition and management of type 2 diabetes. Molecules 21(10):1374
Lipinski CA, Lombardo F, Dominy BW, Feeney PJ (1997) Experimental and computational approaches to estimate solubility and permeability in drug discovery and development settings. Adv Drug Deliv Rev 23(1–3):3–25

Liu K, Rossi PG, Ferrari B, Berti L, Casanova J, Tomi F (2007) Composition, irregular terpenoids, chemical variability and antibacterial activity of the essential oil from Santolina corsica Jordan et Fourr. Phytochemistry 68(12):1698–1705

Lyu X, Lee J, Chen WN (2019) Potential natural food preservatives and their sustainable production in yeast: terpenoids and polyphenols. J Agric Food Chem 67(16):4397–4417

Maaliki D, Shaito AA, Pintus G, El-Yazbi A, Eid AH (2019) Flavonoids in hypertension: a brief review of the underlying mechanisms. Curr Opin Pharmacol 45:57–65

Machado LR, Ottolini B (2015) An evolutionary history of defensins: a role for copy number variation in maximizing host innate and adaptive immune responses. Front Immunol 6:115

Malik EM, Müller CE (2016) Anthraquinones as pharmacological tools and drugs. Med Res Rev 36(4):705–748

Mason TL (1987) Inactivation of red beet β-glucan synthase by native and oxidized phenolic compounds. Phytochemistry 26(8):2197–2202

Matos MJ, Santana L, Uriarte E, Abreu OA, Molina E, Yordi EG (2015) Coumarins—an important class of phytochemicals. In: Phytochemicals-isolation, characterisation and role in human health, pp 113–140

Mignani S, Rodrigues J, Tomas H, Jalal R, Singh PP, Majoral JP, Vishwakarma RA (2018) Present drug-likeness filters in medicinal chemistry during the hit and lead optimization process: how far can they be simplified? Drug Discov Today 23(3):605–615

Moller BL, Olsen CE, Motawia MS (2016) General and stereocontrolled approach to the chemical synthesis of naturally occurring cyanogenic glucosides. J Nat Prod 79(4):1198–1202

Muller WE, Singer A, Wonnemann M (2001) Hyperforin-antidepressant activity by a novel mechanism of action. Pharmacopsychiatry 34(Sup. 1):98–102

Nagren K (2003) PET and knockout mice in drug discovery. Drug Discov Today 8(19):876–877

Nazhand A, Durazzo A, Lucarini M, Romano R, Mobilia MA, Izzo AA, Santini A (2020) Human health-related properties of chromones: an overview. Nat Prod Res 34(1):137–152

Negi JS, Bisht VK, Singh P, Rawat MSM, Joshi GP (2013) Naturally occurring xanthones: chemistry and biology. J Appl Chem 2013(1):1–9

Newman DJ, Cragg GM (2016) Natural products as sources of new drugs from 1981 to 2014. J Nat Prod 79(3):629–661

Ntalli N, Koliopoulos G, Giatropoulos A, Menkissoglu-Spiroudi U (2019) Plant secondary metabolites against arthropods of medical importance. Phytochem Rev 18(5):1255–1275

O'Connor SE (2010) 1.25-Alkaloids. In: Comprehensive natural products II, vol 1, pp 977–1007

Odintsova TI, Slezina MP, Istomina EA (2018) Plant thionins: structure, biological functions and potential use in biotechnology. Vavilovskii Zhurnal Genetiki I Selektsii 22(6):667–675

Othman L, Sleiman A, Abdel-Massih RM (2019) Antimicrobial activity of polyphenols and alkaloids in middle eastern plants. Front Microbiol 10:911

Pedraza-Alva G, Pérez-Martínez L, Valdez-Hernández L, Meza-Sosa KF, Ando-Kuri M (2015) Negative regulation of the inflammasome: keeping inflammation under control. Immunol Rev 265(1):231–257

Pereira DM, Valentão P, Pereira JA, Andrade PB (2009) Phenolics: from chemistry to biology. Molecules 2009(14):2202–2211. https://doi.org/10.3390/molecules14062202

Quinn RJ, Carroll AR, Pham NB, Baron P, Palframan ME, Suraweera L, Pierens GK, Muresan S (2008) Developing a drug-like natural product library. J Nat Prod 71(3):464–468

Rattan RS (2010) Mechanism of action of insecticidal secondary metabolites of plant origin. Crop Prot 29(9):913–920

Reinisalo M, Karlund A, Koskela A, Kaarniranta K, Karjalainen RO (2015) Polyphenol stilbenes: molecular mechanisms of defence against oxidative stress and aging-related diseases. Oxidative Med Cell Longev 2015:340520. https://doi.org/10.1155/2015/340520

Rengasamy KR, Khan H, Gowrishankar S, Lagoa RJ, Mahomoodally FM, Khan Z, Suroowan S, Tewari D, Zengin G, Hassan ST, Pandian SK (2019) The role of flavonoids in autoimmune diseases: therapeutic updates. Pharmacol Ther 194:107–131

Rentzsch M, Wilkens A, Winterhalter P (2009) Non-flavonoid phenolic compounds. In: Wine chemistry and biochemistry. Springer, New York, pp 509–527

Ruszkowska J, Wróbel JT (2003) Tryptophan-derived sulfur-containing phytoalexins—a general overview. In: Developments in tryptophan and serotonin metabolism. Springer, Boston, MA, pp 629–636

Saeed A (2016) Isocoumarins, miraculous natural products blessed with diverse pharmacological activities. Eur J Med Chem 116:290–317

Saghyan AS, Langer P (2016) Non-proteinogenic α-amino acids, natural origin, biological functions. In: Saghyan AS, Langer P (eds) Asymmetric synthesis of non-proteinogenic amino acids. https://doi.org/10.1002/9783527804498.ch1

Saki K, Bahmani M, Rafieian-Kopaei M (2014) The effect of most important medicinal plants on two importnt psychiatric disorders (anxiety and depression)-a review. Asian Pac J Trop Med 1 (7):S34–S42

Sato F (2014) Plant secondary metabolism. In: eLS. Wiley. https://doi.org/10.1002/9780470015902.a0001812.pub2

Schmidt U, Lieberknecht A, Haslinger E (1985) Peptide alkaloids. In: The alkaloids: chemistry and pharmacology, vol 26. Academic Press, pp 299–326

Seca AM, Pinto DC (2018) Plant secondary metabolites as anticancer agents: successes in clinical trials and therapeutic application. Int J Mol Sci 19(1):263

Sell CS (2007) A fragrant introduction to terpenoid chemistry. R Soc Chem. https://doi.org/10.1039/9781847550019

Semwal RB, Semwal DK, Combrinck S et al (2020) Health benefits of chromones: common ingredients of our daily diet. Phytochem Rev 19:761–785. https://doi.org/10.1007/s11101-020-09681-w

Shahidi F, Ambigaipalan P (2015) Phenolics and polyphenolics in foods, beverages and spices: antioxidant activity and health effects—a review. J Funct Foods 18:820–897

Sharma A, Sharma P, Tuli HS, Sharma AK (2018) Phytochemical and pharmacological properties of flavonols. In: eLS, pp 1–12

Sharma A, Tuli HS, Sharma AK (2019) Chemistry and synthetic overview of flavonoids. In: Current aspects of flavonoids: their role in cancer treatment. Springer, Singapore, pp 23–38

Shehadi M, Awada F, Oleik R, Chokr A, Hamze K, Abou Hamdan H, HARB A, KOBAISSI A (2014) Comparative analysis of the anti-bacterial activity of four plant extracts. Int J Curr Res Aca Rev 2(6):83–94

Stevenson DE, Hurst RD (2007) Polyphenolic phytochemicals–just antioxidants or much more? Cell Mol Life Sci 64(22):2900–2916

Takshak S, Agrawal SB (2019) Defense potential of secondary metabolites in medicinal plants under UV-B stress. J Photochem Photobiol B Biol 193:51–88

Teponno RB, Kusari S, Spiteller M (2016) Recent advances in research on lignans and neolignans. Nat Prod Rep 33(9):1044–1092

Thangapazham RL, Sharad S, Maheshwari RK (2013) Skin regenerative potentials of curcumin. Biofactors 39(1):141–149

Turner JV, Agatonovic-Kustrin S (2007) In silico prediction of oral bioavailability. In: Comprehensive medicinal chemistry II ADME Tox approaches. Elsevier Ltd, pp 699–724

Uddin SN, Akond MA, Mubassara S, Yesmin MN (2008) Antioxidant and antibacterial activities of Trema cannabina. Middle East J Sci Res 3(2):105–108

Uddin M, Kabir M, Niaz K, Jeandet P, Clément C, Mathew B, Rauf A, Rengasamy KR, Sobarzo-Sánchez E, Ashraf GM, Aleya L (2020) Molecular insight into the therapeutic promise of flavonoids against Alzheimer's disease. Molecules 25(6):1267

Van Sumere CF (1989) Phenols and phenolic acids. In: Methods in plant biochemistry, vol 1. Academic Press, pp 29–73

Veber DF, Johnson SR, Cheng HY, Smith BR, Ward KW, Kopple KD (2002) Molecular properties that influence the oral bioavailability of drug candidates. J Med Chem 45(12):2615–2623

Velderrain-Rodríguez GR, Palafox-Carlos H, Wall-Medrano A, Ayala-Zavala JF, Chen CO, Robles-Sánchez M, Astiazaran-García H, Alvarez-Parrilla E, González-Aguilar GA (2014) Phenolic compounds: their journey after intake. Food Funct 5(2):189–197

Velu G, Palanichamy V, Rajan AP (2018) Phytochemical and pharmacological importance of plant secondary metabolites in modern medicine. In: Bioorganic phase in natural food: an overview. Springer, Cham, pp 135–156

Vetter J (2000) Plant cyanogenic glycosides. Toxicon 38(1):11–36

Wang S, Alseekh S, Fernie AR, Luo J (2019) The structure and function of major plant metabolite modifications. Mol Plant 12(7):899–919

Ward RS (2000) Recent advances in the chemistry of lignans. In: Studies in natural products chemistry, vol 24. Elsevier, pp 739–798

Wellington KW (2015) Understanding cancer and the anticancer activities of naphthoquinones—a review. RSC Adv 5(26):20309–20338

Wink M (2016) Role in plant diversification of the secondary metabolites. In: Encyclopedia of evolutionary biology, pp 1–9

Xu R, Fazio GC, Matsuda SP (2004) On the origins of triterpenoid skeletal diversity. Phytochemistry 65(3):261–291

Yang W, Chen X, Li Y, Guo S, Wang Z, Yu X (2020) Advances in pharmacological activities of terpenoids. Nat Prod Commun 15(3):1934578X20903555

Yerer MB, Dayan S, Han MI, Sharma A, Tuli HS, Sak K (2020) Nanoformulations of coumarins and the hybrid molecules of coumarins with anticancer effects. Anti Cancer Agents Med Chem 20(15):1797–1816. https://doi.org/10.2174/1871520620666200310094646

Yoshida K, Oyama KI, Kondo T (2012) Chemistry of flavonoids in color development. Recent Adv Polyphenol Res 3:99–129

Young RJ (2014) Physical properties in drug design. Top Med Chem 9:1–68

Zeng T, Liu Z, Liu H, He W, Tang X, Xie L, Wu R (2019) Exploring chemical and biological space of terpenoids. J Chem Inf Model 59(9):3667–3678

# Natural Sources of Plant Secondary Metabolites and the Role of Plant Polyphenols in the Green Photosynthesis of Metallic Nanoparticles

2

Ioana Catalina Fierascu, Irina Fierascu, Radu Claudiu Fierascu, Bruno Stefan Velescu, and Cristina Elena Dinu-Pirvu

**Abstract**

The present chapter has been aimed to offer a glimpse into the fascinating domain of nanoparticle phytosynthesis, instead of being the exhaustive inventory of all the possibilities and compounds involved in the process. Different authors suggest different phenolic compounds being responsible for the reduction and capping of the nanoparticles. This is one of the aspects that needs future research, in order to completely elucidate the process and contribution of particular compounds to the phytosynthesis. Another aspect that should be continuously explored is the recovery of plant secondary metabolites (PSM) from plant wastes resulting from various industrial applications. This circular approach could lead

I. C. Fierascu
University of Medicine and Pharmacy "Carol Davila", Bucharest, Romania

Zentiva Romania S.A., Bucharest, Romania

I. Fierascu
University of Agronomic Sciences and Veterinary Medicine from Bucharest, Bucharest, Romania

Emerging Nanotechnologies Group, National Institute for Research & Development in Chemistry and Petrochemistry – ICECHIM, Bucharest, Romania

R. C. Fierascu (✉)
Emerging Nanotechnologies Group, National Institute for Research & Development in Chemistry and Petrochemistry – ICECHIM, Bucharest, Romania

Politehnica University of Bucharest, Bucharest, Romania

B. S. Velescu
University of Medicine and Pharmacy "Carol Davila", Bucharest, Romania

C. E. Dinu-Pirvu
University of Medicine and Pharmacy "Carol Davila", Bucharest, Romania

University of Agronomic Sciences and Veterinary Medicine from Bucharest, Bucharest, Romania



A. K. Sharma, A. Sharma (eds.), *Plant Secondary Metabolites*,
https://doi.org/10.1007/978-981-16-4779-6_2

not only to a decrease in the overall waste amounts but could also provide the necessary compounds for the nanotechnological processes.

**Keywords**

Plant secondary metabolites (PSM) · Natural sources · Plant polyphenols · Metallic nanoparticles · Phytosynthesis

## 2.1 Plant Secondary Metabolites: From Plants to Industrial Products

Plant secondary metabolites (PSM) represent the basis of phytochemistry discipline. At the same time, these compounds play a critical role in the adaptation of plants to external factors, having multiple actions, including antifungal, antibacterial, antiviral, anti-germinative, anti-feeding or even UV-blocking properties (Li et al. 1993; Zaynab et al. 2018). The role of PSM in plant's defence mechanisms represented the subject of several review papers (Singer et al. 2003; Zaynab et al. 2018). The production of PSM can be enhanced using biotic and abiotic elicitation (Verma and Shukla 2015; Thakur et al. 2019). With the advances in both separation techniques and analytical procedures (Zhang et al. 2011), PSM were found to be useful in a large variety of applications, especially related to human health (Vacca et al. 2016; Singh et al. 2018). According to Bourgaud et al. (2001), PSM generally belong to several classes of compounds: phenolics, terpenes, steroids and alkaloids, respectively. If the phenolic compounds are widespread among higher plants, PSM belonging to the alkaloid family are much more specific and sparsely distributed. In the last decades, several PSM passed the border between laboratory research and industrial applications, resulting in commercially available products, with important health benefits (Table 2.1).

The above presented examples represent only a small peek into the very complex field of PSM uses in industrial applications. In fact, according to Lubbe and Verpoorte (2011), more than 25% of the pharmaceuticals contain or are derived from PSM.

## 2.2 Plant Polyphenols

In the last decade, interest from the research community has risen in what polyphenols are concerned. This increase can be attributed to the very important antioxidant role of polyphenols but also to more recent discoveries in the prevention of several degenerative diseases (Manach et al. 2004; Hu et al. 2013). Another reason for the progressively higher interest concerning the medical potential of plant polyphenols can be attributed to the abundance of these compounds in the natural sources such as vegetables, fruits and whole grains. From the perspective of the number of phenol rings in their componence, polyphenols have been classified into four distinct categories, phenolic acids, lignans, stilbenes and flavonoids, all of which will be further discussed in more details (Hu et al. 2013).

**Table 2.1** Pharmaceutical relevant PSM in commercial products

| Molecule | Class | Structure | Source | Action | Product | References |
|---|---|---|---|---|---|---|
| Morphine | Alkaloid | 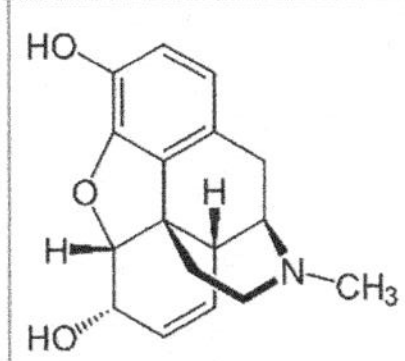 | *Papaver somniferous* L. | Analgesic | Morphine | Stuart-Harris et al. (2000) |
| Quinine | Alkaloid | 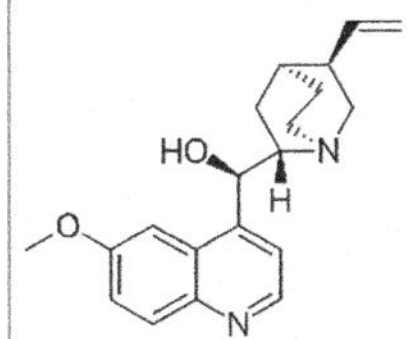 | *Cinchona officinalis* L. | Antimalarial | Quinine | Barennes et al. (1996) |
| Artemisinin | Terpene | 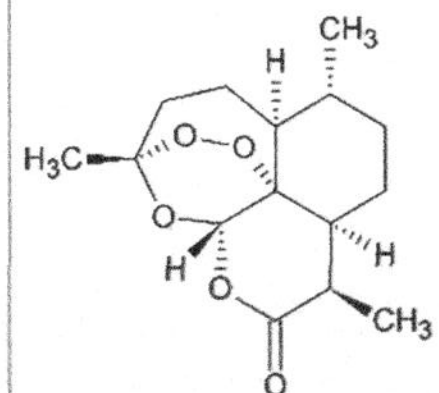 | *Artemisia annua* L. | Antimalarial | Artemisinin | Brown (2006) |
| Taxol | Terpene | 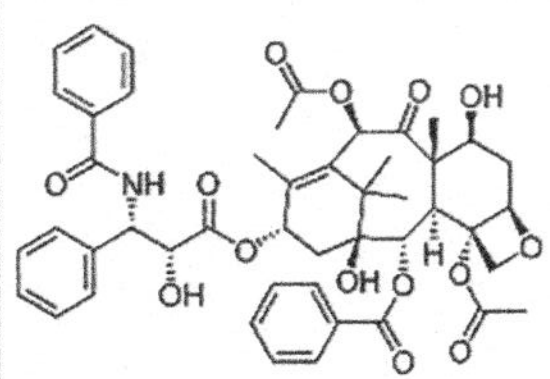 | *Taxus brevifolia* Nutt. | Anticancer | Paclitaxel | Beth (2014) |

(continued)

**Table 2.1** (continued)

| Molecule | Class | Structure | Source | Action | Product | References |
|---|---|---|---|---|---|---|
| Podophyllotoxin | Lignane | 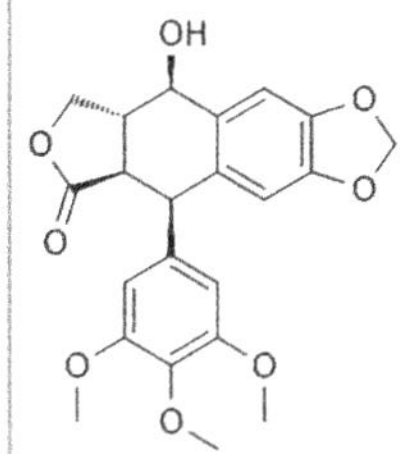 | *Podophyllum peltatum* L. | Anticancer | Condyline | Ardalani et al. (2017) |
| Crofelemer | Oligomeric proanthocyanidin | 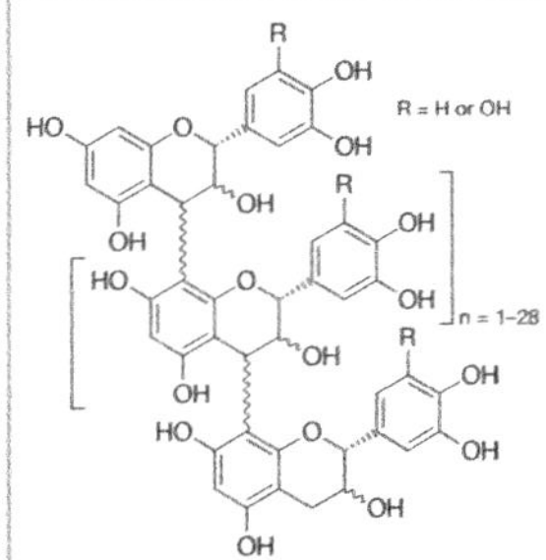 | *Senegalia catechu* (L.f.) P.J.H. Hurter & Mabb. | Antidiarrhoeal (HIV associated) | Fulyzaq | Kleindl et al. (2017) |
| Scopolamine | Alkaloid | 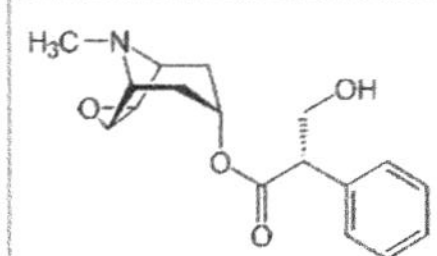 | *Atropa belladonna* L. | Sedative | Scopolamine | Sramska et al. (2017) |

| | | | | | | |
|---|---|---|---|---|---|---|
| Omacetaxine mepesuccinate | Alkaloid | 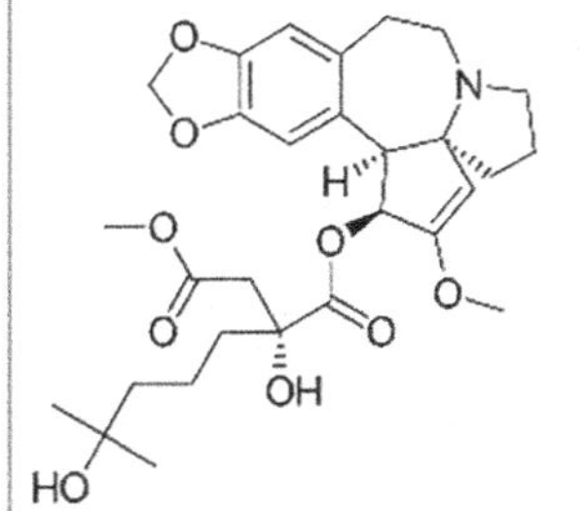 | *Cephalotaxus fortunei* hook. | Leukaemia treatment | Synribo | Atanasov et al. (2015) |
| Digoxin | Steroidal glycosides | 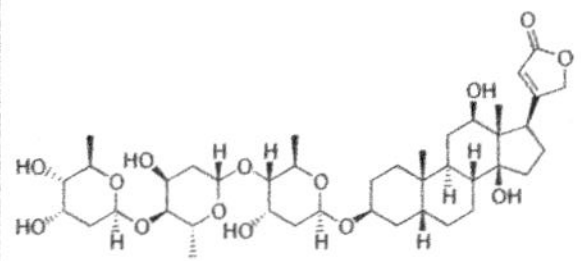 | *Digitalis lanata* Ehrh., 1792 | Heart disorders | Lanoxin | Lubbe and Verpoorte (2011) |

### 2.2.1 Flavonoids

As a result of intense research, flavonoids have gained recognition for their properties, some of high interest in the medical fields, such as prevention of cognitive diseases (Spencer et al. 2017; Panickar and Jang 2013). Considering their chemical structure, approximately 5000 flavonoids have been categorized in the following subgroups: flavones, isoflavones, flavonones, flavonols, flavanols and anthocyanidins (Ververidsi et al. 2007; van de Glind et al. 2013). This group of polyphenols has gained continuous interest due to some important factors, such as low toxicity and low cost, but also their benefits towards human health and the rare occurrence of adverse effects (Ververidsi et al. 2007) (Fig. 2.1).

One of the most studied examples of the effects of a flavonoid on the cognitive deficits is that of luteolin. Yoo et al. (2013) conducted a study on the potential benefits of luteolin on scopolamine-induced cognitive deficits. In a different study, Lin et al. (2016) carried out a study in order to assess the role of luteolin acting as a barrier against kainic acid-induced brain damage on rats.

Other flavonoids, such as quercetin, hesperitin, phloretin and naringenin, have proven their effect on improving the memory function during the studies performed on rodents, after exposing the subjects to chronic stress, isoflurane or streptozotocin (Ghumatkar et al. 2015; Sun et al. 2016; Kheradmand et al. 2018). Although there is a shortage in scientific data, the existing clinical studies endorse the benefits of flavonoid-rich aliments.

Mancini et al. (2017) have assessed in their paper the literature regarding the improved brain functions of subjects consuming green tea, rich in flavonoids, but

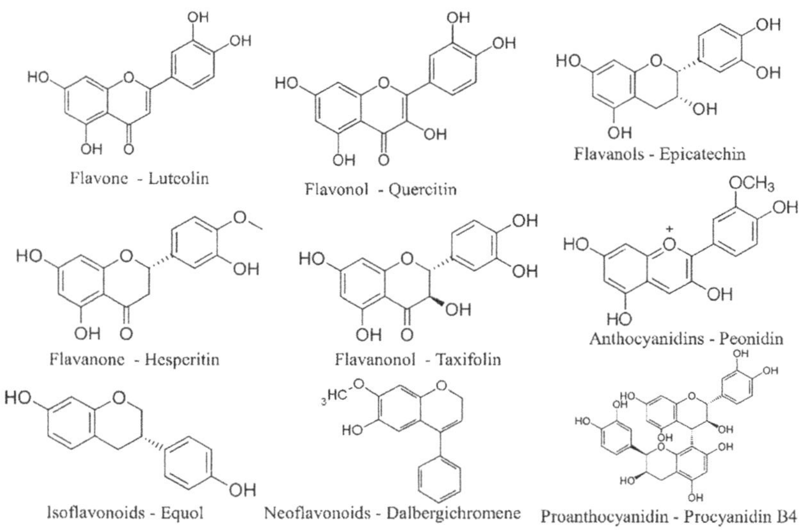


**Fig. 2.1** Chemical structures of biological relevant flavonoids

could not relate to point out a direct connection between the flavonoids' contribution and the positive results on neurological functions.

### 2.2.2 Lignans

As the number of patients suffering from chronic diseases has risen by each year, so has the interest of medical researchers in studying potential new treatments. Adlercreutz (2007) have discussed in their paper the influence of a diet rich in plants over the occurrence of several types of cancer and other chronical diseases as a result of high concentration of phytoestrogens. Phytoestrogens are defined as natural compounds which present similarities with the mammalian oestrogens. From this big class, an important group of compounds can be differentiated as lignans, characterized by a 2–3-dibenzylbutane structure. Lignans are found in fruits and have a vast pharmacological application, starting from antioxidant, nervous system protector, anticancer and anti-inflammatory (Whitten et al. 1997; Zhang et al. 2014) (Fig. 2.2).

The potential effect on neurotoxicity and cognitive impairment of lignans extracted from the fruit of *Schisandra chinensis* have been studied in several studies. Jeong et al. managed to realize the extraction using n-hexane:ethanol (9:1), observing through behavioural tests the cognitive impairment, suggesting a possible breakthrough in the Alzheimer's treatment (Jeong et al. 2013).

López-Biedma et al. (2016) analysed in their review paper the current literature concerning the major lignans found in olives and virgin olive oils, their chemical structures and distribution but also the available information regarding the benefits upon human health of olive- and virgin olive oil-extracted lignans. Two of the most important lignans found in virgin olive oils are 1-acetoxypinoresinol and

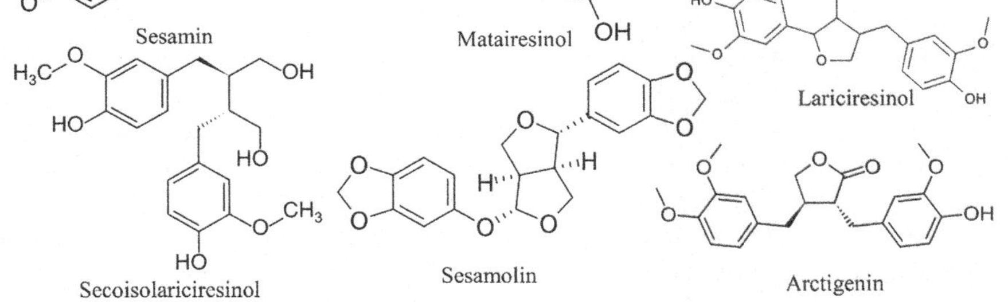


**Fig. 2.2** Chemical structure of important lignans

(+)-pinoresinol. Menendez et al. (2008) highlighted through their study the potential of these lignans in the treatment of breast carcinomas and also the antifungal properties of (+)-pinoresinol.

Previous studies have investigated the anticancer properties of sesamol which is the lignan extracted from sesame seeds (*Sesamum indicum* L.). Majdalawieh and Mansour (2019) have analysed the current literature, providing information towards the anti-inflammatory, antioxidant and chemopreventive action of sesamol.

### 2.2.3 Stilbenes

One other important category of polyphenols are the stilbenes, which can be described as natural compounds with the specificity of 1,2-diphenylethylene backbone. From this category, the most studied stilbene is resveratrol, a natural phytoalexin, which presents a high occurrence in blueberries, cranberries and peanuts and most often grapes, more specifically, the skin of the grapes (Chong et al. 2009). Vestergaard and Ingmer (2019) have discussed the currently known medical applications of resveratrol as an antimicrobial and antifungal, subsequently discussing the disadvantage given by the modest availability for oral administration of stilbene. In a different paper, the cell protective ability, antioxidative properties, the ability to induce cancer cells' inhibition and the cardioprotective action are discussed. The authors' attention to the potential cardioprotective action was attributed to the multitude of previous studies where the consumption of red wine has been connected with a lower incidence of heart diseases (Das and Maulik 2006). Sun et al. (2006) investigated in their paper the possibility of using resveratrol as a proliferating agent against multiple myeloma human cells, concluding that the presented stilbene could have a significant role in the future cancer treatments.

### 2.2.4 Phenolic Acid

Also a big category of the polyphenol family, the phenolic acid can be distinguished by the chemical structure, which includes a phenolic ring and an organic carboxylic acid. Regarding their occurrence, the published literature reports the presence of phenolic acid with a higher incidence in some species of mushrooms and horse gram (Kawsar et al. 2008; Barros et al. 2009).

Previous studies have shown that phenolic acids have managed to gain an important role in medical treatments. Beginning from a cardioprotective function of caffeic acid and ellagic acid, potential treatment of Alzheimer's disease can be done through the usage of *hydroxycinnamic acids*, having an important antibacterial potential, such as inhibiting *Listeria* genes (Chao et al. 2009; Oboh et al. 2013; Stojković et al. 2013).

## 2.3 Phytosynthesis of Metallic Nanoparticles

Nowadays metal nanoparticles have managed to earn a significant role in several human-related industries, leading to major efforts in order to develop synthesis methods with less damaging effects. Furthermore, it has been stated that the option of using biosynthesis for metal nanoparticle production, especially phytosynthesis, can bring advantages never obtained until now through other methods of synthesis (Arunachalama et al. 2012; Fierascu et al. 2019b). Therefore, in the following paragraphs, we will make a short introduction in this topic, not an exhaustive presentation of the domain, in order to present the reader some of the achieved advances in this topic.

In their paper, Li et al. (2011) have discussed about the biosynthesis of nanoparticles using plants, and also, they have analysed how several parameters and characteristics of both the plant extract and the used metal can have a great influence upon the metal nanoparticle production.

In the last decades, the antibacterial property of silver nanoparticles has been intensively studied and has managed to draw the attention of medical researchers, to create more suitable methods of obtaining these nanoparticles (Fierascu et al. 2019a, 2020). In their review paper, Kulkarni and Muddapur (2014) stated the necessity of developing more eco-friendly methods of synthesis of metallic nanoparticles. The most studied property of the metallic nanoparticles is their antimicrobial effect (Fierascu et al. 2017; Sutan et al. 2018, 2019). Prabhu and Poulose (2012) analysed in their paper the toxicity of silver nanoparticles in both the environment and humans. They managed to emphasize the possible adverse effects known until now from current published literature beginning with the impossibility of silver nanoparticles of distinguishing favourable from harmful bacteria (Fig. 2.3).

Dewanjee et al. (2007) have studied in their paper the aspects concerning the phytosynthesis of silver nanoparticles using the extracts from the leaves of *Coccinia*

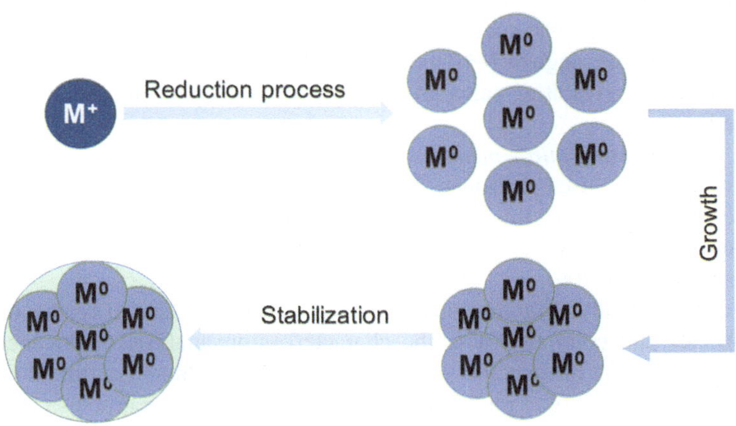


**Fig. 2.3** Mechanism of metallic nanoparticle synthesis

*grandis* L., a worldwide-located herb. They managed to describe the synthesis method to assess their results and concluded that the obtained silver nanoparticles present a powerful photocatalytic property. Other studies have engaged in obtaining silver and gold nanoparticles using extracts from leaves belonging to *Polyalthia longifolia* Sonn. and *Cinnamomum camphora* (L.) J. Presl. The results from the before mentioned studies were encouraging regarding the antibacterial potential of the obtained nanoparticles, a property which was attributed to the compounds in the plants' structure, such as terpenoids, polysaccharides, flavones and phenols (Huang et al. 2007; Mittal et al. 2013).

As it was previously mentioned, silver nanoparticles possess significant antibacterial properties, and having this potential considered, Jyoti et al. (2016) proposed a newly developed synthesis method. They managed to obtain spherical silver nanoparticles, with a size range of 20–30 nm, using *Urtica dioica* Linn. extract as a reducing agent in their phytosynthesis. Upon their characterization of the resulting nanoparticles, which confirmed the formation of the silver nanoparticles, they went further with the investigation in order to confirm the antibacterial potential by testing them against several human pathogenic bacteria. Their study confirmed the antibacterial potential, and together with the characterizations performed previously, the authors concluded that the proposed method of obtaining silver nanoparticles is an eco-friendly and cost-effective method, with significant potential of using it at an industrial scale.

In a different study, Mazariegos et al. (2019) conducted a study which investigated a synthesis method in order to obtain silver nanoparticles, by using the extract obtained from *Justicia spicigera* leaves as a reduction agent. Their characterization analysis presented the formation of spherical nanoparticles with the size range of 86–100 nm and most importantly with significant antibacterial and antifungal properties, making the proposes synthesis method an excellent candidate for prospective studies regarding a scale up for the medical industry.

The extract obtained from the leaves of *Pelargonium graveolens* L'Hér. was used as bioreduction agent in the synthesis process of obtaining silver nanoparticles. Safaepour et al. (2009) succeeded in their study to obtain spherical silver nanoparticles, in the range of 1–10 nm. The authors went further in their experiments and proved that the resulting nanoparticles were efficient in the inhibition of a cancer cell line in vitro.

The leaves from *Datura metel* L. were used to prepare an extraction solution which was later on used to reduce silver nanoparticles. The authors reported their results and concluded that this method can present a novel phytosynthesis method of silver nanoparticles, with a good stability and a diameter size of 16–40 nm (Kesharwani et al. 2009).

In a different study, *Lippia citriodora* Kunth extract was used to synthesize silver nanoparticles and continued with the characterization analysis of the mentioned nanoparticles. Using the specific analysis methods, the authors confirmed the presence of spherical, in the range of 10–45 nm, silver nanoparticles. They also investigated the possibility of using the said nanoparticles in antimicrobial applications. Thus, by comparison with standard drug, the antimicrobial activity of the silver nanoparticles obtained using the plant extract as a reduction agent proved to be more efficient against selected pathogens (Elemike et al. 2017).

Silver nanoparticles were obtained by using phytosynthesis with the extract from *Piper longum* L. leaves. The authors managed to obtain spherical nanoparticles, 18–41 nm, which, based on further investigations, were proven to present cytotoxic activity against a selected cell line (Ali et al. 2011; Jacob et al. 2011).

As it was said, the potential of using plant extracts in the synthesis of metallic nanoparticles has been intensively under research in the last years; thus, there are numerous published articles reporting the formation of nanoparticles with extracts from plants acting as reduction and stabilizing agents. Moreover, the antimicrobial, antifungal and antibacterial activities of the obtained metallic nanoparticles were reported. Silver nanoparticles were obtained by phytosynthesis using *Mentha x piperita* L. (1753) and *Ocimum sanctum* L. leaf extract, respectively. The authors of these two studies continued with the analysis of the nanoparticles and reported obtaining spherical nanoparticles in the range of 4–30 nm. Both studies had tested the antibacterial activity, providing results which confirmed that the silver nanoparticles were effective against isolated human pathogens, *S. aureus* and *E. coli* (Mallikarjun et al. 2011; Singhal et al. 2011).

The antioxidant potential was also investigated by other authors. Banerjee and Narendhirakannan (2011) reported obtaining silver nanoparticles by phytosynthesis using the extract from *Syzygium cumini* (L.) Skeels seeds. Also, concerning the antifungal activity, there have been published articles investigating this potential exposed by the silver nanoparticles synthesized using *Musa x paradisiaca* L. peels as a reduction agent. They have reported a significant activity against several yeasts but also managed to prove that they presented antibacterial potential against well-known pathogens (Bankar et al. 2010).

Balashanmugam and Kalaichelvan (2015) investigated the synthesis of silver nanoparticles using *Cassia roxburghii* DC., with the size range around 35 nm, which through further investigation proved enhanced antifungal activity.

In the published literature up to present, there have also been reports of other metallic nanoparticles obtained using plant extracts. *Calotropis gigantea* (L.) Dryand. extract was used to synthesize zinc nanoparticles. Vidya et al. (2013) managed to obtain zinc nanoparticles in the range of 30–35 nm, using the plant extract as a reducing and stabilizing agent without discussing potential application of the obtained nanoparticles. Their proposal for future research was to investigate further the exact mechanism of synthesis to control the shape and size of the nanoparticles, rather than their potential.

## 2.4 Polyphenols in Nanoparticle Phytosynthesis

As a result of a growing necessity for new materials and methods of obtaining said materials, the nanotechnology field is currently one of the main interests for all categories of researchers. Due to their particle size, from 1 to 100 nm, which comes along with all new properties, the resulting materials in the nano-spectrum are gaining a lead position in most of the human-related industries, from cosmetics and healthcare to the heavier industries such as mechanics, chemical and energy fields (Ahmed et al. 2016).

### 2.4.1 Nanoparticle Synthesis Methods

Even though the interest in nanoparticles is widely expressed, for the biomedical field especially, silver nanoparticles have managed to become one of the most prosper materials, due to their structural properties and potential activities such as antifungal, antioxidant and antibacterial (Joerger et al. 2001). As it was previously mentioned, the interest is not only in obtaining the nanoparticles but also in the development of new synthesis methods. Further, the main categories of synthesis methods, together with their subcategories, will be shortly presented, for a better understanding.

As it can be seen in the below scheme, there are two major synthesis methods, differentiated by their starting material, as follows: in the bottom-up method, the starting point is represented by anatomical or molecular-size components, and for the top-down method, a larger component is subsequently decreased until nanoparticles are obtained (La Rosa et al. 2012; Rajput 2015).

In the biological class of synthesis methods, the "green chemistry" method is drawing increasing interest. This method implies using plants and polysaccharides as reducing agents in the nanoparticle formation reactions. As a classic example, when synthesizing silver and gold nanoparticles with starch, the starch is used as a capping agent and glucose as a reducing agent. In the phytosynthesis of silver and gold nanoparticles, plant extracts and salt of noble metals are used (Fig. 2.4).

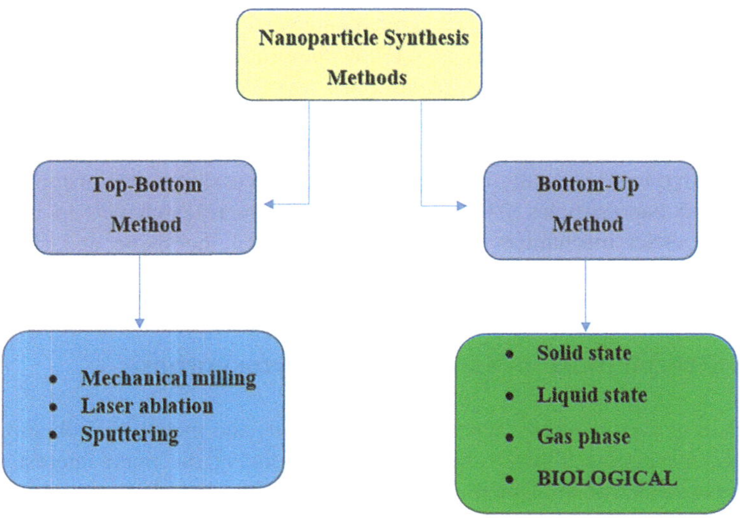


**Fig. 2.4** Schematic representation of the metal nanoparticle synthesis methods

### 2.4.2 Characterization and Properties

When considering metallic nanoparticles, in order to characterize them, several analytical methods can be used, and they will be further mentioned and shortly described.

**Spectrometry Methods**

- UV-Vis spectrometry is mentioned as one of the most used methods of characterization. Light is defined as a flux of photons, and each of these photons is carrying an energy equal to the energy of quantum. The frequency and energy of the photon are inversely proportional to the wavelength. The analysed materials have a specific colour as a result of the absorption of the wavelength and brightness due to the emission of the wavelength of the visible light (Jantschi and Bolboacă 2003).
- Fourier-transform infrared spectroscopy (FTIR) is a measuring technique for the acquisition of infrared spectra. It is based on the detection of molecular vibration, and it is recommended for samples containing pigments as it can detect both organic and inorganic functional groups. The IR radiation is guided by an interferometer, and after passing the sample, the measured signal will represent the interferogram. Finally, the Fourier transform is applied, thus obtaining a spectrum identical to the dispersive form of IR spectroscopy (Scoog et al. 2007).
- X-ray fluorescence spectrometry (XRF) represents a non-destructive method, in which the emission of fluorescence X-ray from a material which has been excited by a bombardment with high-energy X-ray or gamma radiation. This analytical method is commonly preferred due to some advantages such as simple pre-sample preparation, fast and most important non-destructive multi-element analysis and also the ability of providing an overview for the unknown samples. This method has found its application is research as well as in the industrial area (Ron 1999).
- X-ray diffraction (XRD) is a method in which a heated cathode is accelerated to the anode by a potential difference. When there is a sufficiently large potential difference, an X-radiation is emitted. This method allows the experimental determination of the wavelength and the spectrum (Friedrich et al. 1912).
- Inductively coupled plasma atomic emission spectrometry (ICP-AES) is a widely used analysis method, in which a plasma source is used in order to dissociate the sample into the constituent atoms or ions, while exciting them on a higher energy level. The constituents will return to their original form by emitting a characteristic energy photon. The intensity of the radiation is proportional to the concentration of each element in the sample. This analysis method is considered to be advantageous as it is a multi-element analysis method and it allows the direct analysis of liquid samples (Beauchemin and Berman 1989).
- The system for measuring particle size, molecular weight and Z potential through dynamic light diffusion (DLS) is a method in which the Brownian motion is measured and correlated with the particle size. In order to achieve it, the particles are illuminated using a laser, and the intensity of the scattered light fluctuations is further analysed. Of the most important features of DLS with regard to Brownian movement is that small particles move quickly, while the larger ones are slower. The zeta potential of the analysed sample will indicate whether or not the particles in a liquid tend to associate (Dumitriu et al. 2010).

**Thermal Methods**

- Thermogravimetric analysis (TG) is an analysis method mostly used for organic and inorganic chemicals, based on tracking the variation in weight of the substances depending on the temperature (Budrugeac 2001).
- Derived thermogravimetric analysis (DTG). In some situations, due to high speed and close temperatures, the weight variation based on temperature can't be accurately registered. Thus, beside the usual curve, $m = f(T)$, the derived curve obtained from the first derivative, $m'=\frac{dm}{dT}$, must be included in the diagram, in order to increase the sensitivity (Paulik 1995).
- Differential thermal analysis (DTA) is an analysis method in which the temperature difference ($\Delta T$) between a sample ($P$) and a reference material ($R$) is measured as a function of temperature ($T$) of the sample. This method is mostly recommended for qualitative analysis of mixtures of organic substances or mixtures of inorganic substances as well as to the analysis of metals and metal alloys (Ion et al. 2010).

**Microscopy Methods**

- Atomic force microscopy (AFM) is a technique in which the topography of the surfaces is described, and this is performed by using the interaction of a peak with the surface of the analysed sample.
- Scanning electron microscopy (SEM) is a method where the image produced is based on surface processes, as the microscope is able to visualize larger samples, with a greater penetration depth, with the final result of an image as a three-dimensional representation of the sample being subjected to analysis.
- Transmission electron microscopy (TEM) is a microscopy characterization method in which a beam of accelerated electrons provides information regarding the analysed material. For this method, the resulting image can be directly recorded on a photographic film or can be captured by an optical system from a digital camera and afterwards transmitted to a computer screen (Fierascu et al. 2009).

As it has been previously mentioned, the biosynthesis of metal nanoparticles can be realized using different microorganisms, such as fungi, bacteria and plants. Earlier studies have reported that polyphenols are widely encountered in plants; thus, they will be further assessed regarding their role in the phytosynthesis of metal nanoparticles, with examples from the published literature for all the main categories of polyphenols.

### 2.4.3 Flavonoids in Nanoparticle Phytosynthesis

#### 2.4.3.1 *Platycladi cacumen* Extract in Silver Nanoparticle Synthesis

The authors from the following described study proposed a synthesis method for silver nanoparticles using *Platycladi cacumen* (dried twigs and leaves of *Platycladus orientalis* (L.) Franco). The synthesis process began with the extract preparation;

thus, the plant underwent subsequently phases of milling, screening of the obtained powder and dispersing for 9 min in 20 mL deionized water. Finally, the extract solution was centrifuged and deposited al low temperatures. From the obtained extract solution, 10 mL were heated while using a magnetic stirrer for 10 min, during which quantities of aqueous solution of $AgNO_3$ 0.1 M were added to obtain the reaction.

The study went further with the characterization of the obtained silver nanoparticles, starting with UV-Vis analysis. The authors wanted to compare the obtained result based on samples with different concentrations of $AgNO_3$, at different temperatures. Thus, they observed that at the lower considered temperature of 30 °C, peaks corresponding to the silver nanoparticles were observed, independent of the concentration of $AgNO_3$ used. Going further with the analysis, the temperature used was risen to 60 °C and 90 °C, respectively, a tendency of decrease referring to the size distribution of the nanoparticles. These observations were furthermore sustained by the TEM analysis, in which the same comparison was done, leading to the same conclusion that with the increase in temperature of the extract solution, the size distribution will decrease.

In order to investigate the correlation between the formation of silver nanoparticles and the components of extract, the authors decided to analyse the main compounds before but also after the reaction. Thus, they manage to highlight that the quantity of flavonoids decreased after the reaction, resulting in a proposed conclusion that the flavonoids were responsible of the silver ion reduction. This was consistent with other published articles which have mentioned the high concentration of flavonoids in the discussed plant (Zhoud et al. 2004). From the FTIR analysis, another observation was made regarding the biomolecular components of the extract that the spectrum band corresponding to the flavonoids decreased after the reduction, due to the involvement in the bioreduction of the silver ions.

As a brief conclusion, Huang et al. (2011) managed to obtain silver nanoparticles using *Platycladi cacumen* extract as a reducing agent. They also studied the antibacterial potential of the silver nanoparticles with encouraging results, thus proposing the possibility of using the biosynthesis method proposed for an industrial application such as materials with antibacterial properties.

#### 2.4.3.2 *Selaginella bryopteris* (L.) Baker (1884) Extract in Silver Nanoparticle Synthesis

*Selaginella bryopteris* (L.) Baker (1884), commonly named *sanjeevani*, is a plant often used in natural therapies and well known for its biocompounds such as flavonoids, lignans, benzenoids and alkaloids (Dwi 2011). Dakshayani et al. (2019) investigated the synthesis of silver nanoparticles using the extract from *S. bryopteris*. The process began with preparing the plant extract, drying and storing at 40–60 °C for 24 h and continued with the adding of a methanol-water solution and finally boiling for 30 min. Furthermore, a quantity of extract was added to an aqueous solution of $AgNO_3$ (1.0 mmol) while stirring, and finally sonication was performed at 80 °C. Using UV-Vis, the change in colour of mixture was observed, and it was concluded that after 30 min the silver nanoparticles were obtained, with a peak at around 430 nm, as it was expected based on literature (Balavijayalakshmi and Ramalakshmi 2017).

For the characterization of the silver nanoparticles, the authors performed SEM and TEM analysis, which indicated the formation of spherical, well-dispersed nanoparticles, with a diameter between 5 and 10 nm.

The authors wanted to investigate the antibacterial and antifungal potentials, against *S. aureus* and *E. coli* and *A. niger*, respectively. Their conclusions after this analysis were encouraging, as the silver nanoparticles phytosynthesised using *S. bryopteris* extract presented antibacterial activity against *S. aureus*.

Silver nanoparticles were obtained by the authors using *S. bryopteris* extract as a reducing agent; methods proved to be an advantageous, as it is an eco-friendly process, fast and economically convenient. The synthesis led to the development of spherical nanoparticles in the range of 5–10 nm which were well dispersed.

With regard to its potential in the medical field, it was indicated in the results that the obtained silver nanoparticles presented significant antimicrobial and antifungal properties, suggesting an opportunity to be used as a therapeutic agent (Dakshayani et al. 2019).

#### 2.4.3.3 *Salvia officinalis* L. Extract in Silver Nanoparticle Synthesis

In a study conducted by Bunghez et al. (2012), their purpose was to synthesize silver nanoparticles using the extract from *Salvia officinalis* L. In order to obtain the extract, they boiled 50 grams of plant in 250 mL of distilled water for 5 min. The silver nanoparticles were obtained using 10 mL plant extract and 10 mL aqueous solution 1 mM $AgNO_3$ and were extracted from the solution by centrifugation. The resulted nanoparticles were further analysed using UV-Vis, FTIR, SEM and DLS.

The UV-Vis spectrum of the aqueous *Salvia* extract was compared to the spectrum of the absorption of the AgNP-*Salvia* in order to highlight the formation of the silver nanoparticle peak. The *Salvia* extract did not present any absorption between 400 and 600 nm, while the absorption peak specific for the AgNP-*Salvia* was observed at 445 nm. Regarding the absorption bands observed at 345 nm, they were attributed to the flavonoids present in the *Salvia* extract. This was considered by the authors as a possible indication that the flavonoids in the colloidal suspension could have a capping agent role in the bioreduction process of the silver and a stabilization role for the silver nanoparticles.

In the FTIR spectrum, the authors managed to observe intense bands at 1351 $cm^{-1}$ at AgNP-*Salvia*, attributed to the methyl groups, at 1261.25 $cm^{-1}$, specific to the etheric bounds, and at 1074.22 $cm^{-1}$, there is a band specific to the metallic nanoparticles biosynthesized by flavonoids and terpenoids existing in the studied plant.

From the SEM analysis, the authors obtained valuable information regarding the existence of nanoparticles or nanoparticle agglomeration which are dispersed in the matrix. The authors mentioned obtaining high-density silver nanoparticles, synthesized using the extract from *Salvia* leaves, thus confirming the development of silver nanostructure based on uniform plants, with the particle diameter below 100 nm.

In order to finalize their characterization, the authors considered the distribution profile of the silver nanoparticles, at different ultrasound exposure times. After the first ultrasound exposure time, the analysis showed a 397 nm diameter, and at the second exposure time (15 min), the diameter observed had a lower value of 138 nm. Finally, after 30 min of ultrasound exposure, the nanoparticle diameter observed was at 75 nm.

The authors managed to draw several conclusions with high significance for future studies. First of all, the obtained nanoparticles have both antioxidant and antimicrobial potential; they present small dimensions, chemical and thermal stability and low toxicity. Their silver nanoparticles presented a satisfying stability in the aqueous suspension even after several months which indicated the lack of nanoparticle aggregation (Bunghez et al. 2012).

#### 2.4.3.4 *Chenopodium murale* L. Leaf Extract in Silver Nanoparticle Synthesis

In the following part, the possibility of using *Chenopodium murale* L. extract for synthesizing silver nanoparticles will be assessed. The abovementioned plant is commonly found in Egypt and is predisposed to grow in soils which are moist (Kosinova 1975). The authors of the study which will be further discussed proposed a method to synthesize silver nanoparticles using the extract from *C. murale* but also to analyse the obtained nanoparticles. The synthesis process began with the extract preparation. The leaves from *C. murale* were washed and boiled in distilled water and finally freeze dried and filtered. Afterwards the extract solution was added in an aqueous solution of $AgNO_3$ $5 \times 10^{-3}$ M and left until the coloration indicated that the silver nanoparticles had formed (Abdel-Aziz et al. 2014).

After using specific characterization methods, several interesting observations were made. From the UV-Vis analysis, a peak at 440 nm was observed corresponding to plant-AgNP, which is consistent with previous studies from literature. Another important observation was regarding the diameter of the obtained nanoparticles, which was in the range of 30–50 nm, observation made by using TEM analysis.

The authors were interested in investigating the content in phenolic compounds, by comparison, of the plant-silver nanoparticles and the extract solution. From this analysis they managed to demonstrate that the silver nanoparticles obtained by synthesis using the extract from *C. murale* was richer in flavonoids than the extract solution alone.

Regarding the antimicrobial potential, the authors considered an analysis in which they investigated by comparison the antimicrobial activity of the plant extract, the silver nitrate and of course, the plant-silver nanoparticles, concluding based on the obtained results that the highest potential against *S. aureus* was observed in the case of the plant-silver nanoparticles. This was consistent with the findings of other authors which have stated this potential for both silver nanoparticles and plant-silver nanoparticles (Govindaraju et al. 2010).

In this study, the high phenolic content of *C. murale* may be responsible for the synthesis of silver nanoparticles. Several other studies have assessed the antioxidant capacity and the possibility of using the phenolics found in plants for the phytosynthesis of silver nanoparticles (Awika et al. 2003). The capacity of improving the reduction of silver ions by the phenolic compounds found in *C. murale* has been previously demonstrated (Chang et al. 2001). In a different study, Emam (2011) demonstrated that the Chenopodiaceae family, in which *C. murale* is a part of, has a high content of phenolic acids and flavonoids.

### 2.4.4 Phenolic Acids in Nanoparticle Phytosynthesis

#### 2.4.4.1 *Camellia sinensis* (L.) Kuntze Extract in Silver Nanoparticle Synthesis

In the following study, Vilchis-Nestor et al. (2008) investigated the phytosynthesis of silver nanoparticles using *Camellia sinensis* (L.) Kuntze extract as a reducing agent. The study started with the plant extract preparation, realized by boiling 1.5 g of the plant in deionized water. The process continued with the formation of silver nanostructures by adding ammonia solution in $AgNO_3$ $10^{-3}$ M, which was proceeded by the previously extracted solution, and finally adding deionized water for volume adjustments. Visually, the samples were monitored, and the colouring change was assessed, until the solution became dark brown, as it was expected. It has been mentioned before in literature that *C. sinensis* has a high content of polyphenols which have proven antioxidant potential (Scampicchio et al. 2006). Based on published articles, it has been stated that the phenolic compounds are responsible for the reduction processes with high influence regarding the size and stability of the nanoparticles obtained. In this particular study, the *C. sinensis* extract was proven to contain phenolic acids, one example being the gallic acid which has been demonstrated to be successful as a reduction agent in the synthesis of silver nanoparticles (Rababah et al. 2004).

After the synthesis of silver nanoparticles was achieved, the study moved forward with the characterization of the formerly mentioned. The UV-Vis confirmed the formation of the nanoparticles, as a peak was observed around 430 nm.

An important achievement for the discussed study is that the nanoparticles are formed after 4 h of reaction, which can be considered an impressively fast method for obtaining silver nanoparticles using plant extract as a reducing agent.

Their analysis also revealed an important aspect regarding the connection between the volume of extract used and the presence of anisotropy, which is present in the case of large volumes of extract.

Finally, the authors have summarized the information regarding the size and morphology of the silver nanoparticles obtained being sustained by other published studies. They have concluded that phenolic compounds, such as phenolic acids present in *C. sinensis*, can be connected to the modification in the size of the nanoparticles and are directly connected to the bioreduction of silver ions.

In the present study, the authors managed to achieve the formation of silver nanoparticles using *C. sinensis* as a reduction agent, which can lead to further breakthroughs in the medical field due to the advantages brought by this green synthesis method (Vilchis-Nestor et al. 2008).

#### 2.4.4.2 *Parkia speciosa* Hassk. Extract in Silver Nanoparticle Synthesis

*Parkia speciosa* Hassk. is a plant which has gained the interest of researchers due to its antioxidant potential, which was related to the phenolic acids in its componence (Abdalrahim et al. 2012). The authors out of the next study have proposed an eco-friendly silver nanoparticle synthesis method, using the extract from *P. speciosa* leaves.

*P. speciosa* leaves underwent sterilization and distillation before being boiled and finally filtrated and centrifugated. The obtained extract solution was added to a silver nitrate solution 0.01 M which was further completed with deionized water. Besides the visual observation of colour change which suggested the formation of silver nanoparticles, the authors continued with the characterization analysis. By UV-Vis, a peak at approximatively 463 nm was observed, which, based on previously mentioned studies, is specific to the plant-silver nanoparticle. With the information from other studies, the phenolic acids in the extract componence were considered to be responsible for the reduction process (Gan and Aishah 2011).

For a more complete description of the resulted nanoparticles, SEM and TEM analyses were performed, demonstrating the formation of silver nanoparticles with a spherical shape and a diameter size in the range of 22–43 nm.

The authors continued their analysis in order to establish the antibacterial and antimicrobial activity. Thus, based on the dimensions of the inhibition zone, they have stated that the silver nanoparticles obtained with the *P. speciosa* leaf extract presented a high antibacterial potential against *S. aureus*, *B. subtilis* and *E. coli*.

Besides the antibacterial potential which was demonstrated in the discussed study, upon photocatalytic analysis, it was concluded that the plant-silver nanoparticles presented the capability of degrading methylene blue dye. Based on the conclusions drawn from each analysis performed, the authors encouraged more research to be performed upon this subject, as it can find applications in the medical field, due to its high antibacterial potential, but also in the purification of waste water methods (Ravichandran et al. 2019).

### 2.4.5 Lignans in Nanoparticle Phytosynthesis

#### 2.4.5.1 *Forsythia suspensa* (Thunberg) Vahl Fruit Extract in Silver Nanoparticle Synthesis

In the study which will be further discussed, the synthesis of silver nanoparticles using *Forsythia suspensa* (Thunberg) Vahl fruit extract was performed. The plant was selected based on previously published information, which have stated its therapeutic potential (Qu et al. 2012), but also based on their interest to evaluate the potential of the lignans, which are the major component in the respective extract, to act as a reducing agent in the synthesis of silver nanoparticles (Guo et al. 2007).

The fruit from *F. suspensa* was boiled, centrifuged, filtrated and finally freeze dried. For the final step of obtaining silver nanoparticles, the extract was treated with an aqueous solution of silver nitrate and stirred. Although the colour change was observed, suggesting the formation of the silver nanoparticles, this result was also sustained by the UV-Vis, TEM and SEM analyses, with the presence of spherical silver nanoparticles with diameters around 45 nm. The authors used the disk diffusion method in order to demonstrate that the obtained nanoparticles presented antibacterial potential.

The effect of AgNP on the integrity of the membrane of *S. aureus* and *V. parahaemolyticus* was analysed, concluding that they could damage it. Thus, summarizing all the results from the characterization analysis supports the idea that future studies should concentrate on the better understanding of the antibacterial mechanism and on possible applications as a preservative agent (Du et al. 2019).

#### 2.4.5.2 *Sesamum indicum* L. Extract in Silver Nanoparticle Synthesis

In the following study (Bokaeian et al. 2014), the authors investigate the possibility of using *Sesame indicum* L. seeds to synthesis silver nanoparticles. Regarding the selected plant, it has been previously studied, and it has been stated that it has one of the highest contents in lignans (Milder et al. 2005).

The experimental process began with the preparation of the extract from the *S. indicum* seeds, which were firstly sterilized, rinsed, soaked and filtered. After the filtration, a volume of water was added, and finally the solution was deposited for 7 days at room temperature. The silver nanoparticles were synthesized by adding the diluted extract solution to a silver nitrate solution. Upon observations regarding the colour change, the authors concluded that the silver nanoparticles had formed and proceeded with the characterization analysis.

Using UV-Vis analysis and corroborating the results with the TEM images obtained, their assumption of obtaining the desired nanoparticles was confirmed. Although the authors did not continue the analysis in order to establish the potential of the silver nanoparticles, other studies have confirmed that sesamol which is the major lignan in the *S. indicum* seed has anti-inflammatory and antioxidant potential.

#### 2.4.5.3 *Streblus asper* Lour. Extract in Silver Nanoparticle Synthesis

*Streblus asper* Lour. was selected as a subject for the next study due to numerous known applications of the extract obtained from plants (Rastogi et al. 2006). Thus, leaves from *S. asper* were selected, washed, distilled and dried at room temperature. From the dried plant, 10 g of cut leaves were boiled in Millipore water at 80 °C while stirring. The next step was to prepare the silver nanoparticles, which began with adding the extract solution to an aqueous solution of silver nitrate, and this was done while stirring. Finally, the solution was centrifugated and purified, which led to the next phase of characterizing the obtained silver nanoparticles.

In order to confirm their assumption that the silver nanoparticles had formed, which was made based only on the colour of the solution, UV-Vis and TEM analyses were performed. From these characterizations, the final conclusion was that spherical silver nanoparticles, around 13 nm, were obtained by using extract from *Sesamum indicum* seeds.

The authors were also interested in the potential applications of these nanoparticles; thus, they analysed the antibacterial activity of AgNP against *K. pneumonia* and managed to conclude that all the isolated samples had exhibited inhibition upon minimum inhibitory concentration (MIC) analysis (Das et al. 2018).

### 2.4.6 Stilbenes in Nanoparticle Phytosynthesis

Rodríguez-León et al. (2013) proposed in their study the development of silver nanoparticles by phytosynthesis using *Rumex hymenosepalus* Torr. extract, commonly known as wild rhubarb, as a reducing agent. The authors prepared the extract by using 15 g of *R. hymenosepalus* with 100 mL solution of ethanol and water and stored it at room temperature for several days until the extraction was considered complete, based on the colour change. A sample from the extract solution, which will eventually be needed for the NMR analysis, was used at 37 °C in a rotary evaporator and dried afterwards under vacuum. Finally, the obtained sample was washed with tetrahydrofuran being subjected to purifying with a glass filter, which was later on prepared using deuterated dimethyl sulfoxide. The process was finalized with a filtration, resulting in the plant extract further used as a reducing agent in the nanoparticle synthesis. In order to obtain nanoparticles, a solution of $AgNO_3$ 0.1 M was prepared, and it was further divided in several samples with variable volumes which were mixed with a constant volume of plant extract. Finally, samples were adjusted using ethanol in order to obtain different $AgNO_3$ concentrations and left for 96 h so that the reaction could be observed.

The authors performed a UV-Vis analysis of all the samples with different concentrations of $AgNO_3$, confirming that silver nanoparticles were obtained, with the visualization of a peak around 425 as was expected for AgNP. Also, one other conclusion was drawn from the UV-Vis that the peak is more pronounced as the concentration of $AgNO_3$ is higher. The polyphenol peak was observed at around 278 nm as it was also expected, and for a more extended view, the authors also added the analysis of the $AgNO_3$ solution which confirms the visibility of the silver ions at 217 nm. The authors have stated also that from the NMR analysis, peaks were identified as corresponding to polyphenols, more specifically to stilbenes, based on the absorbance peak observed and other studies (Joseph et al. 2007).

The authors went further with the characterization and performed TEM analysis, and they could observe that the size of the obtained nanoparticles is dependent on the concentration of the $AgNO_3$. They performed the analysis on samples with different $AgNO_3$ concentration at 24 h from the beginning of the reaction and at 96 h as well. The size of the nanoparticles was observed to be increasing as the $AgNO_3$ increased. The TEM analysis at 96 h confirmed that the mentioned trend observed in the first TEM analysis is consistent at 96 h as well. In this case, the authors also mentioned that both small and large nanoparticles were observed in the same sample, and with

**Table 2.2** Some examples on the use of plant polyphenols for nanoparticle phytosynthesis

| Plant | Polyphenol | Potential | Nanoparticle diameter | References |
|---|---|---|---|---|
| *Platycladi cacumen* | Flavonoids | Antibacterial | 50–100 nm | Huang et al. (2011) |
| *Selaginella bryopteris* (L.) Baker, 1884 | Flavonoids | Antifungal | 5–10 nm | Dakshayani et al. (2019) |
| *Salvia officinalis* L. | Flavonoids | Antibacterial Antioxidant | 75–100 nm | Bunghez et al. (2012) |
| *Chenopodium murale* L. | Flavonoids | Antimicrobial Antioxidant | 30–50 nm | Abdel-Aziz et al. (2014) |
| *Camellia sinensis* (L.) Kuntze | Phenolic acids | Antibacterial | <100 nm | Vilchis-Nestor et al. (2008) |
| *Parkia speciosa* Hassk. | Phenolic acids | Antibacterial | 22–43 mm | Ravichandran et al. (2019) |
| *Eucalyptus hybrida* | Flavonoids | Antibacterial | 50–100 nm | Dubey et al. (2009) |
| *Forsythia suspensa* (Thunberg) Vahl | Lignans | Antibacterial | 45 nm | Du et al. (2019) |
| *Cinnamomum camphora* (L.) J. Presl. | Phenolic acids | Antibacterial | 16 nm | Huang et al. (2007) |
| *Sesamum indicum* L. | Lignans | Anti-inflammatory | 13 nm | Bokaeian et al. (2014) |
| *Streblus asper* Lour. | Lignans | Antibacterial | 13 nm | Das et al. (2018) |
| *Rumex hymenosepalus* Torr. | Stilbenes | Antimicrobial | 2–40 nm | Rodríguez-León et al. (2013) |

the increase in concentration of $AgNO_3$, the larger nanoparticles tend to increase, while the smaller ones will decrease.

Summarizing the obtained results in the performed analysis, the authors concluded that they manage to achieve their proposed goal of synthesizing silver nanoparticles, with diameters between 2 and 40 nm, while using *R. hymenosepalus* extract as a reducing agent. The proposed method is a promising solution as it has been proven to be an easy-to-perform method, which presents fast kinetics and which does not necessitate the usage of harmful chemicals. From both UV-Vis and NMR, the authors have stated that the studied plant is rich in polyphenols, more specifically in stilbenes, which are known to possess antioxidant potential (Rodríguez-León et al. 2013).

As it was previously mentioned, the interest in developing new eco-friendly biological synthesis methods for obtaining silver nanoparticles has risen due to the possible medical applications and also considering the financial aspects.

Based on the information gathered from the studies which have been discussed in this chapter, and by adding data from other published articles, a summary of some of the plants which can be used in the silver nanoparticle synthesis are presented in Table 2.2.

## 2.5 Concluding Remarks and Perspectives

Plant secondary metabolites represent a wide area of research, with implications in multiple aspects related to our day-to-day life. Although their potential use in applications related to the human health (especially related to the separation of new potentially active compounds) remains on the top of researchers' interest all over the world, having the potential towards new applications (including nanotechnology) (Shavandi et al. 2018) should be continuously explored, as it could lead to new and exciting discoveries.

The nanotechnology (especially the "green chemistry"-driven methods) could find in a relatively small period of time the path from laboratory research to practical, industrial applications. Most of the literature regarding the phytosynthesis of metallic nanoparticles does not intend to elucidate the role of PSM in the phytosynthesis process, but to simply present the possibilities to obtain the nanostructures. However, it is a well-known fact that the possible applications of the nanoparticles strongly depend on the structural and morphological characteristics. Those, in turn, are influenced (among other factors) by the composition of the natural extracts used for phytosynthesis (by the presence and concentration of PSM).

The literature data provides multiple methods to enhance the concentration of PSM (including biotechnological methods) (Halder et al. 2019). Their application could lead to the development of tuned nanoparticles, with tailored characteristics.

Another aspect that should be continuously explored is the recovery of PSM from plant wastes (resulting from various industrial applications). This circular approach could lead not only to a decrease in the overall waste amounts but could also provide the necessary compounds for the nanotechnological processes. The present chapter was aimed at offering a glimpse into the fascinating domain of nanoparticle phytosynthesis, and not an exhaustive inventory of all the possibilities and compounds involved in the process. As can be seen from the examples presented, different authors suggest different phenolic compounds as the ones responsible for the reduction and capping of the nanoparticles. This is one of the aspects that needs future research, in order to completely elucidate the process and contribution of particular compounds to the phytosynthesis.

**Acknowledgements** IF, RCF and CEDP gratefully acknowledge the support obtained through the projects SusMAPWaste, SMIS 104323, contract no. 89/09.09.2016, from the Operational Program Competitiveness 2014–2020, project co-financed by the European Regional Development Fund. IF and RCF also gratefully acknowledge the financial support obtained through the grants of the Romanian National Authority for Scientific Research and Innovation, CNCS/CCCDI – UEFISCDI, project GreenCatOx, project code PN-III-P2-2.1-PED-2019-3166, contract 299PED/2020, and project PREMISE, project code PN-III-P2-2.1-PED-2019-4018, contract 524PED/2020, within PNCDI III.

## References

Abdalrahim A, Khalid MAS, Salman AA, Zhari I, Amin MSAM (2012) Evaluation of antiangiogenic and antioxidant properties of *Parkia speciosa* Hassk extracts. Pakistan J Pharm Sci 25:7–14

Abdel-Aziz MS, Shaheen MS, El-Nekeety AA, Abdel-Wahha MA (2014) Antioxidant and antibacterial activity of silver nanoparticles biosynthesized using *Chenopodium murale* leaf extract. J Saudi Chem Soc 18:356–363

Adlercreutz H (2007) Lignans and human health. Crit Rev Clin Lab Sci 44:483–525

Ahmed S, Ahmad M, Swami BL, Ikram S (2016) A review on plants extract mediated synthesis of silver nanoparticles for antimicrobial applications: a green expertise. J Adv Res 7:17–28

Ali DM, Thajuddin N, Jeganathan K, Gunasekaran M (2011) Plant extract mediated synthesis of silver and gold nanoparticles and its antibacterial activity against clinically isolated pathogens. Colloid Surf B Biointerfaces 85:360–365

Ardalani H, Avan A, Ghayour-Mobarhan M (2017) Podophyllotoxin: a novel potential natural anticancer agent. Avicenna J Phytomed 7:285–294

Arunachalama R, Dhanasingha S, Kalimuthua B, Uthirappana M, Rosea C, Mandalb AB (2012) Phytosynthesis of silver nanoparticles using *Coccinia Grandis* leaf extract and its application in the photocatalytic degradation. Colloid Surf B 94:226–230

Atanasov AG, Waltenberger B, Pferschy-Wenzig EM, Linder T, Wawrosch C, Uhrin P, Temml V, Wang L, Schwaiger S, Heiss EH, Rollinger JM, Schuster D, Breuss JM, Bochkov V, Mihovilovic MD, Kopp B, Bauer R, Dirsch VM, Stuppner H (2015) Discovery and resupply of pharmacologically active plant derived natural products: a review. Biotechnol Adv 33:1582–1614

Awika JM, Rooney LW, Wu X, Prior RL, Zevallos LC (2003) Screening methods to measure antioxidant activity of sorghum (*Sorghum bicolor*) and sorghum products. J Agric Food Chem 51:6657–6662

Balashanmugam P, Kalaichelvan PT (2015) Biosynthesis characterization of silver nanoparticles using *Cassia roxburghii* DC. aqueous extract, and coated on cotton cloth for effective antibacterial activity. Int J Nanomedicine 10:87–97

Balavijayalakshmi J, Ramalakshmi V (2017) *Carica papaya* peel mediated synthesis of silver nanoparticles and its antibacterial activity against human pathogens. J Appl Res Technol 15: 413–422

Banerjee J, Narendhirakannan RT (2011) Biosynthesis of silver nanoparticles from *Syzygium cumini* (L.) seed extract and evaluation of their in vitro antioxidant activities. Dig J Nanomat Biostruct 6:961–968

Bankar A, Joshi B, Kumar AR, Zinjarde S (2010) Banana peel extract mediated novel route for the synthesis of silver nanoparticles. Colloid Surf A 368:58–63

Barennes H, Pussard E, Mahaman Sani A, Clavier F, Kahiatani F, Granic G, Henzel D, Ravinet L, Verdier F (1996) Efficacy and pharmacokinetics of a new intrarectal quinine formulation in children with Plasmodium falciparum malaria. Br J Clin Pharmacol 41:389–395

Barros L, Dueñas M, Ferreira I, Baptista P, Santos-Buelga C (2009) Phenolic acids determination by HPLC–DAD–ESI/MS in sixteen different Portuguese wild mushrooms species. Food Chem Toxicol 47:1076–1079

Beauchemin D, Berman SS (1989) Determination of trace metals in reference water standards by inductively coupled plasma mass spectrometry with on-line preconcentration. Anal Chem 61: 1857–1862

Beth AW (2014) How Taxol/paclitaxel kills cancer cells. Mol Biol Cell 25:2677–2681

Bokaeian M, Sheikh M, Hassanshahian M, Saeidi S, Sahraei S (2014) The antibacterial activity of silver nanoparticles produced in the plant *Sesamum indicum* seed extract: a green method against multi-drug resistant *Escherichia coli*. Int J Enteric Pathog 2:e17928

Bourgaud F, Gravot A, Milesi S, Gontier E (2001) Production of plant secondary metabolites: a historical perspective. Plant Sci 161:839–851

Brown G (2006) Artemisinin and a new generation of antimalarial drugs. Educ Chem 43:97–99

Budrugeac P (2001) The effect of oxygen pressure on the rate of non-isothermal thermo-oxidation of low density polyethylene. J Mat Sci 36:2999–3001

Bunghez IR, Bărbînţă Pătraşcu ME, Badea N, Doncea SM, Popescu A, Ion RM (2012) Antioxidant silver nanoparticles green synthesized using ornamental plants. J Optoelectron Adv Mat 1411: 1016–1022

Chang ST, Wu JH, Wang SY, Kang PL, Yang NS, Shyur LF (2001) Antioxidant activity of extracts from *Acacia confusa* bark and heartwood. J Agric Food Chem 49:3420–3424

Chao PC, Hsu CC, Yin MC (2009) Anti-inflammatory and anti-coagulatory activities of caffeic acid and ellagic acid in cardiac tissue of diabetic mice. Nutr Metabol 6:33

Chong J, Poutaraud A, Hugueney P (2009) Metabolism and roles of stilbenes in plants. Plant Sci 177:143–155

Dakshayani SS, Marulasiddeshwara MB, Sharath Kumar MN, Ramesh G, Raghavendra Kumar P, Devaraja S, Hosamani R (2019) Antimicrobial, anticoagulant and antiplatelet activities of green synthesized silver nanoparticles using *Selaginella* (Sanjeevini) plant extract. Int J Biol Macromol 131:787–797

Das DK, Maulik N (2006) Resveratrol in cardioprotection: a therapeutic promise of alternative medicine. Mol Interv 6:36–47

Das S, Das A, Maji A, Beg M, Singha A, Hossain M (2018) A compact study on impact of multiplicative *Streblus asper* inspired biogenic silver nanoparticles as effective photocatalyst, good antibacterial agent and interplay upon interaction with human serum albumin. J Mol Liq 259:18–29

Dewanjee S, Kundu M, Maiti A, Majumdar R, Majumdar A, Mandal SC (2007) *In vitro* evaluation of antimicrobial activity of crude extract from plants *Diospyros peregrine*, *Coccinia grandis* and *Swietenia macrophylla*. Trop J Pharmaceut Res 6:773–778

Du J, Hu Z, Yu Z, Li H, Pan J, Zhao D, Bai Y (2019) Antibacterial activity of a novel *Forsythia suspensa* fruit mediated green silver nanoparticles against food-borne pathogens and mechanisms investigation. Mater Sci Eng C 102:247–253

Dubey M, Bhadauria S, Kushwah B (2009) Green synthesis of nanosilver particles from extract of *Eucalyptus hybrida* (safeda) leaf. Dig J Nanomater Biostruct 4:537–543

Dumitriu I, Fierăscu RC, Ion RM, Bunghez RI, Corobea MC (2010) Nanotechnology applied in archaeometry: restoration and conservation. SPIE Proc 7821:78211I

Dwi A (2011) Review: natural products from *genus Selaginella* (Selaginellaceae). Nusantara Biosci 3:44

Elemike EE, Onwudiwe DC, Ekennia AC, Ehiri RC, Nnaji NJ (2017) Phytosynthesis of silver nanoparticles using aqueous leaf extracts of *Lippia citriodora*: antimicrobial, larvicidal and photocatalytic evaluations. Mater Sci Eng C 75:980–989

Emam SS (2011) Bioactive constituents of *Atriplex halimus* plant. J Nat Prod 4:25–41

Fierascu RC, Dumitriu I, Ion ML, Catangiu A, Ion RM (2009) Surface and analytical techniques study of Romanian coins. Eur J Sci Theol 5:17–28

Fierascu I, Georgiev MI, Ortan A, Fierascu RC, Avramescu SM, Ionescu D, Sutan A, Brinzan A, Ditu LM (2017) Phyto-mediated metallic nanoarchitectures via *Melissa officinalis* L.: synthesis, characterization and biological properties. Sci Rep 7:12428

Fierascu I, Fierascu IC, Dinu-Pirvu CE, Fierascu RC, Anuta V, Velescu BS, Jinga M, Jinga V (2019a) A short overview of recent developments on antimicrobial coatings based on phytosynthesized metal nanoparticles. Coatings 9:787

Fierascu RC, Ortan A, Avramescu SM, Fierascu I (2019b) Phyto-nanocatalysts: green synthesis, characterization, and applications. Molecules 24:3418

Fierascu RC, Fierascu I, Lungulescu EM, Nicula N, Somoghi R, Diţu LM, Ungureanu C, Sutan AN, Drăghiceanu OA, Paunescu A, Soare LC (2020) Phytosynthesis and radiation-assisted methods for obtaining metal nanoparticles. J Mater Sci 55:1915–1932

Friedrich W, Knipping P, von Laue M (1912) Interferenz-Erscheinungen bei Röntgenstrahlen. Sitzungsberichte der Mathematisch-Physikalischen Classe der Königlich-Bayerischen Akademie der Wissenschaften zu München 303
Gan CY, Aishah AL (2011) Antioxidant *Parkia speciosa* pod powder as potential functional flour in food application: physiochemical properties characterization. Food Hydrocoll 25:1174–1180
Ghumatkar PJ, Patil SP, Jain PD, Tambe RM, Sathaye S (2015) Nootropic, neuroprotective and neurotrophic effects of phloretin in scopolamine induced amnesia in mice. Pharmacol Biochem Behav 135:182–191
Govindaraju K, Tamilselvan S, Kiruthogs V, Simgaravelu G (2010) Biogenic silver nanoparticles by *Solanum torvum* and their promising antimicrobial activity. J Biopest 3:394–399
Guo H, Liu AH, Ye M, Yang M, Guo DA (2007) Characterization of phenolic compounds in the fruits of *Forsythia suspensa* by high-performance liquid chromatography coupled with electrospray ionization tandem mass spectrometry. Rapid Commun Mass Spectrom 21:715–729
Halder M, Sarkar S, Jha S (2019) Elicitation: a biotechnological tool for enhanced production of secondary metabolites in hairy root cultures. Eng Life Sci 19:880–895
Hu N, Yu J, Tan L, Wang Y, Sun L, Tan L (2013) Nutrition and the risk of Alzheimer's disease. Biomed Res Int 2013:1–12
Huang JL, Li QB, Sun DH, Lu YH, Su YB, Yang X, Wang H, Wang Y, Shao W, He N (2007) Biosynthesis of silver and gold nanoparticles by novel sundried *Cinnamomum camphora* leaf. Nanotechnology 18:105104
Huang J, Zhan G, Zheng B, Sun D, Lu F, Lin Y, Chen H, Zheng Z, Zheng Y, Li Q (2011) Biogenic silver nanoparticles by *Cacumen Platycladi* extract: synthesis, formation, mechanism, and antibacterial activity. Ind Eng Chem Res 50:9095–9106
Ion RM, Ion ML, Fierascu RC, Serban S, Dumitriu I, Radovici C, Bauman I, Niculescu VIR (2010) Thermal analysis of Romanian ancient ceramics. J Thermal Anal Calorim 102:393–398
Jacob S, Finub J, Narayanan A (2011) Synthesis of silver nanoparticles using *Piper longum* leaf extracts and its cytotoxic activity against Hep-2 cell line. Colloid Surf B Biointerfaces 91:212–214
Jantschi L, Bolboacă S (2003) Applied chemical and instrumental analysis (In Romanian: *Analiza chimică şi instrumentală aplicată*). Academic Direct, Cluj Napoca
Jeong EJ, Lee HK, Lee KY, Jeon BJ, Kim DH, Park JH, Song JH, Huh J, Lee JH, Sung SH (2013) The effects of lignan-riched extract of *Schisandra chinensis* on amyloid-β-induced cognitive impairment and neurotoxicity in the cortex and hippocampus of mouse. J Ethnopharmacol 146: 347–354
Joerger TK, Joerger R, Olsson E, Granvist C (2001) Bacteria as workers in the living factory: metal-accumulating bacteria and their potential for materials science. Trends Biotechnol 19:15–20
Joseph CC, Moshi MJ, Innocent E, Nkunya MHH (2007) Isolation of a stilbene glycoside and other constituents of *Terminalia sericeae*. Afr J Tradit Complement Altern Med 4:383–386
Jyoti K, Baunthiyal M, Singh A (2016) Characterization of silver nanoparticles synthesized using *Urtica dioica* Linn. leaves and their synergistic effects with antibiotics. J Rad Res Appl Sci 9: 217–227
Kawsar SMA, Huw E, Nahar N, Ozeki Y (2008) Identification and qualification of phenolic acids in *Macrotyloma uniflorum* by reversed phase-HPLC. Am J Plant Physiol 3:165–172
Kesharwani J, Yoon KY, Hwang J, Rai M (2009) Phytofabrication of silver nanoparticles by leaf extract of *Datura metel*: hypothetical mechanism involved in synthesis. J Bionanosci 3:39–44
Kheradmand E, Hajizadeh Moghaddam A, Zare M (2018) Neuroprotective effect of hesperetin and nano-hesperetin on recognition memory impairment and the elevated oxygen stress in rat model of Alzheimer's disease. Biomed Pharmacother 97:1096–1101
Kleindl PA, Xiong J, Hewarathna A, Mozziconacci O, Nariya MK, Fisher AC, Deeds EJ, Joshi SB, Middaugh CR, Schöneich C, Volkin DB, Forrest ML (2017) The botanical drug substance crofelemer as a model system for comparative characterization of complex mixture drugs. J Pharm Sci 106:3242–3256
Kosinova A (1975) Weed communities of winter crops in Egypt. Preslia 47:58–74

Kulkarni N, Muddapur U (2014) Byosyntheis of metal nanoparticles: a review. J Nanotechnol 14: 510246

La Rosa A, Yan M, Fernandez R, Wang X, Zegarra E (2012) An overview of proton-fountain electric-field-assisted nanolithography (PEN) in the context of top-down and bottom-up approaches to nanotechnology. Revciuni 15:87–129

Li J, Ou-Lee TM, Raba R, Amundson RG, Last RL (1993) Arabidopsis mutants are hypersensitive to UV-B radiation. Plant Cell 5:171–179

Li X, Xu H, Chen ZS, Chen G (2011) Biosynthesis of nanoparticles by microorganisms and their applications. J Nanomater 11:270974

Lin TY, Lu CW, Wang SJ (2016) Luteolin protects the hippocampus against neuron impairments induced by kainic acid in rats. Neurotoxicology 55:48–57

López-Biedma A, Sánchez-Quesada C, Delgado-Rodríguez M, Gaforio JJ (2016) The biological activities of natural lignans from olives and virgin olive oils: a review. J Funct Food 26:36–47

Lubbe A, Verpoorte R (2011) Cultivation of medicinal and aromatic plants for specialty industrial materials. Ind Crop Prod 34:785–801

Majdalawieh AF, Mansour ZR (2019) Sesamol, a major lignan in sesame seeds (*Sesamum indicum*): anti-cancer properties and mechanisms of action. Eur J Pharmacol 855:75–89

Mallikarjun K, Narsimha G, Dillip G, Praveen B, Shreedhar B, Lakshmi S (2011) Green synthesis of silver nanoparticles using *Ocimum leaf* extract and their characterization. Dig J Nanomater Biostruct 6:181–186

Manach C, Scalbert A, Morand C, Remesy C, Jimenez L (2004) Polyphenols: food sources and bioavailability. Am J Clin Nutr 79:727–747

Mancini E, Beglinger C, Drewe J, Zanchi D, Lang UE, Borgwardt S (2017) Green tea effects on cognition, mood and human brain function: a systematic review. Phytomedicine 15:26–37

Mazariegos EB, Valdez-Salas B, Gonzalez-Mendoza D, Abdelmoteleb A, Tzintzun Camacho O, Cecena Duran C, Gutierrez-Miceli F (2019) Silver nanoparticles from Justicia spicigera and their antimicrobial potentialities in the biocontrol of foodborne bacteria and phytopathogenic fungi. Rev Argentina Microbiol 51:103–109

Menendez JA, Vazquez-Martin A, Oliveras-Ferraros C, Garcia-Villalba R, Carrasco-Pancorbo A, Fernandez-Gutierrez A, Segura-Carretero A (2008) Analyzing effects of extra-virgin olive oil polyphenols on breast cancer-associated fatty acid synthase protein expression using reverse-phase protein microarrays. Int J Mol Med 22:433–439

Milder IE, Arts IC, van de Putte B, Venema DP, Hollman PC (2005) Lignan contents of Dutch plant foods: a database including lariciresinol, pinoresinol, secoisolariciresinol and matairesinol. Br J Nutr 93:393–402

Mittal AK, Chisti Y, Banerjee UC (2013) Synthesis of metallic nanoparticles using plant extracts. Biotechnol Adv 31:346–356

Oboh G, Agunloye OM, Akinyemi AJ, Ademiluyi AO, Adefegha SA (2013) Comparative study on the inhibitory effect of caffeic and chlorogenic acids on key enzymes linked to Alzheimer's disease and some pro-oxidant induced oxidative stress in rats' brain-in vitro. Neurochem Res 38:413–419

Panickar KS, Jang S (2013) Dietary and plant polyphenols exert neuroprotective effects and improve cognitive function in cerebral ischemia. Recent Pat Food Nutr Agric 5:128–143

Paulik F (1995) Special trends in thermal analysis. Wiley, New York

Prabhu S, Poulose EK (2012) Silver nanoparticles: mechanism of antimicrobial action, synthesis, medical applications, and toxicity effects. Int Nano Lett 2:32

Qu H, Zhang Y, Chai X, Sun W (2012) Isoforsythiaside, an antioxidant and antibacterial phenylethanoid glycoside isolated from *Forsythia suspensa*. Bioorg Chem 40:87–91

Rababah TM, Hettiarachchy NS, Horax R (2004) Total phenolics and antioxidant activities of fenugreek, green tea, black tea, grape seed, ginger, rosemary, gotu kola, and ginkgo extracts, vitamin E, and tert-butylhydroquinone. J Agric Food Chem 52:5183–5186

Rajput N (2015) Methods of preparation of nanoparticles—a review. Int J Adv Eng Technol 7: 1806–1811

Rastogi S, Kulshreshtha DK, Rawat AKS (2006) *Streblus asper* Lour. (Shakhotaka): a review of its chemical, pharmacological and ethnomedicinal properties. Evid Based Complement Alternat Med 3:217–222

Ravichandran V, Vasanthi S, Shalini S, Adnan S, Shah A, Tripathy M, Paliwal N (2019) Green synthesis, characterization, antibacterial, antioxidant and photocatalytic activity of *Parkia speciosa* leaves extract mediated silver nanoparticles. Res Phys 15:102565

Rodríguez-León E, Iñiguez-Palomares R, Navarro RE, Herrera-Urbina R, Tánori J, Iñiguez-Palomares C, Maldonado A (2013) Synthesis of silver nanoparticles using reducing agents obtained from natural sources (*Rumex hymenosepalus* extracts). Nanoscale Res Lett 8:318

Ron J (1999) X-ray fluorescence spectrometry, 2nd edn. Wiley, New York

Safaepour M, Shahverdi AR, Shahverdi HR, Khorramizadeh MR, Gohari AR (2009) Green synthesis of small silver nanoparticles using geraniol and its cytotoxicity against Fibrosarcoma–Wehi 164. Avicenna J Med Biotechnol 1:111–115

Scampicchio M, Wang J, Blasco AJ, Arribas AS, Mannino S, Escarpa A (2006) Nanoparticle-based assays of antioxidant activity. Anal Chem 78:2060–2063

Scoog DA, Holler FJ, Hrouch SR (2007) Principles of instrumental analysis, 6th edn. Thomson Brooks/Cole

Shavandi A, Bekhit AEDA, Saeedi P, Izadifar Z, Bekhit AA, Khademhosseini A (2018) Polyphenol uses in biomaterials engineering. Biomaterials 167:91–106

Singer AC, Crowley DE, Thompson IP (2003) Secondary plant metabolites in phytoremediation and biotransformation. Trends Biotechnol 21:123–130

Singh AK, Singh SK, Singh PP, Srivastava AK, Pandey KD, Kumar A, Yadav H (2018) Biotechnological aspects of plants metabolites in the treatment of ulcer: a new prospective. Biotechnol Rep 18:e00256

Singhal G, Bhavesh R, Kasariya K, Sharma AR, Singh RP (2011) Biosynthesis of silver nanoparticles using *Ocimum sanctum* (Tulsi) leaf extract and screening its antimicrobial activity. J Nanopart Res 13:2981–2988

Spencer SJ, Korosi A, Layé S, Shukitt-Hale B, Barrientos RM (2017) Food for thought: how nutrition impacts cognition and emotion. Sci Food 1:1–7

Sramska P, Maciejka A, Topolewska A, Stepnowski P, Halinski LP (2017) Isolation of atropine and scopolamine from plant material using liquid-liquid extraction and EXtrelut® columns. J Chromatogr B 1043:202–208

Stojković DS, Zivković J, Soković M, Glamočlija J, Ferreira IC, Janković T, Maksimović Z (2013) Antibacterial activity of *Veronica montana* L. extract and of protocatechuic acid incorporated in a food system. Food Chem. Toxicology 55:209–213

Stuart-Harris R, Joel SP, McDonald P, Currow D, Slevin ML (2000) The pharmacokinetics of morphine and morphine glucuronide metabolites after subcutaneous bolus injection and subcutaneous infusion of morphine. Br J Clin Pharmacol 49:207–214

Sun C, Hu Y, Liu X, Wu T, Wang Y, He W, Wei W (2006) Resveratrol downregulates the constitutional activation of nuclear factor-κB in multiple myeloma cells, leading to suppression of proliferation and invasion, arrest of cell cycle, and induction of apoptosis. Cancer Genet Cytogenet 165:9–19

Sun D, Li N, Zhang W, Zhao Z, Mou Z, Huang D, Liu J, Wang W (2016) Design of PLGA-functionalized quercetin nanoparticles for potential use in Alzheimer's disease. Colloid Surf B Biointerfaces 148:116–129

Sutan NA, Manolescu DS, Fierascu I, Neblea AM, Sutan C, Ducu C, Soare LC, Negrea D, Avramescu SM, Fierascu RC (2018) Phytosynthesis of gold and silver nanoparticles enhance in vitro antioxidant and mitostimulatory activity of *Aconitum toxicum* Reichenb. rhizomes alcoholic extracts. Mater Sci Eng C 93:746–758

Sutan NA, Vilcoci DS, Fierascu I, Neblea AM, Sutan C, Ducu C, Soare LC, Negrea D, Avramescu SM, Fierascu RC (2019) Influence of the phytosynthesis of noble metal nanoparticles on the cytotoxic and genotoxic effects of *Aconitum toxicum* reichenb. leaves alcoholic extract. J Clust Sci 30:647–660

Thakur M, Bhattachary S, Khosla PK, Puri S (2019) Improving production of plant secondary metabolites through biotic and abiotic elicitation. J Appl Res Med Aromat Plant 12:1–12

Vacca RA, Valenti D, Caccamese S, Daglia M, Braidy N, Nabavi SM (2016) Plant polyphenols as natural drugs for the management of Down syndrome and related disorders. Neurosci Biobehav Rev 71:865–877

van de Glind EMM, van Enst WA, van Munster BC, Olde Rikkert MGM, Scheltens P, Scholten RJPM, Hooft L (2013) Pharmacological treatment of dementia: a scoping review of systematic reviews. Dement Geriatr Cogn Disord 36:211–228

Verma N, Shukla S (2015) Impact of various factors responsible for fluctuation in plant secondary metabolites. J Appl Res Med Aromat Plant 2:105–113

Ververidsi F, Trantas E, Douglas C, Vollmer G, Kretzschmar G, Panopoulos N (2007) Biotechnology of flavonoids and other phenylpropanoid-derived natural products. Part I: chemical diversity, impacts on plant biology and human health. Biotechnol J 2:1214–1234

Vestergaard M, Ingmer H (2019) Antibacterial and antifungal properties of resveratrol. Int J Antimicrob Agents 53:716–723

Vidya C, Hiremath S, Chandraprabha MN, Lourdu Antonyraj MA, Gopal IV, Jain A, Bansal K (2013) Green synthesis of ZnO nanoparticles by *Calotropis Gigantea*. Int J Curr Eng Technol 2013:118–120

Vilchis-Nestor AR, Sanchez-Mendieta V, Camacho-Lopez MA, Gomez-Espinosa RM, Camacho-Lopez MA, Arenas-Alatorre JA (2008) Solventless synthesis and optical properties of Au and Ag nanoparticles using *Camellia sinensis* extract. Mater Lett 62:3103–3105

Whitten PL, Kudo S, Okubo KK (1997) Isoflavonoids. In: Handbook of plant and fungal toxicants. CRC Press, Boca Raton, pp 117–137

Yoo DY, Choi JH, Kim W, Nam SM, Jung HY, Kim JH, Won MH, Yoon YS, Hwang IK (2013) Effects of luteolin on spatial memory, cell proliferation, and neuroblast differentiation in the hippocampal dentate gyrus in a scopolamine-induced amnesia model. Neurol Res 35:813–820

Zaynab M, Fatima M, Abbas S, Sharif Y, Umair M, Zafar MH, Bahadar K (2018) Role of secondary metabolites in plant defense against pathogens. Microb Pathog 124:198–202

Zhang HF, Yang XH, Wang Y (2011) Microwave assisted extraction of secondary metabolites from plants: current status and future directions. Trends Food Sci Technol 22:672–688

Zhang C, Mao X, Zhao X, Liu Z, Liu B, Li H, Bi K, Jia Y (2014) Gomisin N isolated from *Schisandra chinensis* augments pentobarbital-induced sleep behaviors through the modification of the serotonergic and GABAergic system. Fitoterapia 96:123–130

Zhoud JJ, Xie GR, Yan XJ (2004) Traditional Chinese medicines: molecular structures, natural sources and applications. Chemical and Industrial Press, Beijing

# 3 Plant Secondary Metabolite Determination Through Analytical Chromatographic Techniques: Recent Trends and Advancement

Rohit Sharma, Ashun Chaudhary, Yash Pal Sharma, and Sunil Kumar

**Abstract**

The present chapter presents a summary of analytical instruments like high-pressure thin-layer chromatography, high-pressure liquid chromatography, and gas chromatography- mass spectroscopy for their potential benefits to drug discovery through chemical fingerprints for pharmaceutical, food, and biotechnology industries. The methods outlined in the current chapter are quite robust, rapid, economical, consistent, and effective tools for qualitative and quantitative analysis of natural compounds, and identification of markers could be exploited at the industrial scale and thus have beneficial aspects for human welfare and drug discovery.

**Keywords**

Analytical instruments · High-pressure thin-layer chromatography · High-pressure liquid chromatography · Gas chromatography-mass spectroscopy · Drug discovery

R. Sharma (✉) · Y. P. Sharma
Department of Forest Products, College Of Forestry, Dr. Yashwant Singh Parmar University Of Horticulture and Forestry, Solan, Himachal Pradesh, India
e-mail: rohitsharma@yspuniversity.ac.in

A. Chaudhary
Department of Plant Sciences, Central University of Himachal Pradesh, Dharamshala, TAB, Shahpur, Kangra, Himachal Pradesh, India

S. Kumar
Department of Animal Sciences, Central University of Himachal Pradesh, Dharamshala, TAB, Shahpur, Kangra, Himachal Pradesh, India



A. K. Sharma, A. Sharma (eds.), *Plant Secondary Metabolites*,
https://doi.org/10.1007/978-981-16-4779-6_3

## 3.1 Introduction

Natural products play a vital role in healthcare for decades. Plants are rich source of natural compounds. Natural products from different sources have given a large number of privileged scaffolds for clinical uses and further optimization succeeded to get lead molecules for drug development. If we look at the statistics, 75% of drug present in the market are inspired from natural products. India contributes nearly 10% of the world's biodiversity wealth. Nearly 7500 plants are reported for their medicinal importance in India. Earlier, crude extracts of plants were used for medicinal purpose. However, in the mid-nineteenth century, sincere efforts were made to isolate bioactive molecules of plants. After which library of biological active molecules were generated, viz., atropine (*Hyoscyamus niger*) (**1**), camptothecin (*Camptotheca acuminate*) (**2**), codeine and morphine (*Papaver somniferum*) (**3**), nicotine (*Nicotiana tabacum*) (**4**), pilocarpine (*Pilocarpus jaborandi*) (**5**), quinine (*Cinchona officinalis*) (**6**), etc. (Fig. 3.1). These compounds are available in the market as drugs to cure various diseases. Still, a large number of plants remain unexplored for their chemical and therapeutic potential. In addition to biological activities, they also show an exceptional diversity in chemical structures.

Plants are able to synthesize secondary metabolites which act as plant defense system against various biotic and abiotic stresses. They are present in less than 1% amount in dry weight. These secondary metabolites have chemical diversity and vary from plant to plant and also vary with climatic variation. They are synthesized from biosynthetic precursors such as phosphoenolpyruvate, pyruvate acetate, amino acids, acetyl-CoA, and malonyl-CoA. The structure of secondary metabolites depends on biosynthetic pathways, and they are classified into phenols, alkaloids, saponins, terpenoids, lipids, carbohydrates, and flavonoids on the basis of their structures (Fig. 3.2).

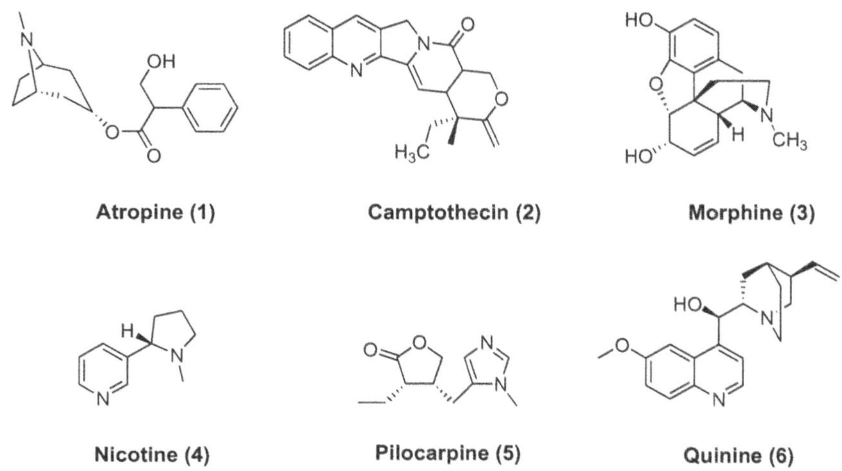


**Fig. 3.1** Natural products with therapeutic potential

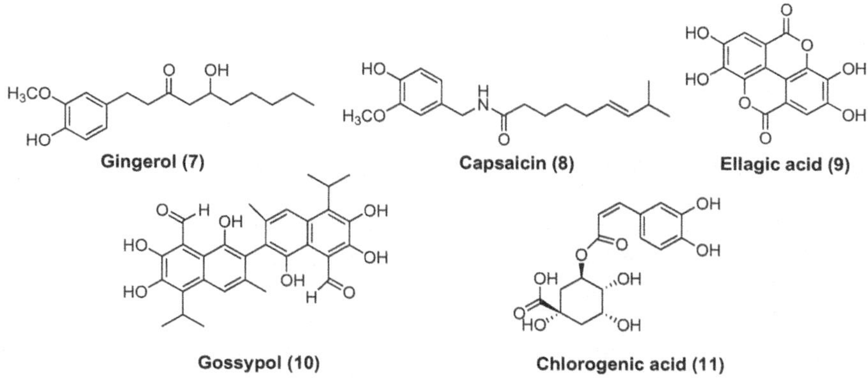


**Fig. 3.2** Various phenol and phenolic acids

Nowadays, pharmaceutical companies and academic institutions are very much interested toward natural product-based drug discovery. In order to validate traditional knowledge of medicinal plants, scientific validation will be required. A large number of secondary metabolites are reported in plants, and these are responsible for various pharmacological activities. Since ages extracts of medicinal plants have been used for curing various diseases rather than active natural compound responsible for that activity. In the last few decades, research has been carried out to identify novel molecules present in plants and further establishing their pharmacological activity. The concentration of active molecules also depends upon the harvesting stage of various medicinal plants as well. So, scientific validation of these medicinal plants is utmost step in drug discovery perspective. The process of scientific validation is done by using various spectroscopic and chromatographic tools like HPLC, HPTLC, NMR, IR, LC-MS, and GC-MS. By validating medicinal plants and scientifically knowing their chemical profiles and concentration of active molecule, they can be further commercialized as medicines.

To unravel various secondary metabolites present in plants, various types of phytochemical analysis were performed such as thin-layer chromatography, high-pressure liquid chromatography, gas chromatography, high-pressure thin-layer chromatography, mass spectroscopy, and nuclear magnetic resonance spectroscopy. To understand the complexity of secondary metabolites, these instruments are boon to organic and medicinal chemist worldwide. The hyphenation of these technologies like liquid chromatography with mass spectroscopy and gas chromatography with mass spectroscopy makes an exciting field of research in analytical chemical, phytochemistry, and pharmacognosy. The chemical profiling of crude extract has become easier by the introduction of UHPLC (ultra-high-pressure liquid chromatography) and provides faster separation of compounds. So, in this chapter, we will summarize recent research carried out by using various analytical instruments.

(a) *High-pressure thin-layer chromatography.*
(b) *High-pressure liquid chromatography.*
(c) *Gas chromatography-mass spectroscopy.*

### 3.1.1 High-Pressure Thin-Layer Chromatography (HPTLC)

Thin-layer chromatography (TLC) is a widely used method for the analysis of natural products before the discovery of high advanced technique like HPTLC. Till today, TLC is used for preliminary identification and semiquantitative analysis of natural compounds. The technique is based on the principle of partition and adsorption. HPTLC is extension to TLC and is robust, rapid, and simple and a tool for quantitative analysis of natural compounds. HPLTLC is also known as flat-bed chromatography or planar chromatography. Quantification of natural molecules at very low concentration can be analyzed with better accuracy and sensitivity.

As we know phenolic compounds are utmost important secondary metabolites present in plant, possessing both nutritional and medicinal importance. In recent times researchers around the globe have utilized HPTLC technique to unravel phenolic compounds from various plants. Srinivasan et al. (2016) carried HPTLC analysis of *Indigofera tinctoria* Linn. for chemical fingerprinting at 254 and 366 nm and revealed the presence of various phenolic compounds in the plant extract. Dinakaran et al. (2019) carried out an analysis of ayurvedic polyherbal drugs, named DBC and DMV, known for antidiabetic potential. DBC and DMV are comprised of 34 and 22 polyherbal materials, respectively, on the basis of HPTLC analysis they concluded in which the presence of phenolic compounds is attributed to the pharmacological effect of DBC and DMV. They further insisted that HPTLC can be utilized as validation method for quality control analysis of herbal formulations. Satija et al. (2020) developed a validated HPTLC method according to ICH guidelines for the identification of berberine (**12**) and rutin (**15**) isolated from *Tinospora cordifolia*. They further carried out antioxidant activity using HPTLC and found berberine (**12**) holds high antioxidant potential than rutin (**15**). Bubueanu and Pavaloiu (2016) carried out HPTLC chemical fingerprinting analysis of *Allium sativum*, *Allium ursinum*, *Malus pumila*, and *Pyrus communis* and found the presence of flavonoid glycosides and hydroxycinnamic acid compounds. Narayanan and Marimuthu (2016) carried out HPTLC analysis of *Cyathea nilgirensis*, *Cyathea* gigantea, and *Cyathea crinita*. They reported the presence of quercetin, (**13**) flavonoids, and gallic acid in all the *Cyathea* species. Better separation of all secondary metabolites was observed, and this method can be further utilized for qualitative and quantitative analyses. Rathore et al. (2016) developed a new validated HPTLC methods for identification of corilagin, (**16**) ellagic acid (**9**), epicatechin (**18**), and gallic acid from *Euphoria longana*; further these polyphenols were confirmed by mass spectroscopy. Ali et al. (2020) carried out HPTLC analysis of ethanolic extract of *Morus alba* and found that the HPTLC fingerprint of the leaf extract consist of phenols and flavonoids responsible for antioxidant activity. Nile and Park (2014) carried out HPTLC analysis of six medicinal plants *Asparagus*

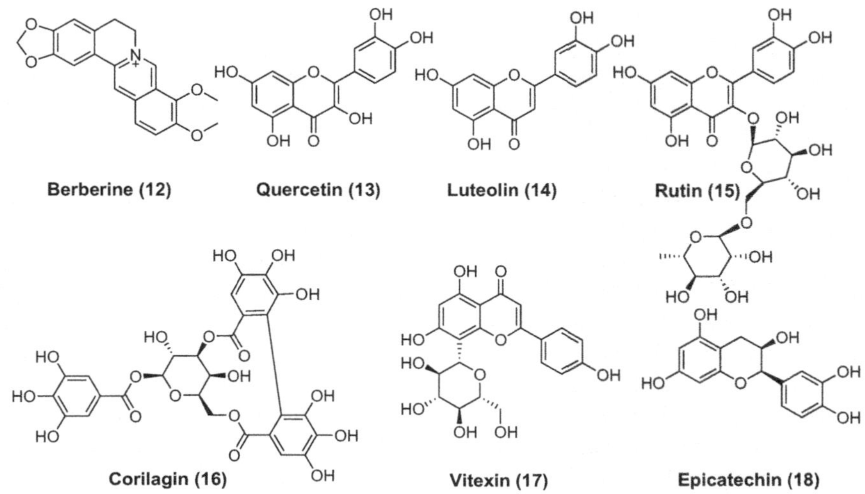


**Fig. 3.3** Structure of compounds

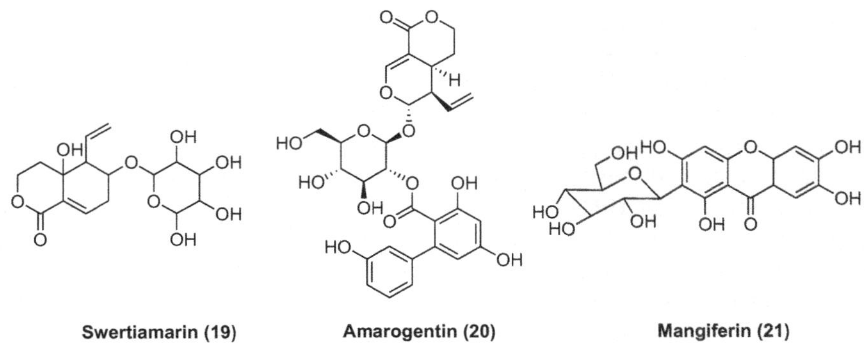


**Fig. 3.4** Structure of compounds

*racemosus*, *Withania somnifera*, *Tephrosia purpurea*, *Plumbago zeylanica*, *Butea monosperma*, and *Vitex negundo* and revealed the presence of quercetin (**13**), rutin (**15**), luteolin (**14**), and vitexin (**17**) in all plants. Rutin (**15**) was present in highest and quercetin (**13**) in lowest concentration in all plant species (Fig. 3.3).

Mundengara et al. (2019) had validated HPTLC protocol for the evaluation of mangiferin (**21**) isolated from *Canscora perfoliata*. On quantification 1.407 ± 0.1% of mangiferin (**21**) was present. Kaur et al. (2019) carried out HPTLC fingerprinting on five different species of *Swertia*, namely, *S. chirata*, *S. paniculata*, *S. cordata*, *S. nervosa*, and *S. angustifolia*, considering specific polyphenols, viz., swertiamarin (**19**), amarogentin (**20**), and mangiferin (**21**). On the basis of the HPTLC study, they concluded that *S. angustifolia and S. paniculata* can be substituted by *S. chirata* (Fig. 3.4).

Pedan et al. (2018) evaluated the HPTLC method for metabolic fingerprinting of cocoa beans for identification of polyphenols and anthocyanins. Nair et al. (2017) developed a new HPTLC-based process for the instantaneous quantification of rutin (**15**), quercetin (**13**), and gallic acid found in *Psidium guajava* and *Aegle marmelos* (L.) Correa. The developed process could be explored for quality control purpose of herbal drugs, and the process is simple, robust, and reproducible. Rocamora et al. (2020) carried out HPTLC analysis combined with effect-directed analysis of essential oils derived from Lamiaceae family for various pharmacological activities. Further, the presence of phenolics was confirmed by derivatizations. They concluded that the AChE activity of these essential oils is due to the presence of phenolics and terpenoids. Conceicao et al. (2020) demonstrated the usefulness of instrument HPTLC coupled with desorption electrospray ionization multistage mass spectroscopy for the analysis of isoquinoline alkaloids from a plant extract. They worked on *Ocotea spixiana* and successfully identified 13 aporphine and 4 benzylisoquinoline-type alkaloids and established acaricide activity of ethyl acetate extract due to the presence of alkaloids. Sridhar et al. (2018) developed a new HPTLC-based quantification method for conophylline (**22**), an indole alkaloid from the leaves of *Tabernaemontana divaricata* collected in different months, viz., August, November, February, and May. The highest content of conophylline (**22**) was found in the month of August. Misra et al. (2020) developed a new HPTLC-UV-based method for the instantaneous assessment of colchicine (**23**) and gloriosine (**24**) alkaloid isolated from *Gloriosa superba*. The content of colchicine (**23**) and gloriosine (**24**) varied from 0.005 to 0.275% and 0.003 to 0.119%. The highest amount of colchicine (**23**) and gloriosine (**24**) was found in the plant material collected from Tirupati, Andhra Pradesh, and Panchgani hill, Maharashtra, respectively. Perala and Ciddi (2018) carried out an HPTLC analysis of nitidine, (**25**) chelerythrine (**26**), and sanguinarine (**27**) from the callus extract of *Zanthoxylum rhetsa*. Further, the developed method was efficient, accurate, and simple and authenticated according to ICH guidelines (Fig. 3.5).

Cortes et al. (2017) developed a bioguided coupled method for the evaluation of cholinesterase inhibitory activity for drug discovery purpose. They evaluated seven different plant species belonging to the family of Amaryllidaceae. The combination of HPTLC along with data obtained by densitometry was followed after enzymatic bioautography with different AChEs and BEAChEs. The desired spots of interest were subjected to GC-MS analysis for the detection of bioactive molecules. Gupta et al. (2019) carried out HPTLC-based chemical fingerprinting of *Careya arborea* bark, leaves, and seeds and concluded that alkaloids were present in the bark and absent in the leaves and seeds. Further they also confirmed the presence of various flavonoids, glycosides, essential oils, saponins, anthracene derivatives, arbutin derivatives, and coumarin derivatives. Hakim et al. (2018) carried out HPTLC fingerprinting analysis considering alkaloids and glycosides of methanol extract of three different species, viz., *Terminalia arjuna*, *T. bellerica*, and *T. chebula* of *Terminalia*, and found that all the plant parts consist of alkaloids and glycosides responsible for cardioprotective activity. Kadam et al. (2018) developed a validated HPTLC method as per ICH guidelines for five medicinally important β-carboline

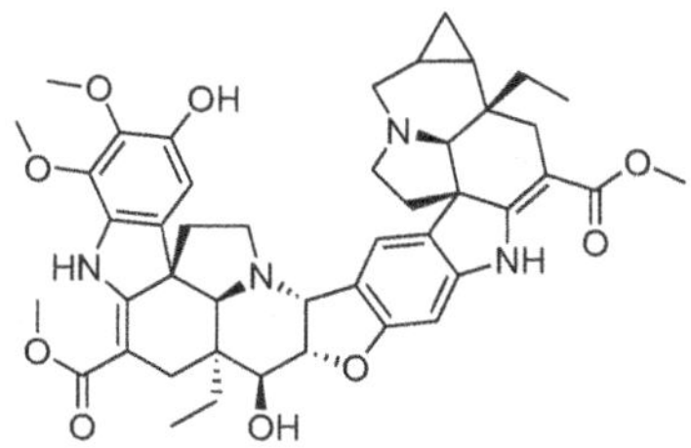

**Conophylline (22)**

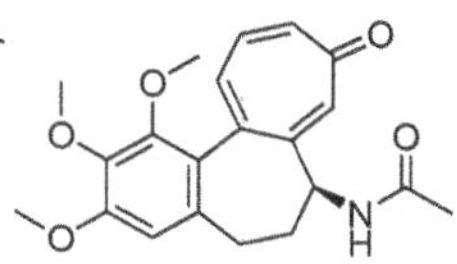

**Colchicine (23)**

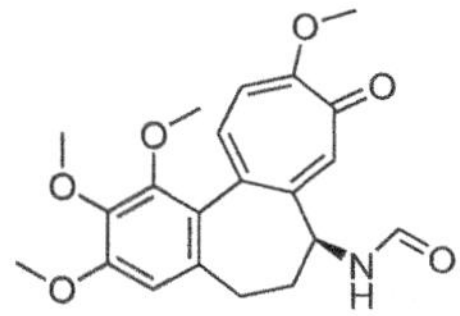

**Gloriosine (24)**

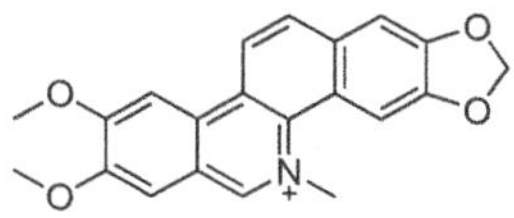

**Nitidine (25)**

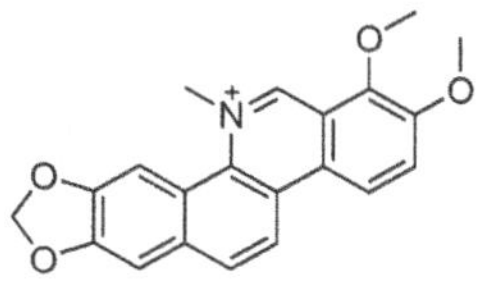

**Chelerythrine (26)**

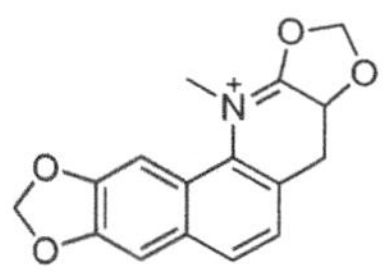

**Sanguinarine (27)**

**Fig. 3.5** Structure of compounds

alkaloids, viz., harmalol (**28**), harmaline (**29**), harmine, harmane (**30**), and norharmane. The method obtained is easy, precise, specific, and reproducible and could be explored for quality control purpose of various herbal drugs. Poornima et al. (2017) carried out HPTLC analysis of the ethanolic extract of *Tabernaemontana divaricata*. The presence of alkaloids was established by derivatizing with Dragendorff's reagent. The pharmacological activity is due to the presence of alkaloids, flavonoids, glycosides, polyphenols, steroids, and tannins. Smyrska-Wieleba et al. (2017) carried out an analysis of pyrrolizidine alkaloids and their *N*-oxides found in the flowers of *Tussilago farfara* and roots of *Arnebia euchroma*. Initial HPTLC analysis confirmed the presence of saturated and unsaturated pyrrolizidine alkaloids in both plants. They concluded that HPTLC analysis is quick for comparative analysis of these specific plants. Further HILIC/ESI-QTOF-MS analysis allowed them for identification of various alkaloids, viz., senkirkine, (**34**) retronecine (**33**), and otonecine types, or *N*-oxides in rapid manner. Senguttuvan and Subramaniam (2016) carried out HPTLC analysis study on various secondary metabolites present in methanolic leaf and root extract of *Hypochaeris radicata* collected from Nilgiris, in the Western Ghats of India. The HPTLC analysis of the leaf methanolic extract revealed the presence of three alkaloids, four flavonoids, one glycoside, and one terpenoids, while the root extract revealed the presence of one alkaloid, one flavonoid, three glycosides, one saponin, and one terpenoid. Geethika and Sunoj Kumar (2018) carried out an HPTLC analysis of methanol, ethanol, aqueous, and chloroform extracts of three species of *Leucas*, viz., *Leucas stelligera*, *Leucas eriostoma*, and *Leucas ciliata*, and found the presence of terpenes, tannins, flavonoids, glycosides, and phenolic compounds. The highest concentration of phenolics, flavonoids, and tannins were found in *L. ciliata*. Jesioneka et al. (2018) developed a new HPTLC method for the analysis of essential oil fractions of

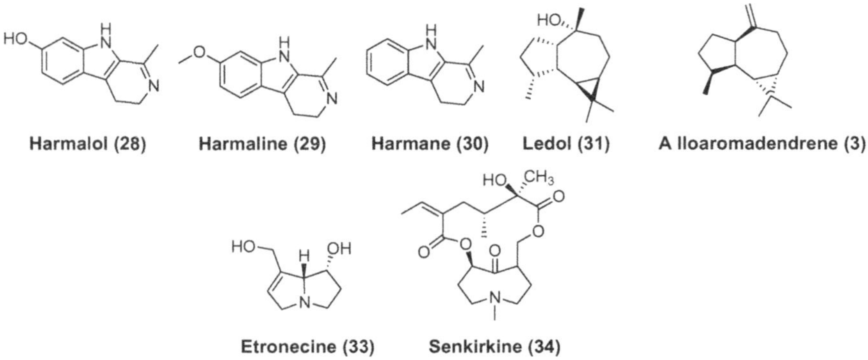


**Fig. 3.6** Structure of compounds

*Rhododendron tomentosum* for ledol (**31**) and alloaromadendrene (**32**). They further carried out the quantification of ledol (**31**) and alloaromadendrene using the newly developed HPTLC method. The mixture of vanillin and phosphoric acid was used for derivatization. Further, HPTLC was utilized for antioxidant analysis of essential oils. GC-MS analysis reveals the detection of specific compounds (Fig. 3.6).

### 3.1.2 High-Pressure Liquid Chromatography

High-pressure liquid chromatography is simple and a commonly available analytical instrument used for qualitative and quantitative analysis of organic compounds. HPLC is a hassle-free technique used for the separation of natural compounds from crude extract efficiently. HPLC consists of the following components: (a) solvent reservoir, (b) pumps, (c) sample injector, (d) column, (e) detector, and (f) work station for data analysis (Fig. 3.7).

Every component is attached to each other in series by steel tubing. The pump maintains the flow of solvents, after which the solvent enters the injector, allowing sample to dissolve in solvent and then passes through the column where the separation of compounds takes place and finally through the detector. The signal of compounds can be recorded in the form of peaks of individual compounds (Fig. 3.8). The injector allows introduction of sample into the column in a controlled and convenient way. Please find below the HPLC chromatogram of andrographolide isolated from *Andrographis paniculata* with retention time of 5.66 min with peak area of 100%.

Bodalska et al. (2020) carried out a study on 23 commercial peppermint samples purchased from different pharmacies in Poland. HPLC (coupled with diode array detector) analysis was performed on nonvolatile fraction and showed the detection of eriocitrin, luteolin-7-*O*-rutinoside, and rosmarinic acid (**36**). Elansary et al. (2020) carried out quantitative HPLC analysis of two species of the leaf extract of *Mentha* for polyphenols located in Saudi Arabia. They found that rosmarinic acid (**36**) was

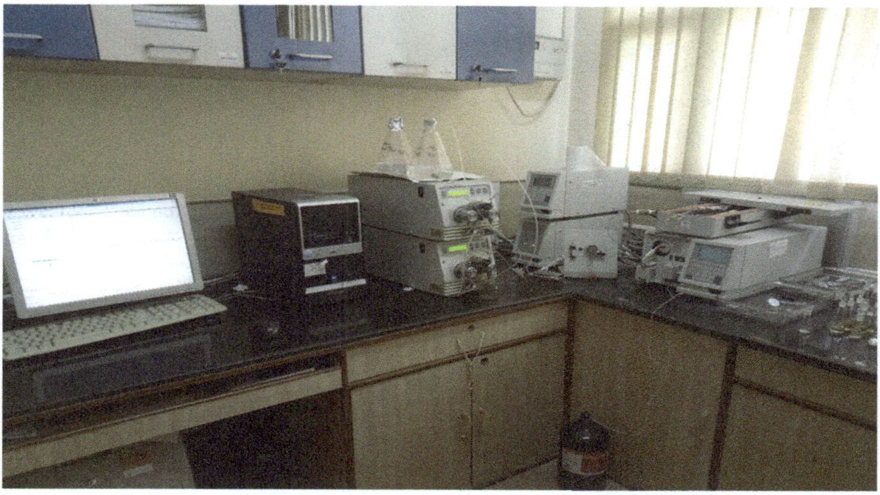

**Fig. 3.7** High-pressure liquid chromatography

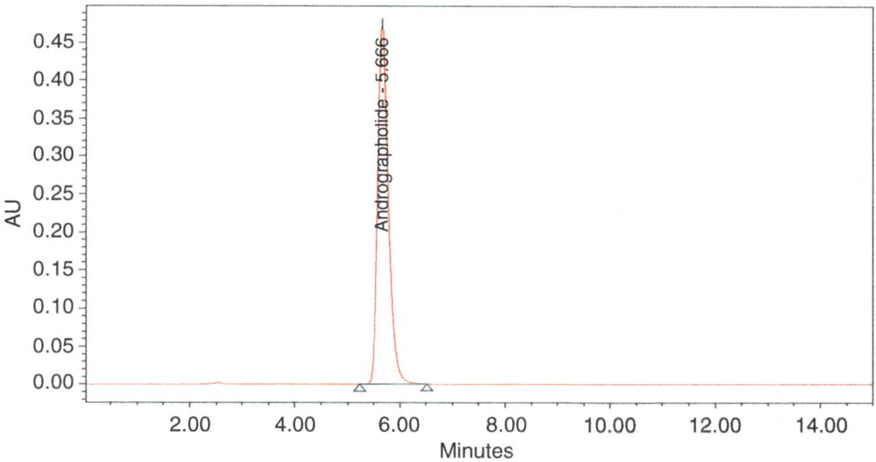


**Fig. 3.8** HPLC chromatogram of andrographolide isolated from *Andrographis paniculata*

present in both species. *Mentha piperita* consists of flavonoids such as naringin (**38**) and cymaroside, while *Mentha longifolia* consists of cryptochlorogenic (**35**), *m*-coumaric acid, and *p*-coumaric acid (**37**) (Fig. 3.9).

Gonzalez-Gonzalez et al. (2019) developed a new HPLC-DAD method for a quick estimation of six phenolic acids (gallic acid, vanillic acid, *p*-hydroxybenzoic acid, caffeic acid, *p*-coumaric acid (**37**), and *trans*-cinnamic acid) in an aqueous extract of *Solanum elaeagnifolium*, *Ampelocissus acapulcensis*, and *Brosimum alicastrum*. Abdelkhalek et al. (2020) analyzed *Eucalyptus camaldulensis* bark

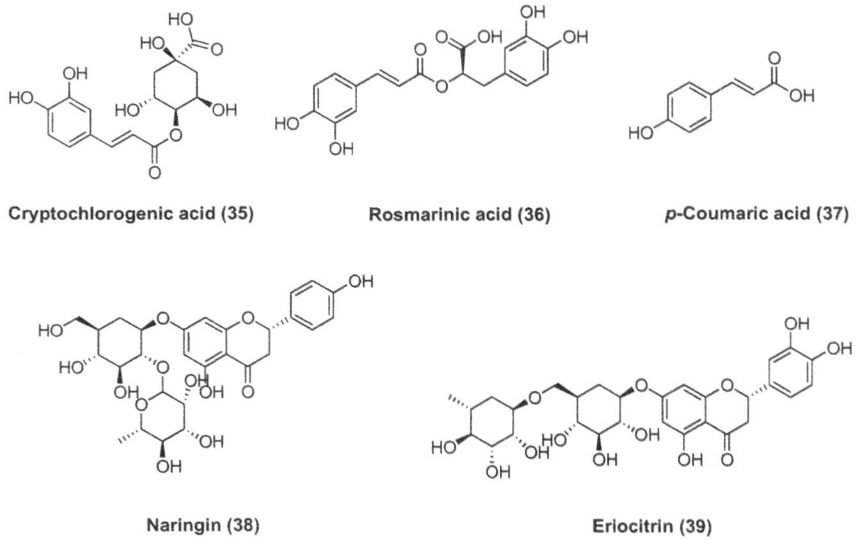


**Fig. 3.9** Structure of compounds

extract for polyphenolic compounds. The HPLC analysis revealed the presence of benzoic acid, salicylic acid, myricetin (**40**), quinol, and rutin (**15**). Belscak-Cvitanovic et al. (2018) developed a "HPLC method for simultaneous determination of phenolic acids and flavonoids in 16 widely spread medicinal plants viz. *Melissa officinalis* L., *Mentha piperita* L., *Salvia officinalis* L., *Lavandula angustifolia* L., *Thymus vulgaris* L., *Rosmarinus officinalis* L., *Taraxacum officinale* L., *Calendula officinalis* L., *Matricaria recutita*, *Achillea millefoiium*, *Tilia cordata*, *Olea europaea* L., *Equisetum arvense*, *Urtica dioica*, and *Plantago lanceolata*." The developed HPLC method was validated by linearity, precision, accuracy, quantification, and limit of detection and able to separate 24 polyphenolic compounds with run time of 30 min. Rosmarinic chicoric acid was present in abundance (Fig. 3.10).

Seal (2016) carried out quantitative RP-HPLC analysis of wild leaves of *Sonchus arvensis* and *Oenanthe linearis*. The newly developed method was validated, and simultaneous estimation of ascorbic acid, phenolic acids, and flavonoids was undertaken in four different solvent extracts. High amount of ascorbic acid and gallic acid was present in 1% aq. acetic acid extract of these plants. Tohma et al. (2016) carried out RP-HPLC analysis for identification and quantification of methanolic extract of three species of *Salvia*, viz., *S. brachyantha* (Bordz.), *S. aethiopis* L., and *S. microstegia* Boiss. and Bal. The HPLC analysis revealed the presence of 18 phenolic acids in *S. brachyantha*. Further, MS analysis confirmed the presence of quinic acid, malic acid, *trans*-aconitic acid, gallic acid, chlorogenic acid (**11**), tannic acid, *p*-coumaric acid (**37**), rutin (**15**), kaempferol (**41**), vanillin (**44**), *trans*-caffeic acid, rosmarinic acid (**36**), apigenin (**45**), etc. Bhatia et al. (2019) developed a validated HPLC method for the determination of seven polyphenols,

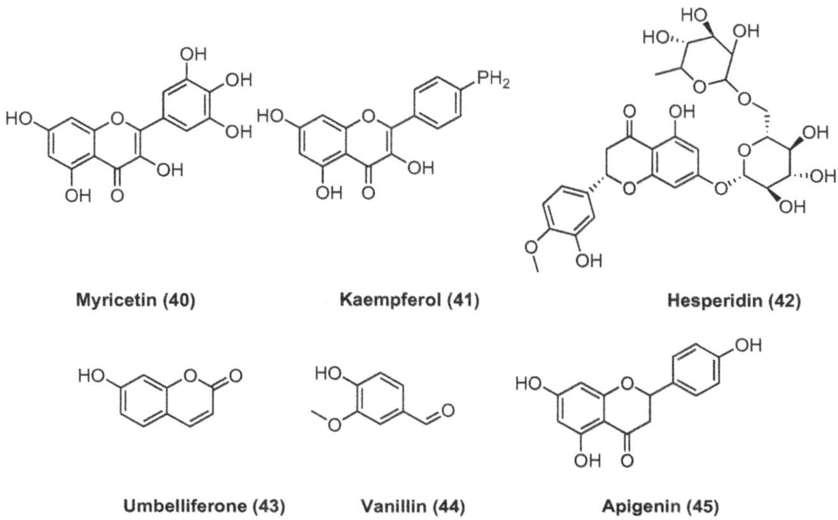


**Fig. 3.10** Structure of compounds

viz., galic acid, catechin, ellagic acid (**9**), quercetin (**13**), kaempferol (**41**), umbelliferone (**43**), and rutin (**15**), found in *Cornus capitata* Wall. (leaves), *Clematis grata* Wall. (whole plant), *Evolvulus nummularius* (L.) L. (whole plant), and *Roylea cinerea* (D. Don) Baill (leaves and stem). The refined method was tested for accuracy, linearity, repeatability, LoD, and LoQ and achieves separation of all seven polyphenols under study.

Çelik et al. (2017) developed a new validated UPLC method coupled with DAD diode array detector and mass spectroscopy. Three aromatic plant species belonging to *Lamiaceae* family, viz., *Origanum majorana* L., *Lavandula officinalis*, and *Mentha pulegium* L., were undertaken for their phenolic profile. UPLC with mass spectroscopy leads to the identification of gallic acid, catechin, vanillic acid, caffeic acid, rutin (**15**), hesperidin (**42**), quercetin (**13**), kaempferol (**41**), apigenin (**45**), etc.

Ekin et al. (2016) investigated the anticholinesterase potential of ethanolic extract of the 34 taxa of *Amelanchier*, *Cotoneaster*, *Pyrus*, and *Sorbus* genera. The active species, viz., *Cotoneaster meyeri* and *Sorbus subfusca*, and two samples of *Sorbus umbellata* were then subjected for HPLC analysis of some specific phenolic acids and flavonoids, out of which chlorogenic acid (**11**) and rutin (**15**) were quantified in all active species (Fig. 3.11).

Viapiana et al. (2016) developed a validated HPLC method for quality analysis of 19 chamomile samples obtained from different manufacturers for comparative fingerprinting. Further, simultaneous quantification of gallic acid, caffeic acid, syringic acid (**46**), *p*-coumaric acid (**37**), ferulic acid (**47**), rutin (**15**), myricetin (**40**), quercetin (**13**), and keampferol (**41**) was carried out. Jiao et al. (2018) developed a new validated HPLC method for identification and quantification of nine polyphenols, viz., gallic acid, chlorogenic acid (**11**), caffeic acid, hyperoside (**48**),

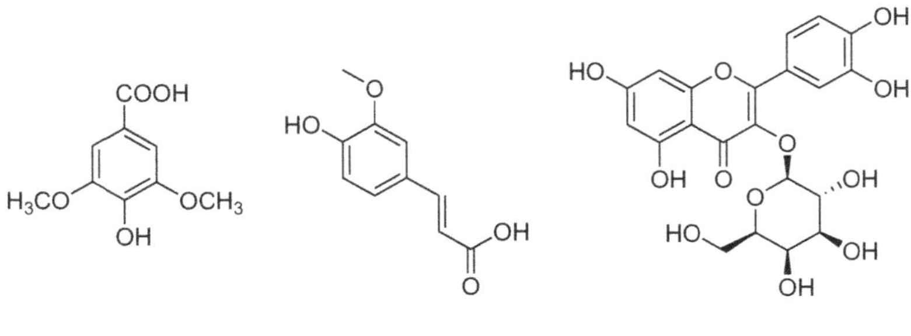


**Fig. 3.11** Structure of compounds

luteoside, myricetin (**40**), quercetin 3-rhamnoside, quercetin (**13**), and kaempferol (**41**) found in *Polygoni avicularis herba*. The method was successfully applied for the analysis of herb samples, and variation in polyphenols contents were seen on the basis of locations. Li et al. (2018) developed a new validated HPLC method for simultaneous identification and determination of quercetin (**13**) and erianin found in *Dendrobium officinale*. The developed method is precise, reliable, and reproducible and can be used to access the quality of *Dendrobium officinale* plant material. Marchelak et al. (2020) developed a new HPLC method for the simultaneous quantification of 30 polyphenols found in *Prunus spinosa* L. (a European herbal medicine) in a 35-min analysis time. The developed method is suitable for the analysis of raw material and extract obtained from different solvents. The results obtained are suitable for quality control and chemical analysis. Migues et al. (2020) carried out comparative analysis using HPLC-DAD and spectrometric methods for polyphenolics (gallic acid, epicatechin) (**18**), catechin, quercetin (**13**), chlorogenic acid (**11**) and rutin (**15**) content in the bark extract of *Schinus terebinthifolius* used as herbal medicine in Central and South America. Further, the quantification of polyphenols was carried out. Results show that both methods were similar and can be used for the quality control purpose of herbal drugs. O Elansary et al. (2019) carried out chemical profiling of bark extract of three species of *Quercus*, viz., Q. *macrocarpa*, *Q. robur*, and *Q. acutissima*, using HPLC. The analysis revealed the presence of polyphenols in high amount. This is the first study to reveal the anticancer potential and presence of ellagic acid (**9**) in all three species of *Quercus*. Rivera-Mondragón et al. (2019) developed a new validated method as per ICH guidelines for the analysis of total flavonoids, flavonolignans, and chlorogenic acid (**11**) found in the leaves of *Cecropia* species. Further identification and quantification of chemical markers were carried out. The developed method is accurate, precise, reproducible, and efficient. This method can be used to check the quality aspects of herbal medicines. Köksal et al. (2017) carried out an HPLC-MS-based analysis of polyphenols present in *Nepeta trachonitica*. The HPLC-MS analysis revealed the presence of 11 major polyphenols, and chlorogenic acid (**11**), quinic acid, and rosmarinic acid (**36**) are present in higher concentration. Tewari et al.

(2017) carried work on ancient medicinal plant *Saraca asoca* used in Ayurvedic medicines. The plant material collected from different geographical regions of India was analyzed for pinitol, flavonols, and polyphenols using HPLC. The results revealed that the bark collected from Uttar Pradesh region were rich in the above phytoconstituents. The presence of pinitol and flavonoids are responsible for the antioxidant activity of *Saraca asoca*. Zhao et al. (2020) developed a new method involving the combination of DNA barcoding, UHPLC, and HPLC-QTOF-MS/MS for chemical profiles, differentiation, and evaluation of quality for *Stephania* species. They concluded that this method is efficient for chemical profiling and quality evaluation of *Stephania* species. Barny et al. (2020) carried out HPLC-MS analysis of pyrrolizidine alkaloids in Apocynaceae family. Results showed the presence of pyrrolizidine alkaloids in tribe *Echiteae*, *Malouetieae*, *Eucorymbia alba*, *Galactophora schomburgkiana*, *Holarrhena pubescens*, and *Kibatalia macrophylla*. In *Wrightia* R. Br. and *Marsdenia* R. Br., pyrrolizidine alkaloids were reported for the first time. Jo et al. (2020) developed a new HPLC-PDA method for the simultaneous determination of four marker compounds, palmatine, geniposide, berberine (**12**), and paeoniflorin, present in Haedoksamul-tang, a traditional medicinal drug in Japan. The developed method was validated as per ICH guidelines. Geniposide was present in highest concentration in Haedoksamul-tang. Wang et al. (2020) carried out quantitative and qualitative HPLC-MS analyses of *Senecio scandens* and related formulations for pyrrolizidine alkaloids. The analysis was utilized for quality control analysis of *Senecio scandens* species. Further safety studies and toxicity were analyzed for *Senecio scandens* growing in various habitats.

### 3.1.3 GC-MS

Hyphenated techniques have gained a lot of importance in analytical chemistry. Gas chromatography was the first technique to introduce with various detectors for the analysis of various natural compounds. GC-MS analysis is one of the most widely used techniques for the detection of molecular mass of volatile compounds. Compounds which are volatile and tolerate high temperature can be analyzed by GC-MS. GC-MS requires a very small amount of sample, and precision is very high in comparison to other analytical techniques. GC-MS is comprised of mainly the injector port, column, and mass spectrometer. Helium, argon, and nitrogen are used as carrier gas for driving the sample to the column. GC-MS has wide application and used in pesticide analysis, flavor and fragrance analysis, and metabolite analysis (Fig. 3.12).

Adejoke et al. (2020) carried out a phytochemical analysis of *Acalypha wilkesiana* used for the treatment of skin disorder. The crude extract was prepared by using 50% methanol and subjected to fractionation with three solvents, viz., hexane, ethyl acetate, and butanol. Phytochemical analysis showed the detection of cardiac glycoside, alkaloid, saponins, flavonoids, and anthraquinones. The essential oil was extracted through hydrodistillation process and responsible for curing skin disorder. The GC-MS analysis of essential oil showed the presence of 4-hexen-2-

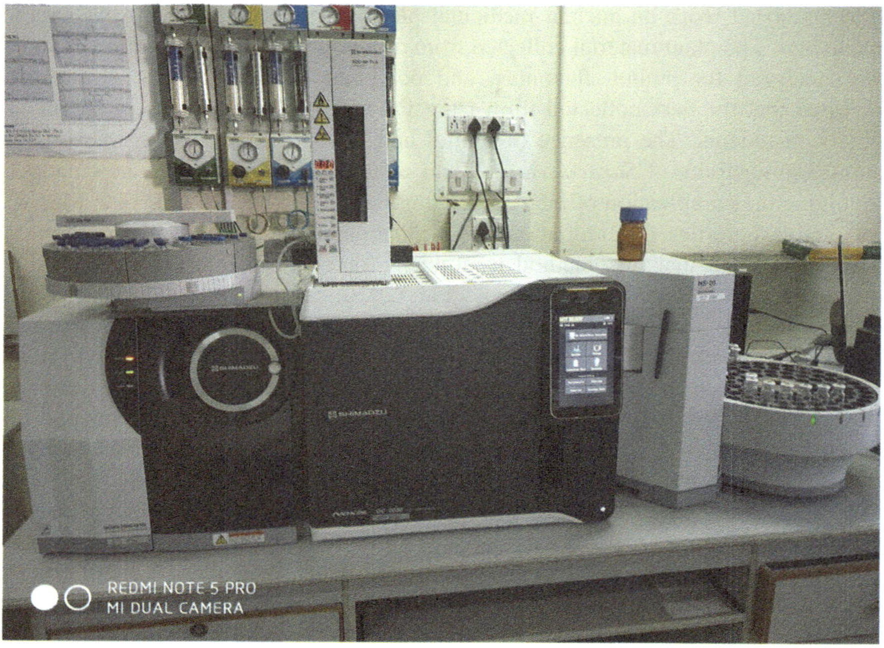


**Fig. 3.12** Gas chromatography-liquid chromatography

one-3-methyl, pyrrole, n-hexadecanoic, and 6-benzamido-4-benzoyl-1,2,4-triazine-3,5. Bhardwaj et al. (2020) carried out the pharmacological evaluation and GC-MS analysis of *Codonopsis clematidea*, a medicinal plant found in the trans-Himalayan cold desert region. The plant is used for various diseases by local tribes. Soxhlet extraction was carried out using methanol as solvent and further fractionation with various organic solvents. Antioxidant potential was accessed by using various assay and revealed the presence of high amount of polyphenolics and flavonoids. The volatile constituents were analyzed using GC-MS analysis. Further luteolin (**14**) was isolated using column chromatography. The plant holds antibacterial and antioxidant potential. Wojciechowska and Gołębiowski (2020) studied the effect of various after effects of using insecticides, viz., deltamethrin and cyfluthrin, on *T. molitor* and acetamiprid and thiamethoxam on *L. decemlineata* after 24, 48, and 72 h. SPME fiber was used to prepare the sample and analyzed using GC-MS which revealed the presence of 20% aldehydes and 8–41% of alkanes in the case of *T. molitor*, while in *L. decemlineata* fatty acids were found in range of 8–65%. Trujillo-Chacón et al. (2020) carried out a qualitative and quantitative analysis of alkaloids in in vitro propagated *Rhodophiala pratensis*. GC-MS analysis was performed and showed alternation in the presence of alkaloids between wild bulbs, in vitro blubs, and callus. The in vitro bulbs have high concentration of alkaloids being provided with supplements. Benamar et al. (2020) for the first time identified nine pyrrolizidine alkaloids using GC-MS analysis of the crude extract of *Pardoglossum cheirifolium*.

Mir et al. (2020) carried antibacterial activity and GC-MS analysis of *Myrtus communis*. Ethanolic leaf extract was prepared using Soxhlet extraction. The crude leaf extract was then subjected to GC-MS analysis to identify compounds responsible for antibacterial activity which revealed the presence of 50 known compounds, and 1,1,8a-trimethyloctahydro-2,6-naphthalenedione (27.6%) and pyrogallol (9.1%) are new compounds. Naz et al. (2020) studied pharmacological evaluation and GC-MS analysis of the methanolic leaf extract of *Jacaranda mimosifolia*. The crude extract shows good antimicrobial, antioxidant, cyctotoxic, and antilipoxygenase activity. GC-MS analysis shows the presence of 15 natural molecules identified as phenolic, fatty acids, and alcoholic compounds. Polyphenolics were measured by using HPLC, while antioxidant and anti-inflammatory activities by spectrometry. Lee et al. (2020) carried out quantitative determination of quinolizidine alkaloids such as angustifoline, sparteine, and 13-hydroxylupanine present in lupin and lupin products. Theses alkaloids are known to be toxic to humans and animals. These alkaloids were determined in lupin cookies, lupin beans, and lupin drinks. GC-MS analysis using flame ionization detectors was carried out, and these products were found to be safe for consumption of human beings. The developed GC-MS analysis method can be used for quality control analysis. Suleiman (2020) studied antimicrobial activity and GC-MS analysis of selected six plants, viz., *Cinnamomum camphora*, *Cichorium intybus*, *Foeniculum vulgare*, *Spartium junceum*, *Nerium oleander*, and *Commiphora myrrha*. Extraction was carried out by using a supercritical fluid extractor. GC-MS analysis was performed to determine active principle present in these plants. Out of all plants, *F. vulgare* has the highest potent against *E. faecalis*, *C. albicans, and S. typhimurium*; this might due to the presence of eugenol, isoeugenol, and squalene in this plant. Ahmet et al. (2020) carried out GC-MS analysis of alkaloids present in *Galanthus fosteri*. The presence of galanthamine was reported in this plant which is why the plant is used for the treatment of Alzheimer's diseases. GC-MS analysis was performed to known alkaloid profiles of this plant and revealed the presence of 22 compounds in the crude extract.

## 3.2 Conclusion

The present chapter has summarized analytical instruments like high-pressure thin-layer chromatography, high-pressure liquid chromatography, and gas chromatography-mass spectroscopy for their potential benefit to drug discovery through chemical fingerprint for pharmaceutical, food, and biotechnology industries. The method outlines are robust, rapid, economical, and consistent and a tool for qualitative as well as quantitative analysis of natural compounds and identification of marker which could be industrial exploitation and thus beneficial for human welfare and drug delivery program.

## References


Abdelkhalek A, Salem MZM, Ali HM, Kordy AM, Salemf AZM, Behiry SI (2020) Antiviral, antifungal, and insecticidal activities of *Eucalyptus* bark extract: HPLC analysis of polyphenolic compounds. Microb Pathog 147:104383

Adejoke OE, Paulinus NS, Omonigho IH (2020) Chemical and phytochemical analyses of extracts from the leaves of *Acalypha wilkesiana*, "an herbal plant used for the treatment of various skin disorders". Ann Sci Technol 5(2):40–48

Ahmet EMİR, Ceren EMİR, Buket Bozkurt NÜS (2020) GC/MS analysis of alkaloids in *Galanthus fosteri* baker and determination of its anticholinesterase activity. Turkish J Pharm Sci 17(1):36

Ali M, Rathaur B, Nishad U (2020) Phytochemical screening, physicochemical and antioxidant activity, tlc & finger print of HPTLC, *Morus Alba* ethanol extraction. Int J Innov Sci Res Technol 5(3):1–14

Barny LA, Tasca JA, Sanchez HA, Smith CR, Koptur S, Livshultz T, Minbiole KP (2020) Chemotaxonomic investigation of Apocynaceae for retronecine-type pyrrolizidine alkaloids using HPLC-MS/MS. bioRxiv

Belscak-Cvitanovic A, Valinger D, Benkovic M, Tusek AJ, Jurina T, Komes D, Kljusuric JG (2018) Integrated approach for bioactive quality evaluation of medicinal plant extracts using HPLC-DAD, spectrophotometric, near infrared spectroscopy and chemometric techniques. Int J Food Prop 20(3):S2463–S2480

Benamar H, Tomassini L, Frezza C, Marouf A, Bennaceur M, Nicoletti M (2020) First study on the pyrrolizidine alkaloids of *Pardoglossum cheirifolium* (L.) E. Barbier & Mathez.: GC-MS analysis of their volatile components in the whole plant. Nat Prod Res:1–6

Bhardwaj P, Naryal A, Thakur MS, Aggarwal NK, Saxena S, Chaurasia OP, Kumar R (2020) Comparative antioxidant, antibacterial, and GC-MS analysis of methanol extract's fractions and isolation of luteolin from leaves of trans-Himalayan *Codonopsis clematidea*. Ind Crop Prod 144: 112046

Bhatia A, Kaur T, Singh B, Arora R, Arora S (2019) Reverse phase HPLC method validation for estimation of polyphenols in medicinal plants and their possible role in reticence of xanthine oxidase activity. Sep Sci Plus 2(7):237–244

Bodalska A, Kowalczyk A, Włodarczyk M, Fecka I (2020) Analysis of polyphenolic composition of a herbal medicinal product-peppermint tincture. Molecules 25(69):1–19

Bubueanu C, Pavaloiu R (2016) HPTLC chromatographic polyphenolic fingerprints of plant species from Eastern Europe. Malays J Med Biol Res 3:41–44

Çelik SE, Tufan AN, Bekdeşer B, Özyürek M, Güçlü K, Apak R (2017) Identification and determination of phenolics in *Lamiaceae* species by UPLC-DAD-ESI-MS/MS. J Chromatogr Sci 55(3):291–300

Conceicao RS, Reis IMA, Cerqueira APM, Perez CJ, Santos Junior MC, Branco A, Ifa DR, Botura MB (2020) Rapid structural characterisation of benzylisoquinoline and aporphine alkaloids from *Ocotea spixiana* acaricide extract by HPTLC-DESI-MS. Phytochem Anal 2020:1–11

Cortes N, Sierra K, Alzate F, Osorioc EH, Osorio E (2017) Alkaloids of amaryllidaceae as inhibitors of cholinesterases (AChEs and BChEs): an integrated bioguided study. Phytochem Anal 29(2):217–227

Dinakaran SK, Chelle S, Avasarala H (2019) Profiling and determination of phenolic compounds in poly herbal formulations and their comparative evaluation. J Tradit Complement Med 9:319–327

Ekin HN, Gokbulut A, Aydin ZU, Donmez AA, Orhan IE (2016) Insight into anticholinesterase and antioxidant potential of thirty-four *Rosaceae* samples and phenolic characterization of the active extracts by HPLC. Ind Crop Prod 91:104–113

Elansary HO, Szopa A, Kubica P, Ekiert H, Klimek-Szczykutowicz M, El-Ansary DO, Mahmoud EA (2020) Polyphenol profile and antimicrobial and cytotoxic activities of natural *Mentha* × *piperita* and *Mentha longifolia* populations in Northern Saudi Arabia. Processes 8(479):1–17. https://doi.org/10.3390/pr8040479

Geethika K, Sunoj Kumar P (2018) Phytochemical screening and high-performance thin-layer chromatography fingerprint profile of three species of *Leucas* (Lamiaceae). Ancient Sci Life 37:102–107

Gonzalez-Gonzalez RM, Barragan-Mendoza L, Peraza-Campos AL, Muniz-Valencia R, Ceballos-Magana SG, Parra-Delgado H (2019) Validation of an HPLC-DAD method for the determination of plant phenolics. Braz J Pharmacogn 29(5):689–693

Gupta P, Patil D, Patil A (2019) Qualitative HPTLC phytochemical profiling of *Careya arborea* Roxb. bark, leaves and seeds. 3Biotech 9:311

Hakim M, Rathod D, Trivedi DA, Panigrahi J, Gantait S, Patel IC (2018) An effective validated method for HPTLC-fingerprinting of alkaloids and glycosides from multiple plant parts of three *Terminalia* spp. Israel J Plant Sci 65(1–2):109–117

Jesioneka A, Poblocka-Olecha L, Zabiegalab B, Bucinskic A, Krauze-Baranowskaa M, Luczkiewicz M (2018) Validated HPTLC method for determination of ledol and alloaromadendrene in the essential oil fractions of *Rhododendron tomentosum* plants and in vitro cultures and bioautography for their activity screening. J Chromatogr B 1086:63–72

Jiao L, Yan C, Zhang K, Xie J, Zhang Y, Wen Z (2018) Comprehensive determination of nine polyphenols in Polygoni Avicularis Herba with a new HPLC–DAD method and their correlation with the antioxidant activities. J Food Meas Charact 12(3):1593–1600

Jo BG, Kang KH, Yang MH (2020) Development and validation of a HPLC–PDA method for the simultaneous determination of berberine, palmatine, geniposide, and paeoniflorin in Haedoksamul-tang. Appl Sci 10(16):5482

Kadam NS, Naik AA, Doshi PJ, Nikam TD (2018) High-performance thin-layer chromatography method for simultaneous determination of antipsychotic and medicinally important five β-carboline alkaloids. J Chromatogr Sci 2019:1–11

Kaur P, Pandey DK, Gupta RC, Dey A (2019) Simultaneous microwave assisted extraction and HPTLC quantification of mangiferin, amarogentin, and swertiamarin in Swertia species from Western Himalayas. Ind Crops Prod 132:449–459

Köksal E, Tohma H, Kılıç Ö, Alan Y, Aras A, Gülçin I, Bursal E (2017) Assessment of antimicrobial and antioxidant activities of *Nepeta trachonitica*: analysis of its phenolic compounds using HPLC-MS/MS. Sci Pharm 85(2):24

Lee HW, Hwang IM, Lee HM, Yang JS, Park EJ, Choi JW et al (2020) Validation and determination of quinolizidine alkaloids (QAs) in lupin products by gas chromatography with flame ionization detection (GC-FID). Anal Lett 53(4):606–613

Li J, Huang HY, Wang YZ (2018) Optimized determination of phenolic compounds in *Dendrobium officinale* stems by reverse-phase high performance liquid chromatography. J Liq Chromatogr Relat Technol 41(9):508–516

Marchelak A, Olszewska MA, Owczarek A (2020) Simultaneous quantification of thirty polyphenols in blackthorn flowers and dry extracts prepared thereof: HPLC-PDA method development and validation for quality control. J Pharm Biomed Anal 184:113121

Migues VH, David JM, David JP (2020) Determination of polyphenols in *Schinus terebinthifolius* Raddi bark extracts and chemometric analysis. Anal Methods 12(11):1478–1485

Mir MA, Bashir N, Alfaify A, Oteef MD (2020) GC-MS analysis of *Myrtus communis* extract and its antibacterial activity against gram-positive bacteria. BMC Complement Med Ther 20(1):1–9

Misra A, Kumar B, Shukla P, Srivastava S (2020) Simultaneous HPTLC-UV quantification of colchicine and gloriosine alkaloids in the natural population of *Gloriosa superba* L., collected from Eastern Ghats of India for the identification of elite chemotypes. J Liq Chromatogr Relat Technol 43(9–10):351–360

Mundengara D, Sulaiman C, Balachandran I, Chandran KPS (2019) Isolation characterization and high-performance thin-layer chromatographic quantification of active xanthonoid mangiferin from *Canscora perfoliata*. J Planar Chromatogr 32(3):231–235

Nair CVJ, Ahamad S, Khan W, Anjum V, Mathur R (2017) Development and validation of high-performance thin-layer chromatography method for simultaneous determination of polyphenolic compounds in medicinal plants. Pharm Res 9:S67–S73

Narayanan J, Marimuthu J (2016) HPTLC fingerprint profile (phenolics) of selected *Cyathea* species from Western Ghats, South India. Chin J Biol 2016:6420371. 1–7

Naz R, Roberts TH, Bano A, Nosheen A, Yasmin H, Hassan MN et al (2020) GC-MS analysis, antimicrobial, antioxidant, antilipoxygenase and cytotoxic activities of *Jacaranda mimosifolia* methanol leaf extracts and fractions. PLoS One 15(7):e0236319

Nile SH, Park SW (2014) HPTLC analysis, antioxidant and antigout activity of Indian plants. Iran J Pharm Res 13(2):531–539

O Elansary H, Szopa A, Kubica P, Ekiert H, Mattar M, Al-Yafrasi MA et al (2019) Polyphenol profile and pharmaceutical potential of *Quercus* spp. bark extracts. Plants 8(11):486

Pedan V, Weber C, Do T, Fischer N, Reich E, Rohn S (2018) HPTLC fingerprint profile analysis of cocoa proanthocyanidins depending on origin and genotype. Food Chem 267:277–287

Perala K, Ciddi V (2018) Simultaneous quantitative determination of nitidine, chelerythrine and sanguinarine using HPTLC from callus extract of *Zanthoxylum rhetsa*. Am J Anal Chem 9:386–396

Poornima K, Palanisamy CP, Sundaram S, Kanniappan GV (2017) Chromatographic fingerprinting analysis of secondary metabolites present in ethanolic extract of *Tabernaemontana divaricata* (L.) R. Br. by HPTLC technique. Anal Chem Lett 7(1):20–29

Rathore AS, Sathiyanarayanan L, Mahadik KR (2016) Determination of major polyphenolic components in *Euphoria longana* Lam. by validated high performance thin-layer chromatography method and direct analysis in real time mass spectrometry. J Chromatogr Sep Tech 7(4):1000330

Rivera-Mondragón A, Broeckx G, Bijttebier S, Naessens T, Fransen E, Kiekens F et al (2019) Ultrasound-assisted extraction optimization and validation of an HPLC-DAD method for the quantification of polyphenols in leaf extracts of *Cecropia species*. Sci Rep 9(1):1–16

Rocamora CR, Ramasamy K, Lim SM, Majeed ABA, Agatonovic-Kustrin S (2020) HPTLC based approach for bioassay-guided evaluation of antidiabetic and neuroprotective effects of eight essential oils of the *Lamiaceae* family plants. J Pharm Biomed Anal 178:112909

Satija S, Tambuwala MM, Pabreja K, Bakshi HA, Chellappan DK, Aljabali AA, Nammi S, Thakur GS, Dureja H, Gupta G, Dua K, Mehta M, Garg M (2020) Development of a novel HPTLC fingerprint method for simultaneous estimation of berberine and rutin in medicinal plants and their pharmaceutical preparations followed by its application in antioxidant assay. J Planar Chromatogr 33:313–319

Seal T (2016) Quantitative HPLC analysis of phenolic acids, flavonoids and ascorbic acid in four different solvent extracts of two wild edible leaves, *Sonchus arvensis* and *Oenanthe linearis* of north-eastern region in India. J Appl Pharm Sci 6(02):157–166

Senguttuvan J, Subramaniam P (2016) HPTLC fingerprints of various secondary metabolites in the traditional medicinal herb *Hypochaeris radicata* L. J Bot 2016:5429625. https://doi.org/10.1155/2016/5429625

Smyrska-Wieleba N, Wojtanowski KK, Mroczek T (2017) Comparative HILIC/ESI-QTOF-MS and HPTLC studies of pyrrolizidine alkaloids in flowers of *Tussilago farfara* and roots of *Arnebia euchroma*. Phytochem Lett 20:339–349

Sridhar SNC, Sengupta P, Paul AT (2018) Development and validation of a new HPTLC method for quantification of conophylline in *Tabernaemontana divaricata* samples obtained from different seasons and extraction techniques: insights into variation of pancreatic lipase inhibitory activity. Ind Crops Prod 111:462–470

Srinivasan S, Wankhar W, Rathinasamy S, Rajan R (2016) Free radical scavenging potential and HPTLC analysis of *Indigofera tinctoria* linn (Fabaceae). J Pharm Anal 6:125–131

Suleiman WB (2020) In vitro estimation of superfluid critical extracts of some plants for their antimicrobial potential, phytochemistry, and GC–MS analyses. Ann Clin Microbiol Antimicrob 19(1):1–12

Tewari R, Gupta M, Ahmad F, Rout PK, Misra L, Patwardhan A, Vasudeva R (2017) Extraction, quantification and antioxidant activities of flavonoids, polyphenols and pinitol from wild and cultivated *Saraca asoca* bark using RP-HPLC-PDA-RI method. Ind Crop Prod 103:73–80

Tohma H, Köksal E, Kılıç Ö, Alan Y, Yılmaz MA, Gülçin I et al (2016) RP-HPLC/MS/MS analysis of the phenolic compounds, antioxidant and antimicrobial activities of *Salvia* L. species. Antioxidants 5(4):38

Trujillo-Chacón LM, Pastene-Navarrete ER, Bustamante L, Baeza M, Alarcón-Enos JE, Cespedes-Acuña CL (2020) In vitro micropropagation and alkaloids analysis by GC–MS of Chilean Amaryllidaceae plants: *Rhodophiala pratensis*. Phytochem Anal 31(1):46–56

Viapiana A, Struck-Lewicka W, Konieczynski P, Wesolowski M, Kaliszan R (2016) An approach based on HPLC-fingerprint and chemometrics to quality consistency evaluation of *Matricaria chamomilla* L commercial samples. Front Plant Sci 7:1561

Wang Y, Li Y, Cang S, Cai Q, Xu H, Wang Y et al (2020) Qualitative and quantitative analysis of pyrrolizidine alkaloids for the entire process quality control from *Senecio scandens* to Senecio scandens-containing preparations by high performance liquid chromatography-tandem mass spectrometry. J Mass Spectrom 55(10):e4532

Wojciechowska M, Gołębiowski M (2020) SPME-GC/MS analysis of volatile compounds contained in the insect larvae of *Tenebrio molitor* and *Leptinotarsa decemlineata* before and after using insecticides. Chem Biodivers 17(4):e1900743

Zhao W, Liu M, Shen C, Liu H, Zhang Z, Dai W et al (2020) Differentiation, chemical profiles and quality evaluation of five medicinal *Stephania* species (Menispermaceae) through integrated DNA barcoding, HPLC-QTOF-MS/MS and UHPLC-DAD. Fitoterapia 141:104453

# Role of Plant Secondary Metabolites as Anticancer and Chemopreventive Agents

# 4

Jeevanjot Kaur, Sonia Mahey, Pankaj Ahluwalia, Rakesh Joshi, and Rakesh Kumar

**Abstract**

Cancer is a life-threatening disease, and its treatment is a challenge for medical science since its inception. This disease is characterized by uncontrolled cell division, and hence its treatment relies on rectifying the functioning of cellular processes. It is estimated that cancer patients will increase by 70% in the coming two decades. Although innumerable drugs are available to treat cancer, the side effects associated with them induces the need to search for nontoxic and efficient anticancer agents. Plants being rich in structurally diverse compounds have immense scope in providing lead bioactive molecules. Plant secondary metabolites have been massively searched for their cancer prevention activities in both in vitro and in vivo models. As a consequence of substantial plant-based research, more than 60% commercial anticancer drugs are derived from plant secondary metabolites. Secondary metabolites are organic molecules essential for the survival of plants where they are engaged in performing numerous functions. This chapter is focused on the chemopreventive and anticancer activity of three major classes of secondary metabolites including terpenoids (monoterpenoids,

J. Kaur
Department of Botanical and Environmental Sciences, Guru Nanak Dev University, Amritsar, India

S. Mahey
Department of Botany, Giani Kartar Singh Memorial Government College, Tanda Urmar, India

P. Ahluwalia
Department of Pathology, Augusta University, Augusta, GA, USA

R. Joshi
Department of Botany and Environmental Science, Sri Guru Granth Sahib World University, Fatehgarh Sahib, India

R. Kumar (✉)
Department of Botany, Doaba College, Jalandhar, India


A. K. Sharma, A. Sharma (eds.), *Plant Secondary Metabolites*,
https://doi.org/10.1007/978-981-16-4779-6_4

diterpenoids, sesquiterpenoids, diterpenoids, and tetraterpenoids), polyphenols (curcumin, quercetin, resveratrol, and flavonoids), and nitrogen-containing secondary metabolites (alkaloids and glucosinolates).

**Keywords**

Anticancer · Secondary metabolites · Chemopreventive · Terpenoids · Polyphenols

## 4.1 Introduction

Cancer is defined as the uncontrolled growth of cells leading to death (Lee et al. 2017). It occurs through transformation, apoptosis dysregulation, multiplication, invasive action, angiogenesis, and finally metastasis of cancer cells. Innumerable methods for the treatment of cancer are known till today, but unfortunately the proportion of cancer patients is increasing at a drastic rate with every passing year. Moreover, from the present scenario, it is estimated that approximately 15.5 million of individuals around the globe will suffer from cancer by the year 2030, and about 11.5 million from this number will not survive (Fig. 4.1) (Amin et al. 2009).

Hence, cancer is the initiating factor of sickness and fatalities globally (Fridlender et al. 2015). This disease occurs on account of improper regulation of cellular processes, viz., gene stability, signaling of growth, stromal microenvironment regulation, signaling of anti-apoptosis, and generation of immune response (Lee et al. 2017; Gali-Muhtasib et al. 2015). Therefore, its treatment is dependent on the reregulation of functioning of cells. For centuries, plants are being used for medical treatment (Fridlender et al. 2015; Gali-Muhtasib et al. 2015). In auxiliary to this, metabolites obtained from plants have been proved to be effective for several biotechnological purposes and therapeutic uses (Korkina and Kostyuk 2012). Plant metabolites possess a myriad of biological activities involving anti-inflammation,

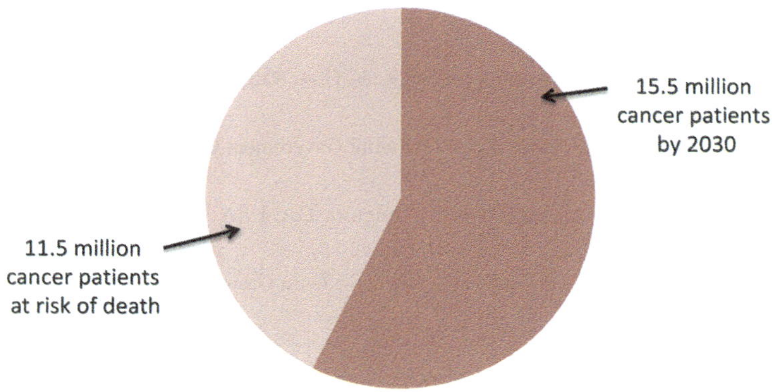


**Fig. 4.1** Projected cancer patients by 2030 and those at risk of death

anticancer, antimicrobial, and analgesic (Fridlender et al. 2015). Approximately 25% of drugs under clinical use are produced by plants (Schmidt et al. 2007). Plants generate about 60% of drugs with anticancer capability. From the above discussion, it is evident that the production of new drugs for treating cancer with lesser side effects and enhanced efficacy is mandatory. Plant-derived metabolites exhibit potent anticancer potential with lessened cytotoxic effect and greater activity (Ijaz et al. 2018). In this chapter, the plant secondary metabolites are categorized with respect to their structure along with the discussion of their anticancer activities.

## 4.2 Side Effects Associated with Cancer Treatment Regimes

Various clinical trials have examined the side effects associated with various cancer treatments such as immunotherapy, chemotherapy, radiations, and antibody treatment (Morrissey et al. 2016). Chemotherapy and radiation possess grievous side effects owing to their cytotoxicity to even healthy cells (Fridlender et al. 2015). Immunotherapy and antibody treatment display extreme specificity potential of cancer targeting; however, they show a limited range of target and are expensive (Morrissey et al. 2016). Furthermore, numerous cancer types obtain resistance following treatment (Fridlender et al. 2015; Morrissey et al. 2016). In the present era, amalgamation of therapies comprising various therapies or drugs is in practice for restricting the demerits of single therapies (Fridlender et al. 2015; Morrissey et al. 2016). Additionally, to overcome the side effects possessed by current anticancer medicines and the need of exploring patient-friendly new natural plant base molecules is on upsurge (Fridlender et al. 2015).

## 4.3 Plant Secondary Metabolites as Anticancer Agents

From the past centuries, a plethora of natural compounds have been obtained from plants, especially the secondary metabolites, that exhibit sufficient variations of their structures and owing to this their synthesis is complicated and thus not accomplished; as well as they possess various bioactivities involving antitumor activity (Nwodo et al. 2016; Habli et al. 2017). Secondary metabolites are defined as natural small organic molecules, which are not integral for the plant to grow, develop, or reproduce. Plant secondary metabolites are composed of mainly three groups such as terpenoids or terpenes (viz., monoterpenoids, diterpenoids, sesquiterpenoids, diterpenoids, and tetraterpenoids), polyphenols (viz., curcumin, quercetin, resveratrol, flavonoids, etc.), and nitrogen-containing compounds (viz., alkaloids and glucosinolates) that are originated by five major precursor pathways (Fig. 4.2).

Numerous plant secondary metabolites exhibiting cytotoxic activity are discovered every year and thus lead to more possibilities for fighting against cancer. These metabolites are outstanding source for the development of anticancer drugs. The modification of their chemical structure results in the enhancement of anticancer activity and selection; improvement of assimilation, distribution, process of

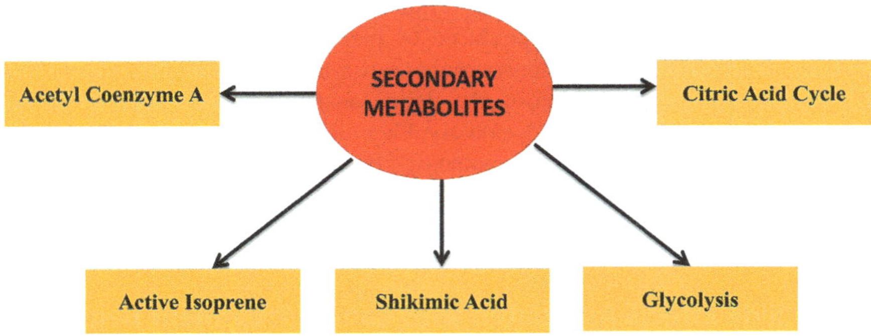


**Fig. 4.2** Five major precursor pathways for producing secondary metabolites

**Table 4.1** Classes of secondary metabolites found in plants

| S. no: | Secondary metabolites | | |
|---|---|---|---|
| 1. | Terpenoids | | |
| | | Monoterpenoids | |
| | | Sesquiterpenoids | |
| | | Diterpenoids | |
| | | Triterpenoids | |
| | | Tetraterpenoids | |
| 2. | Polyphenols | | |
| | | Curcumin | |
| | | Quercetin | |
| | | Resveratrol | |
| | | Flavonoids | |
| 3. | Nitrogen-containing compounds | | |
| | | Alkaloids | Indole alkaloids |
| | | | Isoquinoline alkaloids |
| | | | Phenanthroindolizidine alkaloids |
| | | | Indoquinoline alkaloids |
| | | | Benzophenanthridine alkaloids |
| | | Glucosinolates | |

metabolism and excretory potential; and minimizing their cytotoxicity as well as side effects (Guo 2017; Yao et al. 2017). The forthcoming sections present an overview of the above three mentioned classes of plant secondary metabolites as anticancer agents against various cancer types and also demonstrating their chemopreventive property (Table 4.1).

### 4.3.1 Terpenoids or Terpenes

Terpenoids or terpenes or isoprenoids are small ubiquitous compounds present in daily diet and considerably nontoxic in nature. Therefore, they possess the capability to be utilized in the form of chemopreventive agents for the treatment of cancer (Akihisa et al. 2003). They are synthesized as plant secondary metabolites, and the majority of the natural compounds are terpenoids. Most of the terpenoids exhibit numerous biological potentials and are currently employed for curing several diseases. Structurally, they are composed of five carbon units ($C_5 H_8$) that are also called as isoprene units. They are categorized as monoterpenoids, sesquiterpenoids, diterpenoids, triterpenoids, and tetraterpenoids, dependent on the variations in the number of carbon atoms (Wang et al. 2005a, b). The anticancer properties of the different groups of terpenoids are as follows:

#### 4.3.1.1 Monoterpenoids

A monocyclic monoterpenoid known as limonene belongs to the class of terpenes. Additionally, essential oils obtained from citrus fruits and other plant species contain limonene. D-limonene and L-limonene are two active forms (optically) of limonene that are chemically mirror images of each other. D-limonene is found as a main constituent of various citrus oils such as orange, lime, grapefruit, and mandarin and has acquired focus in the antiproliferative research. D-limonene is well known for its chemopreventive potential against different cancer types. Mammary cancer induced by carcinogens is prevented by D-limonene at the onset of induction and progression stages. It protects from cancer of the liver by enhancing the production of hepatic enzymes which are involved in the detoxification of carcinogens (Sun 2007). Its therapeutic actions are well recognized from the past two decades. It has proved to cause inhibition of progression of stomach, liver, pancreas, skin, and colon cancer under in vivo conditions. The combined effect of cytotoxic agents like docetaxel and fluorouracil (5-FU) and D-limonene is more effective in comparison to a single mode of treatment which takes place through the generation of reactive oxygen species (Rabi and Bishayee 2009). D-limonene is also observed to suppress 3-hydroxy-3-methylglutanyl coenzyme A (HMGCoA) reductase (Clegg et al. 1982) that results in the suppression of isoprenylation of G and p21 proteins along with its localization of the membrane (Kawata et al. 1994). This process is responsible for contributing the efficiency of D-limonene as anticancer and chemopreventive agent; however, this explanation is not appropriate for all types of cancer (Kaji et al. 2001). Studies related to apoptosis demonstrate that D-limonene leads to the upregulation of Bax, cytochrome c release from mitochondria, and cleavage of caspase-9 and caspase-3 but without any effect on caspase-8. The abovementioned reports propose that apoptosis induced by D-limonene usually occurs through mitochondrial death pathway (Ji et al. 2006).

#### 4.3.1.2 Sesquiterpenoid

Artemisinin, an active sesquiterpene, is obtained from *Artemisia annua*. Artemisinin along with its derivatives (ARTs) are employed for suppressing immune response, preventing schistosomiasis, and treating cancer (Meshnick 1998; Burrows et al. 2011; Liu et al. 2011; Yang et al. 2005; Firestone and Sundar 2009; Tan et al. 2011). Its structure is defined as a sesquiterpene trioxane lactone that constitutes a peroxide bridge responsible for its biological activity. ARTs are engaged in suppressing the proliferation of different cancer types, viz., breast cancer, prostate cancer, leukemia, colon cancer, ovarian cancer, lung cancer, hepatoma, melanoma, and gastric cancer (Jiao et al. 2007; Hou et al. 2008; Lu et al. 2008, 2010, 2011; Chen et al. 2009a, b; He et al. 2010; Wang et al. 2010; Li et al. 2009; Efferth et al. 2001, 2003). They possess the potential to circumvent the effect of multidrug resistance in cancerous cells and display comparable anticancer activity both in the cancer cells exhibiting multidrug resistance and parent cancerous cells (Michaelis et al. 2010; Reungpatthanaphong and Mankhetkorn 2002). Their anticancer potency under in vivo conditions is also observed in innumerable xenograft animal models (He et al. 2010; Du et al. 2010; Chen et al. 2009a, b). ARTs in combination with gemcitabine or carboplatin are used as sensitized chemotherapy in models of xenograft tumor (Hou et al. 2008; Chen et al. 2009a, b), suggesting their significant action in the combination chemotherapy. Many reports propose that their anticancer potential is initiated through interference in manifold cellular events. ARTs intervene the arrest of G1 cycle through disturbing the actions of CDK4, cyclin D, p21, NF-kB, CDK2, cyclin E, etc. (Hou et al. 2008; Chen et al. 2009a, b, 2010; Li et al. 2009; Willoughby et al. 2009) and simultaneously initiate the process of apoptosis in different types of cancer cells by activating p38 MAPK, increasing expression of Fas, and activating caspases (Hou et al. 2008; Lu et al. 2008; Chen et al. 2009a, b; Wang et al. 2010; Michaelis et al. 2010; Mu et al. 2008; Handrick et al. 2010). ARTs are also responsible for maintaining the levels of matrix metalloproteinase MMP7, (MMP)2, and MMP9; urokinase plasminogen activator (u-PA); vascular endothelial growth factor (VEGF); and $\alpha v\beta 3$ integrins, therefore, suppressing invasion, angiogenesis, and metastasis (Rasheed et al. 2010; Hwang et al. 2010; Chen et al. 2004).

#### 4.3.1.3 Diterpenoids

Tanshinones are included in the group of diterpenoids and are obtained from *Salvia miltiorrhiza* and are usually employed for treating cardiovascular diseases in China (Zhou et al. 2005). The diterpenoid derivatives obtained from *Salvia miltiorrhiza* have been comprehensively studied. Among them, tanshinone IIA has exhibited anticancer potential both under in vivo and in vitro conditions in different types of cancer such as breast cancer, hepatocellular carcinoma, leukemia, and colon (Sung et al. 1999; Liu et al. 2006; Wang et al. 2005a, b; Su and Lin 2008a, b; Su et al. 2008a, b; Wu et al. 1991; Tang et al. 2003; Yuan et al. 2004). Tanshinone IIA executes its action by binding to the minor groove of DNA which results in damaging of structure of DNA, leading to suppression of binding potential of RNAPII with DNA and RNAPII phosphorylation induction. On the molecular basis, this mechanism is proposed to be responsible for providing anticancer activity to tanshinone IIA. In this regard, the defects in transcription results in the PI3K/AKT pathway suppression, TNF-$\alpha$ upregulation, generation of ROS, erythroblastosis oncogene B downregulation, protein ratio of Bax/Bcl-2 enhancement, and calcium-dependent signaling activation, all these account to the initiation and propagation of the anticancerous mechanisms involved in the case of tanshinone IIA (Liu et al. 2006; Wang et al. 2005a, b; Su and Lin 2008a, b; Su et al. 2008a, b; Yuan et al. 2004; Dai et al. 2011; Won et al. 2010; Chiu and Su 2010; Cheng and Su 2010). Tanshinone IIA also causes differentiation in different types of cancer cells (Zhang et al. 2010; Wang et al. 2007). It also suppresses invasive and metastatic activities of cancer cells by decreasing the levels of MMP2, NF-kB, u-PA, and MMP9 and enhancing the level of TIMP1 and TIMP2 (matrix metalloproteinase protein) tissue suppressor (Yuxian et al. 2009; Shan et al. 2009). Other diterpenoids like cryptotanshinone, tanshinone I, and dihydrotanshinone also possess anticancer property and demonstrate comparable mechanisms with respect to tanshinone IIA (Lee et al. 2008, 2009; Su et al. 2008a, b; Nizamutdinova et al. 2008a, b; Liu et al. 2010; Park et al. 2010). Other diterpenoids such as triptolide, pseudolaric acid B, andrographolide, and oridonin have also demonstrated anticancer activity in various studies.

#### 4.3.1.4 Triterpenoids

Celastrol, a triterpenoid, is obtained from *Tripterygium wilfordii* and demonstrate different biological properties such as anti-inflammatory, antioxidant, and anticancer (Calixto et al. 2004). The researchers are familiar with the anticancer potential of celastrol, but the molecular basis of this activity is not totally recognized. Celastrol affect different signaling pathways which are responsible for its anticancer activity. Celastrol perform its action through different ways: (1) suppresses β kinases and IKKA˜, (2) suppresses the functioning of proteasomes, (3) suppresses the activities of p23 proteins and Cdc37, (4) initiates the activity of HSF1 and simultaneously initiating the heat shock reaction, and (5) suppresses AKT/mTOR/P70S6K signaling which initiates the inhibition of the growth of tumor and process of angiogenesis. The anticancer activity of other triterpenoids, viz., cucurbitacins, alisol, and pachymic acid, have also been reported in several studies.

#### 4.3.1.5 Tetraterpenoids

**Lycopene** (a tertaterpenoid) is an open-chain hydrocarbon and is primarily obtained from tomatoes. Its structure is composed of nonconjugated (2) and conjugated (11) double bonds linearly arranged. Since numerous natural components from diet have exhibited anticancer activity, therefore, the potential of lycopene was also examined in clinical research studies. The efficiency of lycopene combining with soy isoflavones or as a single agent was tested in phase II clinical trial on the concentration of serum PSA in prostate cancer patients. This combination procrastinated the development of both types of prostate cancer including hormone-sensitive and hormone-refractory (Vaishampayan et al. 2007). There are innumerable evidences describing the molecular basis contributing for the marvelous anticancer and chemopreventive actions of lycopene. It displays the antioxidant property by acting as ROS scavenger, which provides assistance to lycopene for inhibiting peroxidation of lipids as well as suppressing damage of DNA. Lycopene is also recognized for its antiangiogenic action, that is, on account of the decreased potentials of u-PA and MMP2, which takes place by increasing the expression of proteins of suppressors of tissues of plasminogen activator inhibitor-1 and MMP (Elgass et al. 2012; Huang et al. 2012; Chen et al. 2012). Despite the cumulative evidences describing the antiproliferative activity of lycopene, the molecular mechanisms responsible for this property are still indefinable. Findings of the proteomic analysis elucidated that lycopene was involved in transforming the expression of a large number of proteins, such as heat shock proteins and cell cycle proteins (Uppala et al. 2013).

The anticancer potential displayed by terpenoids demonstrates their considerable role for developing new anticancer drugs and also improvising already employed modes of treatment. Simultaneously, the biological potentials of several terpenoids still necessitate comprehensive and painstaking research. The natural terpenes consist of divergent classes of molecules, which are suitable for providing enhanced opportunities for the discovery of new anticancer drugs with reduced ill effects. The terpenoids are largely used in practice for treating numerous ailments in Asian nations; however, their pleiotropic effects and biogenesis are still not in competition in comparison to Western medicines. Their chemoprotective and chemopreventive actions propose their possibility in the concurrent anticancerous treatments.

### 4.3.2 Polyphenols

Polyphenols are the secondary metabolites that exhibit great diversity and are abundantly present in plants. In the plant kingdom, they are involved in performing several functions including metabolism and regulation of growth and providing protection against the effects of different pathogens and UV radiation. Approximately 8000 of these compounds are recognized in different species of plants. The requirement of new nontoxic and effectual chemopreventive agents create interest among researchers for analyzing the activities of these metabolites. Their anticarcinogenic activities including modulation of proliferation of cells, growth of tumor, process of angiogenesis, apoptosis, and others have been investigated (Ramos 2008; Fantini et al. 2015). Dietary polyphenols play a pivotal role in affecting and modulating several pathways and biochemical actions engaged in cancer (Han et al. 2007). Additionally, they function as modifiers of biological response and work by maintaining immune system and providing protection from the adverse effects of free radicals. Polyphenols affect numerous cancer-preventing mechanisms such as xenobiotics detoxification, oxidation inhibition, apoptosis initiation, immune system stimulation, anti-inflammation, and cellular signaling system activation. They also affect nuclear factors like AP-1 (activator protein 1) and NF-κB that are engaged in performing different activities such as proliferation of cells, cellular signaling cascades, expression of genes with respect to various stimuli, regulation of transcription of DNA, and finally survival of individuals (Shen et al. 2007; Chen et al. 2000). The individual polyphenols have undergone major testing for understanding the cellular mechanisms and biological processes responsible for their anticancer efficiency. The activity of these compounds is primarily dependent on their interacting ability with other naturally occurring compounds. The following section elucidates the anticancer potential of individual polyphenols.

#### 4.3.2.1 Curcumin

Curcumin obtained from the rhizome of turmeric is well recognized for its antioxidant quality as well as several health benefits. The divergent models of cancer have shown that curcumin possesses the ability to suppress angiogenesis and proliferation of cells, trigger apoptosis, and inhibit progression of cell cycle in cells of tumor (Kunnumakkara et al. 2007; Collett and Campbell 2004; Anto et al. 2002). The growth and development of glioma cell xenograft tumor was also suppressed by curcumin (Aoki et al. 2007). Curcumin has displayed remarkable results not only in the modulation of multiplication of cancerous cells, initiating apoptosis and angiogenesis, but also plays a vital role in affecting the invasion of cancer as well as metastasis (Anand et al. 2008). Different mechanisms are involved behind the anticancer activity of curcumin. In nude mice, curcumin inhibited angiogenesis and proliferation of cancerous cells of the pancreas through suppressing gene products regulated by NF-κB including cmyc, COX-2, Bcl-xL, cyclin D1, apoptosis protein-1, Bcl-2, VEGF (vascular endothelial growth factor), and MMP as depicted in the pancreatic cancer studies (Kunnumakkara et al. 2007). The antitumor potential of curcumin is attributed to its antiangiogenic potency in vivo and its interacting property with the metabolism of arachidonate (Ng et al. 2006). Furthermore, the interaction of curcumin with the receptors of vitamin D elucidates its preventive potential from colon cancer on the account of anticancer activity of vitamin D as proposed in some research studies (Bartik et al. 2010). The synergistic interactions between isoflavones and curcumin result in the inhibition of the generation of PSA (prostate-specific antigen) in the cells of prostate cancer (Ide et al. 2010). The clinical research evidently describes that individual curcumin or combining with other natural compounds exhibited potent anticancer activity against numerous cancers such as breast, colorectal, lung, neck, and head squamous cell carcinoma; prostate; pancreatic; and multiple myeloma (Gupta et al. 2013). Above all, chemopreventive and anticancer actions of curcumin are complicated and multimodal.

#### 4.3.2.2 Quercetin

Quercetin acts as a chemopreventive agent and thus performs different functions such as initiation of arrest of cell cycle as well as carrying out antioxidative and apoptotic processes (Gibellini et al. 2011). Quercetin initiates apoptosis at various stages of cell cycle irrespective of causing any detrimental effects on healthy cells, and this beneficial property has been acknowledged in numerous types of cancer cells under in vitro and in vivo conditions (Gibellini et al. 2011). There are reports elucidating that quercetin possesses the potency to curtail the hazards and progressive development of cancer by its ability to scavenge free radicals (Ekström et al. 2011; Lam et al. 2010). In addition to this, it provides protection to cells from inflammation, damage of DNA, and oxidative stress attributed to its antioxidative action and acts as a modulator of various cancerous cells through inhibiting the proliferation of tumor cells and progression of cell cycle by initiating apoptosis

(Jeong et al. 2009; Yang et al. 2006; Nair et al. 2004; Mu et al. 2007). Moreover, directly injecting quercetin has also resulted in decreased mass of breast tumors (Devipriya et al. 2006). The probability of gastric cancer was decreased by 43% through the consumption of quercetin-rich diet as depicted in epidemiological studies (Ekström et al. 2011), while colon cancer was reduced by 32% (Theodoratou et al. 2007). Intake of quercetin has also caused reduction in the risk of lung cancer by 51% in nonsmokers, while 65% in smokers (Lam et al. 2010). Injecting quercetin intravenously among individuals suffering from various cancer types resulted in the reduced potential of tyrosine kinase enzyme (essential for the growth of tumor) in 9 out of 11 persons (Ferry et al. 1996). Researchers are working in the direction to discover the benefits of synergistic interactions between standard chemotherapeutic drugs and quercetin. It has been demonstrated in vivo and in vitro analyses that quercetin possesses the potency to increase the efficiency of concomitant drugs by different methods such as turning the cancerous cells more sensitive to the effect of chemotherapeutics and increasing the accumulation and bioavailability of these drugs (Miles et al. 2014). Clinically, this potential is beneficial to decrease the concentration of toxic drugs and therefore mitigating their detrimental side effects.

#### 4.3.2.3 Resveratrol

Resveratrol possesses inhibitory effect at every stage of cancer and has exhibited efficacy in treating different cancers such as breast, thyroid, prostate, lung, stomach, pancreas, and colon (Udenigwe et al. 2008). Resveratrol has demonstrated efficiency in the prevention and treatment of esophageal, colon, intestinal, and skin tumors in vivo (Kukreja et al. 2014). Numerous cellular mechanisms responsible for providing anticancer activity to resveratrol are proposed such as suppression of metastasis and angiogenesis as well as initiation of apoptosis (Devipriya et al. 2006). Furthermore, the effect of resveratrol analyzed in healthy individuals confirmed its tolerance ability and evidenced that it executes its action through modulation of enzyme systems engaged in activating and detoxifying carcinogens (Chow et al. 2010).

#### 4.3.2.4 Flavonoids

Flavonoids are categorized under the group of polyphenolic compounds. They are mostly obtained from seeds, fruits, flowers, beverages, and vegetables. Chemically, flavonoids are diphenylpropane ($C_6$-$C_3$-$C_6$) constituting centrally located three carbon rings and aromatic rings (two) on the sides, thus resulting in the formation of oxygenated heterocyclic (Bonfili et al. 2008). They possess different beneficial biological activities such as bone resorption suppression, protection from cancer, hormonal action, and cardioprotection (Tait et al. 2006). They exhibit the ability to join to the ATP-binding sites of various proteins such as calcium plasma membrane ATPase, mitochondrial ATPase, topoisomerase, protein kinase C, and protein kinase A (Bonfili et al. 2008).

The contemporary therapies of cancer such as radiation and chemotherapy are connected with considerable side effects, thus, leading to the urgent requirement of adjuvant or alternative treatments. Polyphenolic compounds, present in abundance in the diet, exhibit remarkable results in treating cancer without causing any adverse effects, i.e., they are completely safe to use. Moreover, they possess the capability to transform numerous biological mechanisms engaged in the initiation and progression of cancer and hence have a broad range of therapeutic activities in comparison to single drugs. The properties of curing and preventing diseases along with the bioavailability of polyphenols can be increased and extended with the assistance of combination therapies that comprise of natural nutrients of similar or diversified chemical classes. Combining micronutrients (required in the maintenance of consistency and reliability of extracellular matrix) with polyphenols provide enhanced anticancer effects. These processes constitute interference with the metabolic pathways essential in the reduction of invasion and metastasis of cancerous cells. Thus, future research is based upon employing natural components in amalgamation with other natural compounds as sources of safe, beneficial, and inexpensive therapeutic medium for treating cancer.

### 4.3.3 Nitrogen-Containing Compounds

A huge proportion of plant secondary metabolites contain nitrogen as the main constituent of their chemical structure. This category is composed of alkaloids, glucosinolates, and others, and majority of them are produced from the commonly occurring amino acids. In this section, the anticancer activity of alkaloids and glucosinolates will be described.

#### 4.3.3.1 Alkaloids

Alkaloids are commonly obtained from medicinal plants and exhibit a range of beneficial effects. They possess greater diversity in their structure and are composed of nitrogen atom along with a ring structure. The position of the nitrogen atom is in the interior of the heterocyclic ring structure in majority of the cases. The biosynthetic pathways are usually employed for the classification of alkaloids into different categories. Moreover, they are widely distributed within the plant kingdom and are usually found in higher plants, especially in families, viz., *Menispermaceae, Ranunculaceae, Loganiaceae, Papaveraceae, and Leguminosae* (Lee 2011; Benyhe 1994; Li et al. 2007). They have demonstrated profound anticancer potential against several cancers. The anticancer effects of some of different categories of alkaloids are discussed as follows:

**Indole Alkaloids**

Medicinal plants produce numerous indole alkaloids. Alkaloid vincristine is isolated from *Vinca rosea*, while vinblastine is from *Catharanthus roseus* which are employed for treating different cancers such as lymphomas, breast cancer, leukemias, lung cancer, Kaposi's sarcoma, and advanced testicular cancer in amalgamation with other anticancerous drugs (Cragg and Newman 2005). The mechanism of action involves inhibiting cell proliferation by bringing alterations in the dynamics of addition of tubulin and causing loss of mitotic spindle microtubules from their ends despite causing their depolymerisation (Jordan et al. 2010; Jordan and Wilson 2004). Camptothecin obtained from *Camptotheca acuminata* possesses a broad range of anticancer potential under in vivo and in vitro conditions. Its first-generation analogues, primarily two of them, are engaged in the treatment of colorectal, lung, and ovarian cancer, while various second-generation analogues are under clinical trials (Li and Zhang 1996). Naucleaoral A and B are isomeric indole alkaloids obtained from *Nauclea orientalis* roots. They exhibit cytotoxicity on κB cells and HeLa cells (Sichaem et al. 2010). A dimeric indole alkaloid with distinctive characteristic named as montamine is obtained from *Centaurea montana* seeds demonstrated reasonable anticancer potential in the Caco-2 colon cancer cells in the MTT assay. The alkaloid schischkiniin obtained from *C. schischkinii* seeds possesses cytotoxic action on Caco-2 colon cancer cells which was analyzed by MTT cytotoxicity and brine shrimp lethality assays (Shoeba et al. 2005). These alkaloids are highly effective in cancer prevention.

### Isoquinoline Alkaloids

Berberine obtained from *Rhizoma coptidis* exhibits integral chemopreventive action against the formation of colon tumor by suppressing cyclooxygenase-2 (COX2) enzyme production in colon cancer cells which is profusely expressed in them and plays a vital role in the colon tumorigenesis (Fukuda et al. 1999). Sanguinarine is isolated from *Sanguinaria canadensis* and initiates the process of apoptosis in pancreatic carcinoma Bxpc-3 and Aspc-1 and human lung cancer (A549) (Ahsan et al. 2007). It also exhibits significant effect with respect to multidrug resistance in the cells of cervical of humans (De Stefano et al. 2009; Jang et al. 2009; Ding et al. 2002). Liriodenine, another isoquinoline alkaloid obtained from *Cananga odorata*, has displayed imperative antiproliferative, cytotoxic, and apoptosis-initiating actions on lung cancer cells and different other cancer cells of humans (Chang et al. 2004). It also suppressed the action of topoisomerase II under in vivo and in vitro conditions. The isoquinoline alkaloids are involved in the antiproliferative activity of cancerous cells.

### Phenanthroindolizidine Alkaloids

Antofine a phenanthroindolizidine alkaloid which is isolated from *Cynanchum paniculatum* possesses antiproliferative and antitumor potential against various human cancers (Min et al. 2010). It also initiates the inhibition of G2/M phase in Col2 cells (Lee et al. 2003; Yin et al. 2005). A benzophenanthridine alkaloid named as 6-methoxydihydrosanguinarine (6ME) is obtained from *Hylomecon* species and displays chemotherapeutic action. It produces inhibitory effect on Hep G2 cells (hepatocellular carcinoma) through apoptosis. Tylophorine (phenanthroindolizidine alkaloid) is isolated from *Tylophora indica* and has demonstrated anticancerous and anti-inflammatory capabilities (Chia-Mao et al. 2009; Wei et al. 2008). Tylophorinidine (TPD) and pergularine (PGL) are alkaloids obtained from *Pergularia pallida* and exhibited potent anticancer action (Narasimha Rao and Venkatachalam 2000). These alkaloids have a major role in anti-inflammation.

#### Indoquinoline Alkaloids

Indoquinoline derivatives named as neocryptolepine and cryptolepine obtained from *Cryptolepis sanguinolenta* roots possess effective cytotoxic action to tumor cells. These compounds get inserted into the DNA and thus affecting the catalytic action of human topoisomerase II. The cytotoxic activity of cryptolepine was observed in cells of human leukemia and murine (Dassonneville et al. 2000). These alkaloids are known for their highly cytotoxic nature against the cancer cells.

#### Benzophenanthridine Alkaloids

A benzophenanthridine alkaloid, 6-methoxydihydrosanguinarine (6ME), obtained from the species of *Hylomecon* possesses potent chemotherapeutic activity. It leads to the cell death by apoptosis in colon carcinoma cells (HT29) (Lee et al. 2004). Benzophenanthridine alkaloids are generally used as chemotherapeutic agents.

Thus, alkaloids play an extremely significant role in the form of anticancer agents by suppressing the action of topoisomerase enzyme that is engaged in the replication of DNA, induction of the process of apoptosis, and p53 gene expression. They undeniably have existence even before the origin of humans, and a number of them possess marvelous similarities with the structures of human neurotransmitters like acetylcholine, serotonin, and dopamine. The familiarity of their medical significance and the analysis of the mechanism of action in suppressing the uncontrolled multiplication of cells would certainly provide assistance in the formulation of anticancer drugs and other lead compounds.

### 4.3.3.2 Glucosinolates

Glucosinolates and their products produced on hydrolysis possess extremely beneficial activities such as antioxidant, anticarcinogenic, antimutagenic, antibacterial, antifungal, and others. These plant secondary metabolites are organic ions composed of sulphonated oxime moieties and b-D-thioglucose. They have marked their presence in approximately 16 families of angiosperms, but a majority of them are present in the plants of Brassicaceae family (Fahey et al. 2001). They constitute roughly 100 recognized natural thioglucosides exhibiting a common structure, having a side chain (R) attached to different aromatic, heteroatomic, and aliphatic carbon skeletons, which are produced from amino acids through oxidation or hydroxylation and process of lengthening of long chains (Hansen et al. 1995). The enzyme myrosinase (thioglucoside glucohydrolase) is also present in plants containing glucosinolates. The reaction between glucosinolates and myrosinase results in the hydrolysis of glucosinolate, and water acts as a mediator in this reaction. The products generated upon hydrolysis are glucose, aglycone moiety, and sulphate. The instability of aglycone moiety results in the rearrangement to produce nitriles, isothiocyanates (ITCs), epithionitriles, thiocyanates, and oxazolidinethiones which rely on the glucosinolate structure and the conditions of reaction.

The chemoprotective attributes of their hydrolysis products are well understood against different chemical carcinogens. They perform their anticarcinogenic action by suppressing the formation of tumors in different tissues such as the bladder, colon, liver, pancreas, small intestine, and others. Methyl sulphonyl glucosinolates upon hydrolysis initiates phase II enzymes which are directly involved in the detoxification of carcinogens. The phase II enzymes minimize the effect of carcinogenesis by direct or indirect manner (Wattenberg 1983). The epidemiological records state that the daily consumption of crucifer vegetables provide protection against several cancers (Verhoeven et al. 1996; Cohen et al. 2000; Ambrosone et al. 2004; Zhang et al. 2000). Isothiocyanates have been reported to provide considerable protecting ability against cancer initiated by several chemical carcinogens in animal studies (Yang et al. 2002; Conaway et al. 2002; Wattenberg 1977; Hecht 2000; Talalay and Fahey 2001; Jiao et al. 1997; Stoner et al. 1991; Morse et al. 1989, 1991). Different research studies have revealed the anticancer effects of SFN (4-methylsulfinylbutyl isothiocyanate). SFN has also reported to provide remarkable defense against tumorigenesis induced by 9,10-dimethyl1,2-benzanthracene in rats (Zhang et al. 1994). It has also suppressed the multiplication of cells of prostate cancer in humans in vitro (Chung et al. 2000). Isothiocyanates (7-methylsulphinylheptyl isothiocyanates and 4-methysulfinylbutyl) obtained from *Rorippa nasturtium-aquaticum* and *Brassica oleracea*, respectively, suppress metalloproteinase-9 and also inhibits the invasive action in human breast cancer cells (MDA-MB-231) under laboratory conditions (Rosea et al. 2005). These components perform action in collaboration against cancerous cells through initiating detoxifying and antioxidative enzymes like UDP-glucuronosyl transferase and glutathione-S-transferases along with suppressing the carcinogen-activating enzymes like cytochrome P450 or bringing alterations in the metabolism of steroid hormones. The commencement of apoptosis by several signal transduction pathways is also their mechanism of action. The induction of apoptosis by SFN is connected with BAX upregulation which is a promoter of the process of apoptosis, BCL-2 downregulation which is an inhibitor of apoptosis and caspases-3, caspases-9, and caspases-8 activation which execute apoptosis.

Glucosinolates are beneficial cancer chemopreventive agents and therefore deserve additional research. They act as chemopreventive agents through soaring apoptosis, inhibiting the progression of cell cycle, and other ways. Some of them have demonstrated excellent results under clinical trials, exhibiting their greater potential in the development of anticancerous drugs.

## 4.4 Conclusion

Cancer has caused uncountable deaths in both developed and developing countries, giving rise to serious challenges to medical science. In this regard, the drugs currently employed in chemotherapy possess drawbacks attributed to their toxicity on the normal cells and consequently causing more health complications. Hence, there is an imperative demand for alternative modes of treatment, and in this regard, the natural-derived anticancerous drugs are considered as the perfect option. The plant secondary metabolites are appropriate anticancer agents that results in the progressive development of new anticancer drugs. Some are already successfully prepared by the pharmaceutical industry. In addition to this, they are beneficial as outstanding lead compounds, and their structural alterations, enhancing the effectiveness of delivery systems, and substitute formulations can increase their pharmacological activity. The latest biotechnological approaches involving the use of nanotechnology contribute to an optimistic approach for cancer therapy. Concurrently, they soar the opportunities for the successful utilization of plant secondary metabolites as cancer therapeutic agents (Gismondi et al. 2015, 2016; Gupta et al. 2017; Raj et al. 2016).

## References

Ahsan H, Reagan-Shaw S, Breur J, Ahmad N (2007) Sanguinarine induces apoptosis of human pancreatic carcinoma AsPC-1 and BxPC-3 cells via modulations in Bcl-2 family proteins. Cancer Lett 249(2):198–108

Akihisa T, Yasukawa K, Tokuda H (2003) Potentially cancer chemopreventive and anti-inflammatory terpenoids from natural sources. Stud Nat Prod Chem 29:73–126

Ambrosone CB, McCann SE, Freudenheim JL, Marshall JR, Zhang Y, Shields PG (2004) Breast cancer risk in premenopausal women is inversely associated with consumption of broccoli, a source of isothiocyanates, but is not modified by GST genotype. J Nutr 134:1134–1138

Amin AR, Kucuk O, Khuri FR, Shin DM (2009) Perspectives for cancer prevention with natural compounds. J Clin Oncol 27:2712–2725

Anand P, Sundaram C, Jhurani S, Kunnumakkara AB, Aggarwal BB (2008) Curcumin and cancer: an 'old-age' disease with an 'age-old' solution. Cancer Lett 267:133–164

Anto RJ, Mukhopadhyay A, Denning K, Aggarwal BB (2002) Curcumin (diferuloylmethane) induces apoptosis through activation of caspase-8, BID cleavage and cytochrome c release: its suppression by ectopic expression of Bcl-2 and Bcl-xL. Carcinogenesis 23:143–150

Aoki H, Takada Y, Kondo S, Sawaya R, Aggarwal B, Kondo Y (2007) Evidence that curcumin suppresses the growth of malignant gliomas *in vitro* and *in vivo* through induction of autophagy: role of Akt and extracellular signal-regulated kinase signaling pathways. Mol Pharmacol 72:29–39

Bartik L, Whitfield GK, Kaczmarska M, Lowmiller CL, Moffet EW, Furmick JK, Hernandez Z, Haussler CA, Haussler MR, Jurutka PW (2010) Curcumin: a novel nutritionally derived ligand of the vitamin D receptor with implications for colon cancer. J Nutr Biochem 21:1153–1161

Benyhe S (1994) Morphine: new aspects in the study of an ancient compound. Life Sci 55:969–979

Bonfili L, Cecarini V, Amici M, Cuccioloni M, Angeletti M, Keller JN, Eleuteri AM (2008) Natural polyphenols as proteasome modulators and their role as anti-cancer compounds. FEBS J 275:5512–5526

Burrows JN, Chibale K, Wells TN (2011) The state of the art in anti-malarial drug discovery and development. Curr Top Med Chem 11:1226–1254

Calixto JB, Campos MM, Otuki MF, Santos AR (2004) Anti-inflammatory compounds of plant origin. Part II. Modulation of pro-inflammatory cytokines, chemokines and adhesion molecules. Planta Med 70:93–103

Chang H-C, Chang F-R, Yang-Chang W, Lai Y-H (2004) Anti cancer effect of liriodenine on human lung cancer cells. Kaohsiung J Med Sci 20(8):365–371

Chen C, Yu R, Owuor ED, Kong ANT (2000) Activation of antioxidant response element (ARE), mitogen-activated protein kinases (MAPKs) and caspases by major green tea polyphenol components during cell survival and death. Arch Pharm Res 23:605–612

Chen HH, Zhou HJ, Wang WQ, Wu GD (2004) Antimalarial dihydroartemisinin also inhibits angiogenesis. Cancer Chemother Pharmacol 53:423–432

Chen H, Sun B, Pan S, Jiang H, Sun X (2009a) Dihydroartemisinin inhibits growth of pancreatic cancer cells in vitro and in vivo. Anti-Cancer Drugs 20:131–140

Chen T, Li M, Zhang R, Wang H (2009b) Dihydroartemisinin induces apoptosis and sensitizes human ovarian cancer cells to carboplatin therapy. J Cell Mol Med 13:1358–1370

Chen H, Sun B, Wang S, Pan S, Gao Y, Bai X, Xue D (2010) Growth inhibitory effects of dihydroartemisinin on pancreatic cancer cells: involvement of cell cycle arrest and inactivation of nuclear factor-κB. J Cancer Res Clin Oncol 136:897–903

Chen ML, Lin YH, Yang CM, Hu ML (2012) Lycopene inhibits angiogenesis both *in vitro* and *in vivo* by inhibiting MMP-2/uPA system through VEGFR2-mediated PI3K-Akt and ERK/ p38 signaling pathways. Mol Nutr Food Res 56:889–899

Cheng CY, Su CC (2010) Tanshinone IIA may inhibit the growth of small cell lung cancer H146 cells by up-regulating the Bax/Bcl-2 ratio and decreasing mitochondrial membrane potential. Mol Med Rep 3:645–650

Chia-Mao W, Yang C-W, Lee Y-Z, Chuang T-H, Wu PL, Chao Y-S, Lee S-J (2009) Tylophorine arrests carcinoma cells at G1 phase by downregulating cyclin A2 expression. Biochem Biophys Res Commun 386(1):140–145

Chiu TL, Su CC (2010) Tanshinone IIA induces apoptosis in human lung cancer A549 cells through the induction of reactive oxygen species and decreasing the mitochondrial membrane potential. Int J Mol Med 25:231–236

Chow HHS, Garland LL, Hsu CH, Vining DR, Chew WM, Miller JA, Perloff M, Crowell JA, Alberts DS (2010) Resveratrol modulates drug-and carcinogen-metabolizing enzymes in a healthy volunteer study. Cancer Prev Res 3:1168–1175

Chung FL, Conaway CC, Rao CV, Reddy BS (2000) Chemoprevention of colonic aberrant crypt foci in Fischer rats by sulforaphane and phenethyl isothiocyanate. Carcinogenesis 21:2287–2291

Clegg RJ, Middleton B, Bell GD, White DA (1982) The mechanism of cyclic monoterpene inhibition of hepatic 3-hydroxy-3- methylglutaryl coenzyme A reductase in vivo in the rat. J Biol Chem 257:2294–2299

Cohen JH, Kristal AR, Stanfordm JL (2000) Fruit and vegetable intakes and prostate cancer risk. J Natl Cancer Inst 92:61–68

Collett GP, Campbell FC (2004) Curcumin induces c-Jun N-terminal kinase-dependent apoptosis in HCT116 human colon cancer cells. Carcinogenesis 25:2183–2189

Conaway CC, Yang YM, Chung FL (2002) Isothiocyanates as cancer chemopreventive agents: their biological activities and metabolism in rodents and humans. Curr Drug Metab 3:233–255

Cragg GM, Newman DJ (2005) Plants as source of anticancer agents. J Ethnopharmacol 100:72–99

Dai ZK, Qin JK, Huang JE, Luo Y, Xu Q, Zhao HL (2011) Tanshinone IIA activates calcium-dependent apoptosis signaling pathway in human hepatoma cells. J Nat Med 66:192–201

Dassonneville L, Lansiaux A, Wattelet A, Wattez N, Mahieu C, Van Miert S, Pieters L, Bailly C (2000) Cytotoxicity and cell cycle effects of the plant alkaloids cryptolepine and neocryptolepine: relation to drug-induced apoptosis. Eur J Pharmacol 409(1):9–18

De Stefano I, Raspaglio G, Zannoni GF, Travaglia D, Prisco MG, Mosca M, Ferlini C, Scambia G, Gallo D (2009) Antiproliferative and antiangiogenic effects of the benzophenanthridine alkaloid sanguinarine in melanoma. Biochem Pharmacol 78(11):1374–1381

Devipriya S, Ganapthy V, Shyamaladevi CS (2006) Suppression of tumor growth and invasion in 9,10 dimethyl benz(a) anthracene induced mammary carcinoma by the plant bioflavonoid quercetin. Chem Biol Interact 162:106–113

Ding Z, Tang S-C, Weerasinghe P, Yang X, Pater A, Liepins A (2002) The alkaloid sanguinarine is effective against multidrug resistance in human cervical cells via bimodal cell death. Biochem Pharmacol 63(8):1415–1421

Du JH, Zhang HD, Ma ZJ, Ji KM (2010) Artesunate induces oncosis-like cell death *in vitro* and has antitumor activity against pancreatic cancer xenografts *in vivo*. Cancer Chemother Pharmacol 65:895–902

Efferth T, Dunstan H, Sauerbrey A, Miyachi H, Chitambar CR (2001) The anti-malarial artesunate is also active against cancer. Int J Oncol 18:767–773

Efferth T, Sauerbrey A, Olbrich A, Gebhart E, Rauch P, Weber HO, Hengstler JG, Halatsch ME, Volm M, Tew KD, Ross DD (2003) Molecular modes of action of artesunate in tumor cell lines. Mol Pharmacol 64:382–394

Ekström AM, Serafini M, Nyren O, Wolk A, Bosetti C, Bellocco R (2011) Dietary quercetin intake and risk of gastric cancer: results from a population-based study in Sweden. Ann Oncol 22:438–443

Elgass S, Cooper A, Chopra M (2012) Lycopene inhibits angiogenesis in human umbilical vein endothelial cells and rat aortic rings. Br J Nutr 108:431–439

Fahey JW, Zalcmann AT, Talalay P (2001) The chemical diversity and distribution of glucosinolates and isothiocyanates among plants. Phytochemistry 56:5–51

Fantini M, Benvenuto M, Masuelli L, Frajese GV, Tresoldi I, Modesti A, Bei R (2015) *In vitro* and *in vivo* antitumoral effects of combinations of polyphenols, or polyphenols and anticancer drugs: perspectives on cancer treatment. Int J Mol Sci 16:9236–9282

Ferry DR, Smith A, Malkhandi J, Fyfe DW, de Takats PG, Anderson D, Baker J, Kerr DJ (1996) Phase I clinical trial of the flavonoid quercetin: pharmacokinetics and evidence for *in vivo* tyrosine kinase inhibition. Clin Cancer Res 2:659–668

Firestone GL, Sundar SN (2009) Anticancer activities of artemisinin and its bioactive derivatives. Expert Rev Mol Med 11:e32

Fridlender M, Kapulnik Y, Koltai H (2015) Plant derived substances with anti-cancer activity: from folklore to practice. Front Plant Sci 6:799

Fukuda K, Hibiya Y, Mutoh M, Koshiji M, Akao S, Fujiwara H (1999) Inhibition by berberine of cyclooxygenase-2 transcriptional activity in human colon cancer cells. J Ethnopharmacol 66 (2):227–233

Gali-Muhtasib H, Hmadi R, Kareh M, Tohme R, Darwiche N (2015) Cell death mechanisms of plant-derived anticancer drugs: beyond apoptosis. Apoptosis 20:1531–1562

Gibellini L, Pinti M, Nasi M, Montagna JP, De Biasi S, Roat E, Bertoncelli L, Cooper EL, Cossarizza A (2011) Quercetin and cancer chemoprevention. Evid Based Complement Alternat Med 2011:591356

Gismondi A, Reina G, Orlanducci S, Mizzoni F, Gay S, Terranova ML, Canini A (2015) Nanodiamonds coupled with plant bioactive metabolites: a nanotech approach for cancer therapy. Biomaterials 38:22–35

Gismondi A, Nanni V, Reina G, Orlanducci S, Terranova ML, Canini A (2016) Nanodiamonds coupled with 5, 7-dimethoxycoumarin, a plant bioactive metabolite, interfere with the mitotic process in B16F10 cells altering the actin organization. Int J Nanomedicine 11:557–574

Guo Z (2017) The modification of natural products for medical use. Acta Pharm Sin B 7:119–136

Gupta SC, Patchva S, Aggarwal BB (2013) Therapeutic roles of curcumin: lessons learned from clinical trials. AAPS J 15:195–218

Gupta C, Prakash D, Gupta S (2017) Cancer treatment with nano-diamonds. Front Biosci (Sch Ed) 9:62–70

Habli Z, Toumieh G, Fatfat M, Rahal ON, Gali-Muhtasib H (2017) Emerging cytotoxic alkaloids in the battle against cancer: overview of molecular mechanisms. Molecules 22:250

Han X, Shen T, Lou H (2007) Dietary polyphenols and their biological significance. Int J Mol Sci 8:950–988

Handrick R, Ontikatze T, Bauer KD, Freier F, Rübel A, Dürig J, Belka C, Jendrossek V (2010) Dihydroartemisinin induces apoptosis by a Bak-dependent intrinsic pathway. Mol Cancer Ther 9:2497–2510

Hansen M, Møller P, Sørensen H, de Trejo MC (1995) Glucosinolates in broccoli stored under controlled atmosphere. J Am Soc Hortic Sci 120:1069–1074

He Q, Shi J, Shen XL, An J, Sun H, Wang L, Hu YJ, Sun Q, Fu LC, Sheikh MS, Huang Y (2010) Dihydroartemisinin upregulates death receptor 5 expression and cooperates with TRAIL to induce apoptosis in human prostate cancer cells. Cancer Biol Ther 9:819–824

Hecht SS (2000) Inhibition of carcinogenesis by isothiocyanates. Drug Metab Rev 32:395–411
Hou J, Wang D, Zhang R, Wang H (2008) Experimental therapy of hepatoma with artemisinin and its derivatives: *in vitro* and *in vivo* activity, chemosensitization, and mechanisms of action. Clin Cancer Res 14:5519–5530
Huang CS, Chuang CH, Lo TF, Hu ML (2012) Anti-angiogenic effects of lycopene through immunomodulation of cytokine secretion in human peripheral blood mononuclear cells. J Nutr Biochem 24:428–434
Hwang YP, Yun HJ, Kim HG, Han EH, Lee GW, Jeong HG (2010) Suppression of PMA-induced tumor cell invasion by dihydroartemisinin via inhibition of PKCα/Raf/MAPKs and NF-κB/AP-1-dependent mechanisms. Biochem Pharmacol 79:1714–1726
Ide H, Tokiwa S, Sakamaki K, Nishio K, Isotani S, Muto S, Hama T, Masuda H, Horie S (2010) Combined inhibitory effect of soy flavones and curcumin on the production of prostate-specific antigen. Prostate 70:1127–1133
Ijaz S, Akhtar N, Khan MS, Hameed A, Irfan M, Arshad MA, Ali S, Asrar M (2018) Plant derived anticancer agents: a green approach towards skin cancers. Biomed Pharmacother 103:1643–1651
Jang B-C, Park J-G, Song D-K, Baek W-K, Yoo SK, Jung K-H, Park G-Y, Lee T-Y, Suh SI (2009) Sanguinarine induces apoptosis in A549 human lung cancer cells primarily via cellular glutathione depletion. Toxicol In Vitro 23(2):281–287
Jeong JH, An JY, Kwon YT, Rhee JG, Lee YG (2009) Effects of low dose quercetin: Cancer cell-specific inhibition of cell cycle progression. J Cell Biochem 106:73–82
Ji J, Zhang L, Wu YY, Zhu XY, Lv SQ, Sun XZ (2006) Induction of apoptosis by d-limonene is mediated by a caspase-dependent mitochondrial death pathway in human leukemia cells. Leuk Lymphoma 47:2617–2624
Jiao D, Smith TJ, Yang CS, Pittman B, Desai D, Amin S, Chung FL (1997) Chemopreventive activity of thiol conjugates of isothiocyanates for lung tumorigenesis. Carcinogenesis 18:2143–2147
Jiao Y, Ge CM, Meng QH, Cao JP, Tong J, Fan SJ (2007) Dihydroartemisinin is an inhibitor of ovarian cancer cell growth. Acta Pharmacol Sin 28:1045–1056
Jordan MA, Thrower D, Wilson L (2010) Mechanism of inhibition of cell proliferation by Vinca alkaloids. Cancer Chemother Pharmacol 70(20):209–224
Jordan MA, Wilson L (2004) Microtubules as a target for anticancer drugs. Nat Rev Cancer 4:253–265
Kaji I, Tatsuta M, Iishi H, Baba M, Inoue A, Kasugai H (2001) Inhibition by d-limonene of experimental hepatocarcinogenesis in Sprague-Dawley rats does not involve p21(ras) plasma membrane association. Int J Cancer 93:441–444
Kawata S, Nagase T, Yamasaki E, Ishiguro H, Matsuzawa Y (1994) Modulation of the mevalonate pathway and cell growth by pravastatin and d-limonene in a human hepatoma cell line (Hep G2). Br J Cancer 69:1015–1020
Korkina L, Kostyuk V (2012) Biotechnologically produced secondary plant metabolites for cancer treatment and prevention. Curr Pharm Biotechnol 13:265–275
Kukreja A, Waddhwa N, Tiwari A (2014) Therapeutic role of resveratrol and piceatannol in disease prevention. J Blood Disord Transfus 5:1–6
Kunnumakkara AB, Guha S, Krishnan S, Diagaradjane P, Gelovani J, Aggarwal BB (2007) Curcumin potentiates antitumor activity of gemcitabine in an orthotopic model of pancreatic cancer through suppression of proliferation, angiogenesis, and inhibition of nuclear factor-κB-regulated gene products. Cancer Res 67:3853–3861
Lam TK, Rotunno M, Lubin JH, Wacholder S, Consonni D, Pesatori AC, Bertazzi PA, Chanock SJ, Burdette L, Goldstein AM, Tucker MA (2010) Dietary quercetin, quercetin-gene interaction, metabolic gene expression in lung tissue and lung cancer risk. Carcinogenesis 31:634–642
Lee MR (2011) The history of Ephedra (ma-huang). J R Coll Phys Edinb 41:78–84

Lee SK, Nam K-A, Heo Y-H (2003) Cytotoxic activity and G2/M cell cycle arrest mediated by antofine, a phenanthroindolizidine alkaloid isolated from Cynanchum paniculatum. Planta Med 69(1):21–25

Lee Y-J, Yin H-Q, Kim Y-H, Li G-Y, Lee B-H (2004) Apoptosis inducing effects of 6-Methoxydihydrosanguinarine in HT29 colon carcinoma cells. Arch Pharm Res 27:1253–1257

Lee CY, Sher HF, Chen HW, Liu CC, Chen CH, Lin CS, Yang PC, Tsay HS, Chen JJ (2008) Anticancer effects of tanshinone I in human non-small cell lung cancer. Mol Cancer Ther 7:3527–3538

Lee WY, Liu KW, Yeung JH (2009) Reactive oxygen species-mediated kinase activation by dihydrotanshinone in tanshinones-induced apoptosis in HepG2 cells. Cancer Lett 285:46–57

Lee CS, Baek J, Han SY (2017) The role of kinase modulators in cellular senescence for use in cancer treatment. Molecules 22:1411

Li Y-F, Zhang R (1996) Reversed-phase high-performance liquid chromatography method for the simultaneous quantitation of the lactone and carboxylate forms of the novel natural productanti cancer agent 10-hydroxycamptothecin in biological fluids and tissues. J Chromatogr B Biomed Sci Appl 686(2):257–265

Li W, Shao Y, Hu L, Zhang X, Chen Y, Tong L, Li C, Shen X, Ding J (2007) BM6, a new semi-synthetic vinca alkaloid, exhibits its potent in vivo anti-tumor activities via its high binding affinity for tubulin and improved pharmacokinetic profiles. Cancer Biol Ther 6:787–794

Li S, Xue F, Cheng Z, Yang X, Wang S, Geng F, Pan L (2009) Effect of artesunate on inhibiting proliferation and inducing apoptosis of SP2/0 myeloma cells through affecting NFκB p65. Int J Hematol 90:513–521

Liu JJ, Lin DJ, Liu PQ, Huang M, Li XD, Huang RW (2006) Induction of apoptosis and inhibition of cell adhesive and invasive effects by tanshinone IIA in acute promyelocytic leukemia cells *in vitro*. J Biomed Sci 13:813–823

Liu JJ, Liu WD, Yang HZ, Zhang Y, Fang ZG, Liu PQ, Lin DJ, Xiao RZ, Hu Y, Wang CZ, Li XD (2010) Inactivation of PI3k/Akt signaling pathway and activation of caspase-3 are involved in tanshinone I-induced apoptosis in myeloid leukemia cells *in vitro*. Ann Hematol 89:1089–1097

Liu R, Dong HF, Jiang MS (2011) Artemisinin: the gifts from traditional Chinese medicine not only for malaria control but also for schistosomiasis control. Parasitol Res 110:2071–2074

Lu JJ, Meng LH, Cai YJ, Chen Q, Tong LJ, Lin LP, Ding J (2008) Dihydroartemisinin induces apoptosis in HL-60 leukemia cells dependent of iron and p38 mitogen-activated protein kinase activation but independent of reactive oxygen species. Cancer Biol Ther 7:1017–1023

Lu JJ, Meng LH, Shankavaram UT, Zhu CH, Tong LJ, Chen G, Lin LP, Weinstein JN, Ding J (2010) Dihydroartemisinin accelerates c-MYC oncoprotein degradation and induces apoptosis in c-MYC-overexpressing tumor cells. Biochem Pharmacol 80:22–30

Lu JJ, Chen SM, Zhang XW, Ding J, Meng LH (2011) The anti-cancer activity of dihydroartemisinin is associated with induction of iron-dependent endoplasmic reticulum stress in colorectal carcinoma HCT116 cells. Investig New Drugs 29:1276–1283

Meshnick SR (1998) Artemisinin antimalarials: mechanisms of action and resistance. Med Trop 58:13–17

Michaelis M, Kleinschmidt MC, Barth S, Rothweiler F, Geiler J, Breitling R, Mayer B, Deubzer H, Witt O, Kreuter J, Doerr HW (2010) Anti-cancer effects of artesunate in a panel of chemoresistant neuroblastoma cell lines. Biochem Pharmacol 79:130–136

Miles SL, McFarland M, Niles RM (2014) Molecular and physiological actions of quercetin: need for clinical trials to assess its benefits in human disease. Nutr Rev 72:720–734

Min HY, Chung HJ, Kim EH, Kim S, Park EJ, Lee SK (2010) Inhibition of cell growth and potentiation of tumor necrosis factor-α (TNF-α)-induced apoptosis by a phenanthroindolizidine alkaloid antofine in human colon cancer cells. Biochem Pharmacol 80:1356–1364

Morrissey KM, Yuraszeck TM, Li CC, Zhang Y, Kasichayanula S (2016) Immunotherapy and novel combinations in oncology: current landscape, challenges, and opportunities. Clin Transl Sci 9:89–104

Morse MA, Wang CX, Stoner GD, Mandal S, Conran PB, Amin SG et al (1989) Inhibition of 4-(methylnitrosamino)-1-(3-pyridyl)-1-butanone-induced DNA adduct formation and tumorigenicity in the lung of F344 rats by dietary phenethyl isothiocyanate. Cancer Res 49:549–553

Morse MA, Eklind KI, Hecht SS, Jordan KG, Choi CI, Desai DH, Amin SG, Chung FL (1991) Structure-activity relationships for inhibition of 4-(methylnitrosamino)-1-(3-pyridyl)-1-butanone lung tumorigenesis by arylalkyl isothiocyanates in A/J mice. Cancer Res 51:1846–1850

Mu C, Jia P, Yan Z, Liu X, Li X, Liu H (2007) Quercetin induces cell cycle G1 arrest through elevating Cdk inhibitors p21 and p27 in human hepatoma cell line (HepG2). Methods Find Exp Clin Pharmacol 29:179–183

Mu D, Zhang W, Chu D, Liu T, Xie Y, Fu E, Jin F (2008) The role of calcium, P38 MAPK in dihydroartemisinin-induced apoptosis of lung cancer PC-14 cells. Cancer Chemother Pharmacol 61:639–645

Nair HK, Rao KVK, Aalinkeel R, Mahajan S, Chawda R, Schwartz SA (2004) Inhibition of prostate cancer cell colony formation by the flavonoid quercetin correlates with modulation of specific regulatory genes. Clin Diagn Lab Immunol 11:63–69

Narasimha Rao K, Venkatachalam SR (2000) Inhibition of dihydrofolate reductase and cell growth activity by the phenanthroindolizidine alkaloids pergularinine and tylophorinidine: the in vitro cytotoxicity of these plant alkaloids and their potential as antimicrobial and anticancer agents. Toxicol In Vitro 14(1):53–59

Ng TP, Chiam PC, Lee T, Chua HC, Lim L, Kua EH (2006) Curry consumption and cognitive function in the elderly. Am J Epidemiol 164:898–906

Nizamutdinova IT, Lee GW, Son KH, Jeon SJ, Kang SS, Kim YS, Lee JH, Seo HG, Chang KC, Kim HJ (2008a) Tanshinone I effectively induces apoptosis in estrogen receptor-positive (MCF-7) and estrogen receptor-negative (MDA-MB-231) breast cancer cells. Int J Oncol 33:4854–4891

Nizamutdinova IT, Lee GW, Lee JS, Cho MK, Son KH, Jeon SJ, Kang SS, Kim YS, Lee JH, Seo HG, Chang KC (2008b) Tanshinone I suppresses growth and invasion of human breast cancer cells, MDA-MB-231, through regulation of adhesion molecules. Carcinogenesis 29:1885–1892

Nwodo JN, Ibezim A, Simoben CV, Ntie-Kang F (2016) Exploring cancer therapeutics with natural products from African medicinal plants, part II: alkaloids, terpenoids and flavonoids. Anticancer Agents Med Chem 16:108–127

Park IJ, Kim MJ, Park OJ, Park MG, Choe W, Kang I, Kim SS, Ha J (2010) Cryptotanshinone sensitizes DU145 prostate cancer cells to Fas (APO1/CD95)-mediated apoptosis through Bcl-2 and MAPK regulation. Cancer Lett 298:88–98

Rabi T, Bishayee A (2009) D-Limonene sensitizes docetaxel-induced cytotoxicity in human prostate cancer cells: generation of reactive oxygen species and induction of apoptosis. J Carcinog 8:9

Raj R, Mongia P, Sahu SK, Ram A (2016) Nanocarriers based anticancer drugs: current scenario and future perceptions. Curr Drug Targets 17:206–228

Ramos S (2008) Cancer chemoprevention and chemotherapy: dietary polyphenols and signaling pathways. Mol Nutr Food Res 52:507–526

Rasheed SA, Efferth T, Asangani IA, Allgayer H (2010) First evidence that the antimalarial drug artesunate inhibits invasion and *in vivo* metastasis in lung cancer by targeting essential extracellular proteases. Int J Cancer 127:1475–1485

Reungpatthanaphong P, Mankhetkorn S (2002) Modulation of multidrug resistance by artemisinin, artesunate and dihydroartemisinin in K562/adr and GLC4/adr resistant cell lines. Biol Pharm Bull 25:1555–1561

Rosea P, Huangb Q, Ongb CN, Whiteman M (2005) Broccoli and watercress suppress matrix metalloproteinase-9 activity and invasiveness of human MDA-MB-231 breast cancer cells. Toxicol Appl Pharmacol 209:105–113

Schmidt BM, Ribnicky DM, Lipsky PE, Raskin I (2007) Revisiting the ancient concept of botanical therapeutics. Nat Chem Biol 3:360–366

Shan YF, Shen X, Xie YK, Chen JC, Shi HQ, Yu ZP, Song QT, Zhou MT, Zhang QY (2009) Inhibitory effects of tanshinone II-A on invasion and metastasis of human colon carcinoma cells. Acta Pharmacol Sin 30:1537–1542

Shen SQ, Zhang Y, Xiang JJ, Xiong CL (2007) Protective effect of curcumin against liver warm ischemia/reperfusion injury in rat model is associated with regulation of heat shock protein and antioxidant enzymes. World J Gastroenterol 13:1953–1961

Shoeba M, Celikb S, Jasparsc M, Kumarasamya Y, MacManusa SM, Nahard L, Thoo-Lin PK, Sarker SD (2005) Isolation, structure elucidation and bioactivity of schischkiniin, a unique indole alkaloid from the seeds of Centaurea schischkinii. Tetrahedron 61(38):9001–9006

Sichaem J, Surapinit S, Siripong P, Khumkratok S, Jong-aramruang J, Tip-pyang S (2010) Two new cytotoxic isomeric indole alkaloids from the roots of Nauclea orientalis. Fitoterapia 81 (7):830–833

Stoner GD, Morrissey DT, Heur YH, Daniel EM, Galati AJ, Wagner SA (1991) Inhibitory effects of phenethyl isothiocyanate on N nitrosobenzylmethylamine carcinogenesis in the rat esophagus. Cancer Res 51:2063–2068

Su CC, Lin YH (2008a) Tanshinone IIA inhibits human breast cancer cells through increased Bax to Bcl-xL ratios. Int J Mol Med 22:357–361

Su CC, Lin YH (2008b) Tanshinone IIA down-regulates the protein expression of ErbB-2 and up-regulates TNF-α in colon cancer cells *in vitro* and *in vivo*. Int J Mol Med 22:847–851

Su CC, Chen GW, Kang JC, Chan MH (2008a) Growth inhibition and apoptosis induction by tanshinone IIA in human colon adenocarcinoma cells. Planta Med 74:1357–1362

Su CC, Chen GW, Lin JG (2008b) Growth inhibition and apoptosis induction by tanshinone I in human colon cancer Colo 205 cells. Int J Mol Med 22:613–618

Sun J (2007) D-Limonene: safety and clinical applications. Altern Med Rev 12:259–264

Sung HJ, Choi SM, Yoon Y, An KS (1999) Tanshinone IIA, an ingredient of *Salvia miltiorrhiza* BUNGE, induces apoptosis in human leukemia cell lines through the activation of caspase-3. Exp Mol Med 31:174–178

Tait S, Salvati AL, Desideri N, Fiore L (2006) Antiviral activity of substituted homoisoflavonoids on enteroviruses. Antivir Res 72:252–255

Talalay P, Fahey JW (2001) Phytochemicals from cruciferous plants protect against cancer by modulating carcinogen metabolism. J Nutr 131:3027–3033

Tan W, Lu J, Huang M, Li Y, Chen M, Wu G, Gong J, Zhong Z, Xu Z, Dang Y, Guo J (2011) Anti-cancer natural products isolated from Chinese medicinal herbs. Chin Med 6:27

Tang Z, Tang Y, Fu L (2003) Growth inhibition and apoptosis induction in human hepatoma cells by tanshinone II A. J Huazhong Univ Sci Technolog Med Sci 23:166–168

Theodoratou E, Kyle J, Cetnarskyj R, Farrington SM, Tenesa A, Barnetson R, Porteous M, Dunlop M, Campbell H (2007) Dietary flavonoids and the risk of colorectal cancer. Cancer Epidemiol Biomarkers Prev 16:684–693

Udenigwe CC, Ramprasath VR, Aluko RE, Jones PJ (2008) Potential of resveratrol in anticancer and anti-inflammatory therapy. Nutr Rev 66:445–454

Uppala PT, Dissmore T, Lau BH, Andacht T, Rajaram S (2013) Selective inhibition of cell proliferation by lycopene in MCF-7 breast cancer cells *in vitro*: a proteomic analysis. Phytother Res 27:595–601

Vaishampayan U, Hussain M, Banerjee M, Seren S, Sarkar FH, Fontana J, Forman JD, Cher ML, Powell I, Pontes JE, Kucuk O (2007) Lycopene and soy isoflavones in the treatment of prostate cancer. Nutr Cancer 59:1–7

Verhoeven DT, Goldbohm RA, Van Poppel G, Verhagen H, Van den Brandt PA (1996) Epidemiological studies on Brassica vegetables and cancer risk. Cancer Epidemiol Biomarkers Prev 5:733–748

Wang G, Tang W, Bidigare R (2005a) Terpenoids as therapeutic drugs and pharmaceutical agents. Nat Prod:197–227

Wang X, Wei Y, Yuan S, Liu G, Lu Y, Zhang J, Wang W (2005b) Potential anticancer activity of tanshinone IIA against human breast cancer. Int J Cancer 116:799–807

Wang J, Wang X, Jiang S, Yuan S, Lin P, Zhang J, Lu Y, Wang Q, Xiong Z, Wu Y, Ren J (2007) Growth inhibition and induction of apoptosis and differentiation of tanshinone IIA in human glioma cells. J Neurooncol 82:11–21

Wang SJ, Gao Y, Chen H et al (2010) Dihydroartemisinin inactivates NF-kappaB and potentiates the anti-tumor effect of gemcitabine on pancreatic cancer both *in vitro* and *in vivo*. Cancer Lett 293:99–108

Wattenberg LW (1977) Inhibition of carcinogenic effects of polycyclic hydrocarbons by benzyl isothiocyanate and related compounds. J Natl Cancer Inst 58:395–398

Wattenberg LW (1983) Inhibition of neoplasia by minor dietary constituents. Cancer Res 43:2448–2453

Wei L, Brossi A, Morris-Natschke SL, Bastow KF, Lee K-H (2008) Antitumor agents 248. Chemistry and antitumor activity of tylophorinerelated alkaloids. Stud Nat Prod Chem 34:3–34

Willoughby JA, Sundar SN, Cheung M, Tin AS, Modiano J, Firestone GL (2009) Artemisinin blocks prostate cancer growth and cell cycle progression by disrupting Sp1 interactions with the cyclin-dependent kinase-4 (CDK4) promoter and inhibiting CDK4 gene expression. J Biol Chem 284:2203–2213

Won SH, Lee HJ, Jeong SJ, Lee HJ, Lee EO, Jung DB, Shin JM, Kwon TR, Yun SM, Lee MH, Choi SH (2010) Tanshinone IIA induces mitochondria dependent apoptosis in prostate cancer cells in association with an inhibition of phosphoinositide 3-kinase/AKT pathway. Biol Pharm Bull 33:1828–1834

Wu WL, Chang WL, Chen CF (1991) Cytotoxic activities of tanshinones against human carcinoma cell lines. Am J Chin Med 19:207–216

Yang YM, Conaway CC, Chiao JW, Wang CX, Amin S, Whysner J, Dai W, Reinhardt J, Chung FL (2002) Inhibition of benzo(a)pyrene-induced lung tumorigenesis in A/J mice by dietary N-acetylcysteine conjugates of benzyl and phenethyl isothiocyanates during the postinitiation phase is associated with activation of mitogen-activated protein kinases and p53 activity and induction of apoptosis. Cancer Res 62:2–7

Yang ZS, Zhou WL, Sui Y, Wang JX, Wu JM, Zhou Y, Zhang Y, He PL, Han JY, Tang W, Li Y (2005) Synthesis and immunosuppressive activity of new artemisinin derivatives. 1. [12(beta or alpha)-Dihydroartemisininoxy]phen(ox)yl aliphatic acids and esters. J Med Chem 48:4608–4617

Yang JH, Hsia TC, Kuo HM, Chou PD, Chou CC, Wei YH, Chung JG (2006) Inhibition of lung cancer cell growth by quercetin glucuronides via G2/M arrest and induction of apoptosis. Drug Metab Dispos 34:296–304

Yao H, Liu J, Xu S, Zhu Z, Xu J (2017) The structural modification of natural products for novel drug discovery. Expert Opin Drug Discovery 12:121–140

Yin H-Q, Kim Y-H, Moon C-K, Lee B-H (2005) Reactive oxygen species-mediated induction of apoptosis by a plant alkaloid 6-methoxydihydrosanguinarine in HepG2 cells. Biochem Pharmacol 70(2):242–248

Yuan SL, Wei YQ, Wang XJ, Xiao F, Li SF, Zhang J (2004) Growth inhibition and apoptosis induction of tanshinone II-A on human hepatocellular carcinoma cells. World J Gastroenterol 10:2024–2028

Yuxian X, Feng T, Ren L, Zhengcai L (2009) Tanshinone II-A inhibits invasion and metastasis of human hepatocellular carcinoma cells *in vitro* and *in vivo*. Tumori J 95:789–795

Zhang Y, Kensler TW, Cho CG, Posner GH, Talalay P (1994) Anticarcinogenic activities of sulforaphane and structurally related synthetic norbornyl isothiocyanates. Proc Natl Acad Sci U S A 91:3147–3150

Zhang SM, Hunter DJ, Rosner BA, Giovannucci EL, Colditz GA, Speizer FE, Willett WC (2000) Intakes of fruits, vegetables, and related nutrients and the risk of non-Hodgkin's lymphoma among women. Cancer Epidemiol Biomarkers Prev 9:477–485

Zhang K, Li J, Meng W, Xing H, Yang Y (2010) C/ EBPβ and CHOP participate in tanshinone IIA-induced differentiation and apoptosis of acute promyelocytic leukemia cells *in vitro*. Int J Hematol 92:571–578

Zhou L, Zuo Z, Chow MS (2005) Danshen: an overview of its chemistry, pharmacology, pharmacokinetics, and clinical use. J Clin Pharmacol 45:1345–1359

# 5 Plant Secondary Metabolites: Natural Compounds as Cosmetic Ingredients and Their Potential Activity in Skin Cancer

Cristina Elena Dinu-Pirvu, Florentina Iuliana Cocos, Valentina Anuta, Mihaela Violeta Ghica, and Lacramioara Popa

**Abstract**

This following chapter represents a brief summary of phytochemicals which presents anticancer properties besides the effects known and for which they have been used since ancient times. Recent studies about phytocompounds have shown that they represent a good alternative to the conventional skin cancer therapy, even metastatic forms, statement sustained through in vitro and in vivo experiments. Modern drug delivery systems, such as liposomes, ethosomes, solid lipid nanoparticles, and nanoemulsions, were developed and tested in order to increase stability and bioavailability of the active compound.

**Keywords**

Phytochemicals · Skin cancer · Melanoma · Drug delivery system · Bioavailability

## 5.1 Introduction

Nowadays, skin cancer is one of the most prevalent diseases. In the last decades, an increasing number of cases of skin cancer were reported (Almeida and Barry 2010; Leiter et al. 2014). The World Health Organization identifies this disease as a priority public health concern because 132,000 new cases are registered every year worldwide.

Skin cancer can be classified in melanoma and non-melanoma skin cancer (basal cell carcinoma and squamous cell carcinoma) (Almeida and Barry 2010; Lomas

C. E. Dinu-Pirvu (✉) · F. I. Cocos · V. Anuta · M. V. Ghica · L. Popa
Department of Physical Chemistry and Colloids, "Carol Davila" University of Medicine and Pharmacy, Bucharest, Romania
e-mail: cristina.dinu@umfcd.ro


A. K. Sharma, A. Sharma (eds.), *Plant Secondary Metabolites*,
https://doi.org/10.1007/978-981-16-4779-6_5

et al. 2012). People with white skin, red hair, and blue eyes can be affected by these types of cancer (Feller et al. 2016). As regards the gender, men are more likely to develop a type of non-melanoma skin cancer than women (Almeida and Barry 2010). A high risk of metastasis is presented by melanoma, followed by squamous cell carcinoma (Feller et al. 2016).

Approximately 30% of White people are likely to develop a non-melanoma skin cancer due to the exposure to the ultraviolet (UV) radiation (Wehner et al. 2012). Basal cell carcinoma can be caused by the exposure to high doses of UVB radiation, while squamous cell carcinoma can be triggered after a chronic exposure to radiation (Leiter et al. 2014). Also, these types of cancer can be initiated and promoted by genetic susceptibility as faulty DNA repair and precursor lesions (dysplastic nevi or atypical moles) (Armstrong and Kricker 2001; Kraemer et al. 1994; Melamed et al. 2017). Moreover, the occupational risk factors can play an important role in the etiology of the melanoma. Populations working in printing and press, telecommunications, and petrochemical industries are more susceptible to the development of this type of skin cancer (Koh et al. 1993). One of the most dangerous occupational factors is the prolonged exposure to arsenic, a class I human carcinogen that is commonly found in the agriculture and medicinal substance industries (Danciu et al. 2018a; Muenyi et al. 2015). PUVA therapy (psoralens, P, and exposure to ultraviolet A radiation, UVA) which represents a treatment for severe skin diseases may induce skin cancer by photomutagenicity and photoinduced immunosuppression. So, it is very important to evaluate the risks before starting the PUVA therapy (Danciu et al. 2018a; Maiorino et al. 2016; Archier et al. 2012).

## 5.2 Skin Cancer Therapy

The common therapeutic options for skin cancer are the surgical excision, radiation therapy, cryosurgery, photodynamic therapy, and topical therapies. In the case of metastasis, chemotherapy and immunotherapy are used.

### 5.2.1 Basal Cell Carcinoma

For this type of skin cancer, the most common therapeutic option is the surgical excision of the skin. The method is applicable only for those lesions which are smaller than 2 cm in diameter. Also, the wounds should be found in the trunk or extremities. One of the advantages of this option is that the tissue can be histologically examined. The presence of the scars and the risk of infection are some of the disadvantages. If the tumor is located on the lips, eyelids, nose, or ears, radiation therapy is the best option. Other therapeutic options are cryosurgery and photodynamic therapy. Also, topical therapies with imiquimod or 5-fluorouracil can be used (Danciu et al. 2018a; Alter et al. 2015).

### 5.2.2 Squamous Cell Carcinoma

For tumors which vary between 4 and 10 mm, surgical excision and cryosurgery are the most common and best options of treatment. As the other type of non-melanoma skin cancer, this one can be treated using nonsurgical effective methods, namely, radiation therapy, photodynamic therapy, and topical applications of imiquimod or 5-fluorouracil (Danciu et al. 2018a; Bahner and Bordeaux 2013). In addition to these two substances, nonsteroidal anti-inflammatory drugs and retinoids can prevent and treat this type of skin cancer (Danciu et al. 2018a; Tran et al. 2001).

### 5.2.3 Melanoma

If the melanoma is early diagnosed, it can be removed surgically along with the surrounding normal structures in order to reduce the tumor recurrence (Almeida and Barry 2010). In the case of metastasis, chemotherapy is used. Substances such as cisplatin, doxorubicin, mitomycin, and temozolomide are frequently administered. Another therapeutic option is the immunotherapy that includes administration of interferon-alpha and interleukin-2 (Leiter et al. 2014; AlQathama and Prieto 2015). The biggest disadvantage of chemotherapy is the number of side effects, such as nausea, vomiting, hair loss, etc. In the last decade, a lot of studies has started in order to develop new drugs with phytochemicals for cancer in order to reduce the side effects of the conventional therapy and to increase the patient adherence to the treatment.

## 5.3 Phytochemicals

In Greek, "phyto" means plant; therefore, phytochemicals can be defined as bioactive, non-nutrient plant compounds. These are especially found in fruits, vegetables, and whole grains (Liu 2004; Block et al. 2001). They have increased interest among researchers, food production, and also for pharmaceutical industry.

Phytochemicals are effective in the chemoprotection and treatment of skin cancer, even metastasis, due to their multi-target activity, being able to interact with proteins and receptors (AlQathama and Prieto 2015; Pal et al. 2016a; Chinembiri et al. 2014).

### 5.3.1 Polyphenolic Compounds

Among the polyphenolic compounds with anticancer activity are flavonoids (quercetin, silymarin, silybin, apigenin, genistein, diosmin, luteolin), resveratrol, curcumin, 6-gingerol, capsaicin, eugenol, caffeic acid and its derivates, and gallic acid and its esters (named catechins) (Chinembiri et al. 2014).

Their main mechanisms of action are antioxidative capacity, photoprotective effect, induction of apoptosis, and inhibition of angiogenesis (Liu 2004; Pal et al.

2016a; Danciu et al. 2018b; Solovchenko and Merzlyak 2003; Liu-Smith and Meyskens 2016).

#### 5.3.1.1 Quercetin

A dietary flavonoid with low toxicity is quercetin, which acts via many mechanisms. It presents antioxidant and anti-inflammatory activities. It promotes the apoptosis of UVB melanoma cells (Liu-Smith and Meyskens 2016; Rafiq et al. 2015). Also, the inhibitory effects of quercetin that interfere with STAT3 phosphorylation and reduce the STAT3 nuclear localization were studied (Cao et al. 2014). The cytotoxic effect is favored by the presence of hydroxyl group at the 3′-position of the ring B in quercetin. Moreover, it presents a proliferative activity on SK-MEL-2 and OCM-1 human melanoma cells. The anticancer effect can be explained by the regulation of cyclin-dependent kinases CDK1 and CDK2 (Kim et al. 2004; Casagrande and Darbon 2001).

#### 5.3.1.2 Apigenin

Another dietary flavonoid with antioxidant, anti-inflammatory, antiviral, and antibacterial properties is apigenin (Liu-Smith and Meyskens 2016; Yan et al. 2017). Since the 1990s, researches have shown that topical administration of apigenin has inhibitory effects on the number and size of chemically and UV-induced skin tumors in hairless mice (Wei et al. 1990; Birt et al. 1997). Hasnat et al. (2015) and Spoerlein et al. (2013) revealed that the reduction of the human melanoma cell number and of the percentage of transwell-migrated cells and cytotoxic effects are dose-dependent.

Apigenin is effective in metastasis by downregulation of STAT3 signaling (Yan et al. 2017; Cao et al. 2016). In addition, it increases terminal differentiation and apoptosis (Pal et al. 2016a; Liu-Smith and Meyskens 2016; Hasnat et al. 2015; Kiraly et al. 2016).

#### 5.3.1.3 Silymarin and Silybin

Silybin is the main active component of silymarin complex, isolate from the seeds of *Silybum marianum* (milk thistle). Its main indication is hepatic ailments, presenting hepatoprotective, antioxidative, and anti-inflammatory effects (Pal et al. 2016a; Ghosh et al. 2010).

Recently, its anticancer activity has been demonstrated at topical administration. Silymarin inhibits carcinogenesis induced by UV radiation, (MEK)-1/2 and RSK-2 kinase activity, and Wnt/β-catenin translocation in melanoma cells (Lee et al. 2013; Gajos-Michniewicz and Czyz 2016). Similar to apigenin, silymarin has beneficial effects in metastasis by downregulation of MMP expression (Cao et al. 2016; Vaid et al. 2015).

In vitro studies revealed that the diminution of oxidative stress caused by UVA radiation is concentration-dependent (Svobodová et al. 2007), while in vivo experiments demonstrated that silymarin application can reduce the UVB-caused sunburn and skin edema (Katiyar et al. 1997).

As regards the oral administration, there were studied several modern formulations including silymarin in order to improve the bioavailability. Therefore, they should become challenges in optimization of drug delivery in the anticancer field.

Self-microemulsifying drug delivery system was tested on male dogs with a single-dose administration. The results reveal a 3-month stability, a faster release of the silymarin, and an increased bioavailability (Li et al. 2010). A significant pharmacokinetic profile was observed in semisolid dispersion binary system using Gelucire 44/14 as carrier (Hussein et al. 2012). A sustained release, followed by an improved pharmacological activity, was proved by the administration of silymarin-loaded chitosan nanoparticles to mice (Gupta et al. 2014a). Xu et al. (2015) formulated silymarin as solid dispersion and silica nanoparticles with slow-release matrix material and release enhancer. For this, a continuous controlled release for 72 h was observed. The liposomal lecithin-based carrier system of phytosomal silymarin has a higher anti-inflammatory effect than the silymarin suspension (Kumar et al. 2014).

#### 5.3.1.4 Diosmin

The main flavonoid administered in cardiovascular disease treatment, such as venous insufficiency, presenting venotonic and vasoprotector effects, is diosmin. Moreover, cyclooxygenases and cytochrome P450 activity is regulated through its antioxidant and anti-inflammatory properties (Alvarez et al. 2008; Conesa et al. 2005). In addition, diosmin is an effective agent against metastatic stages of melanoma, reducing the metastatic nodules compared to rutin and tangeretin (Conesa et al. 2005; Martinez et al. 2005). In order to reduce the adverse reactions of INF-α using a lower dose, diosmin can be associated resulting in a synergic combination with antiproliferative and antimetastatic effects (Alvarez et al. 2008).

#### 5.3.1.5 Fisetin

A neuroprotective effect offers the flavonol named fisetin, being an adjuvant in memory and cognition processes. The primary mechanisms by which it acts is apoptosis, the effect varying with cell type (Pal et al. 2016a; Syed et al. 2014). It was demonstrated that fisetin can be associated with sorafenib. The combination had a positive result when it was tested in vivo, fisetin potentiating the anti-invasive and antimetastatic effects of sorafenib. Therefore, the dose of sorafenib can be reduce, which leads to less side effects (Pal et al. 2016b).

#### 5.3.1.6 Luteolin

Another flavonoid compound known for its antioxidant and anti-inflammatory properties is luteolin. It has beneficial effects in skin cancer therapy by promoting apoptosis and inhibiting cell growth through upregulation of Bax and downregulation of Bcl-3 (Pal et al. 2016a; Iwashita et al. 2000; Nakashima et al. 2010). The apoptosis is induced in a dose- and time-dependent manner, luteolin being capable of increasing the level of intracellular reactive oxygen species (ROS) (Liu-Smith and Meyskens 2016; George et al. 2013; Kim et al. 2016). The antimetastatic activity was proved in a recent in vivo experiment on mice. Luteolin reduced the tumor cells by 50% by increasing the E-cadherin expression, while the β3-integrin and vimentin expressions were reduced (Ruan et al. 2012).

#### 5.3.1.7 Catechins

A subclass of flavonoids is represented by the catechins, which together with their esters, gallic acid, especially epihallocatechin-3-gallatte (EGCG), are studied for their effects in cancer prevention and treatment. EGCG presents photoprotective effect through interleukin-12 and proapoptotic activity through downregulation of inhibitory proteins (Bcl-2, D1, CDK2) and upregulation of Bax (AlQathama and Prieto 2015; Katiyar et al. 2007; Nihal et al. 2005). EGCG is an agonist of 67RL, a cell surface receptor, leading to the suppression of melanoma tumor growth (Umeda et al. 2008). The antimetastatic effect is based on molecular targets, such as COX-2, PGE2 receptors, reduction of angiogenesis promoters, downregulation of matrix metalloproteinase-2 activity, and the impairment of epithelial-to-mesenchymal transmission (AlQathama and Prieto 2015; Yamada et al. 2016; Singh and Katiyar 2011; Chang et al. 2014).

Topical administration of EGCG can reduce skin tumors induced by UV irradiation (Katiyar et al. 2007; Gensler et al. 1996). Moreover, an experiment on rats showed that the local treatment with EGCG prior to UVA irradiation decreased sunburn cell occurrence (Franceschi et al. 2007). This highlights the chemoprevention properties of EGCG.

A double-blind phase randomized clinical trial included 51 volunteers with a precancerous skin disorder, actinic keratosis. Patients had administered locally on one forearm the EGCG formulation and on the other a placebo ointment for 12 weeks. There were no significant differences between the EGCG and the placebo formulations, but the authors hypothesized that EGCG may not be active in the formulation chosen due to its poor bioavailability (Linden et al. 2003).

#### 5.3.1.8 Curcumin

The colored compound of turmeric rhizomes (*Curcuma* sp.), curcumin, is well-known for its anti-inflammatory properties (Pal et al. 2016a). Also, it presents an anticancer activity suppressing cell invasion and inducing autophagy (Zhao et al. 2016). Similar to EGCG, curcumin acts on molecular targets (downregulation of metalloproteinases and collagenases activities, modulation of integrin receptors, interference with STAT3 pathway), mechanisms demonstrated in vivo using mice (Liu-Smith and Meyskens 2016; Chang et al. 2014; Chatterjee et al. 2003; Zhang et al. 2015).

Different formulations with curcumin were tested in order to increase the bioavailability and to improve the stability of the main ingredient of turmeric. An increased oral absorption with a significant increased concentration of curcumin in plasma and liver tissue was observed at nanosuspension and the combination with phosphatidylcholine, both intended for oral administration (Marczylo et al. 2007; Ravichandran 2013). Lipid-based nanoparticles seem to be a potential intravenous delivery system for curcumin's anti-inflammatory properties being selectively delivered to macrophages in the bone narrow and spleen (Sou et al. 2008). Liposomes including curcumin, administered intravenous with oxaliplatin to female mice with cancer, showed inhibition of cell growth and apoptotic activity. The same effects were observed in in vitro studies (Li et al. 2007).

#### 5.3.1.9 Resveratrol

Many of us associate resveratrol, a stilbene polyphenol found in grapes and red wine, with life extension. It has antioxidant, anti-inflammatory, and antiproliferative effects being a strong scavenger for ROS (Pal et al. 2016a; Gupta et al. 2014b; Guthrie et al. 2017; Jagdeo et al. 2010). Resveratrol topical application presents inhibition potential on the different stages of carcinogenesis (Jang et al. 1997; Aziz et al. 2005). Also, it suppresses tumorigenesis and reduces murine epidermal hyperplasia while inhibiting p21, COX-2, Bcl-2 expressions, and enzymes (Kowalczyk et al. 2010). Moreover, studies showed that oral administration of resveratrol is unable to stop proliferation due to its poor bioavailability with a fast liver and intestinal biotransformation (Ndiaye et al. 2011; Niles et al. 2006). Therefore, topical administration is desired for chemoprevention and chemotherapy of skin cancer.

Modern drug delivery systems such as liposomes, nanosponges, resveratrol encapsulated within yeast cell, and zinc pectinate beads were tested in order to improve the bioavailability and the solubility of resveratrol (Coimbra et al. 2011; Ansari et al. 2011; Shi et al. 2008; Das et al. 2010). B-cyclodextrin-based nanosponges presented a good penetration in pigskin, its indication being buccal and topical administration (Ansari et al. 2011).

### 5.3.2 Polysaccharides

Terrestrial and marine organisms are rich in polysaccharides. Similar to polyphenolic compounds, their advantage is low toxicity. Therefore, polysaccharides can be used as immunomodulators in skin cancer prevention and treatment. Through their main mechanisms of action, the induction of inflammatory caspases, downregulation of metalloproteinases expression, and activation of natural killer cells can be enumerated (Imbs et al. 2016).

Beta-glucans extracted from yeast reduce tumor weight and survival rate by activating NK cells. Also, they present no hematopoietic toxicity comparatively with conventional drugs like 5-fluorouracil (Vetvicka and Vetvickova 2015).

Polysaccharides found in raspberry fruits have a synergistic effect with docetaxel in vivo, reducing the hepatic and renal toxicity. Moreover, they inhibit melanoma growth (Yang et al. 2015).

### 5.3.3 Volatile Oils

Essential oils are complexes of monoterpenes, sesquiterpenes, aromatic compounds, and their derivates.

#### 5.3.3.1 Terpin-4-Ol

The volatile oil extracted from tea tree (*Melaleuca alternifolia*) has been studied for its properties in melanoma therapy and chemoprevention. Low concentrations of terpinen-4-ol, the main component of this volatile oil, induce apoptosis (Calcabrini et al. 2004). The topical formulation with 10% tea tree volatile oil proved to inhibit subcutaneous B16-F10 melanoma cell growth (Greay et al. 2010). Tea tree oil in 10% dilution in dimethyl sulfoxide was studied in vivo when gross swelling and dissolution of internal structures of the tumor cells were observed, results which indicate that this formulation can be used for local skin cancer therapy (Ireland et al. 2012).

#### 5.3.3.2 Geraniol

Another phytocompound investigated is geraniol, presented in volatile oils extracted from rose, *Geranium* sp., and lemongrass. It showed a dose-dependent effect on the growth of B16 melanoma cells. A study in which female mice were fed with dietary geraniol for 14 days before and 21 days after tumor transplant revealed the tumor growth suppression for a minimum dose of 6.5 mmol/kg (Yu et al. 1995).

#### 5.3.3.3 α-Pinene

*Schinus terebinthifolius* Raddi. volatile oil presents chemopreventive and chemoprotection in skin cancer. The predominant component of it, α-pinene, acts through the following mechanisms: production of ROS, exposure of phosphatidylserine, disruption of the mitochondrial potential, induction of caspase-3 and DNA fragmentation, and inducing apoptosis of the tumor cells (Matsuo et al. 2011).

#### 5.3.3.4 α-Santalol

The major component of sandalwood oil, α-santalol, represented a traditional treatment of various skin diseases. A 6-h treatment with this compound leads to G2/M phase arrest and depolymerization of microtubules in UACC-62 cells, p53 wild-type human melanoma cells (Zhang et al. 2010).

#### 5.3.3.5 Eugenol

The volatile oil extracted from clove is abundant in eugenol. In vivo and in vitro studies demonstrated the antiproliferative effect of eugenol. S-phase arrest, deregulation and inhibition of transcriptional activity of E2F1, downregulation of oncogenes (c-Myc and H-ras), and change of p53 expression are the mechanisms which lead to tumor growth delay and decrease in tumor size (Ghosh et al. 2005; Pal et al. 2010; Rachoi et al. 2011).

Topical and oral administration of an infusion of cloves into mice with skin cancer reduced the incidence of papilloma development. Moreover, the local application of eugenol reduce inflammation through inhibition of COX-2, induced nitric oxide synthase expression, and decrease of TNF-α, IL-6, and PGE2 levels (Esmaeili et al. 2016). Therefore, a nanoemulsion including 2% eugenol was tested comparatively piroxicam in murine skin for anti-inflammatory effect (Kaur et al. 2010).

#### 5.3.3.6 Boswellic Acids

The frankincense essential oil has shown strong anti-inflammatory activity due to Boswellic acids. Faruck et al. concluded that frankincense volatile oil exhibits in vitro anticancer activity by reducing viability in B16-F10 cell and FM94 cells. Moreover, this can reduce tumor size in C57BL/6 mice melanoma tumor model. Improved hematological biochemical parameters, liver histology, and phase I and phase II drug metabolizing enzymes make frankincense essential oil an effective drug for the prevention of hepatic injury (Hakkim et al. 2019).

### 5.3.4 Alkaloids

Intense pharmacologic effects at low doses present alkaloids. Also, they have a safety profile very different from one of the phytochemicals discussed above. Despite these, mechanisms of action of alkaloids are well studied.

Among the alkaloids with anticancer properties, berberine, harmine, paclitaxel, and glycoalkaloids from *Solanaceae* species may be listed.

#### 5.3.4.1 Berberine

A representative isoquinoline alkaloid is berberine that can be isolated from species of the following families: *Berberidaceae*, *Ranunculaceae*, and *Papaveraceae*. It is well known for its anti-inflammatory, antimicrobial, antihyperlipidemic, and antihyperglycemic activities. Moreover, berberine is one of the most active alkaloids in melanoma by inhibiting PI3K, ERK, and GSK3β kinases (Mantena et al. 2006a; Barbagallo et al. 2015; Song et al. 2015). Its combination with doxorubicin intensely reduces proliferation and increases apoptosis (Mittal et al. 2014).

#### 5.3.4.2 Paclitaxel

There is an approved phytochemical for chemotherapy of several solid tumors with poor prognosis, including skin cancer. The anticancer activity of paclitaxel, a pseudoalkaloid isolated from *Taxus brevifolia*, Pacific yew bark, can be explained though the stabilization of microtubules and disruption of the dynamic equilibrium between free and polymerized tubulin, mechanisms which lead to cell division arrested in the G2/M transition of mitosis (Yue et al. 2010). Paclitaxel antiangiogenic effects are potentiated by COX-2 inhibitors (Merchan et al. 2005). Its extreme hydrophobicity and drug resistance to paclitaxel represent the main disadvantages of this pseudoalkaloid. Lately, modern delivery drug systems have been developed and tested in vitro and in vivo in order to improve the bioavailability of paclitaxel.

The association of paclitaxel and silymarin included in a microemulsion was compared with Taxol, a drug already registered and used for cancer treatment. These two formulations were orally administered to rats. The results showed an increased bioavailability of paclitaxel and a faster absorption for microemulsion compared to the reference drug (Park et al. 2012).

The anticancer activity of paclitaxel was increased by formulating it as a nanoemulsion, combined with curcumin (Ganta et al. 2010).

A combination of paclitaxel with cyclosporine A was tested on patients with cancer. The study revealed that paclitaxel present limited bioavailability by micellar entrapment. Also, it was demonstrated that Cremophor EL increases absorption and metabolism of paclitaxel and reduces the median time to peak concentration (Chu et al. 2008).

The solid nanoemulsion preconcentrate increased the oral bioavailability and had high cumulative in vitro drug release. Also, it presented a significant antiproliferative effect on breast cancer cell line, MCF7 (Gao et al. 2003).

A diffusion-controlled release, low toxicologic profile, and gastrointestinal stability make the niosomal formulation suitable for oral drug delivery (Ahmad et al. 2014). A possible alternative to the conventional therapy is represented by paclitaxel-loaded pegylated ethosomes due to the longer half-life, slower release rate, and increase in cytotoxicity on human melanoma cell line, SK-MEL-3 (Bayindir and Yuksel 2010).

An increased anticancer effect was obtained for solid lipid nanoparticles. The release profile was delayed with the enhancement of lipid concentration. Entrapment efficiency, particle size, and zeta potential value raised as lipid concentration were augmented (Patil and Joshi 2012). The intravenous delivery of solid lipid nanoparticles modified with 2-hydroxypropyl-β-cyclodextrin presented low nephrotoxicity and prolong antiproliferative activity, while intratumorally administration showed a high antitumor efficiency and reduce toxicity to normal organs (Baek et al. 2015). Another group of researchers included the solid lipid nanoparticles into a gel that showed sustained release. Also, the in vivo study confirmed that solid lipid particles entrapping paclitaxel, which were loaded in topical gel, is efficient in the treatment of skin cancer in mice compared to normal paclitaxel-loaded gel (Ritujar et al. 2016).

#### 5.3.4.3 Glycoalkaloids Isolated from *Solanaceae*

Solamargine induces cellular necrosis, inhibiting selectively the growth of WM239 and WM115 melanoma cells (Al Sinani et al. 2016). α-Solanine also induces apoptosis and suppresses melanoma cell invasion trough the reduction of matrix metalloproteinase, MMP-2 and MMP-9 (Lu et al. 2010).

### 5.3.5 Proanthocyanidins

The oligomeric flavonoids, present in grapes seeds, red wine, almonds, cranberries, and cacao, are well-known for their anti-inflammatory, antiarthritic, antioxidant, and antiallergic properties. Additionally, proanthocyanidins prevent skin aging and inhibit UV radiation-induced peroxidation activity exerting anti-skin carcinogenic effects (Pal et al. 2016a; Mittal et al. 2003; Meeran et al. 2009; Nandakumar et al. 2008).

Antitumor-promoting effects of grape seed proanthocyanidins are dose dependent, a fact demonstrated on the SENCAR mouse skin model, on which reduction in tumor incidence, multiplicity, and volume were observed (Sharma et al. 2007).

Researchers studied the mechanisms of action in skin cancer. Therefore, the photoprotective activity can be explained by the inhibition of $H_2O_2$-induced phosphorylation of ERKI/n, JNK, and p38 proteins (Mantena et al. 2006b). The inhibition of skin cell proliferation is mediated through the inhibition of cyclin-dependent kinases and increase of their inhibitors (Meeran and Katiyar 2007). Moreover, the epigenetic mechanisms of the anticancer potential, such as the decrease of the levels of DNA methylation, 5-mehylcytosine, and DNA methyltransferase activity, were explored (Vaid et al. 2012).

### 5.3.6 Caffeic Acid Phenethyl Ester

Besides enzymes, amino acids, and proteins, honey products present a high concentration of caffeic acid phenethyl ester (CAPE). It presents strong in vitro and in vivo inhibitory potentials in many types of cancer (Xiang et al. 2006; Kudugunti et al. 2011; Onori et al. 2009). CAPE inhibits the growth, invasion, and migration of the skin papilloma caused by 12-*O*-tetradecanoylphorbol-13-acetate exposure. Also, this phytocompound can modulate the cell cycle and induces apoptosis by increasing Bax and decreasing Bcl2 expression (Wu et al. 2011; Chen et al. 2001).

### 5.3.7 Allyl Sulfides

Diallyl sulfide, diallyl disulfide, and diallyl trisulfide are the main active compounds extracted from garlic. The chemopreventive and chemotherapeutic effects of allyl sulfides administered topically were proved on murine skin. Diallyl sulfide regulates the p53 expression and reduces tumor by inducing caspase-mediated cell death in mice skin (Page et al. 2016; Arora et al. 2004; Arora and Shukla 2002). Another possible mechanism of action should be the regulation of multiple signaling pathways, such as downregulation of H-ras mRNA, upregulation of p53 and Bax proteins, and downregulation of PI3K/Akt and MAPKs (Arora et al. 2006; Kalra et al. 2006).

Similar activities in skin cancer prevention and therapy are presented by diallyl disulfide (Shan et al. 2016).

Diallyl trisulfide induces apoptosis in human basal cell carcinoma cells through triggering caspase-mediated cell death by upregulating p53 and Bax expression level and lowering Bcl-2 and Bcl-xl expression (Shan et al. 2016).

As regards the potential in inhibiting COX-2 expression in human embryonic cell kidney cells, diallyl trisulfide presents higher effect than the other two allyl sulfides (Wang et al. 2012). Also, trisulfide presents higher chemoprevention and chemotherapeutical potentials than the other sulfides (Wang et al. 2010; Elango et al. 2004).

### 5.3.8 Capsaicin

The bioactive compound extracted from red pepper and red chili pepper is capsaicin. Researchers presented opposite results as regards the chemoprevention and chemotherapeutic effects (Pal et al. 2016a; Bode and Dong 2011). A group of scientists demonstrated that the local application of capsaicin stimulated skin cancer in mouse model (Hwang et al. 2010), while others came up with opposite results and found that topical administration of capsaicin had no substantial increase in the growth of mice skin cancer compared with the control group (Park and Surh 1997). Despite these opposite outcomes, a review highlights the side effects of the phytocompound. Three of the patients develop erythema, burning, and stinging compared to placebo (Mason et al. 2004).

## 5.4 Conclusions

Phytochemicals isolated from medicinal plants and dietary fruits and vegetables have shown great potential in chemoprevention and chemotherapy of different types of skin cancer. Researcher has been interested in discovering and proving mechanisms of action for each potential treatment with less toxicity. The only exception of this toxicological profile is represented by the alkaloids.

There are several studies that illustrate the efficiency of these natural compounds in the metastatic skin cancer, especially metastatic melanoma with a good prognostic.

Moreover, they have been developing and testing modern system drug delivery, such as liposomes, ethosomes, solid lipid nanoparticles, and nanoemulsion, in order to increase the bioavailability of the phytochemical. Additionally, the side effects of some drugs were reduced by associating it with a phytochemical. I think that all these benefits of the new delivery systems can be new start points of developing a new drug with topical administration which can ease the life of the patient and increase their adherence to the treatment.

In conclusion, phytochemicals can be considered an alternative to the conventional skin cancer therapy due to the many advantages and the few disadvantages.

## References

Ahmad J, Mir SR, Kohli K, Chuttani K, Mishra AK, Panda A, Amin S (2014) Solid-nanoemulsion preconcentrate for oral delivery of paclitaxel: formulation design, biodistribution, and gamma scintigraphy imaging. Biomed Res Int 2014:984756

Al Sinani SS, Eltayeb EA, Coomber BL, Adham SA (2016) Solamargine triggers cellular necrosis selectively in different types of human melanoma cancer cells through extrinsic lysosomal mitochondrial death pathway. Cancer Cell Int 16:11

Almeida CA, Barry SA (2010) Cancer: basic science and clinical aspects. Wiley-Blackwell, Chichester, 405 p

AlQathama A, Prieto JM (2015) Natural products with therapeutic potential in melanoma metastasis. Nat Prod Rep 32(8):1170–1182

Alter M, Hillen U, Leiter U, Sachse M, Gutzmer R (2015) Current diagnosis and treatment of basal cell carcinoma. J Dtsch Dermatol Ges 13(9):863–874; quiz 875

Alvarez N, Vicente V, Martınez C (2008) Synergistic effect of diosmin and interferon alpha on metastatic pulmonary melanoma. Cancer Biother Radiopharm 24(3):347–352

Ansari KA, Vavia PR, Trotta F, Cavalli R (2011) Cyclodextrin-based nanosponges for delivery of resveratrol: in vitro characterisation, stability, cytotoxicity and permeation study. AAPS PharmSciTech 12(1):279–286

Archier E, Devaux S, Castela E, Gallini A, Aubin F, Le Maître M, Aractingi S, Bachelez H, Cribier B, Joly P (2012) Carcinogenic risks of psoralen UV-A therapy and narrowband UV-B therapy in chronic plaque psoriasis: a systematic literature review. J Eur Acad Dermatol Venereol 26(Suppl 3):22–31

Armstrong BK, Kricker A (2001) The epidemiology of UV induced skin cancer. J Photochem Photobiol B 63(1–3):8–18

Arora A, Shukla Y (2002) Induction of apoptosis by diallyl sulfide in DMBA-induced mouse skin tumors. Nutr Cancer 44(1):89–94

Arora A, Siddiqui IA, Shukla Y (2004) Modulation of p53 in 7,12-dimethylbenz[a]anthracene-induced skin tumors by diallyl sulfide in Swiss albino mice. Mol Cancer Ther 3(11):1459–1466

Arora A, Kalra N, Shukla Y (2006) Regulation of p21/ras protein expression by diallyl sulfide in DMBA induced neoplastic changes in mouse skin. Cancer Lett 242(1):28–36

Aziz MH, Reagan-Shaw S, Wu J, Longley BJ, Ahmad N (2005) Chemoprevention of skin cancer by grape constituent resveratrol: relevance to human disease? FASEB J 19(9):1193–1195

Baek JS, Kim JH, Park JS, Cho CW (2015) Modification of paclitaxel-loaded solid lipid nanoparticles with 2-hydroxypropyl-betacyclodextrin enhances absorption and reduces nephrotoxicity associated with intravenous injection. Int J Nanomedicine 10:5397–5405

Bahner JD, Bordeaux JS (2013) Non-melanoma skin cancers: photodynamic therapy, cryotherapy, 5-fluorouracil, imiquimod, diclofenac, or what? Facts and controversies. Clin Dermatol 31(6):792–798

Barbagallo CM, Cefalù AB, Noto D, Averna MR (2015) Role of nutraceuticals in hypolipidemic therapy. Front Cardiovasc Med 2:22

Bayindir ZS, Yuksel N (2010) Characterization of niosomes prepared with various nonionic surfactants for paclitaxel oral delivery. J Pharm Sci 99(4):2049–2060

Birt DF, Mitchell D, Gold B, Pour P, Pinch HC (1997) Inhibition of ultraviolet light induced skin carcinogenesis in SKH-1 mice by apigenin, a plant flavonoid. Anticancer Res 17(1A):85–91

Block G, Norkus E, Hudes M, Mandel S, Helzlsouer K (2001) Which plasma antioxidants are most related to fruit and vegetable consumption? Am J Epidemiol 154(12):1113–1118

Bode AM, Dong Z (2011) The two faces of capsaicin. Cancer Res 71(8):2809–2814

Calcabrini A, Stringaro A, Toccacieli L, Meschini S, Marra M, Colone M, Arancia G, Molinari A, Salvatore G, Mondello F (2004) Terpinen-4-ol, the main component of Melaleuca alternifolia (tea tree) oil inhibits the in vitro growth of human melanoma cells. J Invest Dermatol 122 (2):349–360

Cao HH, Tse AKW, Kwan HY, Yu H, Cheng CY, Su T, Fong WF, Yu ZL (2014) Quercetin exerts anti-melanoma activities and inhibits STAT3 signaling. Biochem Pharmacol 87(3):424–434

Cao HH, Chu JH, Kwan HY, Su T, Yu H, Cheng CY, Fu XQ, Guo H, Li T, Tse AKW (2016) Inhibition of the STAT3 signaling pathway contributes to apigenin-mediated anti-metastatic effect in melanoma. Sci Rep 6:21731

Casagrande F, Darbon JM (2001) Effects of structurally related flavonoids on cell cycle progression of human melanoma cells: regulation of cyclin-dependent kinases CDK2 and CDK1. Biochem Pharmacol 61(10):1205–1215

Chang CW, Hsieh YH, Yang WE, Yang SF, Chen Y, Hu DN (2014) Epigallocatechin gallate inhibits migration of human uveal melanoma cells via downregulation of matrix metalloproteinase-2 activity and ERK1/2 pathway. Biomed Res Int 2014:141582

Chatterjee A, Mitra A, Ray S, Chattopadhyay N, Siddiqi M (2003) Curcumin exhibits antimetastatic properties by modulating integrin receptors, collagenase activity, and expression of Nm23 and E-cadherin. J Environ Pathol Toxicol Oncol 22(1):49–58
Chen YJ, Shiao MS, Wang SY (2001) The antioxidant caffeic acid phenethyl ester induces apoptosis associated with selective scavenging of hydrogen peroxide in human leukemic HL-60 cells. Anti-Cancer Drugs 12(2):143–149
Chinembiri TN, Du Plessis LH, Gerber M, Hamman JH, Du Plessis J (2014) Review of natural compounds for potential skin cancer treatment. Molecules 19(8):11679–11721
Chu Z, Chen JS, Liau CT, Wang HM, Lin YC, Yang MH, Chen PM, Gardner ER, Figg WD, Sparreboom A (2008) Oral bioavailability of a novel paclitaxel formulation (Genetaxyl) administered with cyclosporin A in cancer patients. Anti-Cancer Drugs 19(3):275–281
Coimbra M, Isacchi B, Van Bloois L, Torano JS, Ket A, Wu X, Broere F, Metselaar JM, Rijcken CJ, Storm G (2011) Improving solubility and chemical stability of natural compounds for medicinal use by incorporation into liposomes. Int J Pharm 416(2):433–442
Conesa CM, Ortega V, Gascón MJY, Baños MA, Jordana MC, Garcia OB, Castillo JN (2005) Treatment of metastatic melanoma B16F10 by the flavonoids targeretin, rutin and diosmin. J Agric Food Chem 53:6791–6797
Danciu C, Soica C, Antal D, Alexa E, Pavel IZ, Ghiulai R, Ardelean F, Babuta RM, Popescu A, Dehelean CA (2018a) Natural compounds in the chemoprevention of malignant melanoma. Anti Cancer Agents Med Chem 18(5):631–644
Danciu C, Zupko I, Bor A, Schwiebs A, Radeke H, Hancianu M, Cioanca O, Alexa E, Oprean C, Bojin F (2018b) Botanical therapeutics: phytochemical screening and biological assessment of chamomile, parsley and celery extracts against A375 human melanoma and dendritic cells. Int J Mol Sci 19(11)
Das S, Ng KY, Ho PC (2010) Formulation and optimization of zinc-pectinate beads for the controlled delivery of resveratrol. AAPS PharmSciTech 11(2):729–742
Elango EM, Asita H, Nidhi G, Seema P, Banerji A, Kuriakose MA (2004) Inhibition of cyclooxygenase-2 by diallyl sulfides (DAS) in HEK 293T cells. J Appl Genet 45(4):469–471
Esmaeili F, Rajabnejhad S, Partoazar AR, Mehr SE, Faridi-Majidi R, Sahebgharani M, Syedmoradi L, Rajabnejhad MR, Amani A (2016) Anti-inflammatory effects of eugenol nanoemulsion as a topical delivery system. Pharm Dev Technol 21(7):887–893
Feller L, Khammissa R, Kramer B, Altini M, Lemmer J (2016) Basal cell carcinoma, squamous cell carcinoma and melanoma of the head and face. Head Face Med 12:11
Franceschi C, Capri M, Monti D, Giunta S, Olivieri F, Sevini F, Panourgia MP, Invidia L, Celani L, Scurti M (2007) Inflammaging and anti-inflammaging: a systemic perspective on aging and longevity emerged from studies in humans. Mech Ageing Dev 128(1):92–105
Gajos-Michniewicz A, Czyz M (2016) Modulation of WNT/beta-catenin pathway in melanoma by biologically active components derived from plants. Fitoterapia 109:283–292
Ganta S, Devalapally H, Amiji M (2010) Curcumin enhances oral bioavailability and anti-tumor therapeutic efficacy of paclitaxel upon administration in nanoemulsion formulation. J Pharm Sci 99(11):4630–4641
Gao P, Rush BD, Pfund WP, Huang T, Bauer JM, Morozowich W, Kuo MS, Hageman MJ (2003) Development of a supersaturable SEDDS (S-SEDDS) formulation of paclitaxel with improved oral bioavailability. J Pharm Sci 92(12):2386–2398
Gensler HL, Timmermann BN, Valcic S, Wächter GA, Dorr R, Dvorakova K, Alberts DS (1996) Prevention of photocarcinogenesis by topical administration of pure epigallocatechin gallate isolated from green tea. Nutr Cancer 26(3):325–335
George VC, Kumar DRN, Suresh PK, Kumar S, Kumar RA (2013) Comparative studies to evaluate relative in vitro potency of luteolin in inducing cell cycle arrest and apoptosis in HaCaT and A375 cells. Asian Pac J Cancer Prev 14(2):631–637
Ghosh R, Nadiminty N, Fitzpatrick JE, Alworth WL, Slaga TJ, Kumar AP (2005) Eugenol causes melanoma growth suppression through inhibition of E2F1 transcriptional activity. J Biol Chem 280(7):5812–5819

Ghosh A, Ghosh T, Jain S (2010) Silymarin—a review on the pharmacodynamics and bioavailability enhancement approaches. Pharm Sci Tehnol 2(10):348–355
Greay SJ, Ireland DJ, Kissick HT, Heenan PJ, Carson CF, Riley TV, Beilharz MW (2010) Inhibition of established subcutaneous murine tumour growth with topical Melaleuca alternifolia (tea tree) oil. Cancer Chemother Pharmacol 66(6):1095–1102
Gupta S, Singh SK, Girotra P (2014a) Targeting silymarin for improved hepatoprotective activity through chitosan nanoparticles. Int J Pharm Investig 4(4):156–163
Gupta SC, Tyagi AK, Deshmukh-Taskar P, Hinojosa M, Prasad S, Aggarwal BB (2014b) Downregulation of tumor necrosis factor and other proinflammatory biomarkers by polyphenols. Arch Biochem Biophys 559:91–99
Guthrie AR, Chow HS, Martinez JA (2017) Effects of resveratrol on drug- and carcinogen-metabolizing enzymes, implications for cancer prevention. Pharmacol Res Perspect 5(1):e00294
Hakkim FL, Bakshi HA, Khan S, Nasef M, Farzand R, Sam S, Rashan L, Al-Baloshi MS, Hasson SSAA, Al Jabri A, McCarron PA, Tambuwala MM (2019) Frankincense essential oil suppresses melanoma cancer through down-regulation of Bcl-2/Bax cascade signaling and ameliorates hepatotoxicity via phase I and II metabolizing enzymes. Oncotarget 10(37):3479–3490
Hasnat M, Pervin M, Lim JH, Lim BO (2015) Apigenin attenuates melanoma cell migration by inducing anoikis through integrin and focal adhesion kinase inhibition. Molecules 20 (12):21157–21166
Hussein A, El-Menshawe S, Afouna M (2012) Enhancement of the in-vitro dissolution and in-vivo oral bioavailability of silymarin from liquid-filled hard gelatin capsules of semisolid dispersion using Gelucire 44/14 as a carrier. Pharmazie 67(3):209–214
Hwang MK, Bode AM, Byun S, Song NR, Lee HJ, Lee KW, Dong Z (2010) Cocarcinogenic effect of capsaicin involves activation of EGFR signaling but not TRPV1. Cancer Res 70(17):6859–6869
Imbs TI, Ermakova SP, Malyarenko OS, Isakov VV, Zvyagintseva TN (2016) Structural elucidation of polysaccharide fractions from the brown alga Coccophora langsdorfii and in vitro investigation of their anticancer activity. Carbohydr Polym 135:162–168
Ireland DJ, Greay SJ, Hooper CM, Kissick HT, Filion P, Riley TV, Beilharz MW (2012) Topically applied Melaleuca alternifolia (tea tree) oil causes direct anti-cancer cytotoxicity in subcutaneous tumour bearing mice. J Dermatol Sci 67(2):120–129
Iwashita K, Kobori M, Yamaki K, Tsushida T (2000) Flavonoids inhibit cell growth and induce apoptosis in B16 melanoma 4A5 cells. Biosci Biotechnol Biochem 64(9):1813–1820
Jagdeo J, Adams L, Lev-Tov H, Sieminska J, Michl J, Brody N (2010) Dose-dependent antioxidant function of resveratrol demonstrated via modulation of reactive oxygen species in normal human skin fibroblasts in vitro. J Drugs Dermatol 9(12):1523–1526
Jang M, Cai L, Udeani GO, Slowing KV, Thomas CF, Beecher CW, Fong HH, Farnsworth NR, Kinghorn AD, Mehta RG (1997) Cancer chemopreventive activity of resveratrol, a natural product derived from grapes. Science 275(5297):218–220
Kalra N, Arora A, Shukla Y (2006) Involvement of multiple signaling pathways in diallyl sulfide mediated apoptosis in mouse skin tumors. Asian Pac J Cancer Prev 7(4):556–562
Katiyar SK, Korman NJ, Mukhtar H, Agarwal R (1997) Protective effects of silymarin against photocarcinogenesis in a mouse skin model. J Natl Cancer Inst 89(8):556–566
Katiyar S, Elmets CA, Katiyar SK (2007) Green tea and skin cancer: photoimmunology, angiogenesis and DNA repair. J Nutr Biochem 18(5):287–296
Kaur G, Athar M, Alam MS (2010) Eugenol precludes cutaneous chemical carcinogenesis in mouse by preventing oxidative stress and inflammation and by inducing apoptosis. Mol Carcinog 49(3):290–301
Kim YK, Kim YS, Choi SU, Ryu SY (2004) Isolation of flavonol rhamnosides from Loranthus tanakae and cytotoxic effect of them on human tumor cell lines. Arch Pharm Res 27(1):44–47
Kim JK, Kang KA, Ryu YS, Piao MJ, Han X, Oh MC, Boo SJ, Jeong SU, Jeong YJ, Chae S (2016) Induction of endoplasmic reticulum stress via reactive oxygen species mediated by luteolin in melanoma cells. Anticancer Res 36(5):2281–2289

Kiraly AJ, Soliman E, Jenkins A, Van Dross RT (2016) Apigenin inhibits COX-2, PGE2, and EP1 and also initiates terminal differentiation in the epidermis of tumor bearing mice. Prostaglandins Leukot Essent Fatty Acids 104:44–53

Koh HK, Sinks TH, Geller AC, Miller DR, Lew RA (1993) Etiology of melanoma. Cancer Treat Res 65:1–28

Kowalczyk MC, Kowalczyk P, Tolstykh O, Hanausek M, Walaszek Z, Slaga TJ (2010) Synergistic effects of combined phytochemicals and skin cancer prevention in SENCAR mice. Cancer Prev Res (Phila) 3(2):170–178

Kraemer KH, Lee M-M, Andrews AD, Lambert WC (1994) The role of sunlight and DNA repair in melanoma and nonmelanoma skin cancer. The xeroderma pigmentosum paradigm. Arch Dermatol 130(8):1018–1021

Kudugunti SK, Thorsheim H, Yousef MS, Guan L, Moridani MY (2011) The metabolic bioactivation of caffeic acid phenethyl ester (CAPE) mediated by tyrosinase selectively inhibits glutathione S-transferase. Chem Biol Interact 192(3):243–256

Kumar N, Rai A, Reddy ND, Raj PV, Jain P, Deshpande P, Mathew G, Kutty NG, Udupa N, Rao CM (2014) Silymarin liposomes improves oral bioavailability of silybin besides targeting hepatocytes, and immune cells. Pharmacol Rep 66(5):788–798

Lee MH, Huang Z, Kim DJ, Kim SH, Kim MO, Lee SY, Xie H, Park SJ, Kim JY, Kundu JK (2013) Direct targeting of MEK1/2 and RSK2 by silybin induces cell-cycle arrest and inhibits melanoma cell growth. Cancer Prev Res (Phila) 6(5):455–465

Leiter U, Eigentler T, Garbe C (2014) Epidemiology of skin cancer. Adv Exp Med Biol 810:120–140

Li L, Ahmed B, Mehta K, Kurzrock R (2007) Liposomal curcumin with and without oxaliplatin: effects on cell growth, apoptosis, and angiogenesis in colorectal cancer. Mol Cancer Ther 6 (4):1276–1282

Li X, Yuan Q, Huang Y, Zhou Y, Liu Y (2010) Development of silymarin self-microemulsifying drug delivery system with enhanced oral bioavailability. AAPS PharmSciTech 11(2):672–678

Linden KG, Carpenter PM, Mclaren CE, Barr RJ, Rite P, Sun JD, Li K-T, Viner JL, Meyskens FL (2003) Chemoprevention of nonmelanoma skin cancer: experience with a polyphenol from green tea. Recent Results Cancer Res 163(165–171):discussion 264–6

Liu RH (2004) Potential synergy of phytochemicals in cancer prevention: mechanism of action. J Nutr 134(12 Suppl):3479S–3485S

Liu-Smith F, Meyskens FL (2016) Molecular mechanisms of flavonoids in melanin synthesis and the potential for the prevention and treatment of melanoma. Mol Nutr Food Res 60(6):1264–1274

Lomas A, Leonardi-Bee J, Bath-Hextall F (2012) A systematic review of worldwide incidence of nonmelanoma skin cancer. Br J Dermatol 166(5):1069–1080

Lu MK, Shih YW, Chien TTC, Fang LH, Huang HC, Chen PS (2010) alpha-Solanine inhibits human melanoma cell migration and invasion by reducing matrix metalloproteinase-2/9 activities. Biol Pharm Bull 33(10):1685–1691

Maiorino A, De Simone C, Perino F, Caldarola G, Peris K (2016) Melanoma and non-melanoma skin cancer in psoriatic patients treated with high-dose phototherapy. J Dermatolog Treat 27 (5):443–447

Mantena SK, Sharma SD, Katiyar SK (2006a) Berberine inhibits growth, induces G1 arrest and apoptosis in human epidermoid carcinoma A431 cells by regulating Cdki-Cdk-cyclin cascade, disruption of mitochondrial membrane potential and cleavage of caspase 3 and PARP. Carcinogenesis 27(10):2018–2027

Mantena SK, Baliga MS, Katiyar SK (2006b) Grape seed proanthocyanidins induce apoptosis and inhibit metastasis of highly metastatic breast carcinoma cells. Carcinogenesis 27(8):1682–1691

Marczylo TH, Verschoyle RD, Cooke DN, Morazzoni P, Steward WP, Gescher AJ (2007) Comparison of systemic availability of curcumin with that of curcumin formulated with phosphatidylcholine. Cancer Chemother Pharmacol 60(2):171–177

Martinez C, Vicente V, Yanez J, Alcaraz M, Castells Mora MT, Canteras M, Benavente-García García O, Castillo J (2005) The effect of the flavonoid diosmin, grape seed extract and red wine on the pulmonary metastatic B16F10 melanoma. Histol Histopathol 20(4):1121–1129

Mason L, Moore RA, Derry S, Edwards JE, Mcquay HJ (2004) Systematic review of topical capsaicin for the treatment of chronic pain. BMJ 328(7446):991

Matsuo AL, Figueiredo CR, Arruda DC, Pereira FV, JaB S, Massaoka MH, Travassos LR, Sartorelli P, Lago JH (2011) alpha-Pinene isolated from Schinus terebinthifolius Raddi (Anacardiaceae) induces apoptosis and confers antimetastatic protection in a melanoma model. Biochem Biophys Res Commun 411(2):449–454

Meeran SM, Katiyar SK (2007) Grape seed proanthocyanidins promote apoptosis in human epidermoid carcinoma A431 cells through alterations in Cdki-Cdk-cyclin cascade, and caspase-3 activation via loss of mitochondrial membrane potential. Exp Dermatol 16(5):405–415

Meeran SM, Vaid M, Punathil T, Katiyar SK (2009) Dietary grape seed proanthocyanidins inhibit 12-O-tetradecanoyl phorbol-13-acetate-caused skin tumor promotion in 7,12-dimethylbenz[a] anthracene-initiated mouse skin, which is associated with the inhibition of inflammatory responses. Carcinogenesis 30(3):520–528

Melamed RD, Aydin IT, Rajan GS, Phelps R, Silvers DN, Emmett KJ, Brunner G, Rabadan R, Celebi JT (2017) Genomic characterization of dysplastic nevi unveils implications for diagnosis of melanoma. J Invest Dermatol 137(4):905–909

Merchan JR, Jayaram DR, Supko JG, He X, Bubley GJ, Sukhatme VP (2005) Increased endothelial uptake of paclitaxel as a potential mechanism for its antiangiogenic effects: potentiation by Cox-2 inhibition. Int J Cancer 113(3):490–498

Mittal A, Elmets CA, Katiyar SK (2003) Dietary feeding of proanthocyanidins from grape seeds prevents photocarcinogenesis in SKH-1 hairless mice: relationship to decreased fat and lipid peroxidation. Carcinogenesis 24(8):1379–1388

Mittal A, Tabasum S, Singh RP (2014) Berberine in combination with doxorubicin suppresses growth of murine melanoma B16F10 cells in culture and xenograft. Phytomedicine 21(3):340–347

Muenyi CS, Ljungman M, States JC (2015) Arsenic disruption of DNA damage responses-potential role in carcinogenesis and chemotherapy. Biomol Ther 5(4):2184–2193

Nakashima S, Matsuda H, Oda Y, Nakamura S, Xu F, Yoshikawa M (2010) Melanogenesis inhibitors from the desert plant Anastatica hierochuntica in B16 melanoma cells. Bioorg Med Chem 18(6):2337–2345

Nandakumar V, Singh T, Katiyar SK (2008) Multi-targeted prevention and therapy of cancer by proanthocyanidins. Cancer Lett 269(2):378–387

Ndiaye M, Philippe C, Mukhtar H, Ahmad N (2011) The grape antioxidant resveratrol for skin disorders: promise, prospects, and challenges. Arch Biochem Biophys 508(2):164–170

Nihal M, Ahmad N, Mukhtar H, Wood GS (2005) Anti-proliferative and proapoptotic effects of (-)-epigallocatechin-3-gallate on human melanoma: possible implications for the chemoprevention of melanoma. Int J Cancer 114(4):513–521

Niles RM, Cook CP, Meadows GG, Fu YM, Mclaughlin JL, Rankin GO (2006) Resveratrol is rapidly metabolized in athymic (nu/nu) mice and does not inhibit human melanoma xenograft tumor growth. J Nutr 136(10):2542–2546

Onori P, Demorrow S, Gaudio E, Franchitto A, Mancinelli R, Venter J, Kopriva S, Ueno Y, Alvaro D, Savage J (2009) Caffeic acid phenethyl ester decreases cholangiocarcinoma growth by inhibition of NF-kappaB and induction of apoptosis. Int J Cancer 125(3):565–576

Page A, Navarro M, Suarez-Cabrera C, Alameda JP, Casanova ML, Paramio JM, Bravo A, Ramirez A (2016) Protective role of p53 in skin cancer: carcinogenesis studies in mice lacking epidermal p53. Oncotarget 7(15):20902–20918

Pal D, Banerjee S, Mukherjee S, Roy A, Panda CK, Das S (2010) Eugenol restricts DMBA croton oil induced skin carcinogenesis in mice: downregulation of c-Myc and H-ras, and activation of p53 dependent apoptotic pathway. J Dermatol Sci 59(1):31–39

Pal HC, Marchiony Hunt K, Diamond A, A Elmets C, Afaq F (2016a) Phytochemicals for the management of melanoma. Mini Rev Med Chem 16(12):953–979
Pal HC, Diamond AC, Strickland LR, Kappes JC, Katiyar SK, Elmets CA, Athar M, Afaq F (2016b) Fisetin, a dietary flavonoid, augments the anti-invasive and anti-metastatic potential of sorafenib in melanoma. Oncotarget 7(2):1227–1241
Park KK, Surh YJ (1997) Effects of capsaicin on chemically-induced two-stage mouse skin carcinogenesis. Cancer Lett 114(1–2):183–184
Park JH, Park JH, Hur HJ, Woo JS, Lee HJ (2012) Effects of silymarin and formulation on the oral bioavailability of paclitaxel in rats. Eur J Pharm Sci 45(3):296–301
Patil SM, Joshi H (2012) Colloidal drug delivery system for tumor specificity of paclitaxel in mice. Der Pharm Lett 4:961–967
Rachoi BB, Shin SH, Kim UK, Hong JW, Kim GC (2011) S phase cell cycle arrest and apoptosis is induced by eugenol in G361 human melanoma cells. Int J Oral Biol 36(3):129–134
Rafiq RA, Quadri A, Nazir LA, Peerzada K, Ganai BA, Tasduq SA (2015) A potent inhibitor of phosphoinositide 3-kinase (PI3K) and mitogen activated protein (MAP) kinase signalling, quercetin (3, 3′, 4′, 5, 7-Pentahydroxyflavone) promotes cell death in ultraviolet (UV)-B-irradiated B16F10 melanoma cells. PLoS One 10(7):e0131253
Ravichandran R (2013) Pharmacokinetic study of nanoparticulate curcumin: oral formulation for enhanced bioavailability. J Biomater Nanobiotechnol 4:291–299
Ritujar B, Das PJ, Paulomi P, Bhaskr M (2016) Topical delivery of paclitaxel for treatment of skin cancer. Drug Dev Ind Pharm 42(9):1482–1494
Ruan JS, Liu YP, Zhang L, Yan LG, Fan FT, Shen CS, Wang AY, Zheng SZ, Wang SM, Lu Y (2012) Luteolin reduces the invasive potential of malignant melanoma cells by targeting beta3 integrin and the epithelial-mesenchymal transition. Acta Pharmacol Sin 33(10):1325–1331
Shan Y, Wei Z, Tao L, Wang S, Zhang F, Shen C, Wu H, Liu Z, Zhu P, Wang A (2016) Prophylaxis of diallyl disulfide on skin carcinogenic model via p21-dependent Nrf2 stabilization. Sci Rep 6:35676
Sharma SD, Meeran SM, Katiyar SK (2007) Dietary grape seed proanthocyanidins inhibit UVB-induced oxidative stress and activation of mitogen-activated protein kinases and nuclear factor-kappaB signaling in in vivo SKH-1 hairless mice. Mol Cancer Ther 6(3):995–1005
Shi G, Rao L, Yu H, Xiang H, Yang H, Ji R (2008) Stabilization and encapsulation of photosensitive resveratrol within yeast cell. Int J Pharm 349(1–2):83–93
Singh T, Katiyar SK (2011) Green tea catechins reduce invasive potential of human melanoma cells by targeting COX-2, PGE2 receptors and epithelial-to-mesenchymal transition. PLoS One 6(10):e25224
Solovchenko A, Merzlyak M (2003) Optical properties and contribution of cuticle to UV protection in plants: experiments with apple fruit. Photochem Photobiol Sci 2(8):861–866
Song YC, Lee Y, Kim HM, Hyun MY, Lim YY, Song KY, Kim BJ (2015) Berberine regulates melanin synthesis by activating PI3K/AKT, ERK and GSK3beta in B16F10 melanoma cells. Int J Mol Med 35(4):1011–1016
Sou K, Inenaga S, Takeoka S, Tsuchida E (2008) Loading of curcumin into macrophages using lipid-based nanoparticles. Int J Pharm 352(1–2):287–293
Spoerlein C, Mahal K, Schmidt H, Schobert R (2013) Effects of chrysin, apigenin, genistein and their homoleptic copper(II) complexes on the growth and metastatic potential of cancer cells. J Inorg Biochem 127:107–115
Svobodová A, Zdařilová A, Mališková J, Mikulková H, Walterová D, Vostalová J (2007) Attenuation of UVA-induced damage to human keratinocytes by silymarin. J Dermatol Sci 46(1):21–30
Syed DN, Lall RK, Chamcheu JC, Haidar O, Mukhtar H (2014) Involvement of ER stress and activation of apoptotic pathways in fisetin induced cytotoxicity in human melanoma. Arch Biochem Biophys 563:108–117
Tran C, Sorg O, Carraux P, Didierjean L, Saurat JH (2001) Topical delivery of retinoids counteracts the UVB-induced epidermal vitamin A depletion in hairless mouse. Photochem Photobiol 73 (4):425–431

Umeda D, Yano S, Yamada K, Tachibana H (2008) Green tea polyphenol epigallocatechin-3-gallate signaling pathway through 67-kDa laminin receptor. J Biol Chem 283(6):3050–3058
Vaid M, Prasad R, Singh T, Jones V, Katiyar SK (2012) Grape seed proanthocyanidins reactivate silenced tumor suppressor genes in human skin cancer cells by targeting epigenetic regulators. Toxicol Appl Pharmacol 263(1):122–130
Vaid M, Singh T, Prasad R, Katiyar SK (2015) Silymarin inhibits melanoma cell growth both in vitro and in vivo by targeting cell cycle regulators, angiogenic biomarkers and induction of apoptosis. Mol Carcinog 54(11):1328–1339
Vetvicka V, Vetvickova J (2015) Glucan supplementation has strong anti-melanoma effects: role of NK cells. Anticancer Res 35(10):5287–5292
Wang HC, Yang JH, Hsieh SC, Sheen LY (2010) Allyl sulfides inhibit cell growth of skin cancer cells through induction of DNA damage mediated G2/M arrest and apoptosis. J Agric Food Chem 58(11):7096–7103
Wang HC, Hsieh SC, Yang JH, Lin SY, Sheen LY (2012) Diallyl trisulfide induces apoptosis of human basal cell carcinoma cells via endoplasmic reticulum stress and the mitochondrial pathway. Nutr Cancer 64(5):770–780
Wehner MR, Shive ML, Chren M-M, Han J, Qureshi AA, Linos E (2012) Indoor tanning and non-melanoma skin cancer: systematic review and meta-analysis. BMJ 345:e5909
Wei H, Tye L, Bresnick E, Birt DF (1990) Inhibitory effect of apigenin, a plant flavonoid, on epidermal ornithine decarboxylase and skin tumor promotion in mice. Cancer Res 50(3):499–502
Wu J, Omene C, Karkoszka J, Bosland M, Eckard J, Klein CB, Frenkel K (2011) Caffeic acid phenethyl ester (CAPE), derived from a honeybee product propolis, exhibits a diversity of anti-tumor effects in pre-clinical models of human breast cancer. Cancer Lett 308(1):43–53
Xiang D, Wang D, He Y, Xie J, Zhong Z, Li Z, Xie J (2006) Caffeic acid phenethyl ester induces growth arrest and apoptosis of colon cancer cells via the beta-catenin/T-cell factor signaling. Anti-Cancer Drugs 17(7):753–762
Xu X, Yu J, Tong S, Zhu Y, Cao X (2015) Formulation of silymarin with high efficiency and prolonged action and preparation method thereof. US Patent no. US8962017B2
Yamada S, Tsukamoto S, Huang Y, Makio A, Kumazoe M, Yamashita S, Tachibana H (2016) Epigallocatechin-3-O-gallate upregulates microRNA-let-7b expression by activating 67-kDa laminin receptor signaling in melanoma cells. Sci Rep 6:19225
Yan X, Qi M, Li P, Zhan Y, Shao H (2017) Apigenin in cancer therapy: anti-cancer effects and mechanisms of action. Cell Biosci 7:50
Yang YJ, Xu HM, Suo YR (2015) Raspberry pulp polysaccharides inhibit tumor growth via immunopotentiation and enhance docetaxel chemotherapy against malignant melanoma in vivo. Food Funct 6(9):3022–3034
Yu SG, Hildebrandt LA, Elson CE (1995) Geraniol, an inhibitor of mevalonate biosynthesis, suppresses the growth of hepatomas and melanomas transplanted to rats and mice. J Nutr 125(11):2763–2767
Yue QX, Liu X, Guo DA (2010) Microtubule-binding natural products for cancer therapy. Planta Med 76(11):1037–1043
Zhang X, Chen W, Guillermo R, Chandrasekher G, Kaushik RS, Young A, Fahmy H, Dwivedi C (2010) Alpha-santalol, a chemopreventive agent against skin cancer, causes G2/M cell cycle arrest in both p53-mutated human epidermoid carcinoma A431 cells and p53 wild-type human melanoma UACC-62 cells. BMC Res Notes 3:220
Zhang Y, Li Y, Lv Y, Wang J (2015) Effect of curcumin on the proliferation, apoptosis, migration, and invasion of human melanoma A375 cells. Genet Mol Res 14(1):1056–1067
Zhao G, Han X, Zheng S, Li Z, Sha Y, Ni J, Sun Z, Qiao S, Song Z (2016) Curcumin induces autophagy, inhibits proliferation and invasion by downregulating AKT/mTOR signaling pathway in human melanoma cells. Oncol Rep 35(2):1065–1074

# Natural Remedies for a Healthy Heart: The Evidence-Based Beneficial Effects of Polyphenols

6

Denisa Margină, Anca Ungurianu, Carmen Purdel, and George Mihai Nițulescu

**Abstract**

Statistical and international health organizations state that about 31% of all deaths worldwide are caused by cardiovascular disease (CVD), which represents the main cause of death associated with noncommunicable disease. Various researches performed in the last 10 years have shown the beneficial effects induced by polyphenolic compounds in reducing the risk factors for CVD, thus acting either as preventive factors and also, in some cases, as reparative ones, supporting the treatment of patients with chronic cardiovascular impairments. However, larger cohort of studies are still necessary in order to assess and integrate the potential beneficial effects along with data regarding the low bioavailability of the compounds and also with their potential to interfere with the chronic pharmacological treatment of the selected patients.

D. Margină (✉) · A. Ungurianu
Department of Biochemistry, Faculty of Pharmacy, University of Medicine and Pharmacy, UMF Carol Davila Bucharest, Bucharest, Romania
e-mail: denisa.margina@umfcd.ro; anca.ungurianu@umfcd.ro

C. Purdel
Department of Toxicology, Faculty of Pharmacy, University of Medicine and Pharmacy, UMF Carol Davila Bucharest, Bucharest, Romania
e-mail: carmen.purdel@umfcd.ro

G. M. Nițulescu
Department of Pharmaceutical Chemistry, University of Medicine and Pharmacy, UMF Carol Davila Bucharest, Bucharest, Romania
e-mail: george.nitulescu@umfcd.ro





A. K. Sharma, A. Sharma (eds.), *Plant Secondary Metabolites*,
https://doi.org/10.1007/978-981-16-4779-6_6

**Keywords**

Cardiovascular disease (CVD) · Bioavailability · Polyphenolic compounds · Pharmacological treatment

## 6.1 Background

Statistical and international health organizations state that about 31% of all deaths worldwide are caused by cardiovascular disease (CVD), which represents the main cause of death associated with noncommunicable disease (Butler 2011; Santhakumar et al. 2018; Costa et al. 2017). A large body of literature data points out that a diet rich in fruit and vegetables (at least five servings/day) and limiting the salt intake to less than 1 teaspoon/day and engaging in physical activity (minimum of 30 min/day) are important factors contributing to the prevention of cardiovascular pathology, including heart attack and stroke (www.who.org).

Some of the most important mechanisms involved in the development and progression of cardiovascular disease are endothelial dysfunction, dyslipidemia, redox imbalance, inflammation, and platelet function (Cai and Harrison 2000; Libby 2006; Ross 1999).

CVD has, among its determining factors, some that are unmodifiable (age, sex, genetic background) and others that can be influenced in a great manner by lifestyle changes, most of all by diet (endothelial dysfunction, inflammation, plaque formation and rupture, etc.) (Libby 2006; De Caterina et al. 2006).

Atherosclerosis is an inflammatory pathological process initiated by endothelial dysfunction, leading to increased permeation through the endothelium, followed by the accumulation of inflammatory cells (with increased immune local response) and modified fatty molecules inside the artery walls (generating foam cells that contribute to the atherosclerotic plaque formation). All these constitute a vicious circle, being associated with increased reactive oxygen species (ROS) generation and decreased antioxidant protective mechanisms, contributing to the endothelial dysfunction and atherothrombosis through the inactivation of nitric oxide (NO) and the progression of the inflammatory response (through the oxidation of LDL and inducing the pathological signaling pathway associated with it) (Cai and Harrison 2000; Ross 1999; Galkina and Ley 2009; Harrison et al. 2003). In the next step, the lesions are characterized by necrosis of the foam cells accumulated into the intima of the arteries, the release of their fatty contents and plaque rupture due to collagen breakdown by matrix metalloproteinase, and the decreased collagen synthesis by dysfunctional or apoptotic vascular smooth muscle cells (Martinet and De Meyer 2009).

Diet, mainly through its polyphenolic compounds but also through the omega-3 fatty acids or other factors, plays a key role in mitigating oxidative stress and protecting endogenous compounds from oxidative lesions, in counteracting inflammation, in regulating the metabolism, in promoting a protective phenotype, and in improving the endothelial function as well as the platelet function (Kaliora et al. 2006; Morita et al. 2017).

The present chapter aims at gathering literature data (with a focus on the last 10 years of publications) regarding the documented effects of polyphenols and the associated mechanisms involved in the protection/prevention of cardiovascular events. Among these mechanisms, there are some most debated by literature, such as the antioxidant effect, the endothelial protective function, anti-inflammatory action, and the effects on platelet function and on lipid metabolism.

## 6.2 Structure and Bioavailability of Vegetal Secondary Metabolites

Plant secondary metabolites are compounds produced by plants biosynthetically derived from primary metabolites. These metabolites have no role for primary functions as growth, photosynthesis, or reproduction, but play a bioactive role in the adaptation of plants to the environment, as protecting the plants from the attacks or attracting pollinators (Rao and Ravishankar 2002).

More than 250,000 plant secondary metabolites have been identified, and it is demonstrated that each *family* or *specie* produces a specific combination of these compounds. Therefore, secondary metabolite content can be used as a taxonomic character in classifying plants (Liu et al. 2017).

Moreover, plant secondary metabolites exhibit different beneficial effects on humans. Many epidemiological studies show that diets rich in polyphenols or the use of herbal medicines are associated with a lower incidence of cardiovascular disease (Tungmunnithum et al. 2018).

Based on their biosynthetic origins, plant secondary metabolites can be divided into three major groups: terpenes and its derivates, nitrogen and sulfur-containing compounds, and phenolic compounds. As the chapter is focusing on phenolic compounds, especially polyphenols, for the other groups, only major characteristics are mentioned here.

### 6.2.1 Terpenes

Terpenes, the largest group of secondary metabolites, are polymeric isoprene derivatives derived from the five-carbon precursor isopentenyl diphosphate. Approximately 25,000 terpenes and derivates structures are reported, including terpenoids (Gershenzon and Dudareva 2007). Carotenoids from vegetables and fruits (Bohn 2017) or ginkgolide isolated from *ginkgo biloba (*Strømgaard and Nakanishi 2004*)* are well-known terpenoids used in the prevention and treatment of vascular diseases.

### 6.2.2 Nitrogen and Sulfur-Containing Compounds

Members of the group of nitrogen and sulfur-containing compounds are biosynthesized from amino acids (Venditti and Bianco 2018). Alkaloids such as berberine from *Coptis chinensis* (Xia and Luo 2015) or the allyl sulfides and alliin isolated from *Allium sativum* (Gomaa et al. 2018; Bradley et al. 2016) are representative compounds known to exhibit cardioprotective activity.

### 6.2.3 Phenolic Compounds

Phenolic compounds are molecules characterized by having at least one aromatic ring with one or more hydroxyl groups attached. Phenolic compounds are synthesized in plants through the shikimic acid pathway (Ali et al. 2010). Polyphenolic compounds can be easily classified into two groups: non-flavonoids and flavonoids. The group of non-flavonoid compounds includes phenolic acids (benzoic and hydroxycinnamic compounds), lignans, stilbenes, and tannins (De Silva and Alcorn 2019). More than 8000 phenolic compounds have been identified in plants, over 4000 as flavonoids, and the list is still growing (Hollman and Katan 1997).

Phenolic acids, like caffeic and ferulic acid, are abundant in foods. The most frequently encountered is caffeic acid and, to a lesser extent, ferulic acid especially in cereals linked through ester bonds to hemicelluloses (Scalbert and Williamson 2000). Caffeic acid is also found as an ester (chlorogenic acid) in many fruits and vegetables. One cup of Americano coffee from coffee shops contains a mean value of 166 mg of chlorogenic acid (Jeon et al. 2019).

The most abundant polyphenols in our diets are flavonoids (Scalbert and Williamson 2000). Flavonoids consist of a large group of polyphenolic compounds with a phenylbenzo-γ-pyrone base aglycone structure. Based on the level of oxidation, the degree of hydroxylation, methoxylation, or glycosylation of the aglycone structure, flavonoids can be subdivided into various subclasses such as flavones, flavonols, isoflavones, flavan-3-ols, anthocyanidins, catechins, etc. (Kumar and Pandey 2013; Middleton 1998). As an example, flavonols differ from flavanones by a hydroxyl group at the 3 position and a C2–C3 double bond (Kumar and Pandey 2013). It is important to highlight that all modifications on the aglycone structure change the water solubility of flavonoids and directly affect their bioavailability (Ross and Kasum 2002).

The average intake of total flavonoids in Europe corresponds to 428 mg/day, with large regional differences, both in the distribution of intake and the type of flavonoids consumed (Vogiatzoglou et al. 2015). Table 6.1 gives an overview of the main polyphenols' subclasses and the occurrence in common foods.

**Table 6.1** The polyphenol subclasses and main food sources

| Class | Representative compounds (aglycone) | Aglycone chemical structures | Food sources | Reference |
|---|---|---|---|---|
| **Non-flavonoids** | | | | |
| Phenolic acids | Hydroxycinnamic acids (caffeic, ferulic, sinapic acid)<br>Hydroxybenzoic acids and derivatives (gallic acid, syringic acid) | 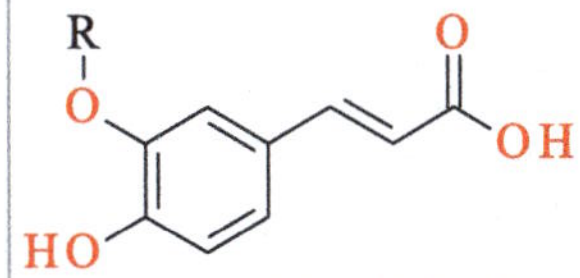<br>caffeic acid (R= H)<br>ferulic acid (R= $CH_3$)<br>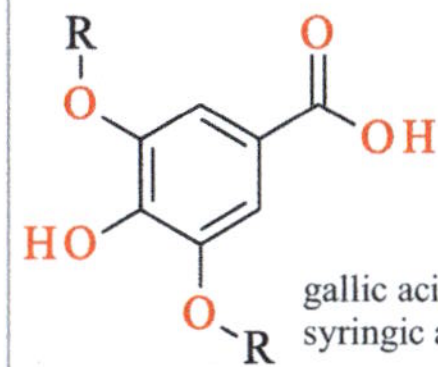<br>gallic acid (R= H)<br>syringic acid (R= $CH_3$) | *Ilex paraguariensis* (mate), *Coffea arabica* (coffee and coffee-related products), *Vitis* spp. (grapevine and grape-derived products) | Ali et al. (2010), De Silva and Alcorn (2019), Scalbert and Williamson (2000), Jeon et al. (2019), Farah et al. (2008) |
| Lignans | Lariciresinol, isolariciresinol<br>Matairesinol<br>Sesamin<br>7-hydroxymatairesinol<br>Pinoresinol, arctigenin, syringaresinol | 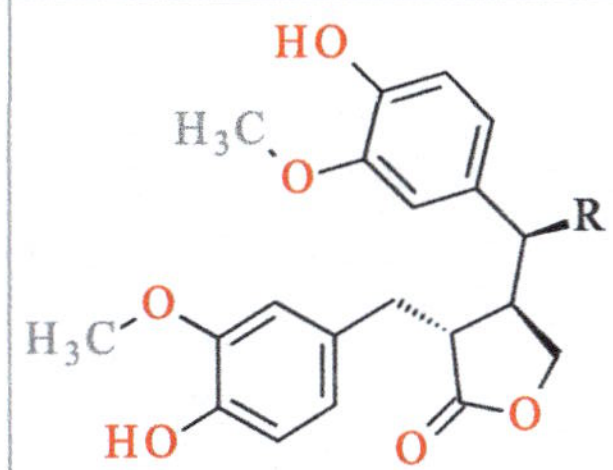<br>matairesinol (R= H)<br>hydroxymatairesinol (R= OH) | *Linum usitatissimum* L. (flaxseed and flaxseed oil), *Poaceae* family (oats, barley, wheat) | De Silva and Alcorn (2019) |

(continued)

**Table 6.1** (continued)

| Class | Representative compounds (aglycone) | Aglycone chemical structures | Food sources | Reference |
|---|---|---|---|---|
| Stilbenes | Resveratrol | 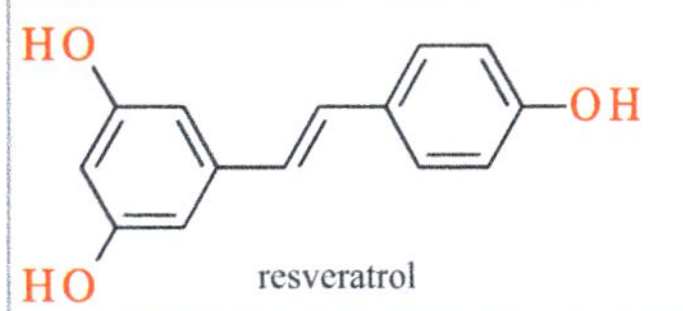 resveratrol | *Vitis* spp. (grapevine and grape-derived products) | Ali et al. (2010) |
| Tannins | Gallotannins, ellagitannins | 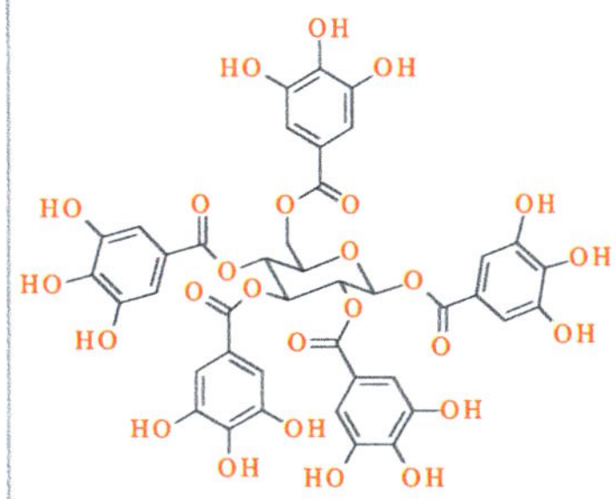 1,2,3,4,6-pentagalloyl glucose | *Camellia sinensis* (green tea)<br>*Rubus* spp. (raspberries, strawberries), *Juglans regia* (walnuts), *Punica granatum* (pomegranate) | Serrano et al. (2009) |
| **Flavonoids** | | | | |
| Flavonols | Kaempferol<br>Quercetin<br>Myricetin | 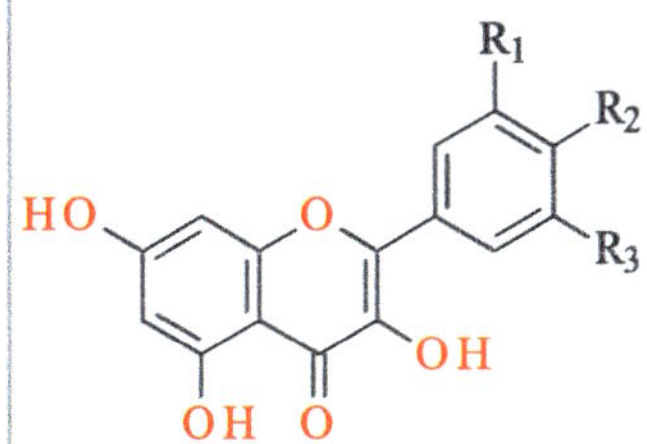 Kaempferol ($R_1$= H, $R_2$= OH, $R_3$= H)<br>Quercetin ($R_1$= OH $R_2$= OH, $R_3$= H)<br>Myricetin ($R_1$= OH, $R_2$= OH, $R_3$ = OH) | *Brassica oleracea* (broccoli, kale)<br>*Vitis* spp. (grapevine)<br>*Allium cepa* (onion)<br>*Camellia sinensis* (green tea)<br>*Sambucus nigra* (elderflower) | Hollman and Katan (1997), Mattivi et al. (2006) |

| | | | | |
|---|---|---|---|---|
| Flavones | Apigenin<br>Luteolin<br>Acacetin<br>Diosmetin | 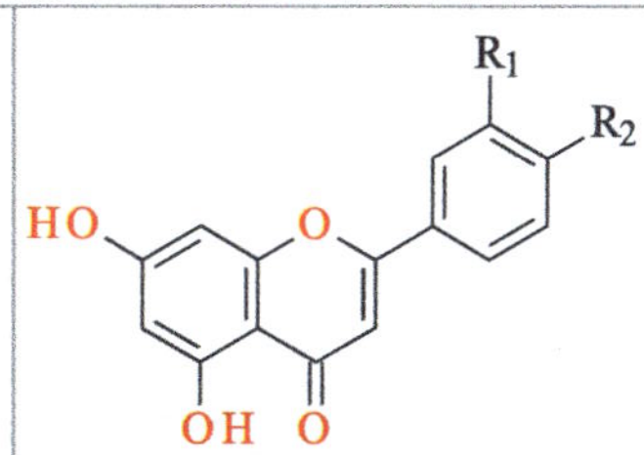<br>Apigenin ($R_1$= H, $R_2$= OH)<br>Acacetin ($R_1$= H, $R_2$= $OCH_3$)<br>Luteolin ($R_1$= OH $R_2$= OH)<br>Diosmetin ($R_1$= OH, $R_2$= OH) | *Origanum vulgare* (oregano)<br>*Rosmarinus officinalis* (rosemary)<br>*Salvia officinalis* (sage)<br>*Citrus* spp.<br>*Petroselinum crispum* (parsley)<br>*Apium graveolens* (celery) | Hostetler et al. (2017) |
| Flavanones | Naringenin<br>Hesperetin | 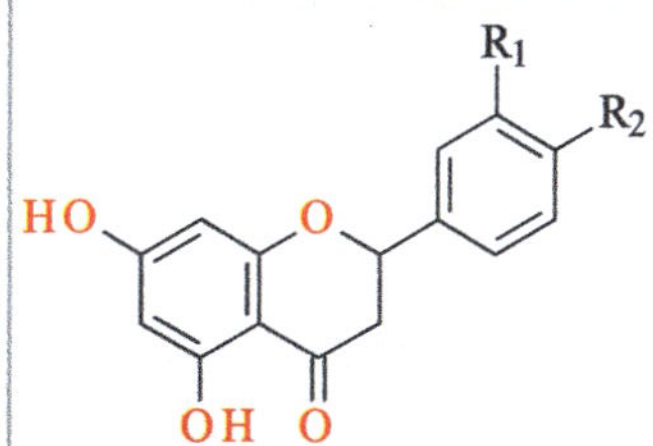<br>Naringenin ($R_1$= H, $R_2$= OH)<br>Hesperetin ($R_1$ = H, $R_2$= $OCH_3$) | *Citrus spp.* (lemon, sweet orange, sour orange, tangerine) | Hollman and Katan (1997), Barreca et al. (2017) |
| Flavanols (catechins and proanthocyanidins) | Catechins (catechin, epicatechin, gallocatechin, epigallocatechin, epicatechin gallate)<br>Proanthocyanidins (oligomers of catechins) | 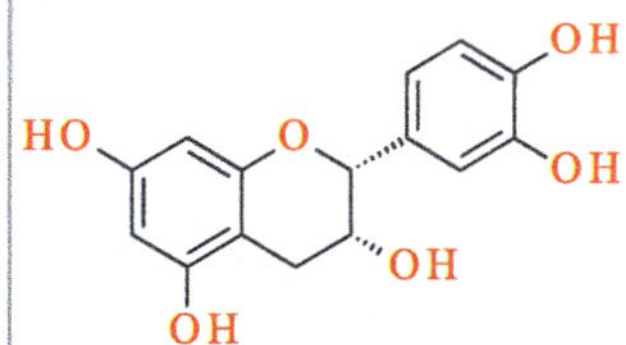<br>(-)-epicatechin | *Camellia sinensis* (green tea)<br>*Vitis* spp. (grapevine and grape-derived products) | Ali et al. (2010) |

(continued)

**Table 6.1** (continued)

| Class | Representative compounds (aglycone) | Aglycone chemical structures | Food sources | Reference |
|---|---|---|---|---|
| Anthocyanidins | Cyanidin<br>Delphinidin<br>Malvidin<br>Pelargonidin<br>Petunidin | 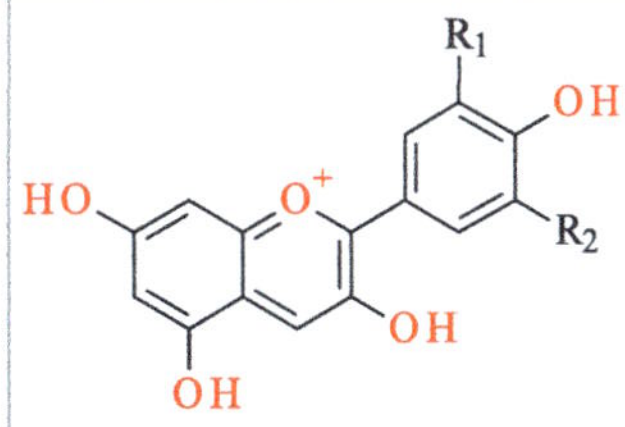<br>Pelargonidin ($R_1$ = H, $R_2$ = H)<br>Cyanidin ($R_1$ = OH, $R_2$ = H)<br>Delphinidin ($R_1$ = OH, $R_2$ = OH) | *Vitis* spp. (grapevine and grape-derived products)<br>*Vaccinium myrtillus* L. (bilberries), *Rubus fruticosus* L. (blackberries),<br>*Ribes nigrum* L. (blackcurrant)<br>*Vaccinium . oxycoccos* L (cranberries) | Ali et al. (2010), Khoo et al. (2017) |
| Isoflavones | Daidzein<br>Genistein<br>Biochanin A<br>Formononetin | 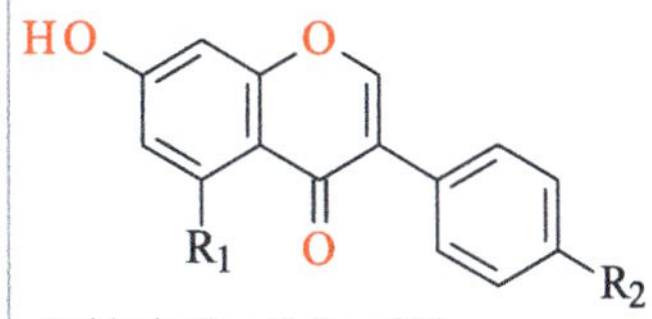<br>Daidzein ($R_1$= H, $R_2$= OH)<br>Formononetin ($R_1$= H, $R_2$= $OCH_3$)<br>Genistein ($R_1$= OH, $R_2$= OH)<br>Biochanin A ($R_1$= OH $R_2$= $OCH_3$) | *Leguminosae family (soyabean,* chickpeas) | Manach et al. (2004) |

### 6.2.4 Polyphenols Bioavailability

There have been published several reviews that have investigated the absorption and bioavailability of flavonoids (Scalbert and Williamson 2000; Ross and Kasum 2002; Manach et al. 2004, 2005; Crozier et al. 2009; Rein et al. 2013), and all concluded that there are significant differences between polyphenols subclasses.

Bioavailability varies in a wide range, from 0.3 to 43% of intake (Manach et al. 2005). Different factors are influencing the bioavailability of polyphenols, such as physicochemical properties of the compound (e.g., molecular size, isomeric configuration, partition coefficients), food processing, or the interindividual variations. Gallic acid is far better absorbed than the other polyphenols. Also, isoflavones in soy are well-absorbed through the gut barrier followed by catechins from green tea, flavanones in citrus fruits, and quercetin glucosides. Proanthocyanidins and anthocyanins are poorly absorbed (Manach et al. 2005). Tannins, especially with high molecular weight, present poor absorption through the gut barrier and have limited metabolism by the intestinal microflora (Serrano et al. 2009).

Flavonoids are present only occasionally in food in the free state (as aglycones) or as methylated derivatives, but mainly as glycosides (Kumar and Pandey 2013). The main sugars to which they are linked are glucose, rhamnose, galactose, and xylose (Ross and Kasum 2002).

It was postulated that only the aglycone can pass through the gut wall, while glycosides are hydrolyzed in the small intestine or colon by microorganisms.

The absorption of aglycone is correlated with its partition coefficient (logP octanol/water). The value of this parameter is indicating whether a compound will passively diffuse across a biological membrane. For example, quercetin has a LogP of 1.2, whereas rutin (quercetin-3-*O*-rhamnoglucoside) has a significantly lower value (−1.1), showing greater hydrophilicity. Other aglycones as luteolin and kaempferol, which have less hydroxy functions than quercetin, possess higher LogP values and are more prone to passive diffusion (Brown et al. 1998).

The isomeric configuration can also affect the absorption. For example, in the case of flavanols diastereoisomers epicatechin and catechin, the bioavailability is greater for (−)-epicatechin isomer and (+)-catechin, in comparison with the other isomers (Ottaviani et al. 2011). The epimerization of native (−)-epicatechin and (+)-catechin occurs due to food processing and reduces the bioavailability, proving that food processing also has a profound effect on the bioavailability of flavonoids (Donovan et al. 2006).

In the case of glycosides derived from glucose, galactose, or xylose, these are hydrolyzed before absorption in the small intestine by *β*-glucosidase and glycosylceramidase, while rhamnose derivates will reach the colon to be converted by rhamnosidases to aglycones and then absorbed (Rein et al. 2013; GutiErrez-Grijalva et al. 2016). Also, in the case of isoflavones, the glycosides are hydrolyzed to aglycones by the small intestine microflora, before its absorption (Rein et al. 2013).

However, in the small intestine, it was observed that the absorption of quercetin glycosides from onions was better compared with quercetin as aglycone (52% vs

24%), thus indicating that these glucosides are better absorbed in the small intestine than the aglycone and the absorption process could imply the sodium-dependent glucose transporter SGLT1 (Hollman et al. 1995).

Anthocyanins, which are also glycosides, appear to be also an exception. Their bioavailability is very low, but the gastric absorption as glycosides occurs quickly. It was suggested the interaction with bilitranslocase as a plausible mechanism for the gastric absorption. Additionally, anthocyanins are absorbed also from the small intestine as aglycones after hydrolysis by various hydrolases (McGhie and Walton 2007).

The different transport mechanisms are involved in the absorption at the gut wall level, such as passive diffusion, facilitated diffusion, or active transport.

As most polyphenols are probably too hydrophilic to penetrate the gut wall by passive diffusion, it was suggested that different membrane carriers are involved in polyphenol absorption, but up to date only Na-dependent saturable transport mechanism involved in hydroxycinnamic and ferulic acids absorption have been identified (Manach et al. 2004). Interestingly, for ellagic acid, a high affinity to organic anion transporters, especially hOAT1, was reported (Rein et al. 2013).

The peak of absorption has been shown to differ markedly among polyphenols. Gallic acid, quercetin glucosides, catechins, free hydroxycinnamic acids, and anthocyanins, which are absorbed in the stomach or the small intestine, reached the plasmatic peak at 1.5 h, whereas rutin, hesperidin, or naringin, which have to be hydrolyzed by the microflora before absorption, reached the plasmatic peak at 5.5 h (Manach et al. 2004).

Interindividual variations of bioavailability of polyphenols have been reported in healthy subjects, possibly because of polymorphism for intestinal enzymes or transporters.

Significant interindividual variations of bioavailability have been identified in the case of grapefruit juice: from 0.012 to 0.37% of naringenin and from 5.0 to 57% of its glucuronides were found in the urine after oral administration of grapefruit juice in healthy adults (Fuhr and Kummert 1995).

Also, in the case of coffee-derived products, after a single serving beverage containing the amount of 412–795μmol total chlorogenic acids, the mean excretion values of their metabolites ranged from 15.7 to 25.2% of intake, with a substantial interindividual variation (Stalmach et al. 2014).

Once absorbed, polyphenols are subjected to phase II reactions such as methylation, sulfation, and glucuronidation. The enzymes involved are catechol-*O*-methyltransferase, sulfotransferases, and glucuronosyl transferases (Rein et al. 2013). Saturation of the conjugation processes is not known. The ratio between sulfation and glucuronidation of polyphenols seems to be influenced, at least in animal studies, by species, sex, and food deprivation. In rats, food deprivation significantly increased plasma isoflavone sulfates with simultaneous decreases in plasma isoflavone glucuronides (Piskula 2000).

The conjugates are then eliminated chiefly in the urine and bile. If the polyphenols conjugates are secreted via the biliary route into the duodenum, they are hydrolyzed by $\beta$-glucuronidase, in the distal segments of the intestine and then reabsorbed. This enterohepatic recycling prolonged significantly the presence of some polyphenols within the body (Manach et al. 2004).

Another factor that could have a profound effect on the bioavailability of flavonoids is food processing. Freezing raspberries at 4 °C for 3 days did not affect anthocyanin levels but increased ellagitannin content (Mullen et al. 2002). Also, in coffee-derived products, the roasting process significantly decreases the content of polyphenols (Dybkowska et al. 2017). Further techniques such as milling, boiling, steaming, microwave heating, or roasting the cereals stimulate the release of bound phenolic compounds and increase their bioavailability (GutiErrez-Grijalva et al. 2016).

Studies regarding the effects exerted by polyphenols as preventing factors interfering with the pathways involved in atherosclerotic plaque formation are based on three types of models:

- Cell line studies
- Preclinical studies using different animal models
- Clinical studies

These researches are performed either with isolated polyphenolic compounds or with extracts from different plant sources.

## 6.3 The General Mechanism Responsible for the Beneficial Effects of Polyphenols in CVD

ROS have been largely involved in endothelial damage, which is followed by progression to atherosclerosis with CVD pathological consequences (Dhalla et al. 2000; Sugamura and Keaney Jr. 2011; Raedschelders et al. 2012). There are a lot of biochemical reactions generating ROS at the vascular level, but the most relevant is catalyzed by mitochondrial enzymes such as NADH/NADPH oxidase and xanthine oxidase (XO) (Cai and Harrison 2000; Paravicini and Touyz 2006).

The impairment of NO-dependent vasorelaxation is the result of either the decreased NO generation (due to reduction in expression of endothelial NOS – eNOS) or the reduction of available NO level (due mainly to degradation by ROS) (Cai and Harrison 2000). So, the imbalance between ROS generation/inactivation and NO generation/catabolism is one of the main factors initiating the endothelial dysfunction.

Dyslipidemia is also directly involved in the development of CVD; high ROS generation oxidizes LDL particles, thus turning them into unrecognizable bodies for endogenous receptors. As a consequence, there is a strong activation of inflammatory phenomena at the vascular level. ROS and inflammatory molecules accentuate the endothelial dysfunction; the increased expression of adhesion molecules (VCAM-1, ICAM-1, E-selectin) favors the adherence of monocytes to the endothelium and the risk of them penetrating within the vascular wall. A plethora of inflammatory molecules are released (TNFα, IFNγ, MCP-1); NFκB is upregulated;

interleukins (IL-6 and IL-8) and gelatinolytic enzymes like metalloproteinases (MMPs) are synthesized contributing to the instability of plaque (Libby 2006).

Studies show that polyphenols have the ability to interfere with the pathological processes induced by redox imbalance, inflammation, and dyslipidemia at the vascular level; the modulated mechanisms are divers—from the inhibition of pro-inflammatory enzymes (COX-2, MAPK, PKC) to reducing the release of inflammatory molecules (TNFα, IL-6, and IL-8 in addition to VCAM-1 and ICAM-1), angiogenic factors to regulating the dynamics of certain microRNAs, stimulating the activity of eNOS and paraoxonase 1 (PON1) or modulating signaling pathways by changing the expression of sirtuin-1 (Sirt1), MAP38 kinase, NFκB, and AP-1 (Hussain et al. 2016; Alvarez-Suarez et al. 2017; Gasparrini et al. 2017; Khurana et al. 2013).

Besides the endothelial dysfunction and the activation of inflammatory pathways at the vascular level, increased platelet aggregation is also involved in the progression of atherosclerosis and rupture. Polyphenols act by diminishing the aggregation risk through targeting platelet NADPH oxidase, the main source of platelet reactive oxygen species and isoprostanes (Violi and Pignatelli 2012), or by alleviating thromboxane synthesis via COX-1 pathway (Santhakumar et al. 2014).

## 6.4 Beneficial Actions of Polyphenols Concerning Redox Homeostasis

Oxidative stress involves a shift from the physiological formation and removal of free radicals, lipid peroxides, and other reactive molecules or reactive oxygen/nitrogen species (ROS/RNS), leading to increased levels of these potential detrimental molecules (Hussain et al. 2016; Khurana et al. 2013; Xia et al. 2017). Physiologically, oxidative species are involved in the regulation of cell homeostasis, affecting a myriad of cellular processes. Pathologically, however, they can initiate and/or accelerate the alteration of key molecules, affecting the integrity and functionality of proteins, lipids, and nucleic acids, causing irreversible damage (Hussain et al. 2016; Ungurianu et al. 2019a).

Polyphenols were proved to exert protective effects on essential molecules against the detrimental actions of ROS and RNS through direct free radical scavenging activity (Hussain et al. 2016), but also via additional direct and indirect effects modulating redox homeostasis, which were lately postulated to be their main mechanism of action (Santhakumar et al. 2018; Amiot et al. 2016).

Some of the most recent data collected from in vitro cell-based and from in vivo animal-based studies are presented in Table 6.2 and in Table 6.3, respectively. However, we included merely a few of the latest reports, without the claim of being exhaustive, in what constitutes a vast and ongoing area of research.

**Table 6.2** Recent reports regarding antioxidant effects in cell-based studies

| Polyphenol/extract | Cell type | Treatment | Mechanism involved | Reference |
|---|---|---|---|---|
| Strawberry | HepG2 cells | 10/50/100μg/mL whole methanolic extract (23.44 ± 0.22 total polyphenols) or 5/10/50 anthocyanin-enriched fraction (531.99 ± 2.01 total polyphenols) | ↓ Intracellular ROS<br>↑ SOD and CAT activities | Forbes-Hernandez et al. (2017a) |
| Green tea catechins | Neutrophils from the peripheral blood of healthy subjects | A mix of 30μM of epigallocatechin-3-gallate, 3μM of epigallocatechin, 2μM of epicatechin gallate, and 1.4μM of epicatechin and each one alone | ↓ ROS and RNS formation<br>↓ MPO activity and HOCl generation<br>↑ SOD, CAT | Marinovic et al. (2015) |
| Mulberry fruit extract | Hippocampal neuronal HT-22 cells | 10, 25, 50, and 100μg/mL mulberry fruit extract | ↑ GSH<br>↑ Cell viability<br>↓ ROS<br>↑ p-Akt, p-TrkB, p-CREB, BDNF<br>↑ Nrf2 | Shin et al. (2019) |
| Olive leaves extract | HUVECs | $H_2O_2$-induced oxidative stress cell model + preconditioning with 100μg extract/mL (23.29 mg GAE/g) | ↓ ROS | De la Ossa et al. (2019) |
| *Lycium ruthenicum* | PC12 neuronal cells | $H_2O_2$-induced oxidative stress cell model + preconditioning with 250, 500 or 1000μg extract/mL | ↑ Cell viability<br>↓ ROS (dose-dependent)<br>↓ Caspase 3, 8, 9 activation | Gao et al. (2020) |
| Tannic acid | C6 rat glioma cells | 6.25–75μM tannic acid | ↓ ROS (>12.5μM)<br>↑ Sulfhydryl content (>12.5μM)<br>↑ CAT, SOD (>25μM) | Bona et al. (2020) |
| Resveratrol | Human lung cells | Pretreatment resveratrol 10μM for 24 h + Diesel exhaust particle-induced damage | ↓ ROS<br>↑ p-AMPK<br>↑ Nrf2 | Zhang et al. (2019) |
| Resveratrol | Human RPE cells | *N*-retinyl-*N*-retinylidene ethanolamine-induced macular degeneration cell model exposed to 25μM resveratrol | ↓ Cytotoxicity<br>↑ SOD activity<br>↓ Mitochondrial $\cdot O_2^-$ | Alaimo et al. (2020) |

(continued)

**Table 6.2** (continued)

| Polyphenol/extract | Cell type | Treatment | Mechanism involved | Reference |
|---|---|---|---|---|
| *Syzygium aromaticum* essential oil extract | Neutrophils | Peptide (formyl-met-leu-phe) or phorbol myristate acetate-induced $^{\cdot}O_2^-$ generation + Pretreatment with 2.5–20µg/mL extract for 30 min | ↓ $^{\cdot}O_2^-$ generation<br>↓ Phosphorylation of MEK1/2 and ERK1/2 | Chniguir et al. (2019) |
| Black chokeberry | Human neutrophils | Crude extract, purified extract standardized to 20 and 40% anthocyanins, and proanthocyanidins + pure compounds (chlorogenic acid, cyanidin-3-*O*-galactoside, epicatechin, rutin, and quercetin) | No effect on spontaneous chemiluminescence<br>Mild effect on PMA-activated chemiluminescence<br>Significant effect on OZP-activated chemiluminescence | Denev et al. (2019) |
| ß-glucogallin | Human retinal epithelial cells ARPE-19 | Methylglyoxal (MG)-induced oxidative stress model + pretreatment with GG (100µM) for 2 h prior to MG treatment | ↑ CAT, SOD, GPx<br>↑ GSH<br>↓ Lipid peroxidation and protein carbonyls | Ma et al. (2019) |
| Lotus seedpod extract | HepG2 cells | Oleic acid NAFLD model + 1–25µg extract/mL | ↓ Lipid accumulation<br>↓ Oxidative stress | Liu et al. (2019) |
| Blueberry | Mouse macrophage RAW 264 and rat hepatoma cells H4IIE | Wild blueberry pomace extract (1176.8 ± 102.9 mg total polyphenols/100 mL) | ↓ ROS and iNOS (in macrophages)<br>↓ Phosphoenolpyruvate carboxykinase (in hepatoma cells) | Hoskin et al. (2019) |

*SOD* superoxide dismutase, *CAT* catalase, *MPO* myeloperoxidase, *p-Akt* phospho-Akt (protein kinase B), *p-TrkB* phospho-tropomyosin receptor kinase B, *p-CREB* phospho-cAMP response element-binding protein, *BDNF* brain-derived neurotrophic factor, *MEK1/2* mitogen-activated protein kinase, *ERK1/2* extracellular signal-regulated kinase, *p-AMPK* phosphorylated AMP kinase, *Nrf2* NF-E2-related factor 2, *NAFLD* nonalcoholic fatty liver disease, *GAE* gallic acid equivalents, *PMA* phorbol 12-myristate 13-acetate, *OPZ* opsonized zymosan particle, *iNOS* inducible nitric oxide synthase

**Table 6.3** Preclinical results on the effects of polyphenols concerning redox homeostasis

| Polyphenol/ extract | Species | Treatment | Observed effects | Reference |
|---|---|---|---|---|
| Cacao | ApoE-deficient (-/-) mice aged 8 weeks ($n = 90$) | 16-week treatment:<br>– Normal mouse chow<br>– Chow with 0.25%<br>– Cacao polyphenols<br>– Chow with 0.40%<br>– Cacao polyphenols | ↓ 4-hydroxynonenal<br>↓ Hexanoyl-lysine<br>↓ Dityrosine | Natsume and Baba (2014) |
| Quercetin | 6-week-old ApoE-deficient mice ($n = 75$, 15/group) | 24-week treatment:<br>– Normal diet<br>– High-fat diet<br>– High fat + low dose of quercetin (25 mg/kg)<br>– High fat + middle dose of quercetin (50 mg/kg)<br>– High fat + high dose of quercetin (100 mg/kg) | ↓ NADPH oxidase expression<br>↓ OxLDL-induced ROS formation | Xiao et al. (2017) |
| Mulberry fruit extract | Swiss CD-1 mice scopolamine memory impairment model | 3-week pretreatment: 100 or 200 mg extract/kg before scopolamine administration | ↓ Lipid peroxidation in brain<br>↑ GSH and GPx activity | Shin et al. (2019) |
| Diosmin | Adult male Wistar rats ($n = 6$/group), benign prostatic hyperplasia model | 28-day treatment:<br>– Control group: normal saline + olive oil i.p. 10 mL/kg/days 19–28<br>– Disease group: testosterone propionate in olive oil 5 mg/kg/days 19–28<br>– Treatment groups: diosmin 20 or 40 mg/kg/days 1–28 + testosterone propionate in olive oil 5 mg/kg/days 19–28<br>– Diosmin group: 40 mg/kg/days 1–28 | ↓ MDA and XO<br>↑ CAT, GSH, GPx, GR | Vafa et al. (2019) |
| Mangostin | Adult male Sprague Dawley rats ($n = 5$/group) | 21-day treatment:<br>– Control<br>– Parkinson's disease model: rotenone 2 mg/kg/day for 21 days<br>– Treatment group: α-mangostin 10 mg/kg/day, i.p. | ↓ Brain MDA<br>↓ Brain nitrite levels | Parkhe et al. (2020) |

(continued)

**Table 6.3** (continued)

| Polyphenol/ extract | Species | Treatment | Observed effects | Reference |
|---|---|---|---|---|
| *Lessonia trabeculate* | C57BL/6J rats | 4 week-treatment:<br>– Diabetic polyphenol-treated group: 200 mg/kg/day (DPG)<br>– Diabetic control group (DCG): saline solution<br>– Normal control group (NCG): saline solution | ↑ CAT, SOD, GSH (DPG versus DCG) | Yuan et al. (2019) |

*ApoE* apolipoprotein E, *MDA* malondialdehyde, *OxLDL* oxidized low-density lipoproteins, *GSH* glutathione, *GPx* glutathione peroxidase, *XO* xanthine oxidase, *CAT* catalase, *GSH* glutathione, *GPx* glutathione peroxidase, *GR* glutathione reductase, *SOD* superoxide dismutase

### 6.4.1 Direct Antioxidant Action

Direct antioxidant effects are dependent, first and foremost, on the particular chemical structure of each polyphenolic compound (Hussain et al. 2016). The number of hydroxyl groups and their position greatly influence molecules' ability to scavenge free radicals or chelate metal ions (Heim et al. 2002; Mishra et al. 2013).

Flavonoids exerted inhibitory effects on ROS-generating enzymatic systems or intercept elements and can promote ROS generation (e.g., metal chelation) (Forbes-Hernandez et al. 2017a; Reis et al. 2016), being able to stop the propagation phase in the generation of free radicals (Reis et al. 2016).

Polyphenols in mulberry (Shin et al. 2019), olive oil extract (De la Ossa et al. 2019), and *Lycium ruthenicum* (Gao et al. 2020) decreased ROS formation, while tannic acid (Bona et al. 2020), catechins (Marinovic et al. 2015), and resveratrol (Xia et al. 2017; Zhang et al. 2019) reduced ROS and RNS formation, and eugenol (terpenic compound) particularly inhibited the generation of superoxide anion (Chniguir et al. 2019). For resveratrol, although its free radical scavenging properties were well-documented (Leonard et al. 2003; Jia et al. 2008; Holthoff et al. 2010; Ungvari et al. 2007; Breuss et al. 2019), its antioxidant properties were not remarkable, and, as such, its noteworthy protective effects against oxidative damage are, most likely, due to the regulation of key genes involved in redox homeostasis (Xia et al. 2017).

*Mimosa pudica* polyphenols proved direct free radical scavenging activity (Ijaz et al. 2019) and also black chokeberry components, especially quercetin and epicatechin (Denev et al. 2019). Saline extract from *Malpighia emarginata* DC leaves (Barros et al. 2019) and pectin-free goji berry extract yielded a significant antioxidant effect (Georgiev et al. 2019). Moreover, the direct action against oxidative species of olive oil was attributed to tyrosol, hydroxytyrosol, and oleuropein, which also reduced LDL oxidation (Khurana et al. 2013; Jemai et al. 2009). Flavonoids were shown to improve membrane function of peripheral blood mononuclear cells (Margina et al. 2012) and inhibit lipid peroxidation (Reis et al. 2016), while the incorporation of anthocyanins was reported in endothelial cells, with the alleviation of ROS and RNS-induced cytotoxicity and reduction of LDL redox impairment (Youdim et al. 2000; Serraino et al. 2003; Forbes-Hernandez et al. 2017b).

Further, polyphenols can interact with nonpolar components of the hydrophobic membrane layer or even permeate and reside in it, conferring protective effects against oxidative damage and interfere with cellular signaling (Oteiza et al. 2005).

### 6.4.2 Indirect Action Supporting Antioxidant Systems

The numerous protective effects reported for polyphenols were, at first, mainly attributed to their antioxidant capacity, followed by reports of their ability to modulate and enhance cellular antioxidant defense systems, increasing the expression of several antioxidant enzymes, such as SOD, CAT, GPx, and GST (Khurana et al. 2013; Bruno and Ghiadoni 2018).

Indeed, catechins and tannic acid stimulated SOD and CAT activities (Forbes-Hernandez et al. 2017a; Marinovic et al. 2015; Bona et al. 2020), while curcumin and B-glucogallin increased GSH while also enhancing CAT, SOD, and GPx (Ma et al. 2019; Aggarwal and Harikumar 2009; Ali et al. 2018). Carnosic acid and carnosol and terpenes from rosemary also increased the cellular level of GSH (Bagetta et al. 2020). Quercetin was reported to support the activity of HO-1 and GST (Khurana et al. 2013) and exhibited a high binding affinity to human serum albumin, resulting in increased protection from glycooxidative damage (Alam et al. 2015).

Interestingly, some polyphenolic compounds, such as myricetin or morin, were shown to exert moderate prooxidant activity and DNA damage through stable complexes with cations, such as $Cu^{2+}$, via Fenton reactions (Jomova et al. 2019). This minor increase in ROS formation stimulated antioxidant defense systems (Jomova et al. 2019).

Resveratrol improved SOD, CAT, and GPx activities, (Alaimo et al. 2020; Breuss et al. 2019; Xia et al. 2010), most likely in a Sirt1-mediated way (Bagetta et al. 2020; Alcendor et al. 2007), as Sirt is a NAD-dependent histone deacetylase with important epigenetic implications, representing one of its most important cellular targets (Xia et al. 2017; Breuss et al. 2019). Also, it determined the upregulation of several genes coding antioxidant proteins, such as thioredoxin-1, HO-1, and γ-glutamylcysteine synthetase (involved in GSH synthesis) (Khurana et al. 2013; Thirunavukkarasu et al. 2007; Yu et al. 2010). Further, it affected, among others, NFκB, cellular tumor antigen p53 (p53), and Forkhead box O (FOXO) transcription factor signaling, resulting in the modified expression of different redox-regulating enzymes (Xia et al. 2017).

Resveratrol also enhanced the pathways of nuclear factor-E2-related factor 2 (Nrf2) (Khurana et al. 2013; Xia et al. 2017; Liu et al. 2019; Breuss et al. 2019; Bagetta et al. 2020) and AMP kinase (AMPK) (Liu et al. 2019). Nrf2 is a redox-sensitive protein which translocates into the nucleus under oxidative stress conditions, activating the ARE, stimulating the expression of several antioxidant enzymes (Xia et al. 2017; Bagetta et al. 2020). Polyphenols in mulberry, besides increasing glutathione levels (Shin et al. 2019), also activated the Nrf2 and protein kinase B (PKB/Akt) pathways, stimulating the expression of antioxidant enzymes and brain-derived neurotrophic factor (BDNF) (Shin et al. 2019), while rosemary polyphenolic compounds reduced the nuclear translocation of NFκB components and ROS formation (Bagetta et al. 2020).

In vivo, cacao polyphenols reduced oxidative impairment—the formation of reactive species and final products such as 4-hydroxynonenal, hexanoyl-lysine and dityrosine (Natsume and Baba 2014), and oxidized LDL-induced ROS generation (Xiao et al. 2017) in apoE-deficient mice. Further, the effects of mulberry polyphenols regarding GSH and GPx were confirmed in a mice model of memory impairment (Shin et al. 2019), while in a prostate hyperplasia model, an increased activity of CAT and GPx, along with increased GSH, was observed (Vafa et al. 2019). In alloxan-induced diabetic rats, the administration of a product based on the fruit powders of bilberry, blackcurrant, sweet pepper, and rose hip enhanced the total serum antioxidant defense (Ungurianu et al. 2019b).

### 6.4.3 Indirect Action Inhibiting Oxidative Stress-Enhancing Systems

Polyphenols were able to interact with nitric oxide synthase (NOS) (Hussain et al. 2016). The inhibitory action of anthocyanins on inducible nitric oxide synthase (iNOS) was reported, with an overall effect of reducing the formation of RNS (Hoskin et al. 2019; Basu et al. 2015). Catechins also reduced iNOS expression and, further, myeloperoxidase (MPO) activity (Marinovic et al. 2015).

NADPH oxidase and XO are major sources of intracellular ROS (Khurana et al. 2013). XO was inhibited by quercetin and luteolin (Hussain et al. 2016), while resveratrol was reported to impede the activity of NADPH oxidase, reducing the generation of ROS (Breuss et al. 2019).

Some enzymes involved in the inflammatory pathway of arachidonic acid, such as COX and LOX, can contribute to the advancement of oxidative damage (Hussain et al. 2016). Green and black tea polyphenols were shown to hinder their activity (Hong et al. 2001).

More recently, a closer look was cast on the ability of polyphenols to alter several cellular pathways including those of Sirt1, mitogen-activated protein kinases (MAPK/ERK), NFκB, or activator protein 1 (AP-1), resulting in a modified expression of certain cytokines and adhesion molecules (Khurana et al. 2013). For example, polyphenols in *Lycium ruthenicum* proved an inhibitory capacity on the activation of caspases 3, 8, and 9 (Gao et al. 2020), while eugenol (terpenic compound) decreased the activation of the MAPK/ERK pathway (Chniguir et al. 2019). Curcumin reduced the activity of XO and MPO, affecting the NFκB, signal transducers and activator of transcription protein (STAT), and hypoxia-inducible factor (HIF) signaling cascades (Aggarwal and Harikumar 2009). Quercetin (Kumar et al. 2014) and olive oil polyphenols (Jemai et al. 2009) also inhibited the NFκB pathway, and, further, olive oil components interfered with Nrf2, Akt, and Sirt1 signaling (Bayram et al. 2012; Samuel et al. 2008).

In vivo, quercetin significantly decreased NADPH oxidase expression in apoE-deficient mice (Xiao et al. 2017) and in a Parkinson's disease rat model; reduced activity of iNOS was observed post-mangostin exposure (Parkhe et al. 2020), while in a prostate hyperplasia model, a decreased XO activity was reported (Vafa et al. 2019).

### 6.4.4 Reports Regarding Antioxidant Effects in Human Studies

The studies we found to be the most relevant concerning the modulation of redox homeostasis by polyphenols are summarized in Table 6.4.

Resveratrol, in a dose of 100 mg/day for 90 days, did not improve the antioxidant capacity in firefighters (Macedo et al. 2015), similar to a 17-day intervention with blueberry and green tea extracts in long-distance runners (Nieman et al. 2013). However, green tea extract administration for 6 weeks determined an increase of total serum antioxidant defense and a limited decrease of lipid oxidative impairment in men with an intense training regimen (Sadowska-Krepa et al. 2019).

Dietary enrichment with strawberries (500 g/day, 30 days) (Alvarez-Suarez et al. 2014), as well as a diet rich in a mix of polyphenols (Annuzzi et al. 2014), reduced urinary isoprostanes. Serum MDA was lowered, and the antioxidant defense was stimulated by increased intake of strawberries (500 g/day for 15 or 30 days) in healthy volunteers (Alvarez-Suarez et al. 2014; Romandinia et al. 2013), green tea extract in sprinters (Jowko et al. 2015), and cacao (Sarria et al. 2012) or *Vaccinium arctostaphylos* fruit extract (Soltani et al. 2014) in hyperlipidemic patients.

As for antioxidant enzymes, 5 weeks of grape powder dietary intervention led to a significant increase of GPx activity in nondiabetic hemodialysis patients (Janiques et al. 2014), and also 90 days of resveratrol (500 mg/day) in T2DM patients, along with SOD and CAT improvement for the latter (Sattarinezhad et al. 2019). Moreover, SOD activity was enhanced by green tea extract in young men following an intensive training routine (Sadowska-Krepa et al. 2019), while soluble mate tea was shown to improve the reduced oxidized glutathione ratio in serum (Panza et al. 2019).

## 6.5 Protective Effects Involving Lipid Metabolism

The ongoing pernicious cycle oxidative stress—low-grade inflammation reported in many of the modern metabolic maladies, such as diabetes mellitus, obesity, or metabolic syndrome—deeply affects the proteome and lipidome of serum lipoproteins, yielding dysfunctional and damaging molecules (Amiot et al. 2016; Ungurianu et al. 2017; Zanfirescu et al. 2019). There are numerous reports of the beneficial actions of polyphenols in preventing and alleviating dyslipidemia,

**Table 6.4** Clinical reports regarding the antioxidant effects of polyphenols

| Source of polyphenols | Design | Population | Intervention | Effects | Reference |
|---|---|---|---|---|---|
| Resveratrol | Placebo-controlled double-blinded study | 60 Brazilian male military firefighters | 90-day treatment: resveratrol 100 mg/day versus placebo | No effect on antioxidant defense systems | Macedo et al. (2015) |
| Water soluble polyphenols from blueberry and green tea extracts | Double-blinded, parallel group design | 35 long-distance runners ages 19–45 years old (18 males and 13 females) | 17-day treatment: 40 g/day extracts (2136 mg gallic acid equivalents) versus placebo | No significant difference between the groups regarding plasma protein carbonyl levels | Nieman et al. (2013) |
| Green tea extract | Randomized controlled trial | 16 young males involved in CrossFit training | 6-week treatment: 250 mg extract (245 mg polyphenols) or placebo | ↑ SOD activity and antioxidant capacity (resting and postexercise) | Sadowska-Krepa et al. (2019) |
| Strawberry | Clinical trial | 33 healthy volunteers (11 males and 12 females) | 30-day intervention: 500 g of strawberries/day ($307.59 \pm 0.01$ anthocyanins/day) | ↓ Serum MDA<br>↓ Urinary isoprostanes | Alvarez-Suarez et al. (2014) |
| Diet constituents (decaffeinated coffee and tea, dark chocolate, blueberry jam, vegetables) | $2 \times 2$ factorial design | 86 overweight/obese individuals (33 males and 45 females for the final assessments) | 8 weeks of different isoenergetic diet:<br>– Low in n-3 polyunsaturated fatty acids and polyphenols<br>– Rich in n-3 polyunsaturated fatty acids<br>– Rich in polyphenols<br>– Rich in n-3 polyunsaturated fatty acids and polyphenols | ↓ Urinary 8-isoprostane | Annuzzi et al. (2014) |

| | | | | | |
|---|---|---|---|---|---|
| Strawberry | Clinical trial | 23 healthy volunteers (11 males and 12 females) | 15-day intervention: 500 g of strawberries/day ($0.216 \pm 0.03$ mg/g total polyphenols and $0.49 \pm 0.14$ mg/g total anthocyanins) | ↑ Total antioxidant capacity of serum<br>↓ Plasma protein carbonyls | Romandinia et al. (2013) |
| Green tea | Double-blind, randomized, placebo-controlled trial | 16 sprinters | 4-week treatment: green tea extract (980 mg of polyphenols/day) versus placebo | ↑ Total antioxidant capacity of serum<br>↓ MDA and SOD postexercise | Jowko et al. (2015) |
| Cacao | Non-controlled, non-randomized, 2-month-long, open-intervention trial | 21 volunteers with hypercholesterolemia (8 males and 13 women) | 2-month intervention: 15 g ×2/day cocoa product (12 g of dietary fiber and 283 mg of soluble polyphenols) | ↓ MDA | Sarria et al. (2012) |
| *Vaccinium arctostaphylos* | Randomized, double-blind, placebo-controlled clinical trial | 50 hyperlipidemic patients (20 males and 30 females) | 4 week intervention: *Vaccinium arctostaphylos* fruit extract ($45 \pm 2$ mg anthocyanins) ×2/day versus | ↓ MDA | Soltani et al. (2014) |
| Grape | Double-blind, placebo-controlled randomized | 32 nondiabetic hemodialysis patients (18 males and 14 females) | 5-week treatment: grape powder (500 mg of polyphenols/day) versus placebo | ↑ GPx | Janiques et al. (2014) |
| Resveratrol | Randomized, double-blind, placebo- | 60 T2DM patients and albuminuria | 90-day treatment: resveratrol (500 mg/day) | ↑ SOD CAT GPx<br>↓ Urinary albumin | Sattarinezhad et al. (2019) |

(continued)

**Table 6.4** (continued)

| Source of polyphenols | Design | Population | Intervention | Effects | Reference |
|---|---|---|---|---|---|
| | controlled clinical trial | | or placebo + losartan (12.5 mg/day) | | |
| Soluble mate tea | Two-phase study | 9 men | 16-day intervention, 8 days/phase: firstly, water (control) and then SMT | ↑ Blood GSH/GSSG | Panza et al. (2019) |
| Extra virgin olive oil | Clinical trial | 10 young and 10 elderly healthy subjects | 12-week intervention: 25 mL raw extra virgin olive oil/day | ↑ PON activity in the young volunteers | Loued et al. (2013) |

*MDA* malonialdehyde, *SOD* superoxide dismutase, *GPx* glutathione peroxidase, *CAT* catalase, *PON* paraoxonase, *T2DM* type 2 diabetes improving serum lipid profile and reducing overall cardiovascular risk (Santhakumar et al. 2018; Garcia et al. 2018; Garcia-Conesa 2017).

Although the main mechanism of action was considered to be their well-known free radical scavenging capacity, several pleiotropic effects have been reported due to their modulation of certain signaling pathways, resulting in reduced LDL/total cholesterol (TC), chelated metal ions, and reduced LDL oxidation, adhesion molecules, and cytokine synthesis/secretion and targeting eNOS, enhancing NO signaling (Santhakumar et al. 2018; Khurana et al. 2013; Bruno and Ghiadoni 2018; Garcia et al. 2018; Garcia-Conesa 2017).

In dietary intervention studies, supplementation with various polyphenol-based extracts determined a reduction of TC (Tokede et al. 2011; Noad et al. 2016), LDL (Basu et al. 2015; Tokede et al. 2011; Anderson et al. 2016; Filip et al. 2015; Hosseini et al. 2016; Ogier et al. 2013; Yubero et al. 2013; Zunino et al. 2014), and, moreover, OxLDL levels (Basu et al. 2015; Khan et al. 2012; Ibero-Baraibar et al. 2014), with an increase of HDL (Basu et al. 2015; Hosseini et al. 2016; Khan et al. 2012; Shema-Didi et al. 2014; Hernaez et al. 2014; Hassellund et al. 2013; Chiva-Blanch et al. 2015; Sarria et al. 2014). However, some studies found no significant changes in lipid profile (Macedo et al. 2015; Broekhuizen et al. 2011; Mathew et al. 2012; Song et al. 2015), while others reported a significant decrease in triglycerides' levels as well (Annuzzi et al. 2014; Ogier et al. 2013; Shema-Didi et al. 2014; Pfeuffer et al. 2013; Kardum et al. 2015; Kusunoki et al. 2015). Although the reported differences in lipid profiles are usually small or very small, they could yield significant long-term protective actions (Garcia-Conesa 2017).

### 6.5.1 Preclinical Reports

Strawberry extract determined a lower expression of proteins involved in fatty acid and cholesterol synthesis, reducing $\beta$-hydroxy-$\beta$-methylglutaryl-CoA (HMG-CoA) activity and stimulating AMPK activation, while overexpressing LDL receptor (Forbes-Hernandez et al. 2017b), resulting in overall improved lipid metabolism and redox status (Forbes-Hernandez et al. 2017a). Mulberry and rosemary extracts also increased AMPK activity (Chang et al. 2013; Tu et al. 2013), while epigallocatechin-3-gallate (EGCG) yielded inhibitory effects on HMG-CoA reductase (Cuccioloni et al. 2011). The mulberry extract decreased the expression of SREBP1 (sterol regulatory element-binding protein 1), FAS ligand, GAPT (GRB2-binding adaptor protein, transmembrane), and HMG-CoA reductase and increased PPAR$\alpha$ (Tu et al. 2013).

Curcumin reduced, via the NF$\kappa$B pathway, the synthesis of several pro-inflammatory molecules (Kim et al. 2007), while caffeic acid was able to regulate energy and redox homeostasis, reducing NADPH levels and increasing AMPK activity (Tyszka-Czochara et al. 2018).

In preclinical studies, polyphenols from chokeberry, strawberry, blueberries, and oregano exerted antihyperlipidemic effects in murine high-fat diet models (Prior et al. 2009; Valcheva-Kuzmanova et al. 2007; Cho et al. 2012). Curcumin reduced lipid peroxidation and improved overall lipid profile and atherogenic indices, also improving leptin levels and PON activity (Khurana et al. 2013).

Cacao polyphenols reduced markers of lipid peroxidation in apoE-deficient mice (Natsume and Baba 2014). In normal Wistar rats, insulin resistance and lipid profiles were ameliorated (Castro et al. 2020), while in diabetic Zucker rats, insulin resistance, body weight gain, and insulin-induced p38 activation were decreased, along with an increase of glucokinase and glucose transporter 2 expression (Cordero-Herrera et al. 2015).

Grape seed extract reduced the expression of SREBP1, microsomal triglyceride transfer protein, and diacylglycerol O-acyltransferase 2 (DGAT2) and improved TG and LDL levels (Quesada et al. 2009), while citrus polyphenols ameliorated insulin resistance, lipid profiles, weight gain, and visceral adiposity in rats fed an obesogenic cafeteria diet (Mayneris-Perxachsab et al. 2019).

### 6.5.2 Clinical Evidence

Numerous systematic reviews and meta-analyses have confirmed some beneficial effects of resveratrol regarding lipid profile, but showed that it failed to significantly reduce LDL and TG and increase HDL (Sahebkar 2013), although it did improve TC (Huang et al. 2016; Haghighatdoost and Hariri 2018), a dose higher than 300 mg/day being most efficient (Huang et al. 2016).

Curcumin is another "high-profile" polyphenol when it comes to managing lipid profiles and long-term cardiovascular protection. Although its protective effects (reducing TC or LDL and augmenting HDL) are well-known for some time now (Soni and Kuttan 1992; Alwi et al. 2008), its effects are still a matter of interest (Adibian et al. 2019).

Besides supplementation with certain isolated polyphenolic compounds, a long-term high-dietary intake of polyphenol-rich food products can yield beneficial effects on lipid profiles and cardiovascular risk. For example, several reports attested the protective effects of regular green tea intake regarding cardiometabolic risk factors (Nagao et al. 2005; Kuriyama et al. 2006; Yang et al. 2004), including a positive effect concerning lipid profiles, decreasing TC and LDL, with an increase in HDL (Imai and Nakachi 1995). Similarly, the polyphenolic compounds found in *Annurca* apples were proven to have a beneficial effect on lipid metabolism (Bagetta et al. 2020).

In healthy individuals, dietary supplementation with strawberries (Alvarez-Suarez et al. 2014) or polyphenol-rich grape extract decreased TC and LDL (Yubero et al. 2013), and cherry juice decreased markers of lipid oxidative damage (Chai et al. 2019).

Although several studies reported blood pressure-lowering effects for various dietary interventions based on increasing the intake of polyphenols, an improvement of lipid profiles was also noticed. As such, in subjects with high-normal blood pressure, chokeberry juice reduced serum cholesterol, triglycerides, and LDL (Kardum et al. 2015), and olive oil supplementation decreased LDL oxidation

(Moreno-Luna et al. 2012), while grape seed extract had no significant effect on serum lipid, although it improved insulin sensitivity (Park et al. 2016). However, chokeberry juice had no significant effect on lipid profiles of patients with untreated mild hypertension (Loo et al. 2016).

In overweight or obese subjects, the short-term use of a polyphenol-rich beverage had little effect on metabolic risk factors (Mullan et al. 2016), although a one-time, meal-associated strawberry drink did improve insulin response (Edirisinghe et al. 2011). Long-term increased dietary polyphenol intake decreased both fasting and postprandial triglycerides (Annuzzi et al. 2014) and increased HDL and GSH to GSSG ratio (Chew et al. 2019).

The cinnamon extract improved glucose and lipid profiles in individuals with impaired fasting glucose (Anderson et al. 2016), while olive leaf extract exerted a positive effect on glucose homeostasis (Wainstein et al. 2012). Black soybean (Kusunoki et al. 2015) and bergamot (Mollace et al. 2019) polyphenols exerted a beneficial effect on serum lipid profile parameters in diabetics, while curcumin also increased adiponectin (Adibian et al. 2019), and a one-time post-meal cacao drink increased postprandial HDL (Basu et al. 2015).

Hyperlipidemic subjects exhibited ameliorated lipid profiles after dietary supplementation with cacao (Sarria et al. 2012) or *Vaccinium arctostaphylos* polyphenol products (Soltani et al. 2014). Further, higher dietary intake of cacao powder increased HDL levels and decreased LDL oxidation (Khan et al. 2012) in patients with high cardiovascular risk, while red wine intake led to augmented levels of HDL, apoA1, and apoA2, with lower lipoprotein(a) (Chiva-Blanch et al. 2015).

Moreover, a 1-year dietary intervention with pomegranate juice led to a significant relative increase of HDL to TG in hemodialysis patients (Shema-Didi et al. 2014). Another long-term study, testing the effect of olive oil extract in postmenopausal women, reported a significant decrease of TC and LDL, along with a protective effect against bone mineral density loss (Filip et al. 2015).

The designs and main findings of the above-cited studies encompassing subjects with different metabolic settings are summarized in Table 6.5.

## 6.6 Polyphenols and Platelet Function: Experimental and Preclinical Results

The effects of polyphenols on platelet function have been reviewed extensively, but most of the experimental and epidemiological studies are older than 10 years (Faggio et al. 2017).

The effect of resveratrol was evaluated on human platelet suspensions. Results show that both low (0.05–0.25μM) and high concentrations of resveratrol (0.8–5.0μM) inhibit platelet aggregation and TXB2 formation in platelets stimulated by collagen or arachidonic acid; other mechanisms involved in the antithrombotic resveratrol effect, effective only for low levels of exposure, are the inhibition of p38 MAP kinase and PKC activity and also the induction of NO release from platelets (Pace-Asciak et al. 1995; Olas and Wachowicz 2005; Shen et al. 2007). By

**Table 6.5** Clinical reports regarding the effects of polyphenols on lipid profiles

| Polyphenol/extract | Design | Population | Intervention | Effects | Reference |
|---|---|---|---|---|---|
| Strawberry | Clinical trial | 33 healthy volunteers | 30-day intervention: 500 g of strawberries/day (307.59 ± 0.01 anthocyanins/day) | ↓ TC, LDL, and TG | Chiva-Blanch et al. (2015) |
| Cherry | Randomized-controlled clinical trial | 37 males and females | 12-week consumption of 480 mL of tart cherry juice/day or control drink | ↑ 8-oxoguanine glycosylase<br>↓ MDA<br>↓ OxLDL | Chai et al. (2019) |
| Grape extract | Randomized, double-blind, placebo-controlled clinical trial | 60 volunteers | 56-day intervention: 700 mg of polyphenol-rich grape extract supplement (2.61 mg/g total flavonols and 2.42 mg/g ellagic acid) versus placebo | ↓ TC and LDL<br>↑ Serum antioxidant capacity<br>↑ Vitamin E | Yubero et al. (2013) |
| Pomegranate | Randomized, controlled crossover trial | 19 young, healthy male subjects | High-fat meal + pomegranate drink (652–948 mg polyphenols) 15 min before the meal or during the meal versus placebo | No differences in TG postprandial levels | Mathew et al. (2012) |
| Chokeberry | Clinical trial | 23 subjects with high normal blood pressure | 4-week intervention: 200 mL of chokeberry juice | ↓ TC, LDL, and TG | Kardum et al. (2015) |
| Olive oil | Double-blind, randomized, crossover dietary intervention study | 24 young women with high-normal blood pressure or stage 1 essential hypertension | 2-month intervention + 4-week washout period: 30 mg/day polyphenol-rich olive oil versus polyphenol-free olive oil | ↓ OxLDL | Moreno-Luna et al. (2012) |
| Grape seed extract | Single-center, randomized, two-arm, double- | 36 middle-aged subjects with prehypertension | 2 weeks of placebo run-in + 6 weeks of treatment: placebo beverage, 333μg/mL total polyphenols (355 mL ×2/ | ↑ Insulin sensitivity<br>No significant changes for fasting glycemia and lipids or OxLDL | Park et al. (2016) |

| | | | | | |
|---|---|---|---|---|---|
| | blinded, placebo-controlled | | day) versus tested beverage with 150 mg of grape seed extract added, resulting in 744μg/mL total polyphenols (355 mL ×2/day) | | |
| Chokeberry | Single-blinded crossover trial (no washout period) | 38 patients with untreated mild hypertension | 16-week intervention, 8-week/phase: 300 mL chokeberry juice and 3 g of chokeberry powder versus placebo | No changes in serum glucose or lipid profiles | Loo et al. (2016) |
| Green tea, grape seed, grape pomace, ruby red grape juice, lemon, and apple extracts | Randomized, double-blind, placebo-controlled | 39 overweight or obese subjects | 4-week intervention: 250 mL of test beverage (361 mg polyphenols and 120 mg ascorbic acid) ×2/day versus placebo | No significant differences in apolipoproteins, insulin, leptin, or adiponectin | Mullan et al. (2016) |
| Strawberry | Crossover study | 24 overweight adults | High-carbohydrate, moderate-fat meal with strawberry beverage (94.66 ± 2.17 total polyphenols/305 mL) versus placebo beverage (2.37 total polyphenols/305 mL) | ↓ Postprandial insulin response | Edirisinghe et al. (2011) |
| Cacao | Randomized, parallel, and double-blind study | 50 overweight and obese subjects | 4 weeks of 15% restriction diet with 1.4 g of cocoa extract (645.3 mg of polyphenols) or placebo | ↓TC, TG, LDL<br>↓ OxLDL<br>↓ MPO | Ibero-Baraibar et al. (2014) |

(continued)

**Table 6.5** (continued)

| Polyphenol/extract | Design | Population | Intervention | Effects | Reference |
|---|---|---|---|---|---|
| Diet constituents (salmon, dentex, anchovies) | 2 × 2 factorial design | 86 overweight/obese individuals (78 for the final assessments) | 8 weeks of isoenergetic diet:<br>– Low in n-3 polyunsaturated fatty acids and polyphenols<br>– Rich in n-3 polyunsaturated fatty acids<br>– Rich in polyphenols<br>– Rich in n-3 polyunsaturated fatty acids and polyphenols | ↓ Fasting TG and VLDL<br>↓ Postprandial TG | Annuzzi et al. (2014) |
| Cranberry | Randomized, double-blind, placebo-controlled, parallel design trial | 78 overweight/obese men and women | 8-week intervention: 450 mL low-calorie, high-polyphenol cranberry extract beverage daily versus placebo | ↑ HDL<br>↑ GSH/GSSG | Chew et al. (2019) |
| Cinnamon | Placebo, double-blind | 137 individuals with impaired fasting glucose | 2-month intervention: 250 mg ×2/day of dried water extract of cinnamon versus placebo | ↓ GLY and HOMA-IR<br>↓ TC and LDL | Anderson et al. (2016) |
| Black soybean | Randomized clinical trial | 36 T2DM patients | 2-month intervention:<br>– BS (n=18): 2.5 g black soybean extract (465 mg of proanthocyanidin and 62 mg of anthocyanin)<br>– F: fenofibrate 80–160 mg/day ($n$ = 11)<br>– BS + FEN ($n$ = 7) | BS: no effect on plasma lipids<br>F: ↓TG (versus baseline)<br>BS + F: ↓TG (versus baseline and F)<br>BS + F: ↓ LDL (versus baseline, effect not observed for F) | Kusunoki et al. (2015) |

| | | | | | |
|---|---|---|---|---|---|
| Cacao | Crossover trial | 18 T2DM subjects | High-fat fast-food-style breakfast with: cocoa beverage (960 mg total polyphenols; 480 mg flavanols) versus flavanol-free placebo (110 mg total polyphenols; <0.1 mg flavanols) | ↑ Postprandial HDL<br>No effect on LDL or TG | Basu et al. (2015) |
| Olive leaf extract | Randomized, placebo-controlled clinical trial | 79 T2DM adults | 14-week intervention:<br>500 mg extract ×2/day versus placebo | ↓ HbA1C<br>↓ Fasting insulin, but not postprandial | Wainstein et al. (2012) |
| Curcumin | Double-blind randomized clinical trial | 44 T2DM patients | 10-week treatment: 1500 mg curcumin or placebo | ↓ TG (versus baseline)<br>↑ Adiponectin (versus control) | Adibian et al. (2019) |
| Bergamot phenolic fraction | Randomized, double blind, placebo-controlled study | 60 T2DM patients | Placebo versus bergamot polyphenolic fraction or its phytosomal formulation | ↓ LDL, TG<br>↑ HDL | Mollace et al. (2019) |
| Cacao | Non-controlled, non-randomized, 2-month-long, open-intervention trial | 21 hypercholesterolemic subjects | 2-month intervention:<br>15 g ×2/day cocoa product (12 g of dietary fiber and 283 mg of soluble polyphenols) | ↓ GLY<br>Slight increase in HDL | Sarria et al. (2012) |
| Cacao | Randomized, crossover feeding trial | 42 high-cardiovascular risk subjects | 4-week intervention: 40 g of cocoa powder with 500 mL of skimmed milk/day versus control (500 mL/day of skimmed milk) | ↑ HDL<br>↓ OxLDL | Khan et al. (2012) |

(continued)

**Table 6.5** (continued)

| Polyphenol/extract | Design | Population | Intervention | Effects | Reference |
|---|---|---|---|---|---|
| Red wine and gin | Crossover study | 67 men with cardiovascular risk | 12-week intervention, 4-week periods: red wine (30 g alcohol/day) or dealcoholized red wine (equivalent amount) versus gin (30 g alcohol/day) | Red wine and gin:<br>↑ HDL, Apo AI, Apo AII<br>Red wine and dealcoholized red wine:<br>↓ Insulin, HOMA-IR<br>Red wine:<br>↓ Lipoprotein(a) | Chiva-Blanch et al. (2015) |
| *Vaccinium arctostaphylos* | Randomized, double-blind, placebo-controlled clinical trial | 50 hyperlipidemic patients | 4-week intervention: fruit extract (45 ± 2 mg anthocyanins) ×2/day versus placebo | ↓ TC, LDL, TG | Soltani et al. (2014) |
| Pomegranate | Double-blind, placebo-controlled, randomized, clinical trial | 101 hemodialysis patients | 1-year intervention, 3×/week, during the first hour of dialysis: 100 mL pomegranate juice (0.7 mM of polyphenols) versus placebo | ↓TG<br>↑ HDL | Shema-Didi et al. (2014) |
| Olive oil | Double-blind study | 64 postmenopausal women with osteopenia | 12-month treatment: 250 mg/day of olive extract + 1000 mg calcium versus placebo (1000 mg calcium) | ↓ TC and LDL<br>↓ Bone mineral density in placebo, but not in treatment group | Filip et al. (2015) |

*TC* total cholesterol, *LDL* low-density lipoproteins, *TG* triglycerides, *MDA* malondialdehyde, *OxLDL* oxidized LDL, *MPO* myeloperoxidase, *VLDL* very-low-density lipoproteins, *HDL* high-density lipoproteins, *GSH* glutathione, *GSSG* oxidized glutathione, *GLY* glycemia, *HOMA-IR* homeostatic model assessment of insulin resistance index, *HbA1C* glycated hemoglobin, *Apo AI* apolipoprotein AI, *Apo AII* apolipoprotein AII derivatization of resveratrol with increasing the lipophilicity, the biological effect on platelet aggregation and NO synthesis was improved (Messina et al. 2015)

In vitro studies demonstrated that cocoa flavanols improve platelet function through reduction of TXA2 synthesis, improvement of NO release, reduction of oxidative stress, inhibition of COX-1, and reduction of ADP-induced expression of the Gp IIb-IIIa receptors (Ludovici et al. 2018; Kathuria et al. 2012; Macakova et al. 2012).

Water dispersible curcuminoids (10–30μg/mL) also act on platelets, inhibiting the aggregation under the effects of collagen, ADP, or arachidonic acid, reducing TXA2, NO, and serotonin level as well as 12-LOX output in rat cells (Maheswaraiah et al. 2015).

The effects of traditional Chinese remedies on vascular function are highly discussed in the literature, but not a lot of studies are available. An experimental study investigated the effect of loureirin A (a natural compound with one OH group) from dragon's blood on platelet function; loureirin A (50–100μM) decreased the collagen, ADP, and thrombin-induced aggregation and P-selectin expression, effects being dose-dependent and mediated through changes in PI3K/Akt signaling (Hao et al. 2015).

Scutellarin is the flavone and a major ingredient of *Erigeron breviscapus*, extensively used by Chinese medicine for coronary heart disease, hypertension, diabetes mellitus, etc. The natural component inhibited ADP-induced platelet aggregation acting as PKC inhibitor, in a dose-dependent manner (Tian et al. 2016).

Studies compared the anticoagulant effects exerted by quercetin-3-O-$\beta$-D-glucoside (isoquercetin) as well as its metabolite quercetin, both in vitro and in vivo in a rat model (thrombin-induced acute thromboembolism), and pointed out comparable effects on platelet function (Choi et al. 2016).

Kaempferol also delayed aggregation both in rats and mice models, through PI3K/Akt pathway, by attenuated phosphorylation of extracellular signal-regulated kinase (Choi et al. 2015).

### 6.6.1 Antiplatelet Effects in Clinical Settings

A clinical study compared the effects induced on smokers and healthy subjects by dark chocolate (patients did not receive any other treatment); results showed that, probably due to the antioxidant effect of the polyphenols in chocolate, there was a reduced platelet ROS and NADPH oxidase 2 (NOX2) activation as well as a decrease of platelet activation via inhibition of platelet 8-iso-PGF2α generation. In vitro incubation of platelets with 0.1–10μM catechin determined the reduction of platelet 8-iso-PGF2α and ROS formation and NOX2 activation but only for cells isolated from smokers and not from healthy controls (Carnevale et al. 2012).

In a study performed on healthy volunteers, who consumed chicory coffee every day for 1 week, results showed that whole blood and plasma viscosity were both significantly decreased, along with serum MIF levels and significant improvements in red blood cell deformability. Authors state that these effects are unlikely to be attributable to a single compound present in chicory coffee; nevertheless, the phenolics, including caffeic acid, are expected to play a substantial role (Schumacher et al. 2011).

In a randomized, double-blind clinical study, athletes received a polyphenol-rich beverage; blood samples were obtained before and after they completed a marathon run. Results show that runners with low intake of polyphenols showed a 2.2-fold increased platelet aggregation directly after completing a marathon and within the following 3 days when compared with baseline values; for the group receiving polyphenols, there was no increase of platelet aggregation (Nickel et al. 2016)

The antithrombotic effect of an anthocyanin-rich Queen Garnet plum juice and anthocyanin-free prune juice was assessed on healthy subjects; results show that Queen Garnet plum juice inhibited platelet aggregation, prolonged the activated-partial thromboplastin clotting time, and reduced the redox stress markers (MDA), the effects being attributed to the polyphenolic components (Santhakumar et al. 2015a). The effect of the same extract was also assessed on the platelet function of sedentary subjects; the extract reduced ADP-induced platelet aggregation and P-selectin expression both with and without oxidative stress induced by aerobic exercise. Concerning the arachidonic acid-induced platelet aggregation, this parameter was reduced only when associated with oxidative stress (Santhakumar et al. 2015b).

The effects of flavanol-rich chocolate were evaluated on endothelial and platelet function in short-term and long-term exposure for congestive heart failure patients. Flow-mediated vasodilatation significantly improved at both evaluated times; platelet aggregation significantly decreased only in the short term, but not after 4 weeks exposure (Flammer et al. 2012). So, the results obtained in vitro showing the ability of flavanol from cocoa/chocolate to modulate aggregation were confirmed in clinical settings; the same type of effects was obtained for grape juice, coffee, and onion supplementation (Table 6.6) (Ludovici et al. 2018).

## 6.7 Polyphenols and Inflammation

There are several studies performed in the last years that withstand the anti-inflammatory effects with positive CVD outcomes of polyphenols.

### 6.7.1 Experimental Studies Supporting the Effect of Polyphenols on Inflammatory Pathways

For example, endothelial cells were exposed to the effect of plasma from subjects receiving 400 mg resveratrol/day for a month, demonstrating that the treatment induced a reduction in the mRNA of inflammatory and adhesion molecules VCAM-1, ICAM-1, and IL-8, thus being a potential candidate for the prevention of atherosclerosis development (Agarwal et al. 2013).

**Table 6.6** Clinical data for platelet effects of polyphenols

| Design (number of patients, age, sex) | Doses/duration | Outcome | Reference |
|---|---|---|---|
| Crossover, randomized single-blind; 20 healthy subjects and 20 smokers | 40 g of dark (cocoa > 85%) or milk chocolate (cocoa < 35%) | – ↓ROS at platelet level in smokers and were not affected in healthy subjects | Carnevale et al. (2012) |
| Double-blind, placebo-controlled crossover study; 60 untreated, mildly hypertensive subjects | Grape polyphenols (polyphenol-rich grape wine extract containing 800 mg of polyphenols), 4 weeks | – ↓Blood pressure during day time<br>– ↓Endothelin-1 level | Draijer et al. (2015) |
| 27 healthy volunteers | 300 mL chicory coffee/day (main component: caffeic acid), 1 week | – ↑Red cell deformability<br>– ↓Plasma viscosity<br>– ↓MIF levels | Schumacher et al. (2011) |
| Randomized, double-blind clinical study; 103 athletes (divided into 2 groups—study and control) | Polyphenol-rich beverage, 3 weeks before a marathon run | – ↓Platelet aggregation induced by the physical effort (marathon run) | Nickel et al. (2016) |
| Double-blind, crossover trial including 21 healthy subjects | Queen Garnet plum juice (QGPJ) and anthocyanin-free prune juice (PJ), 200 mL/day 28 days | – ↓Platelet aggregation induced by ADP, collagen, and arachidonic acid<br>– ↑ Clotting time<br>– ↓ Redox stress markers (MDA) | Santhakumar et al. (2015a) |
| Double-blind, placebo-controlled; including 13 sedentary subjects, before and after aerobic exercise | Queen Garnet plum juice (QGPJ) 200 mL/day 28 days | – ↓ADP-induced platelet aggregation and P-selectin expression both with and without induced oxidative stress induced by the aerobic exercise | Santhakumar et al. (2015a) |
| Double-blind, randomized, placebo-controlled trial clinical study including 20 congestive heart failure patients | Flavanol-rich chocolate, short term (2 h after ingestion of a chocolate bar) and long term (4 weeks, two chocolate bars/day) | – ↑Flow-mediated vasodilatation at both evaluated times<br>– ↓Platelet aggregation only at short term | Flammer et al. (2012) |

In a study exposing human endothelial cells to the effect of Primitivo and Negroamaro polyphenolic extracts or pure polyphenols—such as hydroxycinnamic acids (*p*-coumaric, caffeic, and caftaric acids), flavonols (kaempferol, quercetin, myricetin), or stilbenes (resveratrol)—before stimulation with lipopolysaccharide, the inhibition of adhesion molecules was demonstrated. The anti-inflammatory effect was sustained by a reduction of ICAM-1, VCAM-1 and E-selectin, monocyte chemoattractant protein-1 (MCP-1), and macrophage colony-stimulating factor

(M-CSF), as well as ROS intracellular levels and the activation of NFκB and AP-1; resveratrol significantly reduced also the endothelial expression and release of M-CSF. All tested compounds reduced the adhesion of monocytes to stimulated endothelial cells, this being a key element in the development of atherosclerosis (Calabriso et al. 2016).

The treatment of CCD-18Co myofibroblasts cells with red wine extract decreased mRNA expression of lipopolysaccharide (LPS)-induced inflammatory mediators NFkB, ICAM-1, VCAM-1, and PECAM-1, the anti-inflammatory mechanism being dependent on miR-126 (Angel-Morales et al. 2012).

In a study exposing platelets to the effect of alcohol-free polyphenolic grape extract, alcohol and the polyphenols catechin, epicatechin, resveratrol, and gallic acid, results show that only the total extract induced inhibition of platelet aggregation, through the activation of PECAM-1. This effect was not obtained for any of the isolated tested compounds (de Lange et al. 2007).

Another study tested the effects induced by a mixture containing the four main catechins found in green tea, as well as each one alone, on neutrophils obtained from peripheral blood of healthy individuals; the catechin compounds induced anti-inflammatory effect (reduction of TNFα, IL-1β and IL-6, MPO activity, and HOCl production), together with the induction of antioxidant enzyme activities and Nrf2 mRNA levels, phagocytic capacity, and calcium release, thus proving anti-inflammatory and immunomodulatory actions (Marinovic et al. 2015).

In a study evaluating the effect of platelets' (activated with arachidonic acid) exposure to epicatechin and catechin on platelet-HUVEC (human umbilical vascular endothelial cells) interaction, the tested compounds induced a reduction of sICAM-1, sVCAM-1, and sE-selectin levels as well as an increase of NO bioavailability, proving an anti-inflammatory effect and the ability to counteract endothelial dysfunction only when platelets were harvested from peripheral artery disease patients and not from healthy subjects (Carnevale et al. 2014).

On HUVEC cells stimulated with LPS, exposure to acai (*Euterpe oleracea*) and red muscadine grape phenolics induced an anti-inflammatory effect (reduction of NFκB, IL-6 and IL-8; PECAM-1 and ICAM-1 protein) and a decreased trend in miR-126 expression as a mechanism of regulation for VCAM-1 synthesis (Noratto et al. 2011). The same type of effects was registered also for caffeic acid, which inhibited the LPS activation of NFκB, thus restoring the redox balance as well as the inflammatory one (Kim et al. 2014).

An in vitro study evaluated the effect induced by different flavonoids on the arachidonic acid release from rat neutrophils, kaempferol, luteolin, quercetin, and especially, amentoflavone may reduce the inflammatory response and inhibit the activity of *β*-glucuronidase and lysozyme (Tordera et al. 1994).

The beneficial effects of leafy cabbage (*Brassica oleracea*) extracts (water and methanol, respectively) for CVD risk were examined; for this purpose, HUVECs were preincubated with extracts for 24 h and thereafter stimulated TNFα. Exposure to the extracts from *Brassica oleracea* significantly reduced the TNFα-induced expression of E-selectin, ICAM-1, and VCAM-1 (Kuntz and Kunz 2014).

Another study demonstrated the capacity of extracts from Mango (*Mangifera indica* L.) to counteract TNFα effects on HUVECs, through the inhibition of IL-6, IL-8, COX-2, and ICAM-1 while restoring the expression of eNOS usually downregulated by TNFα (Mura et al. 2015).

Neutrophils exposed to oleacein (one of the abundant compounds from *Olea europaea*) show a reduction of endopeptidase activity, elastase, and MMP-9 and IL-8 release as well as inhibition of CD11b/CD18 expression and increased CD62L expression. This is one explanation for the protective effects of olive oil against endothelial dysfunction (Czerwinska et al. 2014).

An in vitro study on human aortic endothelial cells (HAECs) demonstrated that *Aronia melanocarpa* fruits, due to their polyphenol content, exhibit potent antioxidant properties (decreased intracellular ROS production) and anti-inflammatory effect (inhibited the expression of ICAM-1 and VCAM-1, attenuated the phosphorylation of NFκB p65), as mechanisms involved in their cardioprotective activity (Zapolska-Downar et al. 2012).

Ellagic acid was tested for its ability to inhibit MIF-induced chemotactic migration of PBMCs; results show that ellagic acid at the higher level of exposure failed to inhibit this response when PBMCs were stimulated with MIP-1α but showed high specificity for MIF-induced effect (Sarkar et al. 2015).

The in vitro effects of EGCG on inflammatory response induced by TNFα in adult human retinal pigment epithelial cell line, ARPE-19, were proven; thus, EGCG-reduced ICAM-1, NFκB and IκB expressions, and also reduced ROS level were mediated by TNFα. The results suggest the possible therapeutic involvement of EGCG in blocking TNFα-mediated eye inflammation (Thichanpiang and Wongprasert 2015).

Results were confirmed on vascular endothelial cells derived from human aortas; exposure to low levels of EGCG increased the heme oxygenase-1 (HO-1) expression and blunted the VCAM-1 release stimulated by TNFα (Pullikotil et al. 2012).

Atherosclerotic protective effects were pointed out for an ethanolic propolis extract as well as for its component – pinocembrin – in a cell model of LPS-activated murine macrophages; results show a dose-dependent decrease of MMP-9 activity, mediated by NFκB (as it is the case for inflammatory mediators TNF*α*, IL-1, IL-6, and MCP-1), thus suggesting the involvement of the tested extract in plaque stability through the control of extracellular matrix degradation (Saavedra et al. 2016).

Chlorogenic acid was tested on murine macrophages and microglial cells stimulated with LPS and exerted a dose-depended anti-inflammatory effect showed by the reduction of IL-1β, TNFα, and IL-6 levels as well as the inhibition of NO synthesis and expression of COX-1 and iNOS (Hwang et al. 2014).

Curcumin is also responsible for beneficial effects on inflammatory pathways involved in CVD, acting through the modulation of NFκB, JNK, p38, and STAT-3 in endothelial cells/cardiomyocytes and inducing the reduction of ICAM-1, VCAM-1, and P-selectin. Results for curcumin were confirmed both in experimental studies on endothelial cells but also in animal studies (Table 6.7) (Karimian et al. 2017; Bieganska-Hensoldt and Rosolowska-Huszcz 2017).

**Table 6.7** Anti-inflammatory effects in cell-based experiments

| Plant metabolite/compound | Cell model | Effects | Literature reference |
|---|---|---|---|
| Pure polyphenols (1–25μM) such as *p*-coumaric, caffeic and caftaric acids, kaempferol, quercetin, myricetin, and resveratrol | Human endothelial cells, stimulated with LPS (evaluating the interaction with monocytes) | – ↓ ICAM-1, VCAM-1, E-selectin, MCP-1, and macrophage colony-stimulating factor (M-CSF) | Calabriso et al. (2016) |
| Red wine extract, 0–100μg of gallic acid equivalent/mL | Myofibroblasts cells | – ↓ mRNA expression for mediators NFkB, ICAM-1, VCAM-1, and PECAM-1, dependent of miR-126 | Angel-Morales et al. (2012) |
| Polyphenolic grape extract and the polyphenols catechin, epicatechin, resveratrol, and gallic acid (5, 10, 20, and 25μg/mL) | Platelets | – ↓ Platelet aggregation, through activation of PECAM-1; effects observed for the total extract (25μg/mL level of exposure) and not the isolated polyphenols | de Lange et al. (2007) |
| Green tea components – 30μM of EGCG, 3μM of EGC, 2μM of ECG, and 1.4μM of EC—in mixture as well as each one alone | Neutrophils from healthy individuals | – ↓ of TNFα, IL-1β and IL-6, MPO activity, and HOCl production<br>– ↑ antioxidant enzyme activities | Marinovic et al. (2015) |
| EGCG 2.5μM | Human aortic endothelial cells (HAEC) | – ↓ VCAM-1 release stimulated by TNFα | Pullikotil et al. (2012) |
| Epicatechin (0.1–10μM) and catechin (0.1–10μM) | Platelets (activated with arachidonic acid) from peripheral artery disease patients and healthy subjects | – ↓ sICAM1, sVCAM-1, and sE-selectin<br>– ↑ NO bioavailability | Carnevale et al. (2014) |
| Açaí and red muscadine grape polyphenolics (5–20 mg gallic acid equivalent/L) | HUVEC stimulated with LPS | – ↓ Adhesion molecules and NFκB activation<br>– MicroRNA-126 expression modulated | Noratto et al. (2011) |

(continued)

**Table 6.7** (continued)

| Plant metabolite/compound | Cell model | Effects | Literature reference |
|---|---|---|---|
| Different flavonoids (amentoflavone, quercetagetin-7-O-glucoside, apigenin, fisetin, kaempferol, luteolin, and quercetin) | Rat neutrophils | – ↓ Inflammatory response,<br>– ↓ Arachidonic acid release,<br>– ↓ $\beta$-glucuronidase and lysozyme | Tordera et al. (1994) |
| Methanolic and water extracts from *Brassica oleracea* | HUVECs were preincubated with extracts for 24 h and thereafter stimulated TNFα (10 ng/mL) | – ↓ TNFα-induced expression of E-selectin, ICAM-1, and VCAM-1 | Kuntz and Kunz (2014) |
| Extracts from Mango (*Mangifera indica* L.) | HUVECs, stimulated with TNFα | – ↓ IL-6, IL-8, COX-2, and ICAM-1<br>– ↑ eNOS (counteracting the inhibitory effect of TNFα) | Mura et al. (2015) |
| Oleacein (3,4-dihydroxyphenylethanol-elenolic acid dialdehyde) 50–100μM | Isolated neutrophils | – ↓Neutral endopeptidase activity, elastase, MMP-9, and IL-8 release from neutrophils<br>– ↓ CD11b/CD18 expression and increased CD62L expression | Czerwinska et al. (2014) |
| Chlorogenic acid (0–20μM) | Murine macrophages and microglial cells stimulated with LPS | – ↓ NO production and expression of COX-1 and iNOS without cytotoxicity<br>– ↓ IL-1β, TNFα, and IL-6 | Hwang et al. (2014) |
| *Aronia melanocarpa* fruits, 50μg/mL | Human aortic endothelial cells HAECs preincubated with extract, followed by TNFα stimulation | – ↓ Intracellular ROS production<br>– ↓ Expression of ICAM-1 and VCAM-1,<br>– ↓ The phosphorylation of NFκB p65 | Zapolska-Downar et al. (2012) |
| Ellagic acid (0, 25, and 50μM) | PBMCs | – Concentration-dependent inhibitory effect on MIF-induced chemotactic migration of PBMCs | Sarkar et al. (2015) |

(continued)

**Table 6.7** (continued)

| Plant metabolite/compound | Cell model | Effects | Literature reference |
|---|---|---|---|
| EGCG (10, 50, and 100μM) | Adult human retinal pigment epithelial cell line, ARPE-19, pretreated with EGCG for 1 h followed by exposure to 20 ng/mL of TNFα for 24 h | – ↓ TNFα -induced ICAM-1 expression<br>– 100μM EGCG reversed the intracellular ROS, as induced by TNF-α exposure<br>– 100μM EGCG inhibited the TNF-α-mediated increase of ICAM-1 | Thichanpiang and Wongprasert (2015) |
| Propolis ethanol extract and pinocembrin | Murine macrophages (RAW 264.7) activated with LPS | – ↓ MMP-9 (dose dependent) | Saavedra et al. (2016) |

### 6.7.2 Anti-inflammatory Effects of Polyphenols in Preclinical and Clinical Settings

The effects of caffeic acid (10 mg/kg) were assessed in a mouse model of hyperhomocysteinemia; results show that the tested compound counteracted the effects of high homocysteine levels, reduced the leukocyte adhesion and MCP-1 levels as well as the E-selectin and ICAM-1 expression on cerebrovascular endothelium (Zhao et al. 2012).

The effects of quercetin were assessed on the inflammatory profile of streptozotocin-induced diabetes mellitus; quercetin administration for 6 weeks was associated with the improvement of insulin resistance and decrease of C-reactive protein (CRP) and TNFα, the protective effects on the CVD function being mediated by NFκB signaling (Mahmoud et al. 2013)

Administering 0.25 or 0.40% cacao polyphenols for 16 weeks to apolipoprotein E-deficient mice lead to the reduction of VCAM-1 and ICAM-1 expression, reduction of oxidative stress markers (4-hydroxynonenal in cross-sections of the brachiocephalic trunk), and reduction of cholesterol accumulation in plaque compared to control (Natsume and Baba 2014)

Results are also available regarding the anti-inflammatory effects of polyphenols, generated in human studies.

Resveratrol (20 mg/day) was administered for 60 days, alone or in combination with calcium fructoborate, in a double-blinded, randomized, parallel clinical trial including 166 subjects with stable angina pectoris. Results show the significant reduction of hsCRP level, both alone and in the abovementioned combination after 30-day and 60-day treatment, compared to baseline (Militaru et al. 2013).

A study including men and women with high CVD risk evaluated the effect induced by 330 mL bilberry juice (resveratrol: 1–12 mg/100 g) versus 1 L water/day (control) on biomarkers of inflammation; results show the decrease of CRP, IL-6, and IL-15 (Karlsen et al. 2010).

A purified anthocyanin mixture (640 mg/day, 24 weeks), administered in a double-blind randomized study to 150 hypercholesterolemic subjects, induced the reduction of high-sensitivity CRP, sVCAM-1, and IL-1β. The same type of results was obtained for HepG2 and porcine iliac artery endothelial cells exposed to the purified anthocyanin mixture (reduction of LPS-induced VCAM-1 secretion) (Zhu et al. 2013).

A double-blind, randomized trial including 52 patients with early atherosclerosis tested the effects of 30 mL olive oil or 30 mL of EGCG-supplemented olive oil for 4 months. Inflammatory markers of CVD risk (sICAM, white blood cells, monocytes, lymphocytes, and platelets) were reduced in patients with early atherosclerosis after receiving olive oil, independent of EGCG supplementation (Widmer et al. 2013).

A randomized crossover study tested the effect induced by polyphenols with/without alcohol; 33 male received beer (660 mL/day), the equivalent polyphenol concentration in a nonalcoholic beverage, and gin (the same quantity of alcohol) for 4 weeks. The moderate alcohol intake increased HDL, and adiponectin levels decreased fibrinogen; the nonalcoholic beverage (polyphenols) reduced the monocyte expression of CCR2, TNFβ, and IL-15 levels. The authors concluded that the phenolic component associated in alcoholic drinks contributed to the reduction of inflammatory biomarkers (Chiva-Blanch et al. 2015).

Another study tested the effect of a standardized polyphenolic extract of clove buds (250 mg/day) on inflammatory profile of heavy drinkers. Results were analyzed in a randomized, double-blinded, crossover study with 2 weeks of washout period; the polyphenolic extract induced a decrease of lipid peroxidation, an increase of antioxidant systems (glutathione and superoxide dismutase), and a decrease of inflammatory markers (hsCRP, IL-6) (Mammen et al. 2018).

A parallel, double-blind, controlled, randomized trial tested the effects induced by strawberry and cranberry polyphenols (333 mg/day for 6 weeks) on the metabolic, redox, and inflammatory profile of 41 overweight or obese human patients; the extract improved the insulin sensitivity but did not improve the cardio-metabolic risk factors (Paquette et al. 2017).

By contrast, another randomized, crossover study including obese type 2 diabetes patients pointed out a decrease of IL-6 and TNFα after daily dietary raspberry supplementation. No effect on CRP and plasma lipids were obtained (Schell et al. 2019).

A randomized, placebo-controlled trial including 25 former smokers (vs control) tested the effects induced by *Aronia melanocarpa* extract daily consumption (500 mg/day) on CVD risk markers; results shows a reduction in TC and LDL levels but no improvement of inflammatory and redox markers (Xie et al. 2017).

The effects induced by tart cherry juice (480 mL/day) for 12 weeks was investigated in a controlled study including 37 elderly subjects (65–80 years old); results show that, besides the reduction of blood pressure and LDL level, a decrease of CRP and redox stress markers (MDA) was obtained (Chai et al. 2019).

Ex vivo treatment of fasting blood samples from obese patients with different concentrations (100, 200, and 500μg/mL) of brown, red, and purple rice polyphenol extracts induced the dose-dependent decrease of MDA and TNFα, without an impact on IL-6 concentration (Callcott et al. 2018).

The effects of whole-grain wheat polyphenol supplementation (for 8 weeks) was tested in a placebo-controlled, randomized, parallel-group trial including 80 overweight or obese subjects with high CVD risk (low fruit and vegetable intake, sedentary lifestyle); results show a decrease of TNFα level without significant impact on metabolic parameters (Vitaglione et al. 2015).

## 6.8 Conclusions

Various researches performed in the last 10 years show the beneficial effects induced by polyphenolic compounds in reducing the risk factors for CVD, thus acting either as preventive factors and also, in some cases, as reparative ones, supporting the treatment of patients with chronic cardiovascular impairments. Larger studies are still necessary in order to assess and integrate the potential beneficial effects with data regarding the low bioavailability of the compounds and also with their potential to interfere with the chronic pharmacological treatment of the selected patients.

## References

Adibian M et al (2019) The effects of curcumin supplementation on high-sensitivity C-reactive protein, serum adiponectin, and lipid profile in patients with type 2 diabetes: a randomized, double-blind, placebo-controlled trial. Phytother Res 33(5):1374–1383

Agarwal B et al (2013) Resveratrol for primary prevention of atherosclerosis: clinical trial evidence for improved gene expression in vascular endothelium. Int J Cardiol 166(1):246–248

Aggarwal BB, Harikumar KB (2009) Potential therapeutic effects of curcumin, the anti-inflammatory agent, against neurodegenerative, cardiovascular, pulmonary, metabolic, autoimmune and neoplastic diseases. Int J Biochem Cell Biol 41(1):40–59

Alaimo A et al (2020) Toxic effects of A2E in human ARPE-19 cells were prevented by resveratrol: a potential nutritional bioactive for age-related macular degeneration treatment. Arch Toxicol 94(2):553–572

Alam MM, Ahmad I, Naseem I (2015) Inhibitory effect of quercetin in the formation of advance glycation end products of human serum albumin: an in vitro and molecular interaction study. Int J Biol Macromol 79:336–343

Alcendor RR et al (2007) Sirt1 regulates aging and resistance to oxidative stress in the heart. Circ Res 100(10):1512–1521

Ali K et al (2010) Metabolic constituents of grapevine and grape-derived products. Phytochem Rev 9(3):357–378

Ali BH et al (2018) Curcumin ameliorates kidney function and oxidative stress in experimental chronic kidney disease. Basic Clin Pharmacol Toxicol 122(1):65–73

Alvarez-Suarez JM et al (2014) One-month strawberry-rich anthocyanin supplementation ameliorates cardiovascular risk, oxidative stress markers and platelet activation in humans. J Nutr Biochem 25(3):289–294

Alvarez-Suarez JM et al (2017) Anti-inflammatory effect of Capuli cherry against LPS-induced cytotoxic damage in RAW 264.7 macrophages. Food Chem Toxicol 102:46–52

Alwi I et al (2008) The effect of curcumin on lipid level in patients with acute coronary syndrome. Acta Med Indones 40(4):201–210
Amiot MJ, Riva C, Vinet A (2016) Effects of dietary polyphenols on metabolic syndrome features in humans: a systematic review. Obes Rev 17(7):573–586
Anderson RA et al (2016) Cinnamon extract lowers glucose, insulin and cholesterol in people with elevated serum glucose. J Tradit Complement Med 6(4):332–336
Angel-Morales G, Noratto G, Mertens-Talcott S (2012) Red wine polyphenolics reduce the expression of inflammation markers in human colon-derived CCD-18Co myofibroblast cells: potential role of microRNA-126. Food Funct 3(7):745–752
Annuzzi G et al (2014) Diets naturally rich in polyphenols improve fasting and postprandial dyslipidemia and reduce oxidative stress: a randomized controlled trial. Am J Clin Nutr 99(3):463–471
Bagetta D et al (2020) Mediterranean products as promising source of multi-target agents in the treatment of metabolic syndrome. Eur J Med Chem 186:111903
Barreca D et al (2017) Flavanones: citrus phytochemical with health-promoting properties. Biofactors 43(4):495–506
Barros BRS et al (2019) Saline extract from Malpighia emarginata DC leaves showed higher polyphenol presence, antioxidant and antifungal activity and promoted cell proliferation in mice splenocytes. An Acad Bras Cienc 91(1):e20190916
Basu A et al (2015) Acute cocoa supplementation increases postprandial hdl cholesterol and insulin in obese adults with type 2 diabetes after consumption of a high-fat breakfast. J Nutr 145(10):2325–2332
Bayram B et al (2012) A diet rich in olive oil phenolics reduces oxidative stress in the heart of SAMP8 mice by induction of Nrf2-dependent gene expression. Rejuvenation Res 15(1):71–81
Bieganska-Hensoldt S, Rosolowska-Huszcz D (2017) Polyphenols in preventing endothelial dysfunction. Postepy Hig Med Dosw (Online) 71:227–235
Bohn T (2017) Carotenoids, Chronic Disease Prevention and Dietary Recommendations. Int J Vitam Nutr Res 87(3-4):121–130
Bona NP et al (2020) Tannic acid elicits selective antitumoral activity in vitro and inhibits cancer cell growth in a preclinical model of glioblastoma multiforme. Metab Brain Dis 35(2):283–293
Bradley JM, Organ CL, Lefer DJ (2016) Garlic-derived organic polysulfides and myocardial protection. J Nutr 146(2):403S–409S
Breuss JM, Atanasov AG, Uhrin P (2019) Resveratrol and its effects on the vascular system. Int J Mol Sci 20(7)
Broekhuizen LN et al (2011) Reduction of monocyte chemoattractant protein 1 and macrophage migration inhibitory factor by a polyphenol-rich extract in subjects with clustered cardiometabolic risk factors. Br J Nutr 106(9):1416–1422
Brown JE et al (1998) Structural dependence of flavonoid interactions with $Cu^{2+}$ ions: implications for their antioxidant properties. Biochem J 330(Pt 3):1173–1178
Bruno RM, Ghiadoni L (2018) Polyphenols, antioxidants and the sympathetic nervous system. Curr Pharm Des 24(2):130–139
Butler D (2011) UN targets top killers. Nature 477(7364):260–261
Cai H, Harrison DG (2000) Endothelial dysfunction in cardiovascular diseases: the role of oxidant stress. Circ Res 87(10):840–844
Calabriso N et al (2016) Multiple anti-inflammatory and anti-atherosclerotic properties of red wine polyphenolic extracts: differential role of hydroxycinnamic acids, flavonols and stilbenes on endothelial inflammatory gene expression. Eur J Nutr 55(2):477–489
Callcott ET et al (2018) Coloured rice-derived polyphenols reduce lipid peroxidation and pro-inflammatory cytokines ex vivo. Food Funct 9(10):5169–5175
Carnevale R et al (2012) Dark chocolate inhibits platelet isoprostanes via NOX2 down-regulation in smokers. J Thromb Haemost 10(1):125–132
Carnevale R et al (2014) Epicatechin and catechin modulate endothelial activation induced by platelets of patients with peripheral artery disease. Oxid Med Cell Longev 2014:691015

Castro MC et al (2020) Cacao extract enriched in polyphenols prevents endocrine-metabolic disturbances in a rat model of prediabetes triggered by a sucrose rich diet. J Ethnopharmacol 247:112263
Chai SC et al (2019) Effects of tart cherry juice on biomarkers of inflammation and oxidative stress in older adults. Nutrients 11(2)
Chang JJ et al (2013) Mulberry anthocyanins inhibit oleic acid induced lipid accumulation by reduction of lipogenesis and promotion of hepatic lipid clearance. J Agric Food Chem 61(25):6069–6076
Chew B et al (2019) Chronic consumption of a low calorie, high polyphenol cranberry beverage attenuates inflammation and improves glucoregulation and HDL cholesterol in healthy overweight humans: a randomized controlled trial. Eur J Nutr 58(3):1223–1235
Chiva-Blanch G et al (2015) Effects of alcohol and polyphenols from beer on atherosclerotic biomarkers in high cardiovascular risk men: a randomized feeding trial. Nutr Metab Cardiovasc Dis 25(1):36–45
Chniguir A et al (2019) Eugenol prevents fMLF-induced superoxide anion production in human neutrophils by inhibiting ERK1/2 signaling pathway and p47phox phosphorylation. Sci Rep 9(1):18540
Cho S et al (2012) Carvacrol prevents diet-induced obesity by modulating gene expressions involved in adipogenesis and inflammation in mice fed with high-fat diet. J Nutr Biochem 23(2):192–201
Choi JH et al (2015) Kaempferol inhibits thrombosis and platelet activation. Biochimie 115:177–186
Choi JH, Kim KJ, Kim S (2016) Comparative effect of quercetin and quercetin-3-O-beta-d-glucoside on fibrin polymers, blood clots, and in rodent models. J Biochem Mol Toxicol 30(11):548–558
Cordero-Herrera I et al (2015) Cocoa-rich diet ameliorates hepatic insulin resistance by modulating insulin signaling and glucose homeostasis in Zucker diabetic fatty rats. J Nutr Biochem 26(7):704–712
Costa C et al (2017) Current evidence on the effect of dietary polyphenols intake on chronic diseases. Food Chem Toxicol 110:286–299
Crozier A, Jaganath IB, Clifford MN (2009) Dietary phenolics: chemistry, bioavailability and effects on health. Nat Prod Rep 26(8):1001–1043
Cuccioloni M et al (2011) Epigallocatechin-3-gallate potently inhibits the in vitro activity of hydroxy-3-methyl-glutaryl-CoA reductase. J Lipid Res 52(5):897–907
Czerwinska ME, Kiss AK, Naruszewicz M (2014) Inhibition of human neutrophils NEP activity, CD11b/CD18 expression and elastase release by 3,4-dihydroxyphenylethanol-elenolic acid dialdehyde, oleacein. Food Chem 153:1–8
De Caterina R et al (2006) Nutritional mechanisms that influence cardiovascular disease. Am J Clin Nutr 83(2):421S–426S
De la Ossa JG et al (2019) Waste autochthonous tuscan olive leaves (Olea europaea var. Olivastra seggianese) as antioxidant source for biomedicine. Int J Mol Sci 20(23)
de Lange DW et al (2007) Polyphenolic grape extract inhibits platelet activation through PECAM-1: an explanation for the French paradox. Alcohol Clin Exp Res 31(8):1308–1314
De Silva SF, Alcorn J (2019) Flaxseed lignans as important dietary polyphenols for cancer prevention and treatment: chemistry, pharmacokinetics, and molecular targets. Pharmaceuticals (Basel) 12(2)
Denev P et al (2019) Black chokeberry (Aronia melanocarpa) polyphenols reveal different antioxidant, antimicrobial and neutrophil-modulating activities. Food Chem 284:108–117
Dhalla NS, Temsah RM, Netticadan T (2000) Role of oxidative stress in cardiovascular diseases. J Hypertens 18(6):655–673
Donovan JL et al (2006) (+)-Catechin is more bioavailable than (-)-catechin: relevance to the bioavailability of catechin from cocoa. Free Radic Res 40(10):1029–1034

Draijer R et al (2015) Consumption of a polyphenol-rich grape-wine extract lowers ambulatory blood pressure in mildly hypertensive subjects. Nutrients 7(5):3138–3153
Dybkowska E et al (2017) Assessing polyphenols content and antioxidant activity in coffee beans according to origin and the degree of roasting. Rocz Panstw Zakl Hig 68(4):347–353
Edirisinghe I et al (2011) Strawberry anthocyanin and its association with postprandial inflammation and insulin. Br J Nutr 106(6):913–922
Faggio C et al (2017) Flavonoids and platelet aggregation: a brief review. Eur J Pharmacol 807:91–101
Farah A et al (2008) Chlorogenic acids from green coffee extract are highly bioavailable in humans. J Nutr 138(12):2309–2315
Filip R et al (2015) Twelve-month consumption of a polyphenol extract from olive (Olea europaea) in a double blind, randomized trial increases serum total osteocalcin levels and improves serum lipid profiles in postmenopausal women with osteopenia. J Nutr Health Aging 19(1):77–86
Flammer AJ et al (2012) Cardiovascular effects of flavanol-rich chocolate in patients with heart failure. Eur Heart J 33(17):2172–2180
Forbes-Hernandez TY et al (2017a) Strawberry (cv. Romina) methanolic extract and anthocyanin-enriched fraction improve lipid profile and antioxidant status in HepG2 cells. Int J Mol Sci 18(6)
Forbes-Hernandez TY et al (2017b) Lipid accumulation in HepG2 cells is attenuated by strawberry extract through AMPK activation. Nutrients 9(6)
Fuhr U, Kummert AL (1995) The fate of naringin in humans: a key to grapefruit juice-drug interactions? Clin Pharmacol Ther 58(4):365–373
Galkina E, Ley K (2009) Immune and inflammatory mechanisms of atherosclerosis (*). Annu Rev Immunol 27:165–197
Gao H et al (2020) Profiles and neuroprotective effects of Lycium ruthenicum polyphenols against oxidative stress-induced cytotoxicity in PC12 cells. J Food Biochem 44(1):e13112
Garcia JP et al (2018) The cardiovascular effects of chocolate. Rev Cardiovasc Med 19(4):123–127
Garcia-Conesa MT (2017) Dietary polyphenols against metabolic disorders: how far have we progressed in the understanding of the molecular mechanisms of action of these compounds? Crit Rev Food Sci Nutr 57(9):1769–1786
Gasparrini M et al (2017) Anti-inflammatory effect of strawberry extract against LPS-induced stress in RAW 264.7 macrophages. Food Chem Toxicol 102:1–10
Georgiev KD, Slavov IJ, Iliev IA (2019) Antioxidant activity and antiproliferative effects of Lycium barbarum's (Goji berry) fractions on breast cancer cell lines. Folia Med (Plovdiv) 61(1):104–112
Gershenzon J, Dudareva N (2007) The function of terpene natural products in the natural world. Nat Chem Biol 3(7):408–414
Gomaa AMS, Abdelhafez AT, Aamer HA (2018) Garlic (Allium sativum) exhibits a cardioprotective effect in experimental chronic renal failure rat model by reducing oxidative stress and controlling cardiac Na. Cell Stress Chaperones 23(5):913–920
GutiErrez-Grijalva EP et al (2016) Review: dietary phenolic compounds, health benefits and bioaccessibility. Arch Latinoam Nutr 66(2):87–100
Haghighatdoost F, Hariri M (2018) Effect of resveratrol on lipid profile: an updated systematic review and meta-analysis on randomized clinical trials. Pharmacol Res 129:141–150
Hao HZ et al (2015) Antiplatelet activity of loureirin A by attenuating Akt phosphorylation: in vitro studies. Eur J Pharmacol 746:63–69
Harrison D et al (2003) Role of oxidative stress in atherosclerosis. Am J Cardiol 91(3A):7A–11A
Hassellund SS et al (2013) Effects of anthocyanins on cardiovascular risk factors and inflammation in pre-hypertensive men: a double-blind randomized placebo-controlled crossover study. J Hum Hypertens 27(2):100–106
Heim KE, Tagliaferro AR, Bobilya DJ (2002) Flavonoid antioxidants: chemistry, metabolism and structure-activity relationships. J Nutr Biochem 13(10):572–584
Hernaez A et al (2014) Olive oil polyphenols enhance high-density lipoprotein function in humans: a randomized controlled trial. Arterioscler Thromb Vasc Biol 34(9):2115–2119

Hollman PC, Katan MB (1997) Absorption, metabolism and health effects of dietary flavonoids in man. Biomed Pharmacother 51(8):305–310
Hollman PC et al (1995) Absorption of dietary quercetin glycosides and quercetin in healthy ileostomy volunteers. Am J Clin Nutr 62(6):1276–1282
Holthoff JH et al (2010) Resveratrol, a dietary polyphenolic phytoalexin, is a functional scavenger of peroxynitrite. Biochem Pharmacol 80(8):1260–1265
Hong J et al (2001) Effects of purified green and black tea polyphenols on cyclooxygenase- and lipoxygenase-dependent metabolism of arachidonic acid in human colon mucosa and colon tumor tissues. Biochem Pharmacol 62(9):1175–1183
Hoskin RT et al (2019) Blueberry polyphenol-protein food ingredients: the impact of spray drying on the in vitro antioxidant activity, anti-inflammatory markers, glucose metabolism and fibroblast migration. Food Chem 280:187–194
Hosseini B et al (2016) Effects of pomegranate extract supplementation on inflammation in overweight and obese individuals: a randomized controlled clinical trial. Complement Ther Clin Pract 22:44–50
Hostetler GL, Ralston RA, Schwartz SJ (2017) Flavones: food sources, bioavailability, metabolism, and bioactivity. Adv Nutr 8(3):423–435
Huang H et al (2016) The effects of resveratrol intervention on risk markers of cardiovascular health in overweight and obese subjects: a pooled analysis of randomized controlled trials. Obes Rev 17(12):1329–1340
Hussain T et al (2016) Oxidative Stress and Inflammation: What Polyphenols Can Do for Us? Oxid Med Cell Longev 2016:7432797
Hwang SJ et al (2014) Anti-inflammatory effects of chlorogenic acid in lipopolysaccharide-stimulated RAW 264.7 cells. Inflamm Res 63(1):81–90
Ibero-Baraibar I et al (2014) Oxidised LDL levels decreases after the consumption of ready-to-eat meals supplemented with cocoa extract within a hypocaloric diet. Nutr Metab Cardiovasc Dis 24(4):416–422
Ijaz S et al (2019) HPLC profiling of Mimosa pudica polyphenols and their non-invasive biophysical investigations for anti-dermatoheliotic and skin reinstating potential. Biomed Pharmacother 109:865–875
Imai K, Nakachi K (1995) Cross sectional study of effects of drinking green tea on cardiovascular and liver diseases. BMJ 310(6981):693–696
Janiques AG et al (2014) Effects of grape powder supplementation on inflammatory and antioxidant markers in hemodialysis patients: a randomized double-blind study. J Bras Nefrol 36(4):496–501
Jemai H, El Feki A, Sayadi S (2009) Antidiabetic and antioxidant effects of hydroxytyrosol and oleuropein from olive leaves in alloxan-diabetic rats. J Agric Food Chem 57(19):8798–8804
Jeon JS et al (2019) Contents of chlorogenic acids and caffeine in various coffee-related products. J Adv Res 17:85–94
Jia Z et al (2008) EPR studies on the superoxide-scavenging capacity of the nutraceutical resveratrol. Mol Cell Biochem 313(1-2):187–194
Jomova K et al (2019) A switch between antioxidant and prooxidant properties of the phenolic compounds myricetin, morin, 3′,4′-dihydroxyflavone, taxifolin and 4-hydroxy-coumarin in the presence of copper(II) ions: a spectroscopic, absorption titration and DNA damage study. Molecules 24(23)
Jowko E et al (2015) The effect of green tea extract supplementation on exercise-induced oxidative stress parameters in male sprinters. Eur J Nutr 54(5):783–791
Kaliora AC, Dedoussis GV, Schmidt H (2006) Dietary antioxidants in preventing atherogenesis. Atherosclerosis 187(1):1–17
Kardum N et al (2015) Beneficial effects of polyphenol-rich chokeberry juice consumption on blood pressure level and lipid status in hypertensive subjects. J Med Food 18(11):1231–1238
Karimian MS et al (2017) Curcumin and endothelial function: evidence and mechanisms of protective effects. Curr Pharm Des 23(17):2462–2473

Karlsen A et al (2010) Bilberry juice modulates plasma concentration of NF-kappaB related inflammatory markers in subjects at increased risk of CVD. Eur J Nutr 49(6):345–355

Kathuria A et al (2012) Substrate specificity of acetoxy derivatives of coumarins and quinolones towards Calreticulin mediated transacetylation: investigations on antiplatelet function. Bioorg Med Chem 20(4):1624–1638

Khan N et al (2012) Regular consumption of cocoa powder with milk increases HDL cholesterol and reduces oxidized LDL levels in subjects at high-risk of cardiovascular disease. Nutr Metab Cardiovasc Dis 22(12):1046–1053

Khoo HE et al (2017) Anthocyanidins and anthocyanins: colored pigments as food, pharmaceutical ingredients, and the potential health benefits. Food Nutr Res 61(1):1361779

Khurana S et al (2013) Polyphenols: benefits to the cardiovascular system in health and in aging. Nutrients 5(10):3779–3827

Kim YS et al (2007) Curcumin attenuates inflammatory responses of TNF-alpha-stimulated human endothelial cells. J Cardiovasc Pharmacol 50(1):41–49

Kim SR et al (2014) Caffeic acid regulates LPS-induced NF-kappaB activation through NIK/IKK and c-Src/ERK signaling pathways in endothelial cells. Arch Pharm Res 37(4):539–547

Kumar S, Pandey AK (2013) Chemistry and biological activities of flavonoids: an overview. ScientificWorldJournal 2013:162750

Kumar B et al (2014) Retinal neuroprotective effects of quercetin in streptozotocin-induced diabetic rats. Exp Eye Res 125:193–202

Kuntz S, Kunz C (2014) Extracts from Brassica oleracea L. convar. acephala var. sabellica inhibit TNF-alpha stimulated neutrophil adhesion in vitro under flow conditions. Food Funct 5(6):1082–1090

Kuriyama S et al (2006) Green tea consumption and mortality due to cardiovascular disease, cancer, and all causes in Japan: the Ohsaki study. JAMA 296(10):1255–1265

Kusunoki M et al (2015) Black soybean extract improves lipid profiles in fenofibrate-treated type 2 diabetics with postprandial hyperlipidemia. J Med Food 18(6):615–618

Leonard SS et al (2003) Resveratrol scavenges reactive oxygen species and effects radical-induced cellular responses. Biochem Biophys Res Commun 309(4):1017–1026

Libby P (2006) Inflammation and cardiovascular disease mechanisms. Am J Clin Nutr 83(2):456S–460S

Liu K et al (2017) Novel approach to classify plants based on metabolite-content similarity. Biomed Res Int 2017:5296729

Liu YT et al (2019) Lotus seedpod extracts reduced lipid accumulation and lipotoxicity in hepatocytes. Nutrients 11(12)

Loo BM et al (2016) Consumption of chokeberry (Aronia mitschurinii) products modestly lowered blood pressure and reduced low-grade inflammation in patients with mildly elevated blood pressure. Nutr Res 36(11):1222–1230

Loued S et al (2013) Extra-virgin olive oil consumption reduces the age-related decrease in HDL and paraoxonase 1 anti-inflammatory activities. Br J Nutr 110(7):1272–1284

Ludovici V et al (2018) Polyphenols: anti-platelet nutraceutical? Curr Pharm Des 24(2):146–157

Ma Y, Liu F, Xu Y (2019) Protective effect of beta-glucogallin on damaged cataract against methylglyoxal induced oxidative stress in cultured lens epithelial cells. Med Sci Monit 25: 9310–9318

Macakova K et al (2012) In vitro platelet antiaggregatory properties of 4-methylcoumarins. Biochimie 94(12):2681–2686

Macedo RC et al (2015) Effects of chronic resveratrol supplementation in military firefighters undergo a physical fitness test—a placebo-controlled, double blind study. Chem Biol Interact 227:89–95

Maheswaraiah A, Rao LJ, Naidu KA (2015) Anti-platelet activity of water dispersible curcuminoids in rat platelets. Phytother Res 29(3):450–458

Mahmoud MF et al (2013) Quercetin protects against diabetes-induced exaggerated vasoconstriction in rats: effect on low grade inflammation. PLoS One 8(5):e63784

Mammen RR et al (2018) Clove bud polyphenols alleviate alterations in inflammation and oxidative stress markers associated with binge drinking: a randomized double-blinded placebo-controlled crossover study. J Med Food 21(11):1188–1196
Manach C et al (2004) Polyphenols: food sources and bioavailability. Am J Clin Nutr 79(5):727–747
Manach C et al (2005) Bioavailability and bioefficacy of polyphenols in humans. I. Review of 97 bioavailability studies. Am J Clin Nutr 81(1 Suppl):230S–242S
Margina D, Ilie M, Gradinaru D (2012) Quercetin and epigallocatechin gallate induce in vitro a dose-dependent stiffening and hyperpolarizing effect on the cell membrane of human mononuclear blood cells. Int J Mol Sci 13(4):4839–4859
Marinovic MP, Morandi AC, Otton R (2015) Green tea catechins alone or in combination alter functional parameters of human neutrophils via suppressing the activation of TLR-4/NFkappaB p65 signal pathway. Toxicol In Vitro 29(7):1766–1778
Martinet W, De Meyer GR (2009) Autophagy in atherosclerosis: a cell survival and death phenomenon with therapeutic potential. Circ Res 104(3):304–317
Mathew AS et al (2012) Acute effects of pomegranate extract on postprandial lipaemia, vascular function and blood pressure. Plant Foods Hum Nutr 67(4):351–357
Mattivi F et al (2006) Metabolite profiling of grape: flavonols and anthocyanins. J Agric Food Chem 54(20):7692–7702
Mayneris-Perxachsab JA-HJ, la Hera E, del Bas JM, Arola L, Caimaria A (2019) Supplementation with biscuits enriched with hesperidin and naringenin is associated with an improvement of the metabolic syndrome induced by a cafeteria diet in rats. J Funct Foods 61:103504
McGhie TK, Walton MC (2007) The bioavailability and absorption of anthocyanins: towards a better understanding. Mol Nutr Food Res 51(6):702–713
Messina F et al (2015) Effect of substituted stilbenes on platelet function. Fitoterapia 105:228–233
Middleton E (1998) Effect of plant flavonoids on immune and inflammatory cell function. Adv Exp Med Biol 439:175–182
Militaru C et al (2013) Oral resveratrol and calcium fructoborate supplementation in subjects with stable angina pectoris: effects on lipid profiles, inflammation markers, and quality of life. Nutrition 29(1):178–183
Mishra A, Kumar S, Pandey AK (2013) Scientific validation of the medicinal efficacy of Tinospora cordifolia. ScientificWorldJournal 2013:292934
Mollace V et al (2019) Hypoglycemic and hypolipemic effects of a new lecithin formulation of bergamot polyphenolic fraction: a double blind, randomized, placebo-controlled study. Endocr Metab Immune Disord Drug Targets 19(2):136–143
Moreno-Luna R et al (2012) Olive oil polyphenols decrease blood pressure and improve endothelial function in young women with mild hypertension. Am J Hypertens 25(12):1299–1304
Morita MNY, Yoshikawa T, Niki E (2017) Antioxidant capacity of blueberry extracts: peroxyl radical scavenging and inhibition of plasma lipid oxidation induced by multiple oxidants. J Berry Res 7(1):1–9
Mullan A et al (2016) Effects of a beverage rich in (poly)phenols on established and novel risk markers for vascular disease in medically uncomplicated overweight or obese subjects: a four week randomized placebo-controlled trial. Atherosclerosis 246:169–176
Mullen W et al (2002) Effect of freezing and storage on the phenolics, ellagitannins, flavonoids, and antioxidant capacity of red raspberries. J Agric Food Chem 50(18):5197–5201
Mura M et al (2015) Simultaneous ultrasound-assisted water extraction and beta-cyclodextrin encapsulation of polyphenols from Mangifera indica stem bark in counteracting TNFalpha-induced endothelial dysfunction. Nat Prod Res 29(17):1657–1663
Nagao T et al (2005) Ingestion of a tea rich in catechins leads to a reduction in body fat and malondialdehyde-modified LDL in men. Am J Clin Nutr 81(1):122–129
Natsume M, Baba S (2014) Suppressive effects of cacao polyphenols on the development of atherosclerosis in apolipoprotein E-deficient mice. Subcell Biochem 77:189–198

Nickel T et al (2016) Influence of high polyphenol beverage on stress-induced platelet activation. J Nutr Health Aging 20(6):586–593

Nieman DC et al (2013) Influence of a polyphenol-enriched protein powder on exercise-induced inflammation and oxidative stress in athletes: a randomized trial using a metabolomics approach. PLoS One 8(8):e72215

Noad RL et al (2016) Beneficial effect of a polyphenol-rich diet on cardiovascular risk: a randomised control trial. Heart 102(17):1371–1379

Noratto GD et al (2011) Polyphenolics from acai ( Euterpe oleracea Mart.) and red muscadine grape (Vitis rotundifolia ) protect human umbilical vascular Endothelial cells (HUVEC) from glucose- and lipopolysaccharide (LPS)-induced inflammation and target microRNA-126. J Agric Food Chem 59(14):7999–8012

Ogier N et al (2013) LDL-cholesterol-lowering effect of a dietary supplement with plant extracts in subjects with moderate hypercholesterolemia. Eur J Nutr 52(2):547–557

Olas B, Wachowicz B (2005) Resveratrol, a phenolic antioxidant with effects on blood platelet functions. Platelets 16(5):251–260

Oteiza PI et al (2005) Flavonoid-membrane interactions: a protective role of flavonoids at the membrane surface? Clin Dev Immunol 12(1):19–25

Ottaviani JI et al (2011) The stereochemical configuration of flavanols influences the level and metabolism of flavanols in humans and their biological activity in vivo. Free Radic Biol Med 50(2):237–244

Pace-Asciak CR et al (1995) The red wine phenolics trans-resveratrol and quercetin block human platelet aggregation and eicosanoid synthesis: implications for protection against coronary heart disease. Clin Chim Acta 235(2):207–219

Panza VP et al (2019) Effect of mate tea (Ilex paraguariensis) on the expression of the leukocyte NADPH oxidase subunit p47(phox) and on circulating inflammatory cytokines in healthy men: a pilot study. Int J Food Sci Nutr 70(2):212–221

Paquette M et al (2017) Strawberry and cranberry polyphenols improve insulin sensitivity in insulin-resistant, non-diabetic adults: a parallel, double-blind, controlled and randomised clinical trial. Br J Nutr 117(4):519–531

Paravicini TM, Touyz RM (2006) Redox signaling in hypertension. Cardiovasc Res 71(2):247–258

Park E et al (2016) Effects of grape seed extract beverage on blood pressure and metabolic indices in individuals with pre-hypertension: a randomised, double-blinded, two-arm, parallel, placebo-controlled trial. Br J Nutr 115(2):226–238

Parkhe A et al (2020) Protective effect of alpha mangostin on rotenone induced toxicity in rat model of Parkinson's disease. Neurosci Lett 716:134652

Pfeuffer M et al (2013) Effect of quercetin on traits of the metabolic syndrome, endothelial function and inflammation in men with different APOE isoforms. Nutr Metab Cardiovasc Dis 23(5):403–409

Piskula MK (2000) Soy isoflavone conjugation differs in fed and food-deprived rats. J Nutr 130(7):1766–1771

Prior RL et al (2009) Purified berry anthocyanins but not whole berries normalize lipid parameters in mice fed an obesogenic high fat diet. Mol Nutr Food Res 53(11):1406–1418

Pullikotil P et al (2012) Epigallocatechin gallate induces expression of heme oxygenase-1 in endothelial cells via p38 MAPK and Nrf-2 that suppresses proinflammatory actions of TNF-alpha. J Nutr Biochem 23(9):1134–1145

Quesada H et al (2009) Grape seed proanthocyanidins correct dyslipidemia associated with a high-fat diet in rats and repress genes controlling lipogenesis and VLDL assembling in liver. Int J Obes (Lond) 33(9):1007–1012

Raedschelders K, Ansley DM, Chen DD (2012) The cellular and molecular origin of reactive oxygen species generation during myocardial ischemia and reperfusion. Pharmacol Ther 133(2):230–255

Rao SR, Ravishankar GA (2002) Plant cell cultures: chemical factories of secondary metabolites. Biotechnol Adv 20(2):101–153

Rein MJ et al (2013) Bioavailability of bioactive food compounds: a challenging journey to bioefficacy. Br J Clin Pharmacol 75(3):588–602
Reis JF et al (2016) Action mechanism and cardiovascular effect of anthocyanins: a systematic review of animal and human studies. J Transl Med 14(1):315
Romandinia SML, Giampieria F, Tulipanib S, Gasparrinia M, Forbes-Hernandeza TY, Locorotondoe N, D'Alessandroe M, Mezzettif B, Bompadregand S, Alvarez-Suarez JM (2013) Effects of an acute strawberry (Fragaria×ananassa) consumption on the plasmaantioxidant status of healthy subjects. J Berry Res 3:169–179
Ross R (1999) Atherosclerosis--an inflammatory disease. N Engl J Med 340(2):115–126
Ross JA, Kasum CM (2002) Dietary flavonoids: bioavailability, metabolic effects, and safety. Annu Rev Nutr 22:19–34
Saavedra N et al (2016) Polyphenols from chilean propolis and pinocembrin reduce MMP-9 gene expression and activity in activated macrophages. Biomed Res Int 2016:6505383
Sadowska-Krepa E et al (2019) Effects of medium-term green tea extract supplementation combined with CrossFit workout on blood antioxidant status and serum brain-derived neurotrophic factor in young men: a pilot study. J Int Soc Sports Nutr 16(1):13
Sahebkar A (2013) Effects of resveratrol supplementation on plasma lipids: a systematic review and meta-analysis of randomized controlled trials. Nutr Rev 71(12):822–835
Samuel SM et al (2008) Akt/FOXO3a/SIRT1-mediated cardioprotection by n-tyrosol against ischemic stress in rat in vivo model of myocardial infarction: switching gears toward survival and longevity. J Agric Food Chem 56(20):9692–9698
Santhakumar AB, Bulmer AC, Singh I (2014) A review of the mechanisms and effectiveness of dietary polyphenols in reducing oxidative stress and thrombotic risk. J Hum Nutr Diet 27(1):1–21
Santhakumar AB, Kundur A, Fanning K, Netzel M, Stanley R, Singh I (2015a) Consumption of anthocyanin-rich Queen Garnet plum juice reduces platelet activation related thrombogenesis in healthy volunteers. J Funct Foods 12:11–12
Santhakumar A, Kundur AR, Sabapathy S, Stanley R, Singh I (2015b) The potential of anthocyanin-rich Queen Garnet plum juice supplementation in alleviating thrombotic risk under induced oxidative stress conditions. J Funct Foods 14:747–757
Santhakumar AB, Battino M, Alvarez-Suarez JM (2018) Dietary polyphenols: Structures, bioavailability and protective effects against atherosclerosis. Food Chem Toxicol 113:49–65
Sarkar S et al (2015) Ellagic acid, a dietary polyphenol, inhibits tautomerase activity of human macrophage migration inhibitory factor and its pro-inflammatory responses in human peripheral blood mononuclear cells. J Agric Food Chem 63(20):4988–4998
Sarria B et al (2012) Hypotensive, hypoglycaemic and antioxidant effects of consuming a cocoa product in moderately hypercholesterolemic humans. Food Funct 3(8):867–874
Sarria B et al (2014) Regular consumption of a cocoa product improves the cardiometabolic profile in healthy and moderately hypercholesterolaemic adults. Br J Nutr 111(1):122–134
Sattarinezhad A et al (2019) Resveratrol reduces albuminuria in diabetic nephropathy: a randomized double-blind placebo-controlled clinical trial. Diabetes Metab 45(1):53–59
Scalbert A, Williamson G (2000) Dietary intake and bioavailability of polyphenols. J Nutr 130-(8S Suppl):2073S–2085S
Schell J et al (2019) Raspberries improve postprandial glucose and acute and chronic inflammation in adults with type 2 diabetes. Ann Nutr Metab 74(2):165–174
Schumacher E et al (2011) Thrombosis preventive potential of chicory coffee consumption: a clinical study. Phytother Res 25(5):744–748
Serraino I et al (2003) Protective effects of cyanidin-3-O-glucoside from blackberry extract against peroxynitrite-induced endothelial dysfunction and vascular failure. Life Sci 73(9):1097–1114
Serrano J et al (2009) Tannins: current knowledge of food sources, intake, bioavailability and biological effects. Mol Nutr Food Res 53(Suppl 2):S310–S329
Shema-Didi L et al (2014) Does pomegranate intake attenuate cardiovascular risk factors in hemodialysis patients? Nutr J 13:18

Shen MY et al (2007) Inhibitory mechanisms of resveratrol in platelet activation: pivotal roles of p38 MAPK and NO/cyclic GMP. Br J Haematol 139(3):475–485

Shin SK et al (2019) Mulberry fruit improves memory in scopolamine-treated mice: role of cholinergic function, antioxidant system, and TrkB/Akt signaling. Nutr Neurosci:1–11

Soltani R et al (2014) Evaluation of the effects of Vaccinium arctostaphylos L. fruit extract on serum lipids and hs-CRP levels and oxidative stress in adult patients with hyperlipidemia: a randomized, double-blind, placebo-controlled clinical trial. Evid Based Complement Alternat Med 2014:217451

Song MY et al (2015) Schisandra chinensis fruit modulates the gut microbiota composition in association with metabolic markers in obese women: a randomized, double-blind placebo-controlled study. Nutr Res 35(8):655–663

Soni KB, Kuttan R (1992) Effect of oral curcumin administration on serum peroxides and cholesterol levels in human volunteers. Indian J Physiol Pharmacol 36(4):273–275

Stalmach A, Williamson G, Crozier A (2014) Impact of dose on the bioavailability of coffee chlorogenic acids in humans. Food Funct 5(8):1727–1737

Strømgaard K, Nakanishi K (2004) Chemistry and biology of terpene trilactones from Ginkgo biloba. Angew Chem Int Ed Engl 43(13):1640–1658

Sugamura K, Keaney JF Jr (2011) Reactive oxygen species in cardiovascular disease. Free Radic Biol Med 51(5):978–992

Thichanpiang P, Wongprasert K (2015) Green tea polyphenol epigallocatechin-3-gallate attenuates TNF-alpha-induced intercellular adhesion molecule-1 expression and monocyte adhesion to retinal pigment epithelial cells. Am J Chin Med 43(1):103–119

Thirunavukkarasu M et al (2007) Resveratrol alleviates cardiac dysfunction in streptozotocin-induced diabetes: role of nitric oxide, thioredoxin, and heme oxygenase. Free Radic Biol Med 43(5):720–729

Tian X et al (2016) Delineation of platelet activation pathway of scutellarein revealed its intracellular target as protein kinase C. Biol Pharm Bull 39(2):181–191

Tokede OA, Gaziano JM, Djousse L (2011) Effects of cocoa products/dark chocolate on serum lipids: a meta-analysis. Eur J Clin Nutr 65(8):879–886

Tordera M, Ferrandiz ML, Alcaraz MJ (1994) Influence of anti-inflammatory flavonoids on degranulation and arachidonic acid release in rat neutrophils. Z Naturforsch C J Biosci 49(3-4):235–240

Tu Z et al (2013) Rosemary (Rosmarinus officinalis L.) extract regulates glucose and lipid metabolism by activating AMPK and PPAR pathways in HepG2 cells. J Agric Food Chem 61(11):2803–2810

Tungmunnithum D et al (2018) Flavonoids and other phenolic compounds from medicinal plants for pharmaceutical and medical aspects: an overview. Medicines (Basel) 5(3)

Tyszka-Czochara M et al (2018) Caffeic acid targets AMPK signaling and regulates tricarboxylic acid cycle anaplerosis while metformin downregulates HIF-1alpha-induced glycolytic enzymes in human cervical squamous cell carcinoma lines. Nutrients 10(7)

Ungurianu A et al (2017) Lipoprotein redox status evaluation as a marker of cardiovascular disease risk in patients with inflammatory disease. Mol Med Rep 15(1):256–262

Ungurianu A et al (2019a) Spectrophotometric versus spectrofluorometric assessment in the study of the relationships between lipid peroxidation and metabolic dysregulation. Chem Biol Drug Des

Ungurianu A et al (2019b) Preclinical and clinical results regarding the effects of a plant-based antidiabetic formulation versus well established antidiabetic molecules. Pharmacol Res 150: 104522

Ungvari Z et al (2007) Resveratrol increases vascular oxidative stress resistance. Am J Physiol Heart Circ Physiol 292(5):H2417–H2424

Vafa A et al (2019) Protective role of diosmin against testosterone propionate-induced prostatic hyperplasia in Wistar rats: plausible role of oxidative stress and inflammation. Hum Exp Toxicol:960327119889655

Valcheva-Kuzmanova S et al (2007) Antihyperlipidemic effect of Aronia melanocarpa fruit juice in rats fed a high-cholesterol diet. Plant Foods Hum Nutr 62(1):19–24

Venditti A, Bianco A (2018) Sulfur-containing secondary metabolites as neuroprotective agents. Curr Med Chem

Violi F, Pignatelli P (2012) Platelet oxidative stress and thrombosis. Thromb Res 129(3):378–381

Vitaglione P et al (2015) Whole-grain wheat consumption reduces inflammation in a randomized controlled trial on overweight and obese subjects with unhealthy dietary and lifestyle behaviors: role of polyphenols bound to cereal dietary fiber. Am J Clin Nutr 101(2):251–261

Vogiatzoglou A et al (2015) Flavonoid intake in European adults (18 to 64 years). PLoS One 10(5): e0128132

Wainstein J et al (2012) Olive leaf extract as a hypoglycemic agent in both human diabetic subjects and in rats. J Med Food 15(7):605–610

Widmer RJ et al (2013) Beneficial effects of polyphenol-rich olive oil in patients with early atherosclerosis. Eur J Nutr 52(3):1223–1231

Xia LM, Luo MH (2015) Study progress of berberine for treating cardiovascular disease. Chronic Dis Transl Med 1(4):231–235

Xia N et al (2010) Resveratrol reverses endothelial nitric-oxide synthase uncoupling in apolipoprotein E knockout mice. J Pharmacol Exp Ther 335(1):149–154

Xia N et al (2017) Antioxidant effects of resveratrol in the cardiovascular system. Br J Pharmacol 174(12):1633–1646

Xiao L et al (2017) Quercetin attenuates high fat diet-induced atherosclerosis in apolipoprotein E knockout mice: a critical role of NADPH oxidase. Food Chem Toxicol 105:22–33

Xie L et al (2017) Aronia berry polyphenol consumption reduces plasma total and low-density lipoprotein cholesterol in former smokers without lowering biomarkers of inflammation and oxidative stress: a randomized controlled trial. Nutr Res 37:67–77

Yang YC et al (2004) The protective effect of habitual tea consumption on hypertension. Arch Intern Med 164(14):1534–1540

Youdim KA, Martin A, Joseph JA (2000) Incorporation of the elderberry anthocyanins by endothelial cells increases protection against oxidative stress. Free Radic Biol Med 29(1):51–60

Yu HP et al (2010) Resveratrol prevents endothelial dysfunction and aortic superoxide production after trauma hemorrhage through estrogen receptor-dependent hemeoxygenase-1 pathway. Crit Care Med 38(4):1147–1154

Yuan Y et al (2019) Polyphenol-rich extracts from brown macroalgae lessonia trabeculate attenuate hyperglycemia and modulate gut microbiota in high-fat diet and streptozotocin-induced diabetic rats. J Agric Food Chem 67(45):12472–12480

Yubero N et al (2013) LDL cholesterol-lowering effects of grape extract used as a dietary supplement on healthy volunteers. Int J Food Sci Nutr 64(4):400–406

Zanfirescu A et al (2019) A review of the alleged health hazards of monosodium glutamate. Compr Rev Food Sci Food Saf 18(4):1111–1134

Zapolska-Downar D et al (2012) Aronia melanocarpa fruit extract exhibits anti-inflammatory activity in human aortic endothelial cells. Eur J Nutr 51(5):563–572

Zhang W et al (2019) A multi-scale approach to study biochemical and biophysical aspects of resveratrol on diesel exhaust particle-human primary lung cell interaction. Sci Rep 9(1):18178

Zhao HP et al (2012) Caffeic acid inhibits acute hyperhomocysteinemia-induced leukocyte rolling and adhesion in mouse cerebral venules. Microcirculation 19(3):233–244

Zhu Y et al (2013) Anti-inflammatory effect of purified dietary anthocyanin in adults with hypercholesterolemia: a randomized controlled trial. Nutr Metab Cardiovasc Dis 23(9):843–849

Zunino SJ et al (2014) Dietary grape powder increases IL-1beta and IL-6 production by lipopolysaccharide-activated monocytes and reduces plasma concentrations of large LDL and large LDL-cholesterol particles in obese humans. Br J Nutr 112(3):369–380

# Kauranes as Anti-inflammatory and Immunomodulatory Agents: An Overview of In Vitro and In Vivo Effects

7

Michael R. Mijares, Gricelis P. Martínez, and Juan B. De Sanctis

**Abstract**

Natural products have been used for the prevention and treatment of various pathologies, including acute and chronic inflammatory diseases, autoimmune diseases, and cancer. Kauranes are diterpenes usually found in nature in higher plants, lichens, and liverworts. Recently, various in vitro and in vivo preclinical studies have shown that kauranes affect signaling pathways involved in the production of cytokines, transcription factors, receptors, enzymes, and ligands involved in inflammation. In neutrophils and monocytes/macrophages isolated from humans and rodents, kauranes inhibited the production of reactive oxygen and nitrogen species while increasing the antioxidant defense mechanisms of cells. In LPS-stimulated monocytes/macrophages, some kauranes blocked the NF-κB pathway, involved in the transcription of IL-6, IL-12, and TNF-α, and the pro-inflammatory enzymes NOS-2 and COX-2. Other kauranes affected NLRP3 inflammasome activation and consequently of IL-1β and IL-18 or promote the M2 versus M1 polarization through inhibiting Notch pathway in macrophages.

M. R. Mijares (✉)
Biotechnology Unit, Faculty of Pharmacy, Central University of Venezuela, Caracas, Venezuela

Institute of Immunology, Faculty of Medicine, Central University of Venezuela, Caracas, Venezuela

G. P. Martínez
Institute of Immunology, Faculty of Medicine, Central University of Venezuela, Caracas, Venezuela

J. B. De Sanctis
Institute of Immunology, Faculty of Medicine, Central University of Venezuela, Caracas, Venezuela

Institute of Molecular and Translational Medicine, Faculty of Medicine, Palacky University, Olomouc, Czech Republic


A. K. Sharma, A. Sharma (eds.), *Plant Secondary Metabolites*,
https://doi.org/10.1007/978-981-16-4779-6_7

These compounds inhibited the differentiation and activation of inflammatory subtypes of helper T cells (Th1 and Th17) and their cytokines. These structures favored the differentiation and activation of helper T cell subtypes (Th2 and Treg). Kaurenoic acid, oridonin, stevioside, glaucocalyxin A and B, eriocalyxin B, kamebakaurin, and xylopic acid have shown good therapeutic activity, and these structures are currently used to develop new compounds. In this review article, the immunopharmacological evidence of kaurane-type compound was highlighted through several in vitro and in vivo models, evidencing the molecular and cellular targets of the immune system, opening new expectations for the potential use of these compounds for the treatment of inflammatory and autoimmune diseases in humans.

**Keywords**

Kaurane · Diterpenoid · Inflammation · NF-κB · Nitric oxide · Macrophage · Inflammatory · Natural products

## Abbreviations

| | |
|---|---|
| Aβ | β-amyloid peptide |
| AD | Alzheimer's disease |
| AIM2 | Absent in melanoma 2 |
| ASC | Apoptosis-associated speck-like protein containing a C-terminal caspase recruitment domain |
| BAFF | B cell-activating factor |
| b.i.d. | Twice daily |
| BMDM | Mouse bone marrow-derived macrophages |
| CAR | Carrageenan |
| CD | Crohn's disease |
| CNS | Central nervous system |
| COPD | Chronic obstructive pulmonary disease |
| COX-2 | Cyclooxygenase-2 |
| DAMP | Damage-associated molecular patterns |
| DC | Dendritic cells |
| EriB | Eriocalyxin B |
| fMLP | *N*-formyl-L-methionyl-L-leucyl-L-phenylalanine |
| GCLC | Glutamate-cysteine ligase catalytic |
| GCLM | Glutamate-cysteine ligase modifier |
| GLA | Glaucocalyxin A |
| GLB | Glaucocalyxin B |
| HMGB-1 | High-mobility group box 1 protein |
| HO-1 | Heme oxygenase-1 |
| IBD | Inflammatory bowel disease |
| $IC_{50}$ | Half-maximal inhibitory concentration |

| | |
|---|---|
| $ID_{50}$ | Inhibitory dose 50% |
| IFN-γ | Interferon-gamma |
| IFNR | IFN-γ receptor |
| IκB | Inhibitor of NF-κB |
| IKK | Inhibitor of nuclear factor kappa-B kinase |
| IL | Interleukin |
| iNOS | Inducible nitric oxide synthase |
| KA | Kaurenoic acid |
| KMBK | Kamebakaurin |
| LPS | Lipopolysaccharide |
| LSRM | LPS-stimulated RAW 264.7 macrophages |
| MAPK | Mitogen-activated protein kinase |
| MPO | Myeloperoxidase |
| mRNA | Messenger of ribonucleic acid |
| MSU | Monosodium urate |
| MyD88 | Myeloid differentiation primary response 88 |
| NADPH | Nicotinamide adenine dinucleotide phosphate |
| NF-κB | Nuclear factor-kappa B |
| NGF | Nerve growth factor |
| NIK | NF-κB-inducing kinase |
| NLR | Nucleotide-binding domain leucine-rich repeat containing |
| NLRC4 | NLR family CARD domain containing 4 |
| NLRP3 | NOD-,LRR-and pyrin domain-containing protein 3 |
| NO | Nitric oxide |
| Nrf2 | Nuclear factor erythroid 2-related factor 2 |
| Ori | Oridonin |
| PAMP | Pathogen-associated molecular patterns |
| PD | Parkinson's disease |
| $PGE_2$ | Prostaglandin E2 |
| PHA | Phytohemagglutinin |
| p.o. | Orally (from the Latin "per os," by mouth) |
| poly(I:C) | Polyinosinic:polycytidylic acid |
| PRR | Pattern recognition receptors |
| PYD | Pyrin domain |
| RNS | Reactive nitrogen species |
| ROS | Reactive oxygen species |
| siRNA | Small interfering RNA |
| SLE | Systemic lupus erythematosus |
| SOD | Superoxide dismutase |
| STE | Stevioside |
| TAK1 | Transforming growth factor-beta-activated kinase 1 |
| TGF-β | Transforming growth factor beta |
| TGFβR1 | TGF-β receptor I |

| | |
|---|---|
| TLR | Toll-like receptor |
| TNF-α | Tumor necrosis factor alpha |
| TNFR | Tumor necrosis factor receptor |

## 7.1 Introduction

The inflammatory response involves a complex physiological process developed by the immune system to protect the organism against infection by human pathogens, including bacteria, viruses, or fungi. Also, inflammation protects the body against tissue injury induced by irritants, dust particles, shock, burn, toxic compounds, and other mechanical or physical agents, allowing the tissue to recover (Kishore et al. 2019). Inflammation can be catalogued as microbial, autoimmune, allergic, metabolic, or physical (Hawiger and Zienkiewicz 2019). There are five signs of inflammation that include heat, redness, swelling, pain, and loss of function (Hannoodee and Nasuruddin 2020). The pain associated with inflammation is related to the lowered pH with the consequent tissue acidification (Kuprash and Nedospasov 2016). Three steps of inflammation have been identified. The first step is also called the "silent phase"; at this stage occurs the release of the inflammatory mediators such as histamine, cytokines, prostaglandins, and nitric oxide (NO) from macrophages and mast cells present in the injured tissues. The second stage is called the "vascular phase," and is related to the dilatation of the blood vessels that increase the permeability. The last stage is named the "cellular phase" due to leukocytes and plasma protein infiltration from vessels to the damaged tissue (Kazemi et al. 2018). The selective expression of selectins and other molecules allows the permeation of neutrophils through endothelial cell, avoiding the passage of erythrocytes (Kazemi et al. 2018).

Inflammation can be classified as acute, subacute, and chronic (Hannoodee and Nasuruddin 2020). The acute inflammation is what follows immediately after the tissue damage with the development of the three phases previously mentioned. The resolution of inflammation involves the elimination of the pathogen and the repair of the wound. The period between 2 and 6 weeks is known as subacute inflammation. Chronic inflammation takes place in prolonged cases with no resolution, causing tissue necrosis and degeneration (Kishore et al. 2019). Persistent inflammation becomes destructive and is associated with cytokine dysregulation, which has changed its expression pattern, leading to a new term "inflammaging," which encompassed diseases like atherosclerosis, diabetes, Alzheimer's disease (AD), rheumatoid arthritis, and cancer (Rea et al. 2018).

Ent-kaurane-type diterpenoids are compounds that modulate various biological activities. Moreover, the compounds are involved in antitumor, antibacterial, antihyperglycemic, antioxidant, anti-inflammatory, antidiabetic, cytoprotective, and analgesic responses (González-Burgos et al. 2013a; Woode et al. 2012; Suárez et al. 2009). These activities have been reported in models in vitro and in vivo, by inhibiting the production of proinflammatory cytokines, enzymes, and transcription factors (Ding et al. 2016; Martínez et al. 2019; Zhang et al. 2019; Xu et al. 2018).

Structural modifications and formulations (drug delivery systems) have been performed and analyzed to improve the pharmacokinetics and pharmacodynamics of kauranes and their derivatives in humans (Xu et al. 2018; Zhang et al. 2020). In this review, we address the main anti-inflammatory effects of kauranes and their derivatives at the molecular and cellular level, as well as their pharmacological activity in vivo in animal models of inflammatory diseases.

## 7.2 Chemistry of Kauranes

Kaurane diterpenes are formed by cyclization of the 2E,6E,10E-geranylgeranyl pyrophosphate (GGPP). They are classified as tetracyclic diterpenoids. The perhydrophenanthrene unit (A, B, and C rings) is fused with a fourth ring (D) through a linkage between carbons 8 and 13 (Fig. 7.1). A series of kaurane structures formed by the fusion of rings A and B similarly as occurs in steroids are called "normal," considering its stereochemistry. The prefix "*ent*-" is found in the most common kauranes, and they are the "enantiomeric" corresponding to specular images of the "normal" structures, in which the inverted configurations are related to carbons C-5, C-9, and C-10 (García et al. 2007). In Fig. 7.2 are shown the chemical structures of the most relevant kaurane compounds.

One of the most studied kauranes is the ent-kaur-16-en-19-oic acid (kaurenoic acid, KA), isolated from aerial parts of *Wedelia paludosa* DC. KA is an intermediate molecule in the pathway for the generation of other kauranes. It has been reported that the KA has anti-inflammatory activity, as well as antimicrobial, trypanocidal, and cytotoxic properties (Takahashi et al. 2014).

Some kauranes naturally occur as glycosides. *Stevia rebaudiana* is a terrestrial plant found in South America (Paraguay and Brazil). The leaves of *S. rebaudiana* have the diterpenoid constituents: stevioside (STE), rebaudiosides A and D, and dulcoside A. The STE is a commercial sweetener, which is constituted by a kaurane skeleton and three molecules of glycoside. Rebaudioside A and STE have the same

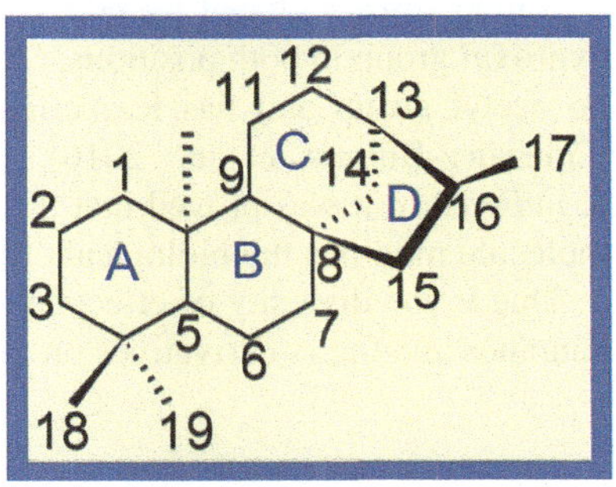


**Fig. 7.1** The carbon skeleton structure of kaurane-type compounds

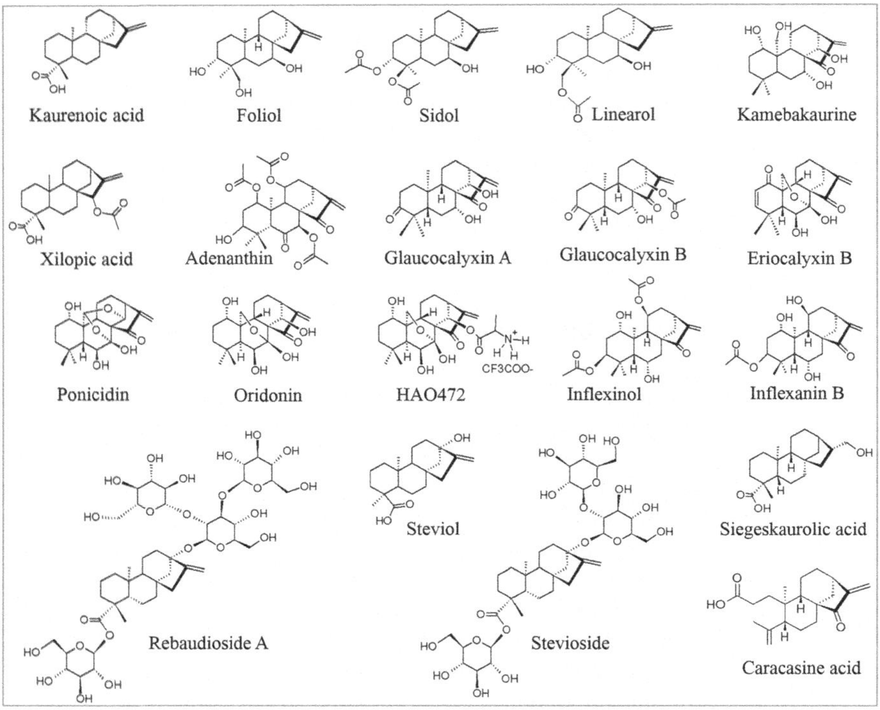


**Fig. 7.2** Kaurane-type compounds with demonstrated anti-inflammatory effects

kaurene structure, but rebaudioside A has an additional glucose moiety (Wang et al. 2018; Gupta et al. 2016). The products of STE hydrolysis are the *ent*-kaurane diterpene steviol (alkaline hydrolysis) or the *ent*-beyerane diterpene isosteviol (acidic hydrolysis) (Wang et al. 2018).

There are some identified features in kaurane compounds that are proven to increase the biological effect. It was reported that the similar compounds linearol and sidol differ in the position of the acetoxy functional group, which is present in position C-18 in linearol and position C-3 in sidol. Both inhibited the mitochondrial oxidative stress induced by $H_2O_2$. Foliol is another similar compound that has a hydroxyl group in both positions, C-3 and C-18. Unlike linearol and sidol, foliol has no acetyl group and has less effectivity, preventing mitochondrial dysfunction (González-Burgos et al. 2016). In a group of *ent*-kauranes isolated from *S. pubescens*, it was proved that the presence of isobutyrate or acetyl groups in the molecule modifies the biological effects.

Due to the diversity of effects, the aim of the review is to focus on the effect of kauranes and theirs derivatives on the immune response.

## 7.3 Anti-inflammatory Activity of Kauranes on Cells of the Immune System

It has been demonstrated that the kaurane-type compounds have effects on various immunocompetent cells, such as neutrophils, peripheral macrophages and microglia, dendritic cells, and T cells, as detailed in Fig. 7.3.

### 7.3.1 Anti-inflammatory Effects on Neutrophils

Neutrophils are the first line of response to inflammation and infection. More than 60% of the hematopoietic production is committed to neutrophil production. These immunocompetent cells are generated in quantities up to a billion every day and have a short life of approximately 19 h (Lawrence et al. 2020). The injured tissues

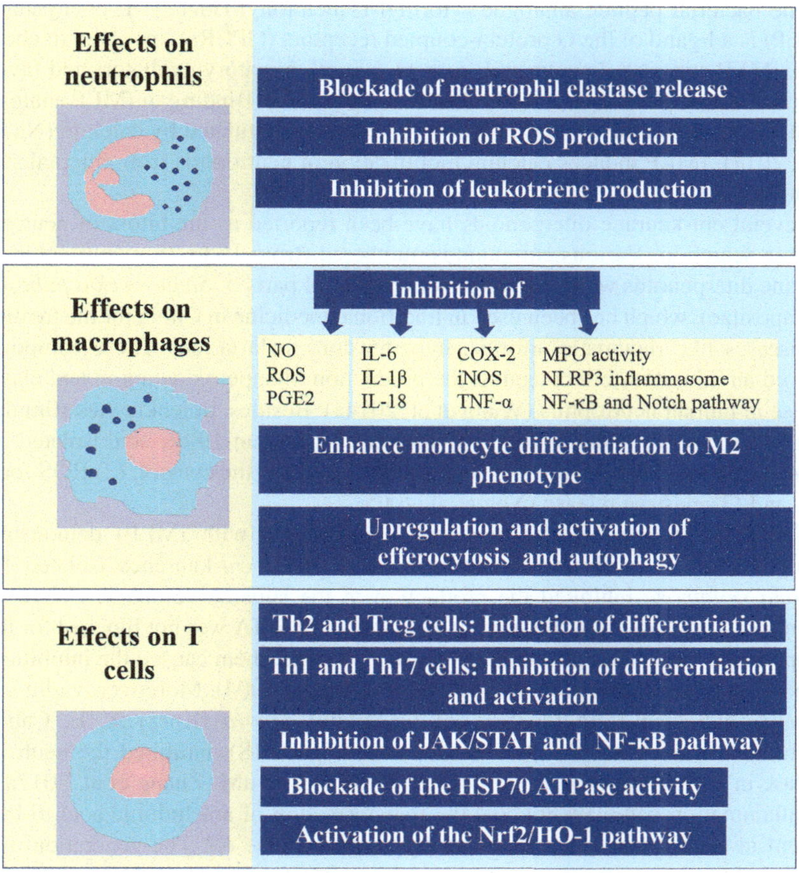


**Fig. 7.3** Effects of kaurane-type compounds in immunocompetent cells

are characterized by the recruitment of a vast number of neutrophils. These cells migrate to the site of inflammation through a process named transendothelial migration induced by cytokines (Papayannopoulos 2018). The neutrophil elastase is released due to stimuli and enhances the inflammatory response (Okada 2017). These enzymes response quickly against pathogens, although they cause tissue destruction (Kubes 2018).

In kauranes, the isobutyrate in position C-17 protects the cells against superoxide generation and elastase secretion. Acetylation at position 18 decreased the inhibitory effect against the production of superoxide radical but increased the inhibition of neutrophil elastase secretion (Wang et al. 2014a).

Based on the in vitro assays, the ent-kauranes with COOH as a functional group in C-19 are more effective in inhibiting the generation of superoxide anion (Yang et al. 2004). Other functional groups to block the reactive oxygen species (ROS) generation are the presence of β-OH and COOH groups at C-16 (Wang et al. 2014a; Yang et al. 2004).

The bacterial peptide analogue n-formyl-L-methionyl-L-leucyl-L-phenylalanine (fMLP) is a ligand of the G protein-coupled receptors (GPCRs), involved in chemotaxis. fMLP-stimulated neutrophils secrete pro-inflammatory mediators and favored chemotaxis and phagocytosis (Winther et al. 2016). The Binding of fMLP analogues to GPCRs induced the production of ROS, a process regulated by calcium (Nguyen et al. 2017). fMLP induces calcium mobilization in neutrophils from internal stores to the cytosol.

Several ent-kaurane diterpenoids have been reported as inhibitors of neutrophil elastase secretion. Various compounds numbered (from 16 to 25) identified as *ent*-kaurane diterpenoids were extracted from the aerial parts of *Siegesbeckia pubescens* (Compositae), which has been used in traditional medicine in China for the treatment of diseases like rheumatic arthritis, hypertension, malaria, etc. These compounds showed an inhibitory effect against the production of superoxide anion and elastase release in human neutrophils (Wang et al. 2014a). Besides, in leukocytes stimulated with fMLP, the *ent*-kaurane 16β,17-dyhidroxy-ent-kauran-19-oic acid isolated from *Annona squamosa* inhibited (1) calcium mobilization to the cytosol, (2) ROS formation, and (3) elastase release (Yeh et al. 2005).

The same in vitro model (neutrophils stimulated with fMLP) demonstrated biological effects in the other two researches. Nine *ent*-kauranes isolated from *Annona squamosa* inhibited the production of the superoxide anion. Conversely, the induced production of the superoxide anion with PMA was not blocked for these compounds, which indicated that the pathway through them caused the inhibition of ROS is not mediated by protein kinase C (Yang et al. 2004). Moreover, various *ent*-kauranes extracted from the bark of *Gochnatia decora* (Kurz) A. L. Cabrera, identified with numbers (1, 2, 4, 5, 6, 7, 12, 15, 17, 18), inhibited the neutrophil elastase in more than 40%, at the concentration of 100 μM (Zhang et al. 2017a).

Inflammation pathways involve the transformation of arachidonic acid to $PGH_2$ present in the lipid bilayer membrane by phospholipase $A_2$. The generation of the $PGH_2$ is due to cyclooxygenase enzymes 1 and 2 (COX-1 and COX-2). COX-1 has many physiological functions as a protection of gastric mucosae and platelet aggregation. The induction of COX-2 in the cyclooxygenase pathway generates large amounts of prostaglandin E2 ($PGE_2$) and other prostaglandins that are responsible for the nerve ending sensitizing and the lowering of the pain threshold (Attiq et al. 2018). The kaurane-type compounds 16αH,17-isovalerate-ent-kauran-19-oic acid and KA have been demonstrated to inhibit the COX-1 activity, with $IC_{50}$ values of 0.21 and 0.15 mM (Kiem et al. 2004).

The leukotriene biosynthesis also takes part in the inflammatory response in neutrophils. The enzyme 5-lipoxygenase oxygenates the arachidonic acid, producing the 5-hydroperoxy-eicosatetraenoic acid (5-HPETE), the precursor of leukotrienes. A kaurane, named methyl ent-7α,9α-dihydroxy-15β-[(2Z)-2-methyl-but-2enoyloxy]-kaur-16-en-19-oate isolated from *Leontopodium alpinum* Cass., inhibited leukotriene production in human neutrophils with a reported half-maximal inhibitory concentration ($IC_{50}$) of 10.4 μM. The KA showed no activity in the same assay (Schwaiger et al. 2004).

### 7.3.2 Anti-inflammatory Effects on Monocytes and Macrophages

Macrophages are innate immune cells that reside in tissues, playing an essential role in the body's defenses against pathogenic microorganisms. Furthermore, they are crucial in the resolution of the inflammatory process, healing wounds, and regulating tissue homeostasis. A proportion of macrophages are generated from the differentiation of monocytes in the blood, which, in turn, come from myeloid precursors produced in the bone marrow (Kratofil et al. 2017). Most tissue-resident macrophages come from embryonic development and can self-maintain through local proliferation. Examples of these kinds of macrophages are alveolar macrophages in the lungs, microglia in the central nervous system (CNS), Kupffer cells in the liver, and Langerhans cells in the skin (Epelman et al. 2014). Additionally, macrophages participate in the adaptive immune response through the process of antigen presentation to B and T cells. Macrophages play important roles in autoimmune diseases, rheumatoid arthritis, inflammatory bowel disease (IBD), systemic lupus erythematosus (SLE), endocrinological diseases, type 1 and 2 diabetes (Espinoza et al. 2012; Siouti and Andreakos 2019), acute lung and liver damage, asthma and chronic obstructive pulmonary disease (COPD) (de Groot et al. 2019; Saradna et al. 2018), cardiovascular diseases (Nasser et al. 2020) and neurological and neurodegenerative diseases such as AD and Parkinson's disease (PD), and multiple sclerosis (Colonna and Butovsky 2017).

On the other hand, two of the most relevant characteristics of macrophages are their plasticity and diversity. These cells are usually classified as M1 and M2, suggesting that they are polarized, that is, they can differentiate into either a pro-inflammatory or anti-inflammatory phenotype (Liu et al. 2014; He et al. 2019). Macrophages M1 (classically activated macrophages) are proinflammatory and play a fundamental role in the defense against pathogenic microorganisms. The sustained or pathological activation of the M1 phenotype can cause significant damage to cells and tissues, thereby promoting the generation of chronic inflammatory and autoimmune diseases, and cancer (Castegna et al. 2020). The M1 macrophages respond to stimulation with microbial components such as LPS and other Toll-like receptor (TLR) ligands, or by cytokines such as TNF-α, IL-1β, or interferon-gamma (IFN-γ) (Castegna et al. 2020). In the presence of these stimuli, M1 macrophages secrete various mediators to eliminate pathogenic elements. Among them are NO and other reactive nitrogen species (RNS), ROS, $PGE_2$, and proinflammatory cytokines such as IL-1β, IL-6, IL-12, IL-15, IL-18, IL-32, and TNF-α, among others (He et al. 2019). Furthermore, $PGE_2$ and NO are produced through the inducible enzymes COX-2 and iNOS, respectively, whose gene expression is regulated by NF-κB (Xue et al. 2018). NF-kB role is associated with inflammatory diseases such as RA, IBD, asthma, COPD, and ulcerative colitis. The Rel family constitutes are NF-kB p50, p52, p65 (Rel A), c-Rel, and Rel B. The most abundant form of NF-κB protein complex is the p50/p65 heterodimer (Mitchell and Carmody 2018). These proteins remain inactive in the cytoplasm as a complex bound to IκB, an inhibitory protein in unstimulated conditions. The activation of NF-kB causes phosphorylation of IκB by IKK that releases and translocates the NF-kB into the nucleus and binds to the kB sites in the promoter of the target genes. The phosphorylated IκB is targeted for ubiquitination and subsequent proteasomal degradation (Zhang et al. 2017b).

M2 macrophages (alternatively activated macrophages) are characterized by the production of anti-inflammatory mediators such as IL-10, as well as other mediators that promote disease resolution and tissue repair (Liu et al. 2014; Kloc et al. 2019). M2 macrophages are activated by anti-inflammatory cytokines IL-4 and IL-13, secreted by various cells of the innate and adaptive immunity.

Various kaurane-type compounds have shown anti-inflammatory effects in vitro, by inhibiting the release of NO and ROS and IL-1β, IL-6, IL-12, and TNF-α, as well as COX-2 expression and iNOS in human monocytes and mouse macrophages stimulated with LPS (Table 7.1). The mechanism behind the inhibition of most of these inflammatory mediators by kauranes has been associated with the inhibition of the NF-κB pathway (Choi et al. 2011; Park et al. 2007; Castrillo et al. 2001).

#### 7.3.2.1 Effect of Kauranes on Monocytes and Peripheral Macrophages

LPS-stimulated RAW 264.7 macrophages (LSRM) are widely used as a model of inflammation to evaluate the potential anti-inflammatory effect of drugs in vitro. In LSRM, a significant number of kauranes inhibited NO production (Table 7.1). In some cases, the inhibition of this inflammatory mediator was accompanied by a reduction of $PGE_2$ and TNF-α release, as well as the downregulation of COX-2 and iNOS (Table 7.1). It has been proved that Ori, KMBK, KA, and other kauranes exert their anti-inflammatory effects in a mechanism which involved blocking the NF-κB activation pathway through increased IκBα degradation and inhibition of nuclear translocation and/or activation of NF-κB (Table 7.1).

Ori, ponicidin, and xindongnin A and xindongnin B affect the binding of the NF-κB p50 and p65 homodimers and heterodimers to DNA. The structures are bound to a different site, through a noncompetitive inhibition (Leung et al. 2005). Besides, EriB blocked NF-κB activation through a mechanism that involved

**Table 7.1** Anti-inflammatory effect of kaurane diterpenoids and their derivatives in LPS-stimulated RAW 264.7 murine macrophages

| | | RAW 264.7-stimulated LPS | | | | | | | | | | |
|---|---|---|---|---|---|---|---|---|---|---|---|---|
| | | NO | TNF-α | $PGE_2$ | COX-2 | | iNOS | | | NF-κB p65 | | |
| Kaurane | Source | $IC_{50}$ (μM) | $IC_{50}$ (μM) | $IC_{50}$ (μM) | Protein | mRNA | Protein | mRNA | IκBα cytosol | Translocation | Activation | References |
| Effusanin C | *Isodon japonicus* | 5.9 | – | – | – | – | – | – | – | – | – | Hong et al. (2008) |
| Effusanin C | *Isodon japonicus* | 1–2[a] | 2–3[a] | – | – | – | – | ↓ | ↑ (β) | ↓ | – | Kim et al. (2013a) |
| Effusanin D | *Isodon japonicus* | 3.3 | – | – | – | – | – | – | – | – | – | Hong et al. 2008) |
| EriB | *Isodon eriocalyx* | – | – | – | ↓ | ↓ | ↓ | ↓ | – | – | – | (Leung et al. (2006) |
| Excisanin A | *Isodon japonicus* | 0.35 | – | 5.37 | – | – | – | – | – | – | ↓ | Hwang et al. (2001) |
| Excisanin B | *Isodon japonicus* | 3.3 | – | – | – | – | – | – | – | – | – | Hong et al. (2008) |
| Excisusin A | *Isodon excisus* | 0.67 | – | – | – | – | – | – | – | – | ↓ | Hong et al. (2007) |
| Excisusin B | *Isodon excisus* | 0.56 | – | – | – | – | – | – | – | – | ↓ | Hong et al. (2007) |
| Excisusin D | *Isodon excisus* | 2.89 | – | – | – | – | – | – | – | – | ↓ | Hong et al. (2007) |
| Excisusin E | *Isodon excisus* | 1.36 | – | – | – | – | – | – | – | – | ↓ | Hong et al. (2007) |
| GLB | *Rabdosia japonica* | 2.5–5 | – | – | – | – | – | – | – | – | – | Gan et al. (2015) |
| Grandiflorolic acid | *Helianthus annuus L* | 5–10 | – | 10–20 | ↓ | ↓ | ↓ | ↓ | – | – | – | Díaz-Viciedo et al. (2008) |

(continued)

**Table 7.1** (continued)

| | | RAW 264.7-stimulated LPS | | | | | | | | | | |
|---|---|---|---|---|---|---|---|---|---|---|---|---|
| | | NO | TNF-α | $PGE_2$ | COX-2 | | iNOS | | | NF-κB p65 | | |
| Kaurane | Source | $IC_{50}$ (μM) | $IC_{50}$ (μM) | $IC_{50}$ (μM) | Protein | mRNA | Protein | mRNA | IκBα cytosol | Translocation | Activation | References |
| Inflexanin A | *Isodon excisus* | 0.63 | – | – | – | – | – | – | – | – | ↓ | Hong et al. (2007) |
| Inflexanin B | *Isodon excisus* | 2.52 | – | – | – | – | – | – | – | – | ↓ | Hong et al. (2007) |
| Inflexarabdonin G | *Isodon excisus* | 0.48 | – | – | – | – | – | – | – | – | ↓ | Hong et al. (2007) |
| Inflexarabdonin I | *Isodon excisus* | 1.24 | – | – | – | – | – | – | – | – | ↓ | Hong et al. (2007) |
| Inflexin | *Isodon excisus* | 0.69 | – | – | – | – | – | – | – | – | ↓ | Hong et al. (2007) |
| Inflexinol | *Isodon excisus* | 3.43 | – | – | ↓ | – | ↓ | – | ↑ | ↓ | ↓ | Lee et al. (2007) |
| Inflexinol | *Isodon excisus* | 0.94 | – | – | – | – | – | – | – | – | ↓ | Hong et al. (2007) |
| Isodojaponin A | *Isodon japonicus* | 6.3 | – | – | – | – | – | – | – | – | – | Hong et al. (2008) |
| Isodojaponin B | *Isodon japonicus* | 10.2 | – | – | – | – | – | – | – | – | – | Hong et al. (2008) |
| Isodojaponin C | *Isodon japonicus* | 8.2 | – | – | – | – | – | – | – | – | – | Hong et al. (2008) |
| Isodojaponin D | *Isodon japonicus* | 8.7 | – | – | – | – | – | – | – | – | – | Hong et al. (2008) |
| Isodojaponin E | *Isodon japonicus* | 20 | – | – | – | – | – | – | – | – | – | Hong et al. (2008) |

| | | | | | | | | | | | | |
|---|---|---|---|---|---|---|---|---|---|---|---|---|
| KA | *Aralia continentalis* | = | – | – | – | = | – | – | = | = | = | Lyu et al. (2011) |
| KA | *Helianthus annuus L* | 5–10 | – | 10–20 | ↓ | ↓ | ↓ | ↓ | – | – | – | Díaz-Viciedo et al. (2008) |
| KA | *Aralia continentalis* | 51.73 | – | 106.09 | ↓ | ↓ | ↓ | ↓ | – | – | ↓ | Choi et al. (2011) |
| Kamebacetal A | *Isodon japonicus* | 0.58 | – | 2.80 | – | – | – | – | – | – | ↓ | Hwang et al. (2001) |
| Kamebanin | *Isodon japonicus* | 0.06 | – | 0.88 | – | – | – | – | – | – | ↓ | Hwang et al. (2001) |
| KMBK | *Isodon japonicus* | 0.15 | – | 2.63 | – | – | – | – | – | – | ↓ | Hwang et al. (2001) |
| KMBK | *Isodon japonicus* | – | <1.0[a] | – | ↓ | ↓ | ↓ | ↓ | – | – | – | Lee et al. (2004) |
| Lasiokaurin | *Isodon japonicus* | 1.6 | – | – | – | – | – | – | – | – | – | Hong et al. (2008) |
| Longikaurin B | *Isodon japonicus* | 0.9 | – | – | – | – | – | – | – | – | – | Hong et al. (2008) |
| Longikaurin B | *Isodon xerophylus* | 0.435 | – | – | – | – | ↓ | ↓ | – | ↓ | ↓ | Aquila et al. (2009) |
| Longikaurin C | *Isodon japonicus* | 1.0 | – | – | – | – | – | – | – | – | – | Hong et al. (2008) |
| Longikaurin D | *Isodon japonicus* | 2.0 | – | – | – | – | – | – | – | – | – | Hong et al. (2008) |
| Loxothyrin A | *Isodon japonicus* | 24.2 | – | – | – | – | – | – | – | – | – | Hong et al. (2008) |

(continued)

**Table 7.1** (continued)

| | | RAW 264.7-stimulated LPS | | | | | | | | | | |
|---|---|---|---|---|---|---|---|---|---|---|---|---|
| | | NO | TNF-α | $PGE_2$ | COX-2 | | iNOS | | | NF-κB p65 | | |
| Kaurane | Source | $IC_{50}$ (μM) | $IC_{50}$ (μM) | $IC_{50}$ (μM) | Protein | mRNA | Protein | mRNA | IκBα cytosol | Translocation | Activation | References |
| Megathyrin A | *Isodon japonicus* | 8.5 | – | – | – | – | – | – | – | – | – | Hong et al. (2008) |
| Ori | *Rabdosia rubescens* | <1 [a] | – | – | – | – | – | ↓ | – | – | – | Zhang et al. (2013) |
| Ori | *Rabdosia rubescens* | – | <30 [a] | – | – | – | – | – | ↑ | ↓ | – | Zhao et al. (2017) |
| Ori | *Isodon rubescens* | – | – | – | ↓ | ↓ | ↓ | ↓ | – | – | – | Leung et al. (2005) |
| Ponicidin | *Isodon rubescens* | – | – | – | ↓ | ↓ | ↓ | ↓ | – | – | ↓ | Leung et al. (2005) |
| Pseurata C | *Isodon japonicus* | 5.4 | – | – | – | – | – | – | – | – | – | Hong et al. (2008) |
| Siegeskaurolic acid | *Siegesbeckia pubescens* | 42.1[a] | >60 [a] | <20[a] | ↓ | ↓ | ↓ | ↓ | ↑ | ↓ | ↓ | Park et al. (2007) |
| STE | *Stevia rebaudiana Bertoni* | >40 | >40 | – | – | – | – | – | – | – | – | Alavala et al. (2019) |
| STE | *Stevia rebaudiana* | – | – | – | – | – | – | – | ↑ | – | – | Fengyang et al. (2012) |
| Xerophilusin A | *Isodon xerophylus* | 0.600 | – | – | – | – | ↓ | ↓ | – | ↓ | ↓ | Aquila et al. (2009) |

| | | | | | | | | | | | | |
|---|---|---|---|---|---|---|---|---|---|---|---|---|
| Xerophilusin B | *Isodon xerophylus* | 0.225 | – | – | – | – | ↓ | ↓ | – | ↓ | ↓ | Aquila et al. (2009) |
| Xerophilusin F | *Isodon xerophylus* | 0.673 | – | – | – | – | ↓ | ↓ | – | ↓ | – | Aquila et al. (2009) |
| Xindongnin A | *Isodon rubescens* | – | – | – | ↓ | ↓ | ↓ | ↓ | – | – | ↓ | Leung et al. (2005) |
| Xindongnin B | *Isodon rubescens* | – | – | – | ↓ | ↓ | ↓ | ↓ | – | – | ↓ | Leung et al. (2005) |

Symbols: [a]The $IC_{50}$ value is expressed in μg/mL, ↓ indicates inhibition/reduction, ↑ indicates increase, (=) indicates no changes, and (–) not determined avoiding the DNA-binding activity of subunits p65 and p50, in a reversible, non-competitive manner, without affecting the nuclear translocation of this transcription factor (Leung et al. 2006). The inhibitory effect of EriB and KMBK on the NF-κB activation pathway was associated with the covalent binding of these kauranes to the cysteine 62 of p50 through the α,β-unsaturated ketone. This interaction prevented the binding of this subunit to the DNA response elements, without affecting the dimerization or the nuclear translocation of p50 and p65 (Lee et al. 2002; Kong et al. 2014). Furthermore, inflexinol inhibited NF-κB activation through direct modification of a cysteine in the p50 subunit of NF-κB (Ban et al. 2009).

In LSRM, effusanin C significantly inhibited the phosphorylation of p38, JNK, and ERK (Kim et al. 2013a). Consequently, effusanin C exhibited an in vitro anti-inflammatory effect in macrophages stimulated by blocking the NF-κB and MAPK signaling pathways (Table 7.1).

It has been proved that the ROS production was inhibited in LSRM treated with Ori (Yang et al. 2019). In the unstimulated cells, Ori protected macrophages against exogenous $H_2O_2$ toxicity (Yang et al. 2019). The effect may be due to the increased expression of nuclear factor erythroid 2-related factor 2 (Nrf2), heme oxygenase-1 (HO-1), and glutamate-cysteine ligase modifier (GCLM) proteins, which are associated with increased antioxidative mechanisms (Yang et al. 2019). The authors of this study demonstrated that Ori regulates Nrf2 expression through activation of Akt and JNK, p38, and ERK MAPK in RAW 264.7 cells (Yang et al. 2019).

Additionally, in LSRM, and the primary peritoneal macrophages of rats stimulated with LPS, treatment with Ori favored a switch in macrophages from classically activated inflammatory (M1, IL-1β, IL-6, and iNOS) into the anti-inflammatory phenotype (M2, IL-10, and CD206) in both cells. In LSRM, the authors of this study demonstrated that Ori induced the phenotypic switch of M1 to M2 through blocking the Notch pathway (reduced the mRNA levels of Jagged-2, Notch1, and Hes-1 in stimulated cells) (Xu et al. 2019).

As part of the anti-inflammatory process of KA in LSRM, this kaurane increased the expression of the messenger of ribonucleic acid (mRNA) of glutamate-cysteine ligase catalytic (GCLC) subunit and HO-1. In stimulated cells, KA increased nuclear translocation and activation of the Nrf2, which controls the gene expression of these two antioxidant mediators (Lyu et al. 2011). Besides, Kim et al. (2017) showed that in LSRM, the mechanism of inhibition of the inflammatory process by KA involved the activation of the signaling process of TGF-β through of the TGF-β receptor 1 (TGFβR1) (Kim et al. 2017).

Furthermore, STE significantly inhibited the secretion of TNF-α and IL-6 and the production of ROS in LSRM (Alavala et al. 2019). The STE anti-inflammatory mechanism in LSRM was associated with the inhibition of NF-κB and MAPK signaling pathways (Fengyang et al. 2012).

In RAW 264.7 and mouse peritoneal macrophages stimulated with LPS/IFN-γ, foliol and linearol significantly inhibited NO production. This study demonstrated that these compounds protected both cell types from LPS-induced apoptosis and further inhibited the phagocytic activity toward zymosan particles (De las Heras et al. 2007).

The *ent*-kaurane diterpenes, foliol, linearol, and KA inhibited the release of NO and TNF-α and protein expression of iNOS and COX- 2 in J774 macrophages (mouse macrophage cell line) stimulated with LPS (Castrillo et al. 2001). These three kauranes inhibited the activation of NF-κB and the inhibitor of NF-κB kinase subunit beta (IKK-β) and decreased phosphorylation of p38, ERK1, and ERK2 mitogen-activated protein kinase (MAPK). Foliol, linearol, and KA inhibited the NF-κB-inducing kinase (NIK) (Castrillo et al. 2001), an enzyme of the MAPK family, which participates in the noncanonical signaling pathway of the NF-κB, in response to the activation of several members of the TNF-α family (Brightbill et al. 2018).

In THP-1 cells stimulated with LPS, STE inhibited the release of NO, IL-1β, and TNF-α, downregulating the expression of NF-κB and IKK-β (Boonkaewwan et al. 2006). Besides, eriocalyxin B (EriB) downregulated the expression of COX-2 mRNA in human THP-1 monocytes stimulated with TNF-α (Leung et al. 2006).

Inflammasomes are complexes of cytosolic proteins that participate in the innate immune response and can be activated by various pathogen-associated molecular patterns (PAMP) and damage-associated molecular patterns (DAMP) (Swanson et al. 2019). After the recognition of PAMP or DAMP, by pattern recognition receptors (PRRs), such as TLR4s, the assembly of the inflammasome occurs through the nucleotide-binding and oligomerization domain (NOD) and nucleotide-binding domain leucine-rich repeat (NLR)-containing protein or pyrin domain (PYD) and HIN domain-containing (PYHIN) family. NLR family members—including NLRP1, NLRP3, NLR family CARD domain containing 4 (NLRC4), and NLRP6—and the PYHIN family member are absent in melanoma 2 (AIM2) (Swanson et al. 2019).

The NLRP3 inflammasome is one of the most studied to date. It is made up of the assembly of (1) the NLRP3 protein, which is an important sensor that detects a wide variety of pathogens and DAMPs; (2) apoptosis-associated speck-like protein containing a C-terminal caspase recruitment domain (ASC), which is an adapter protein; and (3) pro-caspase-1 (Swanson et al. 2019).

The activation of the NLRP3 inflammasome requires two signals. The first signal, also known as the priming signal, is mediated by microbial ligands recognized for PRRs (e.g., TLRs by LPS), TNF-α to its receptor (TNFR), or IFNs to its receptor (IFNR). This receptor-binding process leads to the activation of the NF-κB pathway, which allows the increased expression of various genes that participate in the inflammatory process, including NLRP3 and pro-IL-1β (Zhao and Zhao 2020). The second signal involves the stimulation of PAMP or DAMP receptors with nigericin, extracellular ATP, or the monosodium urate (MSU) crystals. NLRP3 inflammasome assembly leads to the activation of caspase-1, promoting the cleavage of pro-IL-1β and pro-IL-18, to produce biologically active pro-inflammatory cytokines IL-1β and IL-18, as well as gasdermin D that leads to cell death by pyroptosis (Swanson et al. 2019; Liu et al. 2016a). The NLRP3 inflammasome aberrant activation has been associated with a significant number of chronic inflammatory and autoimmune diseases (Swanson et al. 2019; Wang et al. 2020; Ciążyńska et al. 2020; Ding et al. 2019; Guo et al. 2018; Shi et al. 2020; White et al. 2017).

NLRP3 inflammasome is activated as a defense mechanism against different viral elements (Zhao and Zhao 2020).

He et al. (2018) showed that Ori is a specific and covalent inhibitor of the NLRP3 inflammasome. This diterpenoid forms a covalent bond with cysteine 279 of NLRP3 in the NACHT domain, thus preventing the interaction between NLRP3 and NEK7 proteins, and thus, blocking the assembly of this multiple protein complexes (He et al. 2018). The treatment of the LSRM with Ori inhibited the activation of the NLRP3 pathway, as well as the expression of high-mobility group box 1 protein (HMGB-1), TXNIP, and TRX-1 all associated with the inflammatory process (Yang et al. 2019). As will be detailed later in this review, Ori has prophylactic and therapeutic effects in various models of inflammation in rodents, which can be partially explained by the in vivo inhibition of the NLRP3 inflammasome (He et al. 2018).

GLA, an ent-kaurane, isolated from the medicinal plant *Rabdosia japonica* var., inhibited the activation of the canonical and noncanonical pathway of NLRP3 inflammasome in THP-1 cells and mouse bone marrow-derived macrophages (BMDMs) when these cells were primed with LPS or Pam3CSK4 (for noncanonical inflammasome activation) and subsequently stimulated with nigericin, ATP, MSU crystals, or polyinosinic:polycytidylic acid (poly(I:C). Under these stimulating conditions, GLA potently inhibited the activation of caspase-1 and IL-1β secretion in both cell types (Hou et al. 2020). Besides, GLA also blocked the activation of the NLRC4 inflammasome but did not affect the AIM2 inflammasome. The authors of this study showed that GLA inhibited NLRP3 or NLRC4 agonist-induced ASC oligomerization (Hou et al. 2020).

Another anti-inflammatory mechanism of kaurane-type compounds is the activation of the efferocytosis process by macrophages. This phagocytosis mediated for macrophages requires the finding and recognition of the apoptotic cell (Doran et al. 2020). Once the ingestion of apoptotic cells has occurred, macrophages switch the production of proinflammatory cytokines to others that improve tissue repair (Szondy et al. 2017). An impaired efferocytosis can cause pathological inflammation and tissue necrosis and is the cause of various inflammatory diseases, such as neurological disorders, atherosclerosis, and type 2 diabetes (Kawano and Nagata 2018).

Ori upregulated and activated efferocytosis and autophagy. The UV-irradiated apoptotic macrophages 264.7 were co-incubated with live macrophages 264.7 pretreated with Ori, resulting in upregulation and activation of TLR4 and its adaptor proteins TRIF, MyD88, and TRAF6. LPS and Ori induced TLR4 activation and stimulated autophagy. Autophagy was observed by the increased expression of Beclin-1 and LC3II. The process was confirmed by the blockade of TLR4 by small interfering RNA (siRNA) (Zang et al. 2019). The phagocytosis of apoptotic U937 cells induced by Ori was blocked by antihuman TNF-α and IL-1β antibodies, suggesting that these cytokines are important for the efferocytosis process (Liu et al. 2005). Irradiated U937 cells treated with low doses of Ori showed an increase in efferocytosis, and ERK phosphorylation was the signal pathway involved. Phagocytosis was reduced in the presence of the ERK inhibitor PD98059; Beclin-1 and the conversion from LC3I to LC3II were also decreased (Zang et al. 2012a). It was also demonstrated in the cell line U937 that Ori increased NO release, iNOS expression, efferocytosis, and autophagy. Pretreatment with L-NAME blocked the effects of Ori (Zang et al. 2012b).

#### 7.3.2.2 Effects on Macrophages of the Central Nervous System (Microglia)

Neuroinflammation and neurotoxicity reveal excessive inflammatory activation of the microglia (Rajendran and Paolicelli 2018). In vitro, the activation of microglia is induced by LPS, LPS/IFN-γ, or the β-amyloid peptide (**Aβ**). Consequently, it has become a model to assess the effect on neuroinflammation (Gan et al. 2015).

Various kaurane-type compounds decreased the neuro-inflammatory process in activated microglia cells (Dou et al. 2018; Lee et al. 2014). These structures inhibited NO, TNF-α, and $PGE_2$ production in LPS-activated BV-2 murine microglial cells by inhibiting COX-2 and iNOS expression and blocked the activation of the NF-kB pathway (Table 7.2).

In BV-2 cells stimulated with LPS, the treatment with GLA or KMBK inhibited mRNA expression of IL-1β, IL-6, and TNF-α, as well as the phosphorylation of p38 and other MAPKs (Kim et al. 2013b; Kim et al. 2011). Besides, in BV-2 cells, GLA increased Aβ clearance and inhibited tau phosphorylation, through a mechanism that involved inhibition of the mTOR/P70S6K signaling pathway (Zhou et al. 2019a). Furthermore, in BV-2 cells stimulated with LPS, inflexin inhibited the transcription of IL-1β, IL-6, and TNF-α. The inhibition of neuroinflammation in BV-2 cells by this kaurane was due to the inhibition in the NF-κB and Akt activation pathways, without affecting MAPKs pathway (Table 7.2; Ko et al. 2010). In this same model, inflexanin B decreased the release of various inflammatory mediators by inhibiting NF-κB and MAPKs pathways (Table 7.2; Lim et al. 2013).

GLB demonstrated a significant anti-inflammatory effect in BV-2 cells stimulated with LPS by decreasing: (1) the release of IL-1β, IL-6, and ROS; (2) the activity of MPO; and (3) the mRNA levels of TNF-α and IL-1β (Gan et al. 2015; Xu et al. 2017). Nonetheless, GLB significantly increased superoxide dismutase (SOD) activity (Xu et al. 2017). In the LPS-stimulated BV-2 cells, the GLB inhibition of ROS production involved the decreased expression of the gp91phox and p47phox subunits of the nicotinamide adenine dinucleotide phosphate (NADPH) oxidase. As an additional anti-inflammatory mechanism, GLB potently induced HO-1 expression in BV-2 cells. This effect, as in other kauranes, was associated with the inhibition of the NF-κB and p38 MAPK pathways. The neuroprotective effect of GLB prevents the activated microglia from inducing neurotoxicity in the co-culture model (Gan et al. 2015). The anti-inflammatory mechanism of GLB in this model involved the decreased expression of TLR2 and TLR4 and inhibition of activation of the NF-κB pathway, as well as increased nuclear expression of Nrf2 and levels of anti-inflammatory mediators: HO-1, GSTA1, and NQO1 (Xu et al. 2017).

In primary cultures of mouse, microglia stimulated with LPS, the GLA, GLB and KMBK significantly decreased NO production (Gan et al. 2015). Also, GLB inhibited the NO release in LPS-stimulated rat microglia cell line (HAPI) and

**Table 7.2** Effect anti-inflammatory of kauranes diterpenoid and their derivatives on LPS-activated BV-2 murine microglial cells

| | | BV-2 microglia cell-stimulated LPS | | | | | | | | | | |
|---|---|---|---|---|---|---|---|---|---|---|---|---|
| | | NO | $PGE_2$ | TNF-α | COX-2 | | iNOS | | | NF-κB p65 | | |
| Kaurane | Isolated plant | $IC_{50}$ (μM) | $IC_{50}$ (μM) | $IC_{50}$ (μM) | Protein | mRNA | Protein | mRNA | IκBα cytosol | Activation | Translocation | References |
| GLA | *Rabdosia japonica* | 0.1–1 | 1–5 | – | ↓ | ↓ | ↓ | ↓ | ↓ | – | ↓ | Kim et al. (2013b) |
| GLA | *Rabdosia japonica* | 5.23 | – | – | – | – | – | – | – | – | – | Gan et al. (2015) |
| GLB | *Rabdosia japonica* | 4.42 | – | 5–10 | ↓ | ↓ | ↓ | ↓ | ↓ | ↓ | ↓ | Gan et al. (2015) |
| KMBK | *Rabdosia japonica* | 7.35 | – | – | – | – | – | – | – | – | – | Gan et al. (2015) |
| KMBK | *Isodon japonicus* | 0.1–1 | 1–5 | –- | ↓ | ↓ | ↓ | ↓ | – | – | ↓ | Kim et al. (2011) |
| Inflexin | *Isodon excisus* | 1–5 | – | – | ↓ | ↓ | ↓ | ↓ | ↓ | ↓ | ↓ | Ko et al. (2010) |
| Inflexanin B | *Isodon inflexus* | <1–10[a] | 1–10[a] | 1–10[a] | ↓ | – | – | – | – | – | ↓ | Lim et al. (2013) |

Symbols: [a]The $IC_{50}$ value is expressed in μg/mL, ↓ indicates inhibition/reduction, ↑ indicates increase, (=) indicates no changes, and (–) not determined

LPS/IFN-γ-stimulated primary astrocytes (Gan et al. 2015), as well as TNF-α release in primary cultures of mouse microglia, stimulated with LPS (Gan et al. 2015). In primary microglial cells of rats stimulated with LPS, the anti-inflammatory effects of GLA was associated with the inhibition of NO release, as well as downregulation of the iNOS protein and COX-2 (Kim et al. 2013b). In these cells, GLA further inhibited the phosphorylation of IkB-α and p38 (Kim et al. 2013b). Finally, in rat primary microglia stimulated with LPS, pretreatment with Ori inhibited the release of NO, IL-1β, IL-6, TNF-α, and nerve growth factor (NGF), in addition to downregulation of the expression of the iNOS and NGF mRNA, in a mechanism that involved inhibition of NF-κB. Such effects paralleled with inhibition of the DNA binding activity of NF-kB (Xu et al. 2009).

Finally, in primary microglia cultures from mouse, the treatment of cells with EriB exerted a potent anti-inflammatory effect, by blocking the damage induced by MPP+ to dopaminergic (DA) neurons in a microglia co-culture system. EriB exerted its anti-inflammatory effects on cells stimulated with MPP + through (1) inhibition of the production of IFN-γ, TNF-α, IL-1β, and IL-6; (2) shifting the subtype of microglia, from the M1 phenotypic (iNOS and CD86) to M2 phenotypic (CD206 and Arg1); and (3) inhibition of phosphorylation of p65 and IκBα (Dou et al. 2018).

The inhibitory effects of kauranes on microglia allows us to propose the use of several structures for the treatment of various chronic inflammatory and autoimmune diseases.

### 7.3.3 Anti-inflammatory Effects on Dendritic Cells

Dendritic cells (DCs) are professional antigen-presenting cells (APCs) (Qian and Cao 2018). These cells, just like macrophages, represent the link between the innate response and adaptive immunity (Balan et al. 2019). DCs have multiple antigen receptors for pathogens that allow them to recognize a multiplicity of microbial elements, being keys to the initiation of the humoral immune responses, carried out by B cells, and the cellular immune response carried out by CD4 + and CD8 + T cells (Balan et al. 2019). The ability of DCs to interact with other cells of the immune system depends on the subtype and their maturation state. This last event is generally represented by the increased expression of molecules such as MHCII, CD40, CD80, and CD86, as well as the production of various proinflammatory cytokines (IL-1β, IL-6, IL-12, and TNF-α) (Qian and Cao 2018). Beyond its physiological effects on the response to infection, the dysregulation of DCs is part of the onset and perpetuation of various autoimmune diseases and chronic inflammatory diseases. Therefore, its blockade can contribute to the resolution of these pathologies (Qian and Cao 2018).

KMBK showed anti-inflammatory effects on bone marrow-derived dendritic cells (BMDCs) from mice. KMBK inhibited in LPS-treated BMDC (1) NO release; (2) the TNF-α, IL-12, and IL-1β secretion; (3) iNOS, TNF-α, IL- 12, and IL-1β mRNA expression; (4) phosphorylation of p38 and JNK, MEK3/6, and IKK; and (5) the degradation of the total IκBα levels and the nuclear translocation of NF-κB p65 (Kim et al. 2013c). In this study, KMBK directly inhibited the transforming growth factor-beta-activated kinase 1 (TAK1), a protein kinase that participates in the activation process of the NF-κB and MAPK pathways (Kim et al. 2013c). Unlike the effects observed for KMBK in BMDCs, in vitro EriB treatment did not affect the activation of these APCs (Lu et al. 2013).

In LPS-stimulated BMDC, effusanin C showed significant anti-inflammatory activity. The kaurane inhibited (1) NO and TNF-α production; (2) the expression of IL-1β, iNOS, and TNF-α mRNA; (3) the phosphorylation of p38, JNK, and ERK; (4) the degradation of IκBβ; and (5) the nuclear translocation of NF-κB p50 and p65. NF-κB and MAPK signaling pathways (Kim et al. 2013a).

### 7.3.4 Anti-inflammatory Effects on CD4+ Helper T-Cell Activation and Differentiation

CD4+ helper T cells are part of the adaptive response that performs essential functions in defense mechanisms against infection by microbial pathogens (Loo et al. 2018). However, beyond immune defense mechanisms, helper T cells may play a pathophysiological role in the onset and perpetuation of various chronic inflammatory diseases, autoimmune diseases, and cancer (Loo et al. 2018).

Helper T cell subpopulations can be differentiated from naïve T cells into Th1, Th2, Th9, Th17, Th22, follicular helper T (Tfh), peripheral helper T cells (Tph), and regulatory T cells (Tregs) (Gagliani and Huber 2017). Additionally, some helper T cell subsets are originated from other helper T cell subtypes, indicating the existence of plasticity in this type of cell (Cosmi et al. 2014; Sandquist and Kolls 2018). T-bet is the transcription factor involved in Th1 differentiation. Th1 cells release cytokines IL-2, IL-3, IFN-γ, and TNF-α, which are relevant for the inflammatory response. This subpopulation is involved in neuroinflammation observed in neurodegenerative diseases (Yu et al. 2019).

Th2 cells act against helminth infections. GATA-3 is the transcriptional regulator of these subsets of cells, and they release IL-4, IL-5, IL-10, IL-13, and TGF-β. Th2 are usually important in the pathophysiology of various chronic allergic and inflammatory diseases of the airways (Nakayama et al. 2017).

Th17 provides protection against various types of extracellular infections. RORγt is the transcriptional factor that controls the functions of this kind of lymphocyte. IL-17A, IL-17F, IL-22, and TNF-α are the proinflammatory cytokines released by Th17. Nevertheless, uncontrolled production of IL-17 cytokines is observed in chronic diseases, asthma, COPD, ulcerative colitis, multiple sclerosis, and rheumatoid arthritis. The pharmacological inhibition of these cytokines is possible (De Sanctis et al. 2009; Kunkl et al. 2020; Tatiya-Aphiradee et al. 2018; Vyas et al. 2019). Th17 cells have been shown to differentiate into regulatory T cells, depending on the stimulation of TGFβ (Knochelmann et al. 2018).

Tregs are essential in protecting against autoimmune responses while stopping effector responses against exogenous antigens when they can be harmful to the host; under this premise, they preserve the homeostasis of the immune system. Tregs constitutively express the IL-2 receptor α-chain (CD25) and the transcriptional factor Foxp3, which regulates the functionality of these cells (Gagliani and Huber 2017). The balance between Th1/Th2 and Th17/Treg is usually decisive in establishing the degree of inflammation at any given time (Noack and Miossec 2014).

Several studies have shown that kauranes can modulate the differentiation and activation of T cells in vitro and animal models of inflammatory diseases (Chen et al. 2006; Hu et al. 2008; Guo et al. 2013). One of the most relevant mechanisms is the inhibition of the NF-κB activation pathway (Yin et al. 2013; Liu et al. 2016b).

The ent-kaurene diterpenoids isolated from *Isodon serra*, named enmein, nodosin, lasiodonin, and epinodosin, suppressed cellular proliferation of murine splenic T cells (MSTCs) stimulated with concanavalin A (ConA) (Zhang et al. 2005). The inhibitory activity on cell proliferation induced by enmein and nodosin was due to the arrest of the cell cycle in the G1-S stage (Zhang et al. 2005; Li et al. 2010). In ConA-stimulated MSTCs, Ori (5.5; 11 and 22 μg/mL) significantly inhibited the secretion of IL-2, IFN-γ, IL-12p40, and TNF-α (Th1 cytokines) and the expression of the mRNA of these cytokines (Liu et al. 2007), whereas, in the case of nodosin, this kaurane decreased the secretion and expression of the IL-2 mRNA (Li et al. 2010). In non-stimulated murine primary splenocytes, Ori significantly decreased levels of IL-2, IFN-γ, IL-5, and IL-10, and this compound also produced a nonsignificant increase in the Th2/Th1 balance when it was compared to untreated controls (Ku and Lin 2013).

In splenic lymphocytes isolated from rats and stimulated with ConA, Hu et al. (2008) showed that Ori was able to inhibit ConA-induced Th1 polarization, which was evidenced by decreased levels of IL-2 and IFN-γ (Th1 cytokines) and increased levels of cytokines TGF-β and IL-10 (Th2 cytokines) (Hu et al. 2008). In this model, pretreatment with Ori induced differentiation toward $CD4^+/CD25^+$ Tregs (Foxp3+). Additionally, Ori induced the expression of HO-1 protein and its mRNA and increased the activity of this antioxidative enzyme. The results of these authors showed that Ori was able to modulate Th1/Th2 balance by reversing the Th1 polarization in vitro and also promote differentiation toward $CD4^+/CD25^+$ Tregs (Foxp3+) cells. An increased HO-1 expression could explain the anti-inflammatory effects reported in vivo for this kaurane (Hu et al. 2008). In lymphocytes isolated from mesenteric lymph nodes of mice with IBD and stimulated with anti-CD3/CD28 antibodies, Ori and the compound HAO472 (a derivative of Ori water-soluble) significantly decreased the proliferation of $CD4^+$ T cells (Liu et al. 2016b; Wang et al. 2015). Treatment with Ori and HAO472 significantly decreased the amount of IFN-$\gamma^+$ $CD4^+$ and IL-17A+ CD4+ T cells, without significantly affecting the IL-2+ CD4+ T-cell population (Liu et al. 2016b; Wang et al. 2015). The in vitro inhibition of T-cell proliferation and the Th1/Th17 activation by Ori and HAO472 were due to the inhibition of the nuclear translocation of NF-κB p65 (Liu et al. 2016b).

EriB was shown to inhibit the differentiation of Th1 and Th17 inflammatory T-cell populations through the inhibition of activation of Jak/STAT and NF-κB signaling pathways and increased ROS (Lu et al. 2013). This study was able to demonstrate the potential therapeutic effect of EriB for the treatment of multiple sclerosis and to other autoimmune diseases (Lu et al. 2013). Finally, adenanthin, a compound of the kaurane type isolated from the plant *Isodon adenanthus*, decreased Th1 and Th17 cell populations and increased Treg population, through the inhibition of the NF-κB pathway (Yin et al. 2013).

In chronic inflammatory and autoimmune diseases, the activation of NF-κB in T cells has an essential regulatory function, since this transcriptional factor promotes the genetic transcription of various inflammatory mediators. In the Jurkat E6.1 cell line, Ori was shown to block the TNF-α-induced activation of NF-κB. In stimulated cells, Ori completely blocked the binding activity of NF-κB to DNA, without affecting the phosphorylation of IκBα or the nuclear translocation of p65 (Ikezoe et al. 2005). In Jurkat E6.1 cells stimulated with PHA (phytohemagglutinin), caracasine acid, an ent-3,4-seco-kaurane isolated from the plant *Croton micans*, inhibited the activation pathway of NF-κB, by preventing the phosphorylation of the IκBα, IKK-α/β, and NF-κB p65 (Martínez et al. 2019). Furthermore, in this same PMA-stimulated cell line, the treatment with KMBK inhibited the DNA-binding activity of NF-κB (Lee et al. 2002).

Dal Piaz et al. (2013) showed in an elegant study using Jurkat E6.1 cells and chemical proteomics using mass spectrometry that the molecular target of Ori is the stress-inducible heat shock protein 70 1A (HSP70 1A) (Dal Piaz et al. 2013). These authors demonstrated that Ori is capable of bind covalently to Cys267 of HSP70 1A, inhibiting the HSP70 ATPase activity (Dal Piaz et al. 2013).

## 7.4 Other Anti-inflammatory Effects Associated with Kauranes

The human coronary artery smooth muscle cells (HCASMCs) are the primary constituents of the blood vessel wall. When HCASMCs are exposed to oxidative stress and activation by proinflammatory cytokines, the cells change their morphology, lose contractility, and express pro-inflammatory phenotype. This proinflammatory state is named "osteoblastic" transformation of vascular cells (Zhu et al. 2020; Chellan et al. 2018). In peroxide-stimulated HCASMC, GLA inhibited ROS production and secretion of TNF-α, IL-6, and IL-1β induced by p38 and NF-κB pathway. Moreover, GLA induced an increase in SOD and glutathione peroxidase (GPx) activities (Zhu et al. 2020). Ori was able to reduce the oxidative damage and cell death caused by peroxide in the human keratinocyte cell line (HaCaT). These effects were due to a change in the expression of various miRNAs (Bae et al. 2014).

Pretreatment with linearol and sidol inhibited the loss of mitochondrial membrane potential and mitochondrial-induced cell death of two cell lines, human astrocytoma (U373-MG) and rat adrenal pheochromocytoma (PC12) stimulated with peroxide. Thus, the decrease in ATP production was avoided (González-Burgos et al. 2016). Both kauranes reduced the lipid peroxidation generated by ROS production and increased the nuclear expression of Nrf2, which in turn enhanced the expression of antioxidant enzymes such as catalase, superoxide dismutase, glutathione reductase, glutathione peroxidase, and heme oxygenase-1 (González-Burgos et al. 2013b).

Three *ent*-kaurene diterpenes, *ent*-16-kaurene-3β,15β,18-triol, *ent*-3-oxo-16-kaurene-15β,18-diol, and ent-16-kaurene-3β,15β-diol, isolated from *Suregada multiflora*, inhibited the IgE-mediated degranulation in rat basophilic leukemia cells (RBL-2H3) (Cheenpracha et al. 2006; Passante et al. 2009). All three compounds inhibited the release of β-hexosaminidase from RBL-2H3 cell with $IC_{50}$ values from 22.5, 22.9, and 28.7 μM, respectively (Cheenpracha et al. 2006). Besides, kaurane-type diterpenoids, 16α*H*,17-isovaleryloxy-*ent*-kauran-19-oic acid, 16α-hydroxy-17-isovaleryloxy-*ent*-kauran-19-oic acid, and KA, isolated from *Acanthopanax koreanum*, Inhibited trypsin- induced TNF-α release in a human leukemic mast cell line (HMC-1) (Cai et al. 2003).

## 7.5 In Vivo Effects in Animal Models

Various kaurane-type compounds have been shown to exert significant anti-inflammatory activity in models in vivo of acute and chronic inflammation (Chavan et al. 2011; Dalenogare et al. 2019; Osafo et al. 2016; Silva et al. 2015; Wang et al. 2016a). KMBK, KA, and Ori are the most active kaurane-type compounds in these in human inflammatory disease models. The most relevant effects of kaurane compounds in animal models of diseases and syndromes are shown in Fig. 7.4, and the results of these studies in vivo are summarized in Table 7.3.

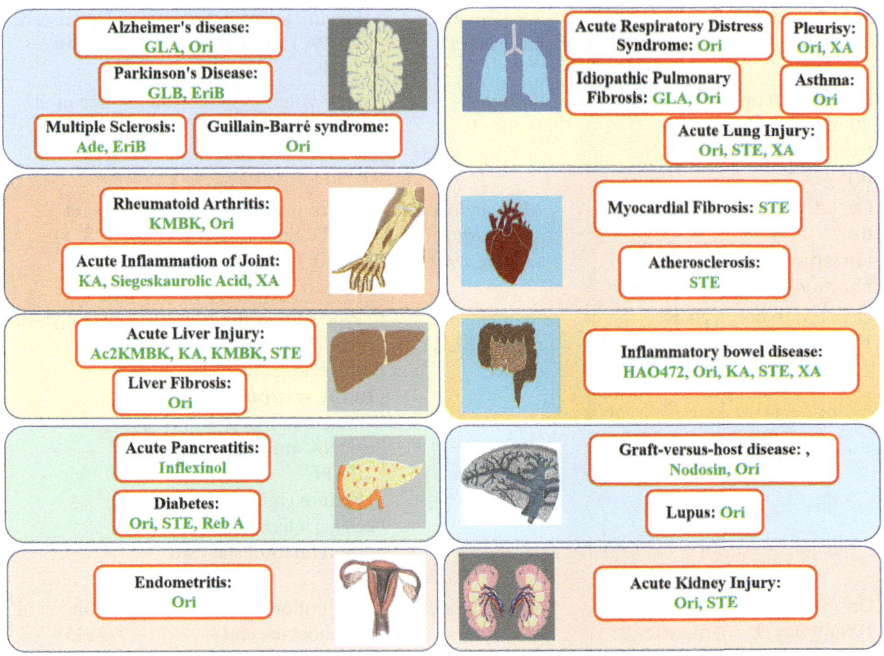


**Fig. 7.4** Effects of kaurane-type compounds in animal models of diseases and syndromes

**Table 7.3** In vivo studies of the anti-inflammatory effect of kauranes in animal models of human diseases and syndromes

| Compound (doses) | Human disease or syndrome | Animal models | Main results | References |
|---|---|---|---|---|
| KMBK (10 or 30 mg/kg, i.p.) | Acute inflammation | Air pouch in mouse | ↓MPO activity; ↓TNF-α and ↓$PGE_2$ in the exudate | Lee et al. (2004) |
| KA (100 mg/ kg, i.p.) | Acute inflammation | CIPO in mice | ↓Paw swelling (inhibition of 71.6%) at 5 h after edema induction | Lim et al. (2009) |
| KA (10 or 20 mg/kg, i.p.) | Acute inflammation | CIPO in rats | ↓Paw swelling | Chavan et al. (2012) |
| SA (20 or 30 mg/kg/day, p.o.) | Acute inflammation | CIPO in rats | ↓Paw swelling (inhibition of 24.8% and 34.4%, respectively) at 3 h after edema induction | Park et al. (2007) |
| XA (10, 30, or 100 mg/kg) | Acute inflammation | Paw edema in mice (induced by CAR, H, $PGE_2$, BK, or S) | ↓Paw swelling<br>A prophylactic and therapeutic effect | Osafo et al. (2018) |
| AEK (12.5 or 25 mg/kg, p. o.) | Acute inflammation | CIPO in rats | ↓Paw swelling (inhibition of 51.56% and 60.93%, respectively) at 2 h after edema induction | Chavan et al. (2011) |
| KA (3.0 μmol/ ear) | Acute inflammation | TPA-mouse ear edema | ↓Ear thickness induced by TPA (inhibition of 71±5%) | Boller et al. (2010) |
| KMBK (10 or 20 mg/kg/day, for 7 days, p. o.) | Arthritis | Adjuvant-induced arthritis in rats | ↓Paw volume (inhibition of 75 ± 15%) | Lee et al. (2004) |
| Ori (20 mg/kg, by intraarticular injection) | Arthritis | MSU-induced gouty arthritis in mice model | ↓Acute joint swelling, ↓production of IL-1β, ↓NLRP3 inflammasome pathway | He et al. (2018) |
| Ori (4.5, 9, or 18 mg/kg, i.p.) | **SLE** | Spontaneous murine SLE model | ↓Rate of B-cell maturation and differentiation<br>**In the serum:**<br>↓creatinine, ↓anti-dsDNA antibody titers; ↓BAFF<br>↓Mature (IgM+IgD+) B cells, ↓intermedia differentiation stages B cells | Zhou et al. (2013) |
| Ori (50 μg or 100 μg/day, i. p.) | Crohn's disease | Mouse model of Crohn's disease | ↓Infiltration of lymphocytes and inflammatory activity; ↓CD4+IL-17+ and | Wang et al. (2015) |

(continued)

**Table 7.3** (continued)

| Compound (doses) | Human disease or syndrome | Animal models | Main results | References |
|---|---|---|---|---|
| | | | CD4+IFN-γ+ T cells (spleen); ↓CD44hiCD62Llow T cells, ↓nuclear translocation of NF-p65 | |
| HAO472 (5 or 7.5 mg/kg) | Colitis | Mouse model of TNBS-induced colitis | **At the colonic level:** ↓TNF-α; ↓IFN-γ; ↓IL-17A; ↓iNOS/COX-2; ↓IFN-γ + CD4 + and IL-17A + CD4 + T cells, ↓expression and activity of NF-κB p65, ↓expression Foxp3 | Liu et al. (2016b) |
| KA (50 and 100 mg/kg, p.o. or rectally) | Colitis | Acetic acid-induced ulcerative colitis in rats | ↓Inflammatory cell infiltration, ↓activity of MPO; ↓MDA levels, ↑areas of surface reepithelialization | Paiva et al. (2002) |
| XA (30 or 100 mg/kg) | Colitis | Acetic acid-induced ulcerative colitis in rats | ↓Mast cell proliferation in the colonic segment; ↓colonic epithelial expression of AgNORs; ↑activity of SOD, CAT, and APx; ↓activity of MPO, ↓MDA levels | Osafo et al. (2019) |
| STE (50 or 100 mg/kg, p. o.) | Colitis | DSS-induced ulcerative colitis in mice | ↓TNF-α, ↓IL-6, ↓COX-2, ↓iNOS, ↑SOD, ↑CAT, ↑GST, ↑HO-1; ↑GSH, ↓activity of MPO, ↓p-IκB, ↓nuclear translocation of NF-p65; ↓p-p38, ↓p-ERK and ↓p-JNK in colon tissues | Alavala et al. (2019) |
| Ori (10 or 20 mg/kg | Asthma | Mouse model of asthma OVA-induced | **In BALF:** ↓eosinophil, ↓total inflammatory cell, ↓IL-4, ↓IL-13, ↓IL-5, and ↓eotaxin levels<br>**In lung tissues**: ↓eosinophilia, ↓mucus production | Wang et al. (2016a) |
| Ori (5, 10, or 20 mg/kg, i.p.) | Pleurisy | Mouse model of CAR-induced pleurisy | **Pleural effusion:** ↓IL-1β, ↓IL-6↓, and TNF-α, ↓MPO activity, ↓MDA, ↓ROS, ↑GSH, ↑SOD<br>**In lung tissues**: ↓p-IκBα; ↓p-NF-κBp65, ↓NLRP3, ↓Caspase-1, ↓IL-1β, | Yang et al. (2020) |

(continued)

**Table 7.3** (continued)

| Compound (doses) | Human disease or syndrome | Animal models | Main results | References |
|---|---|---|---|---|
| | | | ↓TRX-1, ↓TXNIP, ↓NOX-4, ↓KEAP-1, ↑Nrf2, ↑HO-1 | |
| XA (10, 30, or 100 mg/kg, oral route) | Pleurisy | CAR-induced pleurisy model in mice | **In lung tissues**: ↓neutrophil infiltration, ↓alveoli septal thickening, ↓MDA, ↑GSH, ↑SOD activity, ↑CAT activity | Ekuadzi et al. (2018) |
| GLA (10 mg/kg, i.p.) | Idiopathic pulmonary fibrosis | Bleomycin-induced pulmonary fibrosis in mice | **In lung tissues**: ↓infiltration of macrophages; ↓collagen deposition; ↓hydroxyproline, IL-1β, IL-6, TNF-α; and MCP-1 (↓mRNA levels), ↓NF-κ B activation **In BALF:** ↓infiltration of neutrophils, ↓TNF-α, ↓MCP-1 and TGF-β1 (levels of protein) | Yang et al. (2017) |
| STE (25 or 50 mg/kg, i.p.) | Acute lung injury | LPS-induced acute lung injury in mice | **In BALF**: ↓IL-1β, ↓IL-6, and ↓TNF-α, ↓neutrophils, and ↓macrophages **In lung tissues**: ↓MPO activity, ↓nitrate/nitrite content, ↓COX-2 and iNOS, ↓p-IκB-α and p-NF-κBp65 | Yingkun et al. (2013) |
| Ori (20 or 40 mg/kg, i.p.) | Acute lung injury | LPS-induced acute lung injury in mice | **In lung tissues**: ↓MPO activity, ↓MDA, ↑GSH, ↑SOD, ↓NF-kB and NLRP3 pathway activation, ↑Nrf2, ↑HO-1, ↑GCLM, ↑activation of Akt and MAPK | Yang et al. (2019) |
| STE (250 mg/kg, p.o.) | Acute liver injury | LPS-induced acute liver injury in rats | **In the serum**: ↓ALT and ↓AST levels **In the liver tissue**: ↓injury; ↓ALT, ↓AST, ↓IL-1β, ↓TNF-α, and ↓IL-6; ↓oxidative stress; ↑SOD, ↑GSH | Latha et al. (2017) |

(continued)

**Table 7.3** (continued)

| Compound (doses) | Human disease or syndrome | Animal models | Main results | References |
|---|---|---|---|---|
| STE (20 mg/kg, b.i. d., i.p.) | Liver fibrosis, liver cirrhosis | Liver injury induced in rats by TAA | ↓Liver damage; ↓NF-κ Bp65 (expression); ↑Nrf2 ↓IL-17A; ↓IL-1β, ↓TNF-α, ↓IL-6, and ↓IL-10 (mRNA) | Casas-Grajales et al. (2019) |
| KMBK or 1O, 20O-diacetyl KMBK (50 mg/kg/ day, p.o.) | Hepatotoxicity | APAP-induced hepatotoxicity in mice | **In the plasma (levels)**: ↓ALT, ↓AST, ↓TNF-α, ↓IL-6 **In the liver tissue:** ↓MDA,↑GSH, ↓necrosis, ↓RIP1, ↓RIP3, ↓p-JNK1/ 2, ↓nuclear translocation of NF-p65 ↓mRNA expression of Cyp1a2/2e1 (only Ac2KMBK) | Yoshioka et al. (2018) |
| KA (30 mg/ kg) | Hepatotoxicity | APAP-induced hepatotoxicity in mice | **In the plasma (levels)**: ↓ALT and ↓AST, production of IL-1β, IL-33, and TNF-α, and IL-1β, ↓MPO and ↓NAG activity, ↓MDA | Marcondes-Alves et al. (2019) |
| Ori (5 mg/ kg. i.p.) | Liver fibrosis | CCl4-Induced chronic liver injury and fibrosis in mice | **In the serum**: ↓ALT levels **In liver tissues**: ↓injury ↓collagen deposition, ↓recruitment of Kupffer cells, ↓NLRP3, ↓caspase-1, ↓IL-1β | Liu et al. (2020) |
| Ori (10 mg/kg/ day, i.p. for 15 days) | Alzheimer's disease | Aβ1-42-induced mouse model of Alzheimer's disease | **In the hippocampus**: ↓IL-1β, ↓IL-6, ↓TNF-α, ↓MCP-1, ↓iNOS, ↓COX-2, and ↑IL-10 (mRNA expression), ↓iNOS and ↓COX-2 (protein expression), ↓microglia and ↓astrocytes activation, ↑κBα, ↓NF-κ Bp65 nuclear translocation | Wang et al. (2014b) |
| Ori (10 mg/kg/ day, i.p. for 15 days) | Alzheimer's disease | Aβ1-42-induced mouse model of Alzheimer's disease | **In the synaptosomes**: ↑expression PSD-95 and synaptophysin, ↑mitochondrial activity **In the hippocampus**: ↑BDNF/TrkB/CREB signaling pathway **In the Morris water** | Wang et al. (2016b) |

(continued)

**Table 7.3** (continued)

| Compound (doses) | Human disease or syndrome | Animal models | Main results | References |
|---|---|---|---|---|
| | | | **maze test**: ↓latency and searching distance, ↑number of platform crosses | |
| GLB (100, 200, or 400 μg, i.c.v.) | Parkinson's disease | LPS-induced Parkinson's disease in rats | **In the cerebral tissue**: ↓apoptosis (↓cleaved caspase-3, ↓Bax, ↑Bcl-2), ↑TH, ↓GFAP (astrocytes), ↓CD11b (microglia), and ↓(IBA)-1 (activated microglia), ↓NO, ↓TNF-α, ↓IL-1β, ↓IL-6, ↓MPO, ↑SOD,↓ TLR/NF-κB and ↑Nrf2/HO-1 | Xu et al. (2017) |
| Ade (20 mg/kg/day, i.p.) | Multiple sclerosis | EAE in mouse MOG35-55-induced | **In the CNS:** ↓infiltrating CD4+ T cells and ↓F4/80+ macrophages, ↓Th1 and ↓Th17 cell, ↓demyelination, ↑Treg cell<br>**In the serum**: ↓IL-2, ↓IL-6, ↓IL-12, ↓TNF-α, ↓IFN-γ, and ↓IL-17, ↓NF-κB signaling | Yin et al. (2013) |
| Ori (7 or 20 mg/kg/day, p.o.) | Guillain-Barré syndrome (GBS) | Experimental autoimmune neuritis in rats | **In the sciatic nerves**: ↓macrophages (ED1+), ↓pan-T cells (W3/13+), and ↓B cells (OX22+), ↓mRNA levels of IL-1β, IL-6 and iNOS, – IL-10, ↑CD163 (anti-inflammatory-activated phenotype of macrophages), ↓Jagged-2, ↓Notch1, and ↓Hes1 (Notch signaling pathway)<br>**In the spinal roots**: ↓ED1+ cells, ↓W3/13+ cells<br>**In the dorsal horns**: ↓ED1+ cells ↑$CD206^{+}CCD11b^{+}$ cells (anti-inflammatory macrophages) in spleen MNCs | Xu et al. (2019) |

(continued)

**Table 7.3** (continued)

| Compound (doses) | Human disease or syndrome | Animal models | Main results | References |
|---|---|---|---|---|
| STE (10 mg/kg/day, p.o.) | Type 2 diabetes | Mice model via high-fat diet feeding | ↓Macrophage infiltration into adipose tissue, ↓TNF-α, ↓IL-6, ↓IL-10, ↓IL1β, ↓KC, ↓MIP-1α, ↓CD11b, and ↓CD14 (mRNA expression), ↓NF-κB pathway | Wang et al. (2012a) |
| Ori | Type 2 diabetes | T2DM animal model | ↓Infiltrating macrophages in kidney tissues ↓IL-1β, I↓L-6, ↓TNF-α, and ↓MCP-1 (mRNA expression), ↓TLR4 overexpression, ↓p-IκBα, ↓p-NF-κBp65, ↓p-p38 | Li et al. (2018) |

*Ac2KMBK* 1O, 20O-diacetyl KMBK, *AEK* 18-acetoxy-ent-kaur-16-ene, *APAP* acetaminophen, *ALT* alanine aminotransferase, *AgNORs* argyrophilic nucleolar organizer regions, *AST* aspartate aminotransferase, *BK* bradykinin, *BLF* bronchoalveolar lavage fluid, *CCl4* carbon tetrachloride, *CAR* carrageenan, *CIPO* carrageenan-induced paw edema, *D-Gal* D-galactosamine, *DSS* dextran sulphate sodium, *EAE* experimental autoimmune encephalomyelitis, *H* histamine, *H2S* hydrogen sulphide, *i.c.v.* intracerebroventricularly, *MDA* malondialdehyde, *MNCs* mononuclear cells, *MPO* myeloperoxidase, *MCP-1* monocyte chemotactic protein-1, *MOG* myelin oligodendrocyte glycoprotein, *NAG* *N*-acetylglucosaminidase, *Nrf2* nuclear erythroid factor 2, *S* serotonin, *SA* siegeskaurolic acid, *SLE* systemic lupus erythematosus, *TNBS* trinitrobenzene sulfonic acid. Symbols: ↓ indicates inhibition/reduction, ↑ indicates increase, and – indicates no changes

### 7.5.1 Effects of Kauranes in Animal Models of Acute Inflammation

In the air pouch model, KMBK, at 10 mg/kg and 30 mg/kg, administered i.p., caused significant suppression of PMN recruitment (measured as MPO activity) and other CAR-induced inflammatory mediators (Table 7.3; Lee et al. 2004).

CAR induces an early exudative phase of the acute inflammatory process, and the inhibition of this acute phase will eventually increase the resolution of the lesion (Osafo et al. 2018). KA showed important in vivo anti-inflammatory effects in rodents in the CAR-induced paw edema (CIPO), even 5 h after CAR injection (Table 7.3; Choi et al. 2011; Lim et al. 2009; Chavan et al. 2012).

Siegeskaurolic acid, a kaurane isolated from the roots of *Siegesbeckia pubescens*, showed a significant anti-inflammatory effect in the rat model previously described. Kaurane decreased the plantar edema in the animals (Table 7.3; Park et al. 2007). Additionally, this kaurane exhibited antinociceptive effects in the model of acetic acid-induced abdominal constriction and hot plate stimulus in mice (Park et al. 2007).

Xylopic acid (XA), a kaurene obtained from the plant *Xylopia aethiopica*, was able to inhibit the acute inflammatory response in the paw edema in mice model induced by various mediators of the acute inflammation process. XA significantly inhibited maximum plantar edema (maximal edema), showing prophylactic and therapeutic effects (Table 7.3; Osafo et al. 2018).

In the 12-O-tetradecanoylphorbol acetate (TPA)-induced mouse acute ear edema, the topical application of KA and grandiflorolic acid significantly decreased the extent of swelling induced by TPA (Díaz-Viciedo et al. 2008). In- ear biopsies of pretreated mice with KA and grandiflorolic acid, a significant decrease in neutrophil infiltration induced by this irritant was observed (Díaz-Viciedo et al. 2008). The topical application of KA (1.0 and 3.0μmol/ear) significantly reduced TPA-induced ear inflammation (Table 7.3; Boller et al. 2010).

### 7.5.2 Effects of Kauranes in Animal Models of Rheumatoid Arthritis

The treatment with KMBK (10 or 20 mg/kg, p.o.) significantly decreased the process of chronic inflammation in an adjuvant-induced arthritis model in rats. Treatment with KMBK (20 mg/kg) inhibited paw edema (Table 7.3; Lee et al. 2004). In the MSU-induced gouty arthritis in mice model, a type of NLRP3 inflammasome-dependent arthritis, the animal treated with Ori (20 mg/kg, by intraarticular injection) had significantly decreased acute joint swelling, as well as the production of IL-1β in the joint tissue of WT mice, but not in Nlrp3-/- mice. The results of this study indicated that Ori counteracts the inflammatory process and joint damage through inhibition of the NLRP3 inflammasome pathway (He et al. 2018).

### 7.5.3 Effects of Kauranes in Animal Models of SLE

On the other hand, in a spontaneous murine SLE model (MRL (lpr/lpr) mice), treating mice with Ori improved significantly the clinical and serological manifestations of SLE. These effects led to a more significant survival rate of the animals, lower levels of proteinuria, and kidney damage as compared to the animals that were not treated with Ori. Besides, a substantial decrease in serum creatinine and anti-dsDNA antibody titers were observed in mice treated with Ori, which were associated with decreased expression of B-cell activating factor (BAFF) and decreased differentiation and maturation of B cells (Table 7.3; Zhou et al. 2013).

### 7.5.4 Effects of Kauranes in Animal Models of IBD

IBD is the general term to define two chronic inflammatory pathologies of the gastrointestinal tract: Crohn's disease (CD) and ulcerative colitis (UC), whose onset and pathophysiology involves the combination of various environmental, genetic factors, infections, and a dysregulated immune response (Yeshi et al. 2020). On the evolution of IBD, the immune response plays a fundamental role due to the aberrant activation of various subtypes of CD4 helper T cells, including Th1 and Th17 cells, and their cytokines (Liu et al. 2016b). This activation promotes damage to the intestinal mucosa and generates chronic inflammation characteristic of the disease (Fu et al. 2020; Leppkes and Neurath 2020).

Various IBD animal models have been developed to try to understand the pathophysiology of the disease and to perform the preclinical evaluation of new drugs that may emerge as an alternative or complementary therapy for these pathologies (Mizoguchi et al. 2020). Some models use genetically modified animals, while others are based on the use of chemical agents for the induction of the disease (Mizoguchi et al. 2020).

Another promising anti-inflammatory effect in vivo for Ori was observed in mouse models of chemical-induced IBD. Treatment with Ori showed a significant anti-inflammatory effect in a mouse model of trinitrobenzene sulfonic acid (TNBS)-induced colitis (a mouse model of Crohn's disease) (Wang et al. 2015). In these animals, Ori decreased the mortality caused by TNBS and the inflammatory process at the colonic level. Additionally, the bodyweight of the Ori-treated mice was significantly higher than that of the animals with TNBS. Treatment with Ori significantly decreased the frequency of effector/memory T cells in the spleen, as well as the level of pro-inflammatory T cells at the splenic level of mice with TNBS-induced colitis (Table 7.3; Wang et al. 2015). The decrease in Th1/Th17 populations at the splenic and colonic levels by Ori could explain the anti-inflammatory activity of this diterpenoid.

In a mouse model of TNBS-induced colitis, the treatment of mice with the Ori derivative HAO472 significantly decreased (1) the damage to colonic mucosal, (2) the clinical symptoms of the disease, and (3) the severity of the disease while increasing the survival rate of the animals (Liu et al. 2016b). The treatment of mice with HAO472 meaningfully reduced the frequencies of splenic Th1/Th17 cells as compared with the TNBS group. The treatment of mice with HAO472 showed a decreased expression of diverse proinflammatory mediators in the infiltrated inflammatory cells in the colonic mucosa (Table 7.3; Liu et al. 2016b).

In another IBD model, in rat colitis induced by acetic acid, the treatment of animals with KA or XA decreased (1) the severity of mucosal damage; (2) MPO activity to colonic level; and (3) malondialdehyde (MDA) level. The treatment of rats with KA reduced inflammatory cell infiltration and submucosal edema in colon segments (Table 7.3; Paiva et al. 2002; Osafo et al. 2019). Like other kauranes, XA and KA can exert significant anti-inflammatory activity in IBD by blocking pro-inflammatory mediators and suppressing oxidative stress (Table 7.3; Paiva et al. 2002; Osafo et al. 2019).

In the dextran sulfate sodium (DSS)-induced ulcerative colitis model in mice, the treatment with the STE significantly decreased the disease activity index (DAI) score and improved the clinical signs (changes in body weight, stool consistency, and the presence of blood in the stools). This reduction can be due to the inhibition of inflammatory mediators in the colon tissue in comparison to DSS alone (Table 7.3; Alavala et al. 2019). Furthermore, the treatment of animals with STE reduced oxidative stress by various mechanisms (Table 7.3; Alavala et al. 2019). Finally, STE significantly suppressed activation pathways of NF-κB p65 and MAPK in colon tissues (Table 7.3; Alavala et al. 2019).

### 7.5.5 Effects of Kauranes in Animal Models of Lung Conditions

Various animal models have demonstrated the importance of kauranes in the treatment of lung maladies including asthma (Wang et al. 2016a), pleurisy (Yang et al. 2020; Ekuadzi et al. 2018), acute lung injury (Yang et al. 2019; Yingkun et al. 2013; Huang et al. 2018), acute respiratory distress syndrome (Jiang et al. 2017), and idiopathic pulmonary fibrosis (Yang et al. 2017; Fu et al. 2018).

In the murine asthma model, ovalbumin (OVA)-induced airway inflammation, Ori was able to regulate the Th1/Th2 cytokine balance, significantly decreasing airway hyperresponsiveness. In bronchoalveolar lavage fluid (BALF) of mice treated with this compound, there was a significant decrease in the eosinophil number and total inflammatory cell number, as well as a decrease in Th2-type cytokines and eotaxin levels, without affecting Th1 levels (Table 7.3; Wang et al. 2016a).

Ori was able to significantly decrease the effect of LPS-induced acute respiratory distress syndrome (ARDS) in mice by inhibiting NF-κB. Treatment with Ori significantly reduced the lung W/D ratio (wet-to-dry ratio index of lung water accumulation), the serum levels of IL-6 and TNF-α, and their expression in pulmonary tissues (Jiang et al. 2017).

In a mouse model of CAR-induced pleurisy, treatment with Ori decreased the scores of lower lung injury as compared to the CAR group. The compound decreased (1) inflammatory granulocyte infiltration, (2) edema, (3) the disseminated thickening of alveolar septa, (4) the focal hemorrhaging of lung tissues, and (5) the release of various proinflammatory cytokines in pleural effusion (Table 7.3; Yang et al. 2020). Inhibition of NF-κB and TXNIP/NLRP3 pathways, along with the activation of antioxidant mechanisms dependent on Nfr2, was responsible for the effect (Table 7.3; Yang et al. 2020). GLA was shown to decreased pulmonary fibrosis in the bleomycin-induced pulmonary fibrosis (BIPF) mouse model. The treatment of fibrotic animals with GLA significantly increased the survival rate and reduced weight loss and collagen deposition and hydroxyproline content in the bleomycin treated group, an effect mediated by blocking NF-κB signal transduction (Table 7.3; Yang et al. 2017).

Several studies validated the effects of kauranes for the treatment of respiratory ailments, in the LPS, CAR, or TNF-α-induced acute lung injury (ALI) in mice. Treatment with STE, Ori, or XA demonstrated an increase in animal survival and the decreased pulmonary edema, by the inhibition of pro-inflammatory mediators and the NF-kB and/or NLRP3 pathways (Table 7.3; Zhao et al. 2017; Yang et al. 2019; Ekuadzi et al. 2018; Yingkun et al. 2013; Huang et al. 2018). In some cases, the increase of antioxidants at the lung, due to kaurane treatment, contributed to the resolution of inflammation (Table 7.3; Yang et al. 2019; Ekuadzi et al. 2018). The anti-inflammatory effects observed with kauranes in the ALI rodent model gives new perspectives to possible therapies in humans.

### 7.5.6 Effects of Kauranes in Animal Models of Acute Liver Injury, Hepatotoxicity, and Liver Fibrosis

Several kauranes have pharmacological effects on murine models of acute liver damage, drug-induced hepatotoxicity, and liver fibrosis (Yoshioka et al. 2018; Liu et al. 2020; Shi et al. 2019; Yoshioka et al. 2017).

#### 7.5.6.1 Acute Liver Injury

In the LPS-induced acute liver injury model in rats, STE exerted a robust hepatoprotective activity. The treatment of rats with STE decreased LPS-induced hepatic damage, which was evidenced by a significant decrease in transaminase levels in the liver and serum. Histopathology revealed that STE attenuated LPS-induced structural changes by decreasing hepatocellular damage and hepatic necrotic areas. Furthermore, the treatment with STE significantly decreased the levels of oxidative stress and induced the activation of antioxidant mechanisms, when these markers were compared with the control group (Table 7.3; Latha et al. 2017). Finally, this study showed that STE decreased proinflammatory cytokine levels in animals compared to the control group (Table 7.3; Latha et al. 2017).

STE extracted from *Stevia rebaudiana* (Bertoni) was evaluated in the liver injury model in rats induced by long-term thioacetamide (TAA) administration. The treatment of rats with STE reversed liver damage and decreased NF-κB activation and the expression of various cytokines induced by TAA. STE increased the expression of Nrf2 protein and consequently decrease oxidative stress. *In vivo*, STE exhibits antioxidant, anti-inflammatory and immunomodulatory activities, in a model of liver damage, which makes it to be considered a potentially useful compound for treating liver disease in humans (Table 7.3; Casas-Grajales et al. 2019).

#### 7.5.6.2 Hepatotoxicity

A murine in vivo model serves for evaluating the acetaminophen-induced liver (APAP) injury, which depends on the release of various inflammatory mediators including ROS, proinflammatory cytokines, and NF-κB activation. In this model, pretreatment with KMBK or the synthetic derivative of KMBK, 1*O*, 20*O*-diacetyl KMBK (Ac2KMBK), significantly inhibited APAP-induced liver damage (Table 7.3; Yoshioka et al. 2017, 2018). KMBK and Ac2KMBK decreased plasma levels of ALT, AST, TNF-α, and IL-6. These two compounds inhibited lipid peroxidation and liver necrosis and apoptosis while increasing GSH levels. This hepatoprotection process involved the decreased expression of RIP1 and RIP3 proteins, decreased JNK1/2 activation, and decreased nuclear translocation of NF-p65 (Table 7.3; Yoshioka et al. 2017, 2018). Ac2KMBK (but not KMBK) significantly inhibited Cyp1a2 and Cyp2e1 mRNA expression. Since Cyp1a2/2e1 in the liver are the enzymes of cytochrome p450 responsible for transforming APAP to its toxic metabolites (toxic NAPQI). The protective effect of Ac2KMBK was more potent than that of KMBK (Table 7.3; Yoshioka et al. 2018).

In the APAP hepatotoxicity model, the treatment of mice with KA (30 mg/kg) increased animal survival as compared to the vehicle-treated control group (Marcondes-Alves et al. 2019). In these mice, KA decreased (1) APAP-induced hepatic necrosis, (2) ALT and AST plasmatic levels, (3) APAP-induced neutrophils and macrophage recruitment, (4) oxidative stress, and (5) the production of proinflammatory cytokines. As part of the hepatoprotection mechanism, treatment with KA significantly increased GSH and IL-10 levels (Table 7.3; Marcondes-Alves et al. 2019).

KA, KMBK, and Ac2KMBK decreased the lethality and liver damage in the model of APAP-induced hepatotoxicity, through the inhibition of cells recruitment and inflammatory mediators, in a mechanism that depends on the inhibition of the NF-κB activation pathway.

#### 7.5.6.3 Liver Fibrosis

In the mouse model of liver damage and fibrosis carbon tetrachloride (CCl4), treatment with Ori significantly decreased liver injury and reduced ALT levels. Ori decreased collagen deposition and the expression of α-smooth muscle actin in murine livers with fibrosis. Moreover, Ori decreased the activation of NLRP3 in the liver, by reducing the expression of the NLRP3, caspase-1, and IL-1β proteins (Table 7.3; Liu et al. 2020).

### 7.5.7 Effects of Kauranes in Animal Models of Neurological Diseases

#### 7.5.7.1 Alzheimer's Disease

AD is the most frequent neurodegenerative disease in older adults (Tiwari et al. 2019). This degenerative disorder is characterized by neuroinflammation and the accumulation of Aβ plaques and hyperphosphorylated tau protein, generating neurofibrillary tangles (Fakhoury 2018; Nisbet and Götz 2018).

Aβ peptide plays a critical role in synaptic loss, being fundamental for the pathogenesis of AD. In vitro and in vivo studies have shown that inhibition of Aβ peptide formation or its clearance inhibits neuroinflammation and neurodegeneration in AD (Xin et al. 2018). The depositions of Aβ peptides generate neuroinflammation that activates astrocytes and microglia (Fakhoury 2018), releasing IL-1β, IL-6, TNF-α, iNOS, and COX-2, contributing to neuronal damage, cell death, and amyloid toxicity. Like other chronic inflammatory diseases, the activation of the NF-κB and NLRP3 plays an essential role in the onset and perpetuation of the disease (Minter et al. 2016). Consequently, inhibition of the inflammatory process through blocking NF-κB pathways and NLRP3 may be a successful therapeutic strategy for AD (Wang et al. 2014b).

Inhibition of the mammalian target of rapamycin (mTOR) protein is an important pharmacological strategy for the treatment of AD. The inhibition of mTOR increases Aβ clearance and regulates the phosphorylation and degradation of tau through mediating the activity of P70S6K and the activation of the autophagy system process, being, therefore, another pharmacological target of importance for research in AD.

Various kaurane-type compounds have been shown to inhibit the production and/or clearance of Aβ (Zhang et al. 2013; Zhou et al. 2019a). GLA increases Aβ clearance by inhibiting the protein kinase B/mTOR/autophagy signaling pathway and tau phosphorylation inhibition involving the mTOR/70-kDa ribosomal protein S6 kinase pathway (Zhou et al. 2019a).

In the APP/PS1-21 transgenic mouse model, an animal model of cerebral amyloidosis for AD, the administration of Ori, significantly decreased (1) Aβ deposition, (2) plaque-associated amyloid precursor protein (APP expression), and (3) microglial activation in the cortex and hippocampus of mice. Treatment with Ori decreased social interaction deficits in AD mice, henceforth, the potential usefulness of Ori in AD and other neurodegenerative diseases (Zhang et al. 2013).

Wang et al. (2014b) showed in vivo that Ori improved cognitive impairment and attenuate memory deficits (escape latency, searching distance, crossing platform times), inhibited glial activation, and decreased the expression of proinflammatory mediators in the hippocampus of the Aβ1-42-induced mouse model of AD (Table 7.3). In these animals, Ori inhibited the NF-κB pathway and Aβ1-42-induced neuronal tissue apoptosis (Table 7.3; Wang et al. 2014b). In the Aβ1-42-induced mouse model of AD, the treatment of animals with Ori was able to reverse impaired behavior, and hence rescue synaptic loss induced by Aβ1-42 in vivo. The effect of the Ori was through BDNF/TrkB/CREB signaling pathway in the hippocampus of AD mice (Table 7.3; Wang et al. 2016b). Ori attenuated alterations in dendritic structure and spine density observed in the hippocampus of AD mice (Wang et al. 2016b).

#### 7.5.7.2 Parkinson's Disease

Parkinson's disease (PD) is the second prevalent neurodegenerative disease in the old population. This disease is characterized by dopaminergic neuron loss, inflammation, and oxidative stress injury in the substantia nigra. In an LPS-induced Parkinson's disease model in rats, the treatment of animals with GLB attenuated experimental PD symptoms induced by LPS. GLB showed significant anti-inflammatory and antiapoptotic effects in this animal model (Table 7.3; Xu et al. 2017). In 1-methyl-4-phenyl-1,2,3,6-tetrahydropyridine (MPTP)-induced PD mouse model, treatment with EriB decreased damage to dopamine neurons in the substantia nigra region by inhibiting microglia activation. In the open-field activity test used to determine the changes in behavioral activity, EriB attenuated the motor deficit severity and increased locomotion and rearing function. The effects of EriB on mice with PD are associated with decreased activation of microglia, inhibition of proinflammatory cytokines, and NF-κB signaling pathway (Dou et al. 2018).

#### 7.5.7.3 Multiple Sclerosis

Adenanthin (Ade) showed preventive and therapeutic effects for the treatment of experimental autoimmune encephalomyelitis (EAE, animal model for multiple sclerosis). In mice with EAE, Ade significantly inhibited the clinical symptoms, disease severity, and demyelination in CNS. The effects of Ade on induced EAE in mice were associated with decreased infiltration and activation of inflammatory cells, increase in Treg cells, reduced liberation of proinflammatory cytokines, and decreased stimulatory capacity of APCs, through inhibition of NF-κB signaling pathway (Table 7.3; Yin et al. 2013). In cited work, Ade exhibited potent immunomodulatory activity in the treatment of this multiple sclerosis model (Yin et al. 2013).

Similar to Ade, EriB showed significant prophylactic and therapeutic anti-inflammatory activity in the EAE model in mice. The pretreated animals with EriB (10 mg/kg, i.p.), before EAE induction, was associated with a delay in the appearance of clinical symptoms and may lower disease severity. In EriB-treated mice, a reduction in CNS inflammation and demyelination was observed. When MOG-reactive T cells were treated with EriB and subsequently transfer into normal mice, EAE did not occur (Lu et al. 2013). EriB decreased ROS and inhibited Th1 and Th17 differentiation by inhibiting NF-κB and Jak/STAT and pathways, which have been associated with an inflammatory process in macrophages (Lu et al. 2013; Liu et al. 2018). EriB binds directly to STAT3 through a covalent bond between an α,β-unsaturated carbonyl of EriB and the Cys712 of STAT3 (Yu et al. 2015).

#### 7.5.7.4 Guillain-Barré Syndrome

Ori was able to inhibit inflammation induced by the synthetic neurogenic peptide P2 57-81 used to induce experimental autoimmune neuritis (EAN). In this animal model, treatment with Ori exerted a significant prophylactic and therapeutic effect, decreased peak disease severity and disease duration, suppressed paraparesis, and increased latency EAN appearance. These effects were achieved through the inhibition of local inflammatory response and the increase of macrophages of the anti-inflammatory phenotype (M2) in the peripheral nerves. The inhibition involved the Notch pathway (Table 7.3; Xu et al. 2019).

### 7.5.8 Effects of Kauranes in Animal Models of Diabetes

Another relevant effect of kauranes is related to the control of insulin resistance and adipose tissue inflammation. In an insulin resistance mice model via high-fat diet feeding, treatment with STE significantly improved fasting glucose, basal insulin levels, glucose tolerance, and whole-body insulin sensitivity. Mechanistically, these changes were associated with a significant decrease in macrophage infiltration and of inflammatory mediators into adipose tissue (Table 7.3; Wang et al. 2012a). The reduction of the inflammatory process in adipose tissue was associated with less activation of the NF-κB pathway. STE decreased glucose resistance and the inflammatory state in adipose tissue and consequently could become a suitable therapeutic strategy for type 2 diabetes (Wang et al. 2012a).

In the model in streptozotocin (STZ)-induced diabetic rats, the treatment of the animals with the rebaudioside A (Reb A) (200 mg/kg, p.o.) inhibited the lipid peroxidation, hyperglycemia, and hyperlipidemia, and it may contribute to improving plasma insulin level in diabetic rats, consequently lowering blood glucose levels (Saravanan and Ramachandran 2013). The T2DM animal model, treated with Ori, significantly inhibited (1) diabetes-induced renal injury in rats, (2) excretion levels of urinary proteins, (3) serum creatinine and blood urea nitrogen concentrations, (4) infiltration of inflammatory cells in kidney tissues in rats, (5) levels of various inflammatory mediators, (6) TLR4 overexpression, and (7) NF-κB and MAPK activation pathways (Li et al. 2018). The pharmacological effects described for kauranes in animal models of diabetes can be attributed partially to their powerful anti-inflammatory activity.

### 7.5.9 Effects of Kauranes in Other Animal Models of Inflammation In Vivo

Other anti-inflammatory effects in vivo of kaurane diterpenes involved the inhibition of (1) atherosclerosis and myocardial fibrosis in mice by STE (Geeraert et al. 2010; Wang et al. 2019), (2) acute kidney injury by Ori and STE (Potočnjak et al. 2017; Yan et al. 2020), and (3) the skeletal muscle injury by STE (Bunprajun et al. 2012), acute pancreatitis in mice by inflexinol (Ahn et al. 2013), and endometritis process in mice by Ori (Zhou et al. 2019b). Furthermore, Ori prevented graft rejection (Guo et al. 2013) and exhibited protective effects in sepsis (Zhao et al. 2016). GLA treatment increased the survival of mice with induced lethal endotoxic shock (Hou et al. 2020). Besides, nodosin favored orthotopic liver transplantation in rats, by increasing HO-1 levels (Wang et al. 2012b).

## 7.6 Future Perspectives

The structural variety of kaurane compounds makes them an excellent starting point for the development of lead compounds. The HAO472 (a derivative of Ori water-soluble) has led the way for the effective development of more active drugs with a better pharmacokinetic profile. The successful design of kauranes with clinical utility refers to improving the activity, selectivity, solubility, bioavailability, and safety of these metabolites. The existence of high-performance technologies for the detection of natural products and their organic isolation, as well as synthesis techniques, will allow more efficiency in obtaining the quantities of kauranes that are required for large-scale production in the pharmaceutical industry. Given the significant number of plants, microbes, animals, marine organisms, and other sources of metabolites in nature, which have not yet been studied, it would be expected that a wide variety of new diterpenoid kauranes may be discovered soon. Kaurane-type compounds obtained from nature and/or their semisynthetic or synthetic metabolites have the challenge to overcome the preclinical studies until the incorporation of some of these compounds in the therapeutic arsenal for the treatment of pathologies associated with the pathophysiology of inflammation.

## 7.7 Conclusions

For many decades, natural products have been the primary source for the discovery and obtaining of most of the drugs approved for the treatment of various pathologies in humans. Extraordinary efforts have been carried out to identify and develop metabolites derived from nature as potential anti-inflammatory drugs. Diterpenoid kauranes have shown a wide range of anti-inflammatory effects in a variety number of in vitro and in vivo models. These compounds have been shown to inhibit the inflammatory process in various immune and nonimmune cells by suppressing a vast cascade of inflammatory signaling pathways. In neutrophils, they have been shown to inhibit (1) the production of ROS; (2) the release of elastase; and (3) the production of leukotrienes. In mouse macrophages and human monocyte LPS-stimulated, they decreased (1) ROS and RNS production; (2) COX-2 and iNOS expression; (3) the production of inflammatory cytokines; (4) the activation of the NF-κB pathway and the NLRP3 inflammasome; and (5) the activation of MAPKs and JAKs/STAT. In T cells, these compounds inhibited the differentiation and activation of the inflammatory phenotypes, Th1 and Th17, and the cytokines released by inhibiting various activation pathways and transcriptional factors. In animal models of inflammatory diseases, such as rheumatoid arthritis, asthma, lupus, acute liver damage, acute lung damage, acute kidney damage, neuroinflammatory and neurodegenerative diseases, EriB, GLA, GLB, KA, KMBK, Ori, STE, XA, and other kauranes have shown significant inflammatory activity in vivo. A variety of specific molecular targets such as NF-κB, inflammasome NLRP3, STAT3, and HSP70 have already been identified, which allow explaining the amplitude of the observed biological effects for these compounds. Future studies are required to address critical issues concerning the availability of the compound and the effects in several tissues so that the structures may have an impact on acute and chronic inflammation in vivo.

**Conflict of Interest** The authors declare that they have no conflict of interest.

## References

Ahn DW, Ryu JK, Kim J et al (2013) Inflexinol reduces the severity of acute pancreatitis by inhibiting nuclear factor-κB activation in cerulein-induced pancreatitis. Pancreas 42(2):279–284

Alavala S, Sangaraju R, Nalban N et al (2019) Stevioside, a diterpenoid glycoside, shows anti-inflammatory property against dextran sulphate sodium-induced ulcerative colitis in mice. Eur J Pharmacol 855:192–201

Aquila S, Weng ZY, Zeng YQ, Sun HD, Ríos JL (2009) Inhibition of NF-kappaB activation and iNOS induction by ent-kaurane diterpenoids in LPS-stimulated RAW264.7 murine macrophages [published correction appears in J Nat Prod. 2009 Sep;72(9):1724]. J Nat Prod 72(7):1269–1272

Attiq A, Jalil J, Husain K, Ahmad W (2018) Raging the war against inflammation with natural products. Front Pharmacol 9:976

Bae S, Lee EJ, Lee JH et al (2014) Oridonin protects HaCaT keratinocytes against hydrogen peroxide-induced oxidative stress by altering microRNA expression. Int J Mol Med 33(1):185–193

Balan S, Saxena M, Bhardwaj N (2019) Dendritic cell subsets and locations. Int Rev Cell Mol Biol 348:1–68
Ban JO, Oh JH, Hwang BY et al (2009) Inflexinol inhibits colon cancer cell growth through inhibition of nuclear factor-kappaB activity via direct interaction with p50. Mol Cancer Ther 8(6):1613–1624
Boller S, Soldi C, Marques MC et al (2010) Anti-inflammatory effect of crude extract and isolated compounds from Baccharis illinita DC in acute skin inflammation. J Ethnopharmacol 130(2):262–266
Boonkaewwan C, Toskulkao C, Vongsakul M (2006) Anti-inflammatory and immunomodulatory activities of stevioside and its metabolite steviol on THP-1 cells. J Agric Food Chem 54(3):785–789
Brightbill HD, Suto E, Blaquiere N et al (2018) NF-κB inducing kinase is a therapeutic target for systemic lupus erythematosus. Nat Commun 9(1):179
Bunprajun T, Yimlamai T, Soodvilai S, Muanprasat C, Chatsudthipong V (2012) Stevioside enhances satellite cell activation by inhibiting of NF-κB signaling pathway in regenerating muscle after cardiotoxin-induced injury. J Agric Food Chem 60(11):2844–2851
Cai XF, Shen G, Dat NT et al (2003) Inhibitory effect of kaurane type diterpenoids from Acanthopanax koreanum on TNF-alpha secretion from trypsin-stimulated HMC-1 cells. Arch Pharm Res 26(9):731–734
Casas-Grajales S, Ramos-Tovar E, Chávez-Estrada E et al (2019) Antioxidant and immunomodulatory activity induced by stevioside in liver damage: in vivo, in vitro and in silico assays. Life Sci 224:187–196
Castegna A, Gissi R, Menga A et al (2020) Pharmacological targets of metabolism in disease: opportunities from macrophages. Pharmacol Ther 210:107521
Castrillo A, de Las HB, Hortelano S, Rodriguez B, Villar A, Bosca L (2001) Inhibition of the nuclear factor kappa B (NF-kappa B) pathway by tetracyclic kaurene diterpenes in macrophages. Specific effects on NF-kappa B-inducing kinase activity and on the coordinate activation of ERK and p38 MAPK. J Biol Chem 276(19):15854–15860
Chavan MJ, Wakte PS, Shinde DB (2011) Analgesic and anti-inflammatory activities of 18-acetoxy-ent-kaur-16-ene from Annona squamosa L. bark. Inflammopharmacology 19(2):111–115
Chavan MJ, Kolhe DR, Wakte PS, Shinde DB (2012) Analgesic and antiinflammatory activity of kaur-16-en-19-oic acid from Annona reticulata L. bark. Phytother Res 26(2):273–276
Cheenpracha S, Yodsaoue O, Karalai C et al (2006) Potential anti-allergic ent-kaurene diterpenes from the bark of Suregada multiflora. Phytochemistry 67(24):2630–2634
Chellan B, Sutton NR, Hofmann Bowman MA (2018) S100/RAGE-mediated inflammation and modified cholesterol lipoproteins as mediators of osteoblastic differentiation of vascular smooth muscle cells. Front Cardiovasc Med 5:163
Chen ZJ, Li YS, Zhou JY, Wang ZT (2006) Effect of glaucocalyxin A on level of Th1/Th2 type cytokines in mice. Zhongguo Zhong Yao Za Zhi 31(15):1257–1260
Choi RJ, Shin EM, Jung HA, Choi JS, Kim YS (2011) Inhibitory effects of kaurenoic acid from Aralia continentalis on LPS-induced inflammatory response in RAW264.7 macrophages. Phytomedicine 18(8–9):677–682
Ciążyńska M, Bednarski IA, Wódz K, Narbutt J, Lesiak A (2020) NLRP1 and NLRP3 inflammasomes as a new approach to skin carcinogenesis. Oncol Lett 19(3):1649–1656
Colonna M, Butovsky O (2017) Microglia function in the central nervous system during health and neurodegeneration. Annu Rev Immunol 35:441–468
Cosmi L, Maggi L, Santarlasci V, Liotta F, Annunziato F (2014) T helper cells plasticity in inflammation. Cytometry A 85(1):36–42
Dal Piaz F, Cotugno R, Lepore L et al (2013) Chemical proteomics reveals HSP70 1A as a target for the anticancer diterpene oridonin in Jurkat cells. J Proteomics 82:14–26
Dalenogare DP, Ferro PR, De Prá SDT et al (2019) Antinociceptive activity of Copaifera officinalis Jacq. L oil and kaurenoic acid in mice. Inflammopharmacology 27(4):829–844

de Groot LES, van der Veen TA, Martinez FO, Hamann J, Lutter R, Melgert BN (2019) Oxidative stress and macrophages: driving forces behind exacerbations of asthma and chronic obstructive pulmonary disease? Am J Physiol Lung Cell Mol Physiol 316(2):L369–L384

De las Heras B, Hortelano S, Girón N, Bermejo P, Rodríguez B, Boscá L (2007) Kaurane diterpenes protect against apoptosis and inhibition of phagocytosis in activated macrophages. Br J Pharmacol 152(2):249–255

De Sanctis JB, Garmendia JV, Moreno D et al (2009) Pharmacological modulation of Th17. Recent Pat Inflamm Allergy Drug Discov 3(2):149–156

Díaz-Viciedo R, Hortelano S, Girón N et al (2008) Modulation of inflammatory responses by diterpene acids from Helianthus annuus L. Biochem Biophys Res Commun 369(2):761–766

Ding Y, Ding C, Ye N et al (2016) Discovery and development of natural product oridonin-inspired anticancer agents. Eur J Med Chem 122:102–117

Ding S, Xu S, Ma Y, Liu G, Jang H, Fang J (2019) Modulatory mechanisms of the NLRP3 inflammasomes in diabetes. Biomolecules 9(12):850

Doran AC, Yurdagul A Jr, Tabas I (2020) Efferocytosis in health and disease. Nat Rev Immunol 20(4):254–267

Dou F, Chu X, Zhang B et al (2018) EriB targeted inhibition of microglia activity attenuates MPP+ induced DA neuron injury through the NF-κB signaling pathway. Mol Brain 11(1):75

Ekuadzi E, Biney RP, Benneh CK, Osei Amankwaa B, Jato J (2018) Antiinflammatory properties of betulinic acid and xylopic acid in the carrageenan-induced pleurisy model of lung inflammation in mice. Phytother Res 32(3):480–487

Epelman S, Lavine KJ, Randolph GJ (2014) Origin and functions of tissue macrophages. Immunity 41(1):21–35

Espinoza A, Peón AN, Terrazas LI (2012) Alternatively activated macrophages in types 1 and 2 diabetes. Mediators Inflamm 2012:815953

Fakhoury M (2018) Microglia and astrocytes in Alzheimer's disease: implications for therapy. Curr Neuropharmacol 16(5):508–518

Fengyang L, Yunhe F, Bo L et al (2012) Stevioside suppressed inflammatory cytokine secretion by downregulation of NF-κB and MAPK signaling pathways in LPS-stimulated RAW264.7 cells. Inflammation 35(5):1669–1675

Fu Y, Zhao P, Xie Z, Wang L, Chen S (2018) Oridonin inhibits myofibroblast differentiation and bleomycin-induced pulmonary fibrosis by regulating transforming growth factor β (TGFβ)/Smad pathway. Med Sci Monit 24:7548–7555

Fu SH, Chien MW, Hsu CY, Liu YW, Sytwu HK (2020) Interplay between cytokine circuitry and transcriptional regulation shaping helper T cell pathogenicity and plasticity in inflammatory bowel disease. Int J Mol Sci 21(9):E3379

Gagliani N, Huber S (2017) Basic aspects of T helper cell differentiation. Methods Mol Biol 1514: 19–30

Gan P, Zhang L, Chen Y et al (2015) Anti-inflammatory effects of glaucocalyxin B in microglia cells. J Pharmacol Sci 128(1):35–46

García PA, de Oliveira AB, Batista R (2007) Occurrence, biological activities and synthesis of kaurane diterpenes and their glycosides. Molecules 12(3):455–483

Geeraert B, Crombé F, Hulsmans M, Benhabilès N, Geuns JM, Holvoet P (2010) Stevioside inhibits atherosclerosis by improving insulin signaling and antioxidant defense in obese insulin-resistant mice. Int J Obes (Lond) 34(3):569–577

González-Burgos E, Carretero ME, Gómez-Serranillos MP (2013a) Involvement of Nrf2 signaling pathway in the neuroprotective activity of natural kaurane diterpenes. Neuroscience 231:400–412

González-Burgos E, Carretero ME, Gómez-Serranillos MP (2013b) Kaurane diterpenes from Sideritis spp. exert a cytoprotective effect against oxidative injury that is associated with modulation of the Nrf2 system. Phytochemistry 93:116–123

González-Burgos E, Duarte AI, Carretero ME, Moreira PI, Gómez-Serranillos MP (2016) Kaurane diterpenes as mitochondrial alterations preventive agents under experimental oxidative stress conditions. Pharm Biol 54(4):705–711

Guo W, Zheng P, Zhang J, Ming L, Zhou C, Zhang S (2013) Oridonin suppresses transplant rejection by depleting T cells from the periphery. Int Immunopharmacol 17(4):1148–1154

Guo C, Fu R, Wang S et al (2018) NLRP3 inflammasome activation contributes to the pathogenesis of rheumatoid arthritis. Clin Exp Immunol 194(2):231–243

Gupta E, Purwar S, Sundaram S, Tripathi P, Rai G (2016) Stevioside and rebaudioside A—predominant ent-kaurene diterpene glycosides of therapeutic potential: a review. Czech J Food Sci 34:281–299

Hannoodee S, Nasuruddin DN (2020) Acute inflammatory response. StatPearls Publishing, Treasure Island (FL)

Hawiger J, Zienkiewicz J (2019) Decoding inflammation, its causes, genomic responses, and emerging countermeasures. Scand J Immunol 90(6):e12812

He H, Jiang H, Chen Y, Ye J, Wang A, Wang C et al (2018) Oridonin is a covalent NLRP3 inhibitor with strong anti-inflammasome activity. Nat Commun 9(1):2550

He W, Kapate N, Shields CW 4th, Mitragotri S (2019) Drug delivery to macrophages: a review of targeting drugs and drug carriers to macrophages for inflammatory diseases. Adv Drug Deliv Rev. S0169-409X(19)30239-X

Hong SS, Lee SA, Han XH et al (2007) Kaurane diterpenoids from Isodon excisus inhibit LPS-induced NF-kappaB activation and NO production in macrophage RAW264.7 cells. J Nat Prod 70(4):632–636

Hong SS, Lee SA, Han XH et al (2008) ent-Kaurane diterpenoids from Isodon japonicus. J Nat Prod 71(6):1055–1058

Hou X, Xu G, Wang Z, Zhan X, Li H, Li R et al (2020) Glaucocalyxin A alleviates LPS-mediated septic shock and inflammation via inhibiting NLRP3 inflammasome activation. Int Immunopharmacol 81:106271

Hu AP, Du JM, Li JY, Liu JW (2008) Oridonin promotes CD4+/CD25+ Treg differentiation, modulates Th1/Th2 balance and induces HO-1 in rat splenic lymphocytes [published correction appears in Inflamm Res. 2008 Jul;57(7):350]. Inflamm Res 57(4):163–170

Huang W, Huang M, Ouyang H, Peng J, Liang J (2018) Oridonin inhibits vascular inflammation by blocking NF-κB and MAPK activation. Eur J Pharmacol 826:133–139

Hwang BY, Lee JH, Koo TH et al (2001) Kaurane diterpenes from Isodon japonicus inhibit nitric oxide and prostaglandin E2 production and NF-kappaB activation in LPS-stimulated macrophage RAW264.7 cells. Planta Med 67(5):406–410

Ikezoe T, Yang Y, Bandobashi K et al (2005) Oridonin, a diterpenoid purified from Rabdosia rubescens, inhibits the proliferation of cells from lymphoid malignancies in association with blockade of the NF-kappa B signal pathways. Mol Cancer Ther 4(4):578–586

Jiang J, Shan X, Zhu L (2017) Effects and mechanisms of oridonin in the treatment of acute respiratory distress syndrome mice. Int J Clin Exp Med 10:6191–6197

Kawano M, Nagata S (2018) Efferocytosis and autoimmune disease. Int Immunol 30(12):551–558

Kazemi S, Shirzad H, Rafieian-Kopaei M (2018) Recent findings in molecular basis of inflammation and anti-inflammatory plants. Curr Pharm Des 24(14):1551–1562

Kiem PV, Cai XF, Minh CV, Lee JJ, Kim YH (2004) Kaurane-type diterpene glycoside from the stem bark of Acanthopanax trifoliatus. Planta Med 70(3):282–284

Kim BW, Koppula S, Kim IS et al (2011) Anti-neuroinflammatory activity of Kamebakaurin from Isodon japonicus via inhibition of c-Jun $NH_2$-terminal kinase and p38 mitogen-activated protein kinase pathway in activated microglial cells. J Pharmacol Sci 116(3):296–308

Kim JY, Kim HS, Kim YJ et al (2013a) Effusanin C inhibits inflammatory responses via blocking NF-κB and MAPK signaling in monocytes. Int Immunopharmacol 15(1):84–88

Kim BW, Koppula S, Hong SS et al (2013b) Regulation of microglia activity by glaucocalyxin-A: attenuation of lipopolysaccharide-stimulated neuroinflammation through NF-κB and p38 MAPK signaling pathways. PLoS One 8(2):e55792

Kim JY, Kim HS, Kim YJ et al (2013c) Inhibition of TAK1 by kamebakaurin in dendritic cells. Int Immunopharmacol 15(1):138–143
Kim KH, Han JW, Jung SK, Park BJ, Han CW, Joo M (2017) Kaurenoic acid activates TGF-β signaling. Phytomedicine 32:8–14
Kishore N, Kumar P, Shanker K, Verma AK (2019) Human disorders associated with inflammation and the evolving role of natural products to overcome. Eur J Med Chem 179:272–309
Kloc M, Ghobrial RM, Wosik J, Lewicka A, Lewicki S, Kubiak JZ (2019) Macrophage functions in wound healing. J Tissue Eng Regen Med 13(1):99–109
Knochelmann HM, Dwyer CJ, Bailey SR et al (2018) When worlds collide: Th17 and Treg cells in cancer and autoimmunity. Cell Mol Immunol 15(5):458–469
Ko HM, Koppula S, Kim BW et al (2010) Inflexin attenuates proinflammatory responses and nuclear factor-kappaB activation in LPS-treated microglia. Eur J Pharmacol 633(1–3):98–106
Kong LM, Deng X, Zuo ZL, Sun HD, Zhao QS, Li Y (2014) Identification and validation of p50 as the cellular target of eriocalyxin B. Oncotarget 5(22):11354–11364
Kratofil RM, Kubes P, Deniset JF (2017) Monocyte conversion during inflammation and injury. Arterioscler Thromb Vasc Biol 37(1):35–42
Ku CM, Lin JY (2013) Anti-inflammatory effects of 27 selected terpenoid compounds tested through modulating Th1/Th2 cytokine secretion profiles using murine primary splenocytes. Food Chem 141(2):1104–1113
Kubes P (2018) The enigmatic neutrophil: what we do not know. Cell Tissue Res 371(3):399–406
Kunkl M, Frascolla S, Amormino C, Volpe E, Tuosto L (2020) T helper cells: the modulators of inflammation in multiple sclerosis. Cells 9(2):482
Kuprash DV, Nedospasov SA (2016) Molecular and cellular mechanisms of inflammation. Biochemistry (Mosc) 81(11):1237–1239
Latha S, Chaudhary S, Ray R (2017) Hydroalcoholic extract of Stevia rebaudiana bert. leaves and stevioside ameliorates lipopolysaccharide induced acute liver injury in rats. Biomed Pharmacother 95:1040–1050
Lawrence SM, Corriden R, Nizet V (2020) How neutrophils meet their end. Trends Immunol 41(6):531–544
Lee JH, Koo TH, Hwang BY, Lee JJ (2002) Kaurane diterpene, kamebakaurin, inhibits NF-kappa B by directly targeting the DNA-binding activity of p50 and blocks the expression of antiapoptotic NF-kappa B target genes. J Biol Chem 277(21):18411–18420
Lee JH, Choi JK, Noh MS, Hwang BY, Hong YS, Lee JJ (2004) Anti-inflammatory effect of kamebakaurin in in vivo animal models. Planta Med 70(6):526–530
Lee JW, Lee MS, Kim TH et al (2007) Inhibitory effect of inflexinol on nitric oxide generation and iNOS expression via inhibition of NF-kappaB activation. Mediators Inflamm 2007:93148
Lee M, Kim SH, Lee HK, Cho Y, Kang J, Sung SH (2014) ent-kaurane and ent-pimarane diterpenes from Siegesbeckia pubescens inhibit lipopolysaccharide-induced nitric oxide production in BV2 microglia. Biol Pharm Bull 37(1):152–157
Leppkes M, Neurath MF (2020) Cytokines in inflammatory bowel diseases—update 2020. Pharmacol Res 104835
Leung CH, Grill SP, Lam W, Han QB, Sun HD, Cheng YC (2005) Novel mechanism of inhibition of nuclear factor-kappa B DNA-binding activity by diterpenoids isolated from Isodon rubescens. Mol Pharmacol 68(2):286–297
Leung CH, Grill SP, Lam W, Gao W, Sun HD, Cheng YC (2006) Eriocalyxin B inhibits nuclear factor-kappaB activation by interfering with the binding of both p65 and p50 to the response element in a noncompetitive manner. Mol Pharmacol 70(6):1946–1955
Li J, Du J, Sun L, Liu J, Quan Z (2010) Anti-inflammatory function of nodosin via inhibition of IL-2. Am J Chin Med 38(1):127–142
Li J, Bao L, Zha D et al (2018) Oridonin protects against the inflammatory response in diabetic nephropathy by inhibiting the TLR4/p38-MAPK and TLR4/NF-κB signaling pathways. Int Immunopharmacol 55:9–19

Lim H, Jung HA, Choi JS, Kim YS, Kang SS, Kim HP (2009) Anti-inflammatory activity of the constituents of the roots of Aralia continentalis. Arch Pharm Res 32(9):1237–1243

Lim JY, Sul D, Hwang BY, Hwang KW, Yoo KY, Park SY (2013) Suppression of LPS-induced inflammatory responses by inflexanin B in BV2 microglial cells. Can J Physiol Pharmacol 91(2):141–148

Liu YQ, You S, Zhang CL, Tashiro S, Onodera S, Ikejima T (2005) Oridonin enhances phagocytosis of UV-irradiated apoptotic U937 cells. Biol Pharm Bull 28(3):461–467

Liu J, Yang F, Zhang Y, Li J (2007) Studies on the cell-immunosuppressive mechanism of oridonin from Isodon serra. Int Immunopharmacol 7(7):945–954

Liu YC, Zou XB, Chai YF, Yao YM (2014) Macrophage polarization in inflammatory diseases. Int J Biol Sci 10(5):520–529

Liu X, Zhang Z, Ruan J et al (2016a) Inflammasome-activated gasdermin D causes pyroptosis by forming membrane pores. Nature 535(7610):153–158

Liu QQ, Wang HL, Chen K et al (2016b) Oridonin derivative ameliorates experimental colitis by inhibiting activated T-cells and translocation of nuclear factor-kappa B. J Dig Dis 17(2):104–112

Liu X, Yin S, Chen Y et al (2018) LPS-induced proinflammatory cytokine expression in human airway epithelial cells and macrophages via NF-κB, STAT3 or AP-1 activation. Mol Med Rep 17(4):5484–5491

Liu D, Qin H, Yang B, Du B, Yun X (2020) Oridonin ameliorates carbon tetrachloride-induced liver fibrosis in mice through inhibition of the NLRP3 inflammasome. Drug Dev Res 81(4):526–533

Loo TT, Gao Y, Lazarevic V (2018) Transcriptional regulation of CD4+ TH cells that mediate tissue inflammation. J Leukoc Biol 104(6):1069–1085

Lu Y, Chen B, Song JH et al (2013) Eriocalyxin B ameliorates experimental autoimmune encephalomyelitis by suppressing Th1 and Th17 cells. Proc Natl Acad Sci U S A 110(6):2258–2263

Lyu JH, Lee GS, Kim KH et al (2011) ent-kaur-16-en-19-oic Acid, isolated from the roots of Aralia continentalis, induces activation of Nrf2. J Ethnopharmacol 137(3):1442–1449

Marcondes-Alves L, Fattori V, Borghi SM et al (2019) Kaurenoic acid extracted from Sphagneticola trilobata reduces acetaminophen-induced hepatotoxicity through inhibition of oxidative stress and pro-inflammatory cytokine production in mice. Nat Prod Res 33(6):921–924

Martínez GP, Mijares MR, Chávez K et al (2019) Caracasine acid, an ent-3,4-seco-kaurene, promotes apoptosis and cell differentiation through NFkB signal pathway inhibition in leukemia cells. Eur J Pharmacol 862:172624

Minter MR, Taylor JM, Crack PJ (2016) The contribution of neuroinflammation to amyloid toxicity in Alzheimer's disease. J Neurochem 136(3):457–474

Mitchell JP, Carmody RJ (2018) NF-κB and the transcriptional control of inflammation. Int Rev Cell Mol Biol 335:41–84

Mizoguchi E, Low D, Ezaki Y, Okada T (2020) Recent updates on the basic mechanisms and pathogenesis of inflammatory bowel diseases in experimental animal models. Intest Res 18(2):151–167

Nakayama T, Hirahara K, Onodera A et al (2017) Th2 cells in health and disease. Annu Rev Immunol 35:53–84

Nasser MI, Zhu S, Huang H et al (2020) Macrophages: first guards in the prevention of cardiovascular diseases. Life Sci 250:117559

Nguyen GT, Green ER, Mecsas J (2017) Neutrophils to the ROScue: mechanisms of NADPH oxidase activation and bacterial resistance. Front Cell Infect Microbiol 7:373

Nisbet RM, Götz J (2018) Amyloid-β and tau in Alzheimer's disease: novel pathomechanisms and non-pharmacological treatment strategies. J Alzheimer's Dis 64(s1):S517–S527

Noack M, Miossec P (2014) Th17 and regulatory T cell balance in autoimmune and inflammatory diseases. Autoimmun Rev 13(6):668–677

Okada Y (2017) Chapter 8: Proteinases and matrix degradation

Osafo N, Biney RP, Obiri DD (2016) Aqueous ethanol fruit extract of Xylopia aethiopica and xylopic acid exhibit anti-inflammatory activity through inhibition of the arachidonic acid pathway. UK J Pharm Biosci 4:35–41

Osafo N, Obiri DD, Antwi AO, Yeboah OK (2018) The acute anti-inflammatory action of xylopic acid isolated from Xylopia aethiopica. J Basic Clin Physiol Pharmacol 29(6):659–669

Osafo N, Obiri DD, Danquah KO, Essel LB, Antwi AO (2019) Potential effects of xylopic acid on acetic acid-induced ulcerative colitis in rats. Turk J Gastroenterol 30(8):732–744

Paiva LA, Gurgel LA, Silva RM et al (2002) Anti-inflammatory effect of kaurenoic acid, a diterpene from Copaifera langsdorffii on acetic acid-induced colitis in rats. Vascul Pharmacol 39(6):303–307

Papayannopoulos V (2018) Neutrophils stepping through (to the other side). Immunity 49(6):992–994

Park HJ, Kim IT, Won JH et al (2007) Anti-inflammatory activities of ent-16alphaH,17-hydroxy-kauran-19-oic acid isolated from the roots of Siegesbeckia pubescens are due to the inhibition of iNOS and COX-2 expression in RAW 264.7 macrophages via NF-kappaB inactivation. Eur J Pharmacol 558(1–3):185–193

Passante E, Ehrhardt C, Sheridan H, Frankish N (2009) RBL-2H3 cells are an imprecise model for mast cell mediator release. Inflamm Res 58(9):611–618

Potočnjak I, Broznić D, Kindl M, Kropek M, Vladimir-Knežević S, Domitrović R (2017) Stevia and stevioside protect against cisplatin nephrotoxicity through inhibition of ERK1/2, STAT3, and NF-κB activation. Food Chem Toxicol 107(Pt A):215–225

Qian C, Cao X (2018) Dendritic cells in the regulation of immunity and inflammation. Semin Immunol 35:3–11

Rajendran L, Paolicelli RC (2018) Microglia-mediated synapse loss in Alzheimer's disease. J Neurosci 38(12):2911–2919

Rea IM, Gibson DS, McGilligan V, McNerlan SE, Alexander HD, Ross OA (2018) Age and age-related diseases: role of inflammation triggers and cytokines. Front Immunol 9:586

Sandquist I, Kolls J (2018) Update on regulation and effector functions of Th17 cells. F1000Res 7:205

Saradna A, Do DC, Kumar S, Fu QL, Gao P (2018) Macrophage polarization and allergic asthma. Transl Res 191:1–14

Saravanan R, Ramachandran V (2013) Modulating efficacy of Rebaudioside A, a diterpenoid on antioxidant and circulatory lipids in experimental diabetic rats. Environ Toxicol Pharmacol 36(2):472–483

Schwaiger S, Adams M, Seger C, Ellmerer EP, Bauer R, Stuppner H (2004) New constituents of Leontopodium alpinum and their in vitro leukotriene biosynthesis inhibitory activity. Planta Med 70(10):978–985

Shi M, Deng Y, Yu H et al (2019) Protective effects of oridonin on acute liver injury via impeding posttranslational modifications of interleukin-1 receptor-associated kinase 4 (IRAK4) in the toll-like receptor 4 (TLR4) signaling pathway. Mediators Inflamm 2019:7634761

Shi C, Yang H, Zhang Z (2020) Involvement of nucleotide-binding oligomerization domain-like receptor family pyrin domain containing 3 inflammasome in the pathogenesis of liver diseases. Front Cell Dev Biol 8:139

Silva JJ, Pompeu DG, Ximenes NC et al (2015) Effects of kaurenoic acid and arginine on random skin flap oxidative stress, inflammation, and cytokines in rats. Aesthetic Plast Surg 39(6):971–977

Siouti E, Andreakos E (2019) The many facets of macrophages in rheumatoid arthritis. Biochem Pharmacol 165:152–169

Suárez AI, Chavez K, Mateu E et al (2009) Cytotoxic activity of seco-entkaurenes from Croton caracasana on human cancer cell lines. Nat Prod Commun 4(11):1547–1550

Swanson KV, Deng M, Ting JP (2019) The NLRP3 inflammasome: molecular activation and regulation to therapeutics. Nat Rev Immunol 19(8):477–489

Szondy Z, Sarang Z, Kiss B, Garabuczi É, Köröskényi K (2017) Anti-inflammatory mechanisms triggered by apoptotic cells during their clearance. Front Immunol 8:909
Takahashi JA, Gomes DC, Lyra FH, Dos Santos GF, Martins LR (2014) The remarkable structural diversity achieved in ent-kaurane diterpenes by fungal biotransformations. Molecules 19(2):1856–1886
Tatiya-Aphiradee N, Chatuphonprasert W, Jarukamjorn K (2018) Immune response and inflammatory pathway of ulcerative colitis. J Basic Clin Physiol Pharmacol 30(1):1–10
Tiwari S, Atluri V, Kaushik A, Yndart A, Nair M (2019) Alzheimer's disease: pathogenesis, diagnostics, and therapeutics. Int J Nanomedicine 14:5541–5554
Vyas SP, Hansda AK, Goswami R (2019) Rheumatoid arthritis: 'melting pot' of T helper subsets. Int Rev Immunol 38(5):212–231
Wang Z, Xue L, Guo C et al (2012a) Stevioside ameliorates high-fat diet-induced insulin resistance and adipose tissue inflammation by downregulating the NF-κB pathway. Biochem Biophys Res Commun 417(4):1280–1285
Wang CF, Wang ZY, Tao SF et al (2012b) Preconditioning donor liver with nodosin perfusion lessens rat ischemia reperfusion injury via heme oxygenase-1 upregulation. J Gastroenterol Hepatol 27(4):832–840
Wang R, Liu YQ, Ha W et al (2014a) In vitro anti-inflammatory effects of diterpenoids and sesquiterpenoids from traditional Chinese medicine Siegesbeckia pubescens. Bioorg Med Chem Lett 24(16):3944–3947
Wang S, Yang H, Yu L et al (2014b) Oridonin attenuates Aβ1-42-induced neuroinflammation and inhibits NF-κB pathway. PLoS One 9(8):e104745
Wang S, Zhang Y, Saas P et al (2015) Oridonin's therapeutic effect: suppressing Th1/Th17 simultaneously in a mouse model of Crohn's disease. J Gastroenterol Hepatol 30(3):504–512
Wang J, Li F, Ding J et al (2016a) Investigation of the anti-asthmatic activity of oridonin on a mouse model of asthma. Mol Med Rep 14(3):2000–2006
Wang S, Yu L, Yang H et al (2016b) Oridonin attenuates synaptic loss and cognitive deficits in an Aβ1-42-induced mouse model of Alzheimer's disease. PLoS One 11(3):e0151397
Wang M, Li H, Xu F et al (2018) Diterpenoid lead stevioside and its hydrolysis products steviol and isosteviol: biological activity and structural modification. Eur J Med Chem 156:885–906
Wang J, Shen W, Zhang JY, Jia CH, Xie ML (2019) Stevioside attenuates isoproterenol-induced mouse myocardial fibrosis through inhibition of the myocardial NF-κB/TGF-β1/Smad signaling pathway. Food Funct 10(2):1179–1190
Wang Z, Zhang S, Xiao Y et al (2020) NLRP3 inflammasome and inflammatory diseases. Oxid Med Cell Longev 2020:4063562
White CS, Lawrence CB, Brough D, Rivers-Auty J (2017) Inflammasomes as therapeutic targets for Alzheimer's disease. Brain Pathol 27(2):223–234
Winther M, Holdfeldt A, Gabl M, Wang JM, Forsman H, Dahlgren C (2016) Formylated MHC class Ib binding peptides activate both human and mouse neutrophils primarily through formyl peptide receptor 1. PLoS One 11(12):e0167529
Woode E, Ameyaw EO, Boakye-Gyasi E, Abotsi WK (2012) Analgesic effects of an ethanol extract of the fruits of Xylopia aethiopica (Dunal) A. Rich (Annonaceae) and the major constituent, xylopic acid in murine models. J Pharm Bioallied Sci 4(4):291–301
Xin SH, Tan L, Cao X, Yu JT, Tan L (2018) Clearance of amyloid beta and tau in Alzheimer's disease: from mechanisms to therapy. Neurotox Res 34(3):733–748
Xu Y, Xue Y, Wang Y, Feng D, Lin S, Xu L (2009) Multiple-modulation effects of oridonin on the production of proinflammatory cytokines and neurotrophic factors in LPS-activated microglia. Int Immunopharmacol 9(3):360–365
Xu W, Zheng D, Liu Y, Li J, Yang L, Shang X (2017) Glaucocalyxin B alleviates lipopolysaccharide-induced Parkinson's disease by inhibiting TLR/NF-κB and activating Nrf2/HO-1 pathway. Cell Physiol Biochem 44(6):2091–2104
Xu J, Wold EA, Ding Y, Shen Q, Zhou J (2018) Therapeutic potential of oridonin and its analogs: from anticancer and antiinflammation to neuroprotection. Molecules 23(2):474

Xu L, Li L, Zhang CY, Schluesener H, Zhang ZY (2019) Natural diterpenoid oridonin ameliorates experimental autoimmune neuritis by promoting anti-inflammatory macrophages through blocking notch pathway. Front Neurosci 13:272
Xue Q, Yan Y, Zhang R, Xiong H (2018) Regulation of iNOS on immune cells and its role in diseases. Int J Mol Sci 19(12):3805
Yan Y, Tan RZ, Liu P et al (2020) Oridonin alleviates IRI-induced kidney injury by inhibiting inflammatory response of macrophages via AKT-related pathways. Med Sci Monit 26:e921114
Yang YL, Chang FR, Hwang TL, Chang WT, Wu YC (2004) Inhibitory effects of ent-kauranes from the stems of Annona squamosa on superoxide anion generation by human neutrophils. Planta Med 70(3):256–258
Yang F, Cao Y, Zhang J, You T, Zhu L (2017) Glaucocalyxin A improves survival in bleomycin-induced pulmonary fibrosis in mice. Biochem Biophys Res Commun 482(1):147–153
Yang H, Lv H, Li H, Ci X, Peng L (2019) Oridonin protects LPS-induced acute lung injury by modulating Nrf2-mediated oxidative stress and Nrf2-independent NLRP3 and NF-κB pathways. Cell Commun Signal 17(1):62
Yang H, Huang J, Gao Y, Wen Z, Peng L, Ci X (2020) Oridonin attenuates carrageenan-induced pleurisy via activation of the KEAP-1/Nrf2 pathway and inhibition of the TXNIP/NLRP3 and NF-κB pathway in mice. Inflammopharmacology 28(2):513–523
Yeh SH, Chang FR, Wu YC et al (2005) An anti-inflammatory ent-kaurane from the stems of Annona squamosa that inhibits various human neutrophil functions. Planta Med 71:904–909
Yeshi K, Ruscher R, Hunter L, Daly NL, Loukas A, Wangchuk P (2020) Revisiting inflammatory bowel disease: pathology, treatments, challenges and emerging therapeutics including drug leads from natural products. J Clin Med 9(5):E1273
Yin QQ, Liu CX, Wu YL et al (2013) Preventive and therapeutic effects of adenanthin on experimental autoimmune encephalomyelitis by inhibiting NF-κB signaling. J Immunol 191(5):2115–2125
Yingkun N, Zhenyu W, Jing L, Xiuyun L, Huimin Y (2013) Stevioside protects LPS-induced acute lung injury in mice. Inflammation 36(1):242–250
Yoshioka H, Aoyagi Y, Fukuishi N et al (2017) Suppressive effect of kamebakaurin on acetaminophen-induced hepatotoxicity by inhibiting lipid peroxidation and inflammatory response in mice. Pharmacol Rep 69(5):903–907
Yoshioka H, Nonogaki T, Ohnishi H et al (2018) 1O, 20O-diacetyl kamebakaurin protects against acetaminophen-induced hepatotoxicity in mice. Biomed Res 39(5):251–260
Yu X, He L, Cao P, Yu Q (2015) Eriocalyxin B Inhibits STAT3 signaling by covalently targeting STAT3 and blocking phosphorylation and activation of STAT3. PLoS One 10(5):e0128406
Yu S, Liu M, Hu K (2019) Natural products: potential therapeutic agents in multiple sclerosis. Int Immunopharmacol 67:87–97
Zang L, He H, Ye Y et al (2012a) Nitric oxide augments oridonin-induced efferocytosis by human histocytic lymphoma U937 cells via autophagy and the NF-κB-COX-2-IL-1β pathway. Free Radic Res 46(10):1207–1219
Zang L, Xu Q, Ye Y et al (2012b) Autophagy enhanced phagocytosis of apoptotic cells by oridonin-treated human histocytic lymphoma U937 cells. Arch Biochem Biophys 518(1):31–41
Zang L, Wang J, Ren Y et al (2019) Activated toll-like receptor 4 is involved in oridonin-induced phagocytosis via promotion of migration and autophagy-lysosome pathway in RAW264.7 macrophages. Int Immunopharmacol 66:99–108
Zhang Y, Liu J, Jia W, Zhao A, Li T (2005) Distinct immunosuppressive effect by Isodon serra extracts. Int Immunopharmacol 5(13–14):1957–1965
Zhang ZY, Daniels R, Schluesener HJ (2013) Oridonin ameliorates neuropathological changes and behavioural deficits in a mouse model of cerebral amyloidosis. J Cell Mol Med 17(12):1566–1576
Zhang M, Zhao C, Dai W, He J, Jiao S, Li B (2017a) Anti-inflammatory ent-kaurenoic acids and their glycosides from Gochnatia decora. Phytochemistry 137:174–181

Zhang Q, Lenardo MJ, Baltimore D (2017b) 30 Years of NF-κB: a blossoming of relevance to human pathobiology. Cell 168(1-2):37–57

Zhang Z, Xu J, Liu B et al (2019) Ponicidin inhibits pro-inflammatory cytokine TNF-α-induced epithelial-mesenchymal transition and metastasis of colorectal cancer cells via suppressing the AKT/GSK-3β/Snail pathway. Inflammopharmacology 27(3):627–638

Zhang Y, Wang S, Dai M, Nai J, Zhu L, Sheng H (2020) Solubility and bioavailability enhancement of oridonin: a review. Molecules 25(2):332

Zhao C, Zhao W (2020) NLRP3 inflammasome—a key player in antiviral responses. Front Immunol 11:211

Zhao YJ, Lv H, Xu PB et al (2016) Protective effects of oridonin on the sepsis in mice. Kaohsiung J Med Sci 32(9):452–457

Zhao G, Zhang T, Ma X et al (2017) Oridonin attenuates the release of pro-inflammatory cytokines in lipopolysaccharide-induced RAW264.7 cells and acute lung injury. Oncotarget 8(40):68153–68164

Zhou L, Sun L, Wu H et al (2013) Oridonin ameliorates lupus-like symptoms of MRL(lpr/lpr) mice by inhibition of B-cell activating factor (BAFF). Eur J Pharmacol 715(1–3):230–237

Zhou T, Zhuang J, Wang Z et al (2019a) Glaucocalyxin A as a natural product increases amyloid β clearance and decreases tau phosphorylation involving the mammalian target of rapamycin signaling pathway. Neuroreport 30(4):310–316

Zhou M, Yi Y, Hong L (2019b) Oridonin ameliorates lipopolysaccharide-induced endometritis in mice via inhibition of the TLR-4/NF-κB pathway. Inflammation 42(1):81–90

Zhu S, Zhang J, Lv Y (2020) Glaucocalyxin A inhibits hydrogen peroxide-induced oxidative stress and inflammatory response in coronary artery smooth muscle cells. Clin Exp Pharmacol Physiol 47(5):765–770

# Role of Plant Secondary Metabolites in Metabolic Disorders

8

Younis Ahmad Hajam, Raksha Rani, Preeti Sharma, Imran Ali Khan, and Rajesh Kumar

**Abstract**

Living cells are actively engaged in conducting various processes, sustaining the survival of cells, depending on highly orchestrated biochemical reactions. The body of any living being is a beautifully sophisticated factory. However, any alteration in normal cell physiology leads to various cellular, biochemical, metabolic, histological, and molecular alterations. Metabolic disorder includes a group of threatening reasons that significantly increases the prevalence of various diseases such as cardiovascular disease, obesity, diabetes, dyslipidemia (increased triglyceride levels, LDL, and VLDL but decreased HDL), increased blood pressure and irregular glucose metabolism, etc. Lifestyle changes, especially eating behaviors and physical activity, are predicted by two independent risk factors for the development of metabolic syndrome, more than genetic factors (only 10%) for the occurrence of metabolic disorder. These metabolic alterations often lead to organ damage, such as hepatic injury, renal damage, neurodegeneration, and reproductive abnormalities. Various drugs are available for the treatment of these metabolic disorders, but in one way or another, they have several other side defects, such as liver failure, kidney damage, neuronal injury, and several others. Phytopolyphenols such as phenols, alkaloids,

Y. A. Hajam (✉) · R. Rani · R. Kumar
Division Zoology, Department of Biosciences, Career Point University, Hamirpur, Himachal Pradesh, India
e-mail: drkumar83@rediffmil.com

P. Sharma
Department of Zoology, Career Point University, Hamirpur, Himachal Pradesh, India

I. A. Khan
Department of Clinical Laboratory Sciences, College of Applied Medical Sciences, King Saud University, Riyadh, Saudi Arabia
e-mail: imkhan@ksu.edu.sa



A. K. Sharma, A. Sharma (eds.), *Plant Secondary Metabolites*,
https://doi.org/10.1007/978-981-16-4779-6_8

flavonols, flavonoids, flavonoids, terpenes, saponins, etc. have therapeutic effectiveness in the treatment of many diseases. These phytochemicals are versatile and have a great deal of interest from the scientific community due to their numerous biological activities, bioavailability, and bio-accessibility, and these polyphenols are used to prevent various chronic diseases. In addition, polyphenol-rich diets serve as a buffer against various forms of oxidative stress, cancer, type 2 diabetes, osteoporosis, pancreatitis, cardiovascular diseases, gastrointestinal diseases, lung diseases, and neurodegenerative diseases. The use of medicinal herbs can also be an alternative/adjuvant treatment for both metabolic and chronic diseases.

**Keywords**

Phytopolyphenols · Therapeutics · Biological activities · Bioavailability · Bio-accessibility · Polyphenol-rich diet · Medicinal herbs · Metabolic and chronic diseases

## 8.1 Introduction

### 8.1.1 Physiology of Metabolism

Living cells are actively engaged in conducting various processes, sustaining the survival of cells, depending on highly orchestrated biochemical reactions. The body of any living being is a beautifully sophisticated factory. It recognizes unprocessed substances (food), some of these substances are burned to produce energy, some are used to generate complete objects, some are stored, and unused goods are removed from the body to maintain homeostasis in the body. Storage products are used for growth and development. Metabolism is a general concept that constitutes all biochemical changes that are taking place in living organisms to sustain life. Metabolism processes are important for growth and reproduction and enable living organisms to maintain their structure as well as to respond to their environmental needs. Another general term is metabolic processes that describe the sequence of chemical reactions that either break down higher molecular weights and complex compounds into smaller units (catabolism) or the creation of larger molecules from smaller ones (anabolism). For example, eating bread or rice breaks the starch into glucose units in food canal bread or rice. These glucose units are further catabolized to release energy and perform various biological functions, including muscle contractions. In contrast, an anabolic reaction assembles glucose molecules to form glycogen as a storage form. Metabolic processes never stop or are inactive, but these processes are continuous depending on internal or external stimuli.

### 8.1.2 Cell: The Metabolic Processing Center

Cells are the workstation of metabolic processes (both catabolic and anabolic), but the body of living organisms is composed of different types of cells, including hepatic cells, neural cells, renal cells, reproductive cells, and muscle cells. A typical animal cell is the main two parts of the nucleus and the membrane-bounded space called cytoplasm. Cytoplasm is a semifluid substance of a cell that fills the cytoplasm with different organelles inside the cytosol, small units that perform unique metabolic functions. One of these organelles is a capsule-like mitochondria (cell powerhouse), which generates energy through various energy-generating pathways.

Carbohydrates are an essential source of energy for various reactions. Glycolysis is a basic metabolic pathway present in all species through which glucose is converted into energy-release products (ATPs), NADH, and two pyruvate molecules. Glycogen, which is processed as glucose in vertebrates, is synthesized through the glycogenesis process when blood glucose levels increase and break through the glycogenolysis process (glucose deficiency/glucose supply shortness). Glucose is also synthesized from noncarbohydrate sources known as gluconeogenesis under stress conditions. In addition, the pentose phosphate pathway makes a cell capable of switching glucose 6-phosphate (G6P), a glucose derivative, to ribose 5-phosphate (used to synthesize nucleotides as well as nucleic acids) and various other monosaccharides. NADPH, produced via the pentose pathway, is an important reductant present in the cells. In vertebrates, glucose is transported through circulatory blood throughout the body. As the energy level in the body decreases, the glucose contained in the form of glycogen in the liver and muscle is converted through the glycolysis process. Some cells in different organs (brain, red blood cells, and skeletal muscle cells) need a continuous supply of glucose to meet energy requirements.

Protein metabolism consists of different biochemical processes required for protein and amino acid synthesis (anabolism) and protein degradation by catabolism. Necessary amino acids must be supplied or taken from external sources in the form of a diet. Amino acids derived after protein breakdown are accelerated to provide precursors for gluconeogenesis and protein resynthesis, as well as to facilitate the process of DNA replication and cell proliferation, such as immune system cells, wound care, and other processes. Proteins have a particular and complex metabolism role, proteins do not act as energy sources for the growth and development of the organism, but they also make up raw materials for the construction of the body. When the proteins are fully digested, the free amino acids released pass through the portal circulation and are then dispersed across the body and provide substrates for various synthetic transformations associated with the creation of tissue components. Proteins in each tissue are supposed to be synthesized within the same cell locality, but influential factors are responsible for the mixture of amino acids in the correct order as well as in the correct proportions (Rose 1933). The proteins have various important positions in the structure as well as the functions of muscle, hemoglobin, hormones, fibrin, and receptors.

As amino acids are degraded from ketoacidosis through the process of oxidative deamination or transamination (the amino acid group is separated from the amino acid to form urea), acetyl coenzyme A may also be produced. Acetyl coenzyme A enters the citric acid cycle to generate ATP. In standard situations, protein metabolism is suppressed due to the presence of glucose levels in the circulation of the blood. A small amount of glucose is required to increase the level of insulin in circulatory blood to block protein breakdown (Kovacevic and McGivan 1983).

Lipids include a wide collection of water-insoluble (hydrophobic) organic molecules, which are removed from tissues by nonpolar solvents due to their insolubility in aqueous solutions. The body of the organism is divided into various structures, such as the plasma membrane/triglycerols globule, in adipocytes, or into the plasma associated with proteins, as well as in the lipoprotein unit or albumin.

Lipids are the key basis of the power to conduct routine functions in the body, and they also have a hydrophobic shield. In addition to this lipid, it also performs other functions, including fat-soluble vitamins acting as regulators or coenzymes, and prostaglandins as well as steroid hormones that contribute to the maintenance of homeostasis in the body. Lipid undergoes a process of emulsification in the small intestine, particularly in the duodenum. The method of emulsification is mediated by bile salts, as well as by mechanical peristalsis mixing. Bile salts (cholesterol derivatives) are formed in hepatocytes and stored in the gallbladder. Bile salts consist of a sterol ring structure having a side chain in which a glycine or taurine molecule is covalently bound by an amide linkage. Bile salts cooperate with lipid particles found in diet and watery duodenal compounds, which in small intestines stabilize the lipid unit as it grows smaller and thus prevents them from coalescing. Synthesized lipids are used for a number of purposes. Triglycerols are found within chylomicrons and are initially broken down in the skeletal muscle and adipocyte capillaries, as well as in the heart, lungs, kidneys, and liver. Chylomicrons containing triacylglycerol are converted to free fatty acids and glycerol with the aid of lipoprotein lipase. Free fatty acids obtained from TAG hydrolysis are then transported to the circulation of serum albumin in the blood; free fatty acids are oxidized in the cells to generate energy. Released TAG glycerol is primarily used by liver cells to produce glycerol 3-phosphate, entering either glycolysis or gluconeogenesis by oxidation to dihydroxyacetone phosphate. Released TAG fatty acids require hydrolytic action, mediated by hormone-sensitive lipase, and fatty acids are removed from TAG carbon 1/carbon 3. In addition, certain lipases are specified for diacylglycerol, or other fatty acids are extracted by monoacylglycerol. Hormonal regulation of lipolysis involves the degradation of triglycerides by the action of lipases. Epinephrine, glucagon, and insulin regulates the concentration of blood glucose, moreover insulin also decrease fat deposition. However, if some change occurs in the metabolism of these biomolecules, various kinds of metabolic disorders are induced, such as irregular carbohydrate metabolism leading to diabetes, altered lipid metabolism leading to dyslipidemia, and cardiovascular disease and altered protein metabolism leading to malnutrition and several other conditions.

## 8.2 Introduction to Metabolic Disorders

Metabolic syndrome, which originated in 1920, was initially explained as a combination of intestinal obesity and other metabolic deformities in cardiovascular disease and type 2 diabetes mellitus (Alić et al. 2017). Recently, however, the definition of MetS is the product of numerous research reports (Alić et al. 2017; Haffner et al. 1992; Kaplan 1989). Metabolic disorder occurs when normal metabolic functions do not occur, and these changes in these processes are either higher or lower in the quantity of vital substances needed to maintain health. Metabolic disorder thus involves a category of threatening causes that greatly raises the prevalence of multiple diseases such as cardiovascular disease, obesity, diabetes, dyslipidemia (increased triglyceride levels, LDL, and VLDL but decreased HDL), increased blood pressure, abnormal glucose metabolism, etc. in addition to lower body obesity and/or raised insulin resistance (IR). Lifestyle changes, especially eating behaviors and physical activity, are predicted by two independent risk factors for the development of metabolic syndrome, more than genetic factors (only 10%) for the occurrence of metabolic disorder (Weiss et al. 2013; Reilly et al. 2005; Mozaffarian et al. 2011). Some studies have indicated that overall dietary habits instead of nutrients are not responsible for metabolic syndrome. Dietary activities as well as physical activity are expected to be linked to insulin resistance. However, day-to-day study findings that describe the role of dietary patterns such as sugar-sweetened drink intake (Hu 2013) may be more closely linked to the development of metabolic disorder and CVD rather than physical activity or sedentary lifestyle (Casazza et al. 2009). The prevalence of metabolic disorders has increased over the past decades due to an increase in the sedentary lifestyle associated with excess calorie intake. Elevated physical activity and healthy eating habits are the main factors for reducing or reversing the rise in weight as well as its negative effects.

Various research groups are trying to develop a diagnostic condition for the proper identification of metabolic disorders. The World Health Organization (WHO) and the European Community, the American Association of Clinical Endocrinology (AACE), and the National Heart, Lung, and Blood Institute (NHLBI) have tried to research diabetes and insulin resistance (EGIR) (Alić et al. 2017). The National Heart, Lung, and Blood Institute (NHLBI) Concepts are components of the scientific program that has been proposed to classify patients with metabolic disorders due to coronary inhibitory disease. These guidelines are intended to promote the accessibility and ease of use of these requirements in clinical practice (Nikolić et al. 2007). Among the five (5) criteria that have been identified, a patient may have at least three (3) criteria, such as elevated blood glucose levels, hypertension, elevated triglycerides, and low HDL levels, as well as increased waist perimeter.

## 8.3 Association of Significant Dietary Habits with Metabolic Syndrome

The incidence of metabolic syndrome is growing at an accelerated rate, but no treatment has been available to date. Different foods consumed have a wide variety of interactions between nutrients and foods. Rodríguez and Moreno (2006) investigated the relationship between the dietary factors and CVD or MetS or any particular distinguishing parameters of MetS. Historically, dietary guidelines involve lower fat consumption due to higher caloric density relative to proteins or carbohydrates. Current studies have indicated that the form of fat consumed is predicted to be responsible for dehydration. Similarly, the quantity of carbohydrates is not important, as is their quality and their source. Dietary patterns are an attempt to examine the relationship between diet and disease (Kant 2004; Newby and Tucker 2004; Hu 2002). Dietary habits are very useful to prescribe diets, due to the general dietary habits that might be easy for the public to understand. Earlier research indicated that an inappropriate dietary pattern is a key factor linked to MetS components, including obesity, dyslipidemia, hypertension (HTN), and CVD (Hu 2002; Pierce et al. 2002). As a result, the incidence of MetS is growing worldwide, and understanding the association of dietary patterns with MetS and its constituents may help to reduce the incidence of metabolic disorders (el Bilbeisi et al. 2017) (Table 8.1).

## 8.4 Causes of Metabolic Disorders

The basic cause of any metabolic disorder is an abnormal and uncontrolled metabolic process, which can occur at both biochemical and genetic levels. Popular examples are diabetes due to unregulated glucose metabolism, which can cause various effects in the body such as oxidative stress, hormonal imbalance, and hematological deficiency (Kanikarla-Marie and Jain 2016).

## 8.5 Role of Polyphenols in Health

Polyphenols are flexible phytochemicals in the diet and are highly attenuated by the scientific community due to their various biological activities, bioavailability, and bio-accessibility, and these polyphenols are used to avoid various chronic diseases. Epidemiological studies have shown that consumption of food containing high levels of polyphenols can be used to avoid cancer and other chronic diseases, including cardiovascular diseases, neurodegenerative diseases, type 2 diabetes, renal failure, hepatic damage, and obesity (Del Rio et al. 2013). Recent studies have documented that long-term intakes of polyphenol-rich diets serve as a protection mechanism against various forms of cancer, type 2 diabetes, osteoporosis, pancreatitis, cardiovascular diseases, gastrointestinal diseases, lung diseases, and neurodegenerative diseases (Fujiki et al. 2015; Xiao and Hogger 2015;

**Table 8.1** Association of dietary patterns with metabolic syndromes and possible treatments

| Sr. no. | Disease | Defective enzyme/system | Symptoms | Complications | Treatments |
|---|---|---|---|---|---|
| 1. | Diabetes | Beta cells of the pancreas/ insensitive<br>Insulin receptor | Increase in blood<br>Glucose level | Obesity, CVDs, infertility, dyslipidemia, etc. | Metformin/glibenclamide |
| 2. | Dyslipidemia | Defective lipid metabolism (lipase enzyme) | Increase in TG, LDL, cholesterol, and VLDL, decrease<br>In HDL level<br>In serum | Coronary artery diseases (CAD), peripheral artery diseases (PAD), heart attack and stroke, etc. | HMG-CoA reductase inhibitors (the "statins"), The fibrates (gemfibrozil, clofibrate, and fenofibrate), niacin/nicotinic acid, and the bile acid binding resins (colestipol and cholestyramine) |
| 3. | Obesity | Adipose tissue/leptin | Breathlessness, increased sweating, snoring, feeling very tired every day, back and joint pains, low confidence and self-esteem | Weight gain, inflammation, osteoarthritis, rheumatoid arthritis | Orlistat (Xenical, Alli), lorcaserin (Belviq), phentermine-topiramate (Qsymia), naltrexone-bupropion (Contrave), and liraglutide (Saxenda) |
| 4. | Hypothyroidism | Deficiency of $T_3$ and $T_4$ | Low body temperature<br>Anemia, low BMI, heart failure | Obesity, joint pain, infertility, and heart disease | Levothyroxine (Synthroid, Levoxyl, Levothroid, Unithroid, Tirosint), liothyronine (Cytomel, Triostat),<br>Thyroid desiccated (Armour thyroid, nature-Throid, Westhroid) |
| 5. | Hyperthyroidism (overactive thyroid) | Hypersecretion of thyroxin | **Whole body**: Excessive sweating, excessive hunger, fatigue, or heat intolerance<br>**Behavioral:** Hyperactivity, | Arrhythmia (abnormal heart beat, such as atrial fibrillation), cardiac dilation (increase in the size of the | Treatments include radioactive iodine, medication, and sometimes surgery |

(continued)

**Table 8.1** (continued)

| Sr. no. | Disease | Defective enzyme/system | Symptoms | Complications | Treatments |
|---|---|---|---|---|---|
| | | | irritability, or restlessness<br>**Heart:** Abnormal heart rhythm, fast heart rate, or palpitations<br>**Mood:** Mood change, tension, or dread attack<br>**Eyes:** Irregular extension of eyes or swollen eyes<br>**Menstrual:** Menstrual disturbances or short and prolonged bleeding<br>**Sleep:** Trouble lessening asleep or sleeplessness | heart cavities, which actually thins the heart muscle), and congestive heart failure. Sudden cardiac arrest | Medication includes:<br>**Antithyroid:** Methimazole; propylthiouracil (PTU)<br>**Beta blockers:** Propranolol; Inderal |
| 6. | Hyperlipoproteinemia, type 3 | Higher cholesterol and triglycerides | Pancreatitis (type 1)<br>Abdominal pain (types 1 and 5)<br>Enlarged liver or spleen (type 1)<br>Accumulation of lipid or xanthomas (type 1)<br>Previous history of heart ailment (types 2 and 4)<br>Ancestral history of diabetes (types 4 and 5)<br>Heart attack and stroke | Carotid artery disease<br>Coronary heart disease, including angina or heart attack<br>Peripheral artery disease<br>Stroke | Atorvastatin (Lipitor)<br>Fluvastatin (Lescol XL)<br>Pravastatin (Pravachol)<br>Ezetimibe (Zetia) |

| | | | | | |
|---|---|---|---|---|---|
| 7. | Familial hypolipoproteinemia | Lowered levels of any or all lipids and/or lipoproteins in the blood | Abetalipoproteinemia<br>Familial hypobetalipoproteinemia<br>Chylomicron retention disease | Overactive thyroid, anemia, undernutrition, cancer, chronic infection, or impaired absorption of foods from the digestive tract | **Statins, cholesterol absorption inhibitors, omega-3 fatty acid supplements, niacin, fibrates.** The medications fenofibrate (Tricor) and gemfibrozil (Lopid) |
| 8. | Abdominal obesity and the metabolic syndrome | Insulin resistance/lipases | Hypertriglyceridemia,<br>Low HDL cholesterol<br>Insulin resistance<br>Elevated apolipoprotein B, prothrombotic profile, increase in inflammatory indicators<br>Increase in insulin level<br>Increase in small, dense LDL | Increased insulin level | |
| 9. | Osteoporosis | Demineralization of bones/ deficiency of bones | Defective calcium metabolism | Depression, pain<br>Weakening of bones, pain in joints, fractures<br>A stooping posture<br>Persistent *back and neck pain* | **Calcium supplements, vitamin D, estrogen therapy** |
| 10. | Galactosemia | Inherited disorder that prevents a person from processing the sugar galactose, which is found in many foods | Higher risk of infection<br>Galactosemia means too much galactose buildup in the blood<br>Loss of appetite, vomiting, jaundice, fluid building up in the abdomen and swelling, abnormal | Enlarged liver, kidney failure, cataracts in the eyes, or brain damage. If untreated, as many as 75% of infants with galactosemia will die<br>Tremors, speech problems and delays, learning | There is no cure for galactosemia or approved medication to replace the enzymes |

(continued)

**Table 8.1** (continued)

| Sr. no. | Disease | Defective enzyme/system | Symptoms | Complications | Treatments |
|---|---|---|---|---|---|
| | | | bleeding, diarrhea, irritability, fatigue or lethargy, weight loss, weakness | disabilities, fine motor difficulties, low bone mineral density, reproductive problems, premature ovarian insufficiency | |
| 11. | Zellweger syndrome (absence of peroxisomes) | Mutation in the PEX1 gene | Poor muscle tone (hypotonia), poor feeding, seizures, hearing loss, vision loss, distinctive facial features, and skeletal abnormalities | | |

Martín-Peláez et al. 2013; Fraga et al. 2010). These protective dominant activities of polyphenols are based on their "biochemical scavenging theory" which explains that polyphenol compounds neutralize free radicals by forming stabilized chemical complexes and therefore inhibit the formation of free radicals (Sroka and Cisowski 2003). Other evidence supports the anti-stress behavior of polyphenols such as hydrogen peroxide formation, which promotes and controls immune responses, such as cell growth (Sroka and Cisowski 2003; Saeidnia and Abdollahi 2013).

During the digestion process polyphenols show preabsorptive interaction, phenols present in the diet reduce the transport of thiamine as well as folic acid and also alter the activity of drugs through interactions that influence drug transporters or enzymes involved in these drug transformation reactions, therefore inhibition as well as increase in bioavailability depends on the status. For example, the iron chelating and inhibitory effects of polyphenols on iron absorption may be due to poor iron status (Hurrell and Egli 2010). Mechanistic cause may be that people who eat rich food inhibit the absorption of iron, including sorghum, beans, and millet.

Earlier studies indicated that isoflavones found in soy products have a negative effect on the risk of estrogen-sensitive breast cancer and endometrial cancer, suggesting endocrine disrupting activity of these compounds (Yang et al. 2015; Carbonel et al. 2015). In the 1990s, considerable emphasis was placed on the use of polyphenols for various human health-related issues, which also decreases the incidence of many diseases, including CVD, DM, and cancer (Hertog et al. 1993). Polyphenols are elevated against LDL-mediated atherosclerosis due to their antioxidant properties (Lusis 2000).

## 8.6 Polyphenols and Their Role in the Human Body

Polyphenols are plant-based secondary metabolites that are abundantly present in foods that are appropriate for consumption, owing to their high antioxidant capacity, and their ingestion along with diet/diet use reduces the prevalence of various diseases such as coronary heart disease (Hertog et al. 1993; Virgili and Marino 2008). The beneficial effects of polyphenols on cardiovascular disease were due to their antioxidant properties, although the present study stated that they were vasodilatory, suggesting their ability to stabilize lipids and associated variables and to change the mechanism of apoptosis in vascular endothelium (Andújar et al. 2012). Evidence is growing quite rapidly, all of which suggesting that polyphenols are capable of preventing neuronal disorders such as Alzheimer's disease and Parkinson's disease by reducing inflammatory stress signaling and by up-regulation of gene expression that encodes antioxidant enzymes and cytoprotective proteins (Cabrera et al. 2006). Polyphenols are known to function as separate molecules and are also capable of interacting with receptors and enzymes and preventing diseases and enhancing human health (Murakami and Ohnishi 2012). Dietary proteins and polyphenols are both nutritious and dietary surpluses (inhibits various diseases and increases metabolic activity) and contribute to the development of food structure and enhance the functional capacity of food. Polyphenols in the form of capsules is found to be more effective; therefore, encapsulation techniques have been continuously increased due to various factors, including the efficacy of the capsule forming agent, the stabilization of the capsule forming agent, and the release of the capsulated compounds. Over the last decades, polyphenolic compounds have been shown to reduce numerous chronic as well as degenerative diseases such as obesity, cardiovascular diseases, and neurodegenerative diseases (Landgrave-Gómez et al. 2015) and to prevent cancer by influencing bacterial metabolizing enzyme, thereby reducing the overall risk of cancer (dos Reis et al. 2019).

## 8.7 Metabolic Syndrome and Oxidative Stress

Oxidative stress can be defined as an imbalance between the generation of reactive oxygen/reactive nitrogen species (ROS/RNS) and their elimination by antioxidant enzymes. Antioxidant enzymes are found in cells and molecules, which are subdivided into water-/lipid-soluble enzymes, or their origins may be endogenous/exogenous (dietary). Administration of different synthetic pharmaceutical agents enhances the resistance of LDL to oxidation (Brennan et al. 2002). This imbalance between free radicals and the antioxidant system causes oxidative damage to various structural and functional biomolecules such as proteins, fats, nucleic acids, and carbohydrates. Antioxidants shield the body from the harmful effects of free radicals. Supplementation of exogenous antioxidants defends the body against oxidative stress (Azab and Albasha 2018).

Reactive oxygen species (ROS) are mostly formed in mitochondria via the electron transport chain during aerobic respiration. However, most electrons enter complex III of the transport chain of electrons, and approximately 1–3% of electrons react prematurely with oxygen and create superoxide radicals (Kausar et al. 2018). Free radicals play a very important role in various physiological functions in the body, such as the production of O– and NO by neutrophils and macrophages, aiding the progression of phagocytosis and removing bacteria. Superoxide radicals in phagocytic cells are known to function as nonselective antibiotics and destroy any infective bacteria (as well as neutrophils) and cause damage to neighboring tissue cells; these free radicals often cause inflammatory reactions and encourage cell proliferation (mitotic division) of fibroblasts. Endogenous ROS/RNS is produced by a variety of processes, such as inflammation and activation of immune cells, ischemia and anxiety, incidence of cancer and infectious diseases, and aging. However, exogenous ROS/RNS are created due to contamination (water, air) by consuming alcohol, smoking, heavy metals, certain drugs (tacrolimus and cyclosporine), rays, cooking, and various solvents such as benzene. These compounds are decomposed after entering the body (Dasgupta and Klein 2014).

These ROS damage all macromolecules, including proteins, lipids, and nucleic acids, causing protein and nucleic acid modulation. In addition, these free radicals initiate and evolve various diseases such as diabetes, heart attacks, atherosclerosis, hepatic diseases, and cancer (Halliwell 2007). Oxidative stress is also responsible for the development of cancer and stimulation of ontogenesis. DNA breakage and start of ROS to AP-1 and NF-kB signal transduction procedures have been shown to trigger transcription of genes associated with the regulation of cell growth as well as the initiation of cancer.

Lipid peroxidation products are produced when hydrogen atoms are removed from unsaturated fatty acids. Accelerated lipid peroxidation affects membrane fluidity and integrity of biomolecules (membrane-bound proteins or cholesterol). These oxidizable lipids may attack neighboring proteins, resulting in an increased formation of protein carbonyls (Almroth et al. 2005). It has been stated that free radical induces peroxidation of membrane lipids, and even oxidative damage to DNA is correlated with a number of chronic health problems such as cancer, atherosclerosis, weakening of the nervous system, and aging (Finkel and Holbrook 2000).

The increase in the production of reactive oxygen species and a decrease in the activity of the anti-oxidative enzymatic defense system lead to oxidative stress. Oxidative stress causes damage only when the antioxidant system is unable to combat the formation of ROS. Under regulated physiological circumstances, ROS is neutralized/weakened under in vivo conditions by various antioxidant systems. O− is then converted to $H_2O_2$ by the action of superoxide dismutase (SOD), and hydrogen peroxide is reduced by the action of glutathione peroxide (GSH-Px), catalase, and peroxiredoxins (Fridlyand and Philipson 2006; Wood et al. 2000). In addition, ROS may also be neutralized or made less active in vivo by different endogenous molecules (albumin, uric acid) or by various exogenous dietary antioxidants (vitamin C or vitamin E). Attention has been based on the possibility that oxidative stress leads to the development of metabolic syndrome. Experimental evidence has shown that obesity is associated with elevated levels of ROS, increased expression of NAPH oxidase and lower levels control the expression of antioxidant enzymes, and that it induces major disruptions in the production of adiponectin, IL-6, and MCP-1. Previous studies indicated that the administration of NADPH oxidase inhibitor reduces the development of ROS in adipocytes, improves the irregularity of adipocytokines, and reduces the occurrence of diabetes, hyperlipidemia, and hepatic steatosis.

## 8.8 Phytotherapeutic and Metabolic Disorders

Various foods and beverages rich in different polyphenols are used for the treatment of various metabolic disorders, such as green tea, nuts, red wine, grape seeds, berries, and dark chocolate, containing various monomeric flavonols (catechins) and their oligomers (proanthocyanidins), which are a major source of content (Del Rio et al. 2013). Green tea (*Camellia sinensis* L.) plant contains high levels of flavonols (catechins) and alkaloid caffeine has some metabolic effects. Nagao et al. (2007) study reported that the treatment of green tea revealed substantial reductions in body weight, BMI, waist circumference, and body fat mass. It has been found that caffeine has the ability to interfere with the energy balance by rising energy consumption and reducing energy intake. Suliburska et al. (2012) reported that green tea containing epigallocatechingallate (EGCG) decreases BMI and waist circumference in obese patients. This weight-reducing effect of tea depends on the effects of caffeine and catechins on the adrenergic system. Caffeine (1,3,7-trimethylxantine), a purine alkaloid, suppresses the phosphodiesterase enzyme that hydrolyzes cyclic adenosine monophosphate (cAMP) to AMP. Diepvens et al. (2007) reported that cAMP signal is activated by beta-adrenergic stimulation, causing adrenergic effects such as reduced appetite while increasing energy intake and lipolysis. Evans and Bahng (2014) reported that caffeine up-regulates the expressions of uncoupling protein and increases thermogenesis by inhibiting phosphodiesterase and activation of cAMP protein kinase A. Catechins present in green tea affect the inactivation of catecholamines, inhibit the enzyme catechol-*O*-methyltransferase (COMT), and cause overstimulation of adrenergic receptors (Shixian et al. 2006). Similarly, caffeine-induced suppression of phosphodiesterase cause inactivation of COMT by growing energy intake, fat oxidation and lipid breakdown (Diepvens et al. 2007). Previous studies have shown that the administration of green tea enhances the lipid profile by substantially reducing LDL cholesterol (Basu et al. 2010; Suliburska et al. 2012; Chu et al. 2017; Belcaro et al. 2013) and triglycerides. Green tea reduces oxidative stress as well as cardiac alterations in dialysis and reverses neuronal inflammation (Mandel et al. 2008).

Anthocyanin water-insoluble compounds are commonly found in a number of fruits (grapes, cherries, cranberries, strawberries, blueberries, blackberries, currants) and vegetables (beetroot, red cabbage, and red onions). Anthocyanin has antioxidant potential due to its electron/hydrogen atoms that donate and receive free radicals from various hydroxyl groups (Wang et al. 1999). Blueberries (*Vaccinium myrtillus* L.) are packed with anthocyanidins, chlorogenic acid, flavonoids, and stilbenes, including pterostilbene and resveratrol. Blueberries have a variety of preventive and therapeutic potentials, such as ability to reduce oxidative pressure and inflammatory reactions and protection from cardiovascular disorders, hypertension, and diabetes (Naseri et al. 2018). Diet rich in blueberry protects the metabolic syndrome associated with a proinflammatory disorder in obese people elevating adiponectin expression and inhibits the expression of NF-kB in the liver (Vendrame et al. 2013). In addition, blueberries also have the potential to reduce triglycerides, fasting insulin, abdominal fat mass, liver weight, body weight, total fat mass and increase skeletal and adipose muscle PPAR activity. Food products rich in anthocyanins regulate the glycemic level as well as the effect on LDL-C (Yang et al. 2017).

Pomegranate contains polyphenols, primarily ellagitannins and anthocyanins, which can be used for the treatment of various metabolic disorders. In vivo studies have shown that pomegranate has the potential to reduce blood glucose, improve insulin sensitivity, inhibit 5-007-glucosidase, and increase the activity of glucose transporter 4 (GLUT4). In addition to this, pomegranate causes anti-inflammatory effects by altering the peroxisome proliferator-activated receptor pathway expressions, and current research has shown that pomegranate also reduces blood pressure during metabolic syndrome. Cocoa produces many polyphenols, including catechins, anthocyanins and proanthocyanidins. Epidemiological studies have confirmed that cocoa polyphenols cause cardiovascular beneficial effects in humans.

Cocoa polyphenols modulate some primary signaling pathways, such as Toll as receptor 4/NF-a-B signal transduction and transcription activation. Cocoa polyphenols release NO by stimulating the endothelial NO synthase, which is responsible for the dilation of blood vessels and also serves as a cardioprotective agent. Consumption of dark chocolates rich in flavonoids by healthy individuals increase the coronary flow rate reserve but is expected to be free flavonoid white chocolate (Shiina et al. 2009). Blood pressure lowering of cocoa activity can be explained by a number of mechanisms, such as NO elevation, which shows an antihypertensive effect (Napoli and Ignarro 2009). In addition, flavanols are capable of suppressing angiotensin-converting enzyme activity under in vitro conditions.

Curcumin derived from *Curcuma longa* (turmeric) conducts a number of biological activities, in particular turmeric rhizomes. Curcumin analogs are the major bioactive components of curcumin, demethoxycurcumin, and bisdemethoxycurcumin (Teiten et al. 2014). Curcumin and curcuminoids have small oral bioavailability due to their small intestinal absorption and fast metabolism (Liu et al. 2016). Curcumin possesses a range of therapeutic properties, such as anti-inflammatory, anticancer, hypoglycemic, antioxidant, antiviral, and antimicrobial activity, and shows that curcumin can be used as an adjuvant for multiple illnesses (Gupta et al. 2013). Many specific biological functions of curcumins can be explained by their ability to interact directly with various cell signaling molecules associated with inflammation and cancer, etc. In addition, curcumin modulates activities of different transcription factors, growth factors, inflammatory factors, cytokine, protein kinase, and enzymes (Milani et al. 2017). Various clinical studies have shown that curcumin can be used for the treatment of metabolic syndrome.

Yang et al. (2014) estimated the impact of curcumin extract (95% curcumin) on weight, glucose, and lipid profile against metabolic syndrome. Curcumin treatment substantially raises HDL-C levels, but decreases LDL-C levels. De Melo et al. (2018) studied that the administration of curcuminoids and/or curcumin decreases HbA1c without affecting homeostasis. Both alone and in combination, curcuminoids reduce the concentration of fasting blood glucose in people with dysglycemic conditions (pre-diabetes, diabetes, or metabolic syndrome, but not in nondiabetic or euglycemic individuals).

Olive oil is used as a food supplement; it primarily contains more oleic acid, but less of other biologically active ingredients such as vitamins and polyphenols. Olive oil also contains more than 230 compounds such as tocopherols, fatty alcohols, triterpenic alcohols, squalene, plant sterols, and polyphenols such as oleuropein and its metabolites (hydroxytyrosol and tyrosol) (Ruiz-Canelol) Polyphenols, especially hydroxytyrosol and tyrosol, have an anti-inflammatory effect and also affect cell multiplication and apoptosis in cancer cells. Curcumin is used as a treatment to prevent cardiovascular disease; improves BP; regulates glucose levels, endothelial function, and oxidative stress; and reduces triglycerides and total and LDL-C, but increases HDL-C and decreases inflammatory markers such as C-reactive protein and IL-6 (Estruch et al. 2006). Intake of olive oil along with various other phenols at different concentrations increases HDL-C, decreases both total cholesterol and oxidative stress bioindicators, and decreases triglyceride levels in circulatory plasma. Various flavonoids and phenolic acids found in the flower extract, including kaempferol, quercetin, and their glucosides, are capable of raising glucose and oleic uptake in both human skeletal muscle cells and liver cells under in vitro conditions (Ho et al. 2017). Clinical trials have shown that an elderflower extract can be used as a supplementary or functional food for the treatment of diabetes.

## 8.9 Role of Secondary Metabolites in Type 1 and Type 2 Diabetes Mellitus

Secondary metabolites (polyphenols) are known as flavonoids, phenolic acids, stilbenes and lignans. Flavonoids are further subdivided into flavones, flavonols, flavanols, flavanes, isoflavones, and anthocyanins. Several studies have documented the antidiabetic efficacy of some dietary polyphenols used to prevent type 1 and type 2 diabetes (Kim et al. 2016). Different mechanistic approaches are suggested that contribute to the maintenance of glucose homeostasis by the following means: suppression of carbohydrate digestion and absorption of glucose in the intestine; stimulation of insulin secretion from pancreatic β-cells; altering the release of glucose from the liver; and activating insulin receptors and glucose uptake in insulin-sensitive tissue (Hanhineva et al. 2010). Flavonoids include antioxidant action, central nervous system effects, and increased insulin sensitivity (Prasain et al. 2010). Anthocyanins include flavonoids that have an enormous dietary value and are mostly consumed in combination with diets such as fruits and vegetables (Guo and Ling 2015). Quercetin is commonly used in human diets due to its antidiabetic and anti-inflammatory activities (Vinayagam and Xu 2015; Li et al. 2016).

*Momordica charantia* (bitter melon) is a cucurbitane-type triterpenoid carantine (a steroidal glycoside similar to a mixture of stigmasterol glucoside and β-sitosterol glucoside) and also a polypeptide-p, vicine, and a ribosome inactivating momordin protein with hyperglycemic activity (Tan et al. 2008; Joseph and Jini 2013). Previous studies recorded that bitter melon extract suppresses glucose absorption in the intestines (Grover and Yadav 2004; Chaturvedi 2012), inhibits significant glucose metabolism enzymes (Shibib et al. 1993), and decreases gluconeogenesis in the liver (Tsai et al. 2012). Previous studies have confirmed that *M. charantia* increases the activity of AMP-activated protein kinase (AMPK) pathway (essential for the control of lipid and glucose metabolism in cells) and decreases the expression of phosphoenol pyruvate carboxykinase (Shih et al. 2014). Polypeptide-p is also known as "plant insulin," and clinical trials have shown that the treatment of polypeptide-p substantially lowers blood sugar (Baldwa et al. 1977). Current studies have shown that the addition of cucurbitane-type triterpenoid (known as compound K16) lowers both the blood glucose level and the lipid content, thus enhancing glucose tolerance. Compound K16 is also reported to have an up-regulatory effect on insulin signaling pathway-released protein expression (Jiang et al. 2016).

Diosgenin (3b-hydroxy-5-spirostene), 4-hydroxyisoleucine, and the soluble dietary fiber fraction of fenugreek seed are mainly bioactive compounds examined

(Fuller and Stephens 2015). Significant aglycone saponin has been reported to possess glucose-reducing potential by regenerating pancreatic beta-cells as well as stimulating insulin secretion (Kalailingam et al. 2014) and antioxidant potential and helping to differentiate adipocytes and increasing insulin-dependent glucose transport (Uemura et al. 2010). 4-Hydroxyisoleucine is a branched amino acid derivative usually present in plants and serves as a measure of the total free amino acid content in fenugreek seeds (Fuller and Stephens 2015). 4-Hydroxyisoleucine reveals both insulinotropic and antidiabetic properties by stimulating glucose-dependent insulin secretion and also reduces insulin resistance in the muscles and liver (Jetté et al. 2009). Fenugreek seeds are rich in fiber (50–65 g of fiber/100 g of seed), and these soluble dietary fibers increase glycemic control. These hypoglycemic properties of fibers are due to inhibition of lipid hydrolyzing and carbohydrate hydrolyzing enzymes in the digestive system (Hannan et al. 2007). Fenugreek seed is also known as galactomannan because it lowers the rate of glucose absorption (Srichamroen et al. 2009). It can also be concluded that fenugreek seed can be used as an effective remedy for type 2 diabetes, obesity, and dyslipidemia due to hypoglycemic and anti-dyslipidemic properties.

Ivy gourd (*C. grandis*) is suspected to imitate the action of insulin. *C. grandis* has a hypoglycemic effect by insulin secretion by influencing glucose-associated enzymes (Sauvaire et al. 1996). Clinical studies recorded *C. grandis* extract reduces the level of glucose-6-phosphatase and lactase dehydrogenase in glycolytic pathways and also restores the function of lipoprotein lipase in lipolytic pathways. In addition, both animal and human studies have shown that *C. grandis* can be used as a dietary supplement for diabetes.

Earlier studies indicated that extracts of cinnamon species possess antidiabetic properties (Akilen et al. 2012; Chen et al. 2012; Cheng et al. 2012; Verspohl et al. 2005). The presence of procyanidin oligomers in cinnamon is suspected to be responsible for antidiabetic activity (Lu et al. 2011; Chen et al. 2012). Several studies support the hypoglycemic activity of different cinnamon species. Chen et al.'s (2012) study showed that *C. cassia* extract can promote lipid deposition in adipose tissues and the liver, but *Cinnamomum tamala* extract typically increases insulin concentrations in the blood and pancreas. This antidiabetic activity is caused by the presence of procyanidin oligomer components in these extracts (Chen et al. 2012). Cinnamon extract therapy increases insulin resistance, hyperglycemic activity, and lipid metabolism (Sheng et al. 2008).

Secondary metabolites as inhibitors of alpha-glucosidase are used for the treatment of type 2 diabetes.

The role of glycosidases is to catalyze the hydrolysis of glycoside bonds in polysaccharides and glycoconjugates, to contribute to various biological functions such as carbohydrate digestion, glycoconjugate lysosomal breakdown, and post-translational changes in cell glycoproteins (Davies et al. 2005; Vocadlo and Davies 2008). Specifically, the terminal stage of starch digestion and disaccharides present in the human diet at higher concentrations is catalyzed in mammals' 5-007-glucosidase (AG) in the mucosal brush border of the small intestine. 5-007-glucosidase (AG) inhibitors interrupt the degradation of carbohydrates in the small intestine and also decrease the blood volume of postprandial glucose. As a result, glycosidase inhibition greatly affects the metabolism of polysaccharides, the production of glycoproteins, cell interactions, and the extension of the formulation of new therapeutic molecules used against different diseases, including diabetes, obesity, metastatic cancer, and viral infection (Kajimoto and Node 2009; Stutz and Wrodnigg 2011). AGs present in the brush wider cells of the small intestine possess selectively hydrolyzing property at the terminal end (1 → 4)-linked α-glucose residues (starch or disaccharides) and release an individual α-glucose molecule (Bischoff 1994). AG inhibitors have been extensively studied and numerous AG inhibitors have been marketed, such as acarbose, miglitol, voglibose, and 1-deoxynojirimycin (DNJ) and anti-glucosidase anti-type 2 diabetes drugs, a chronic situation in which the body is immune to the normal impact of insulin, resulting in an efficient blood glucose regulation (Olokoba et al. 2012). Plants have different phytochemicals, some of which have a health-promoting effect. Different plants have been tested for different bioactive compounds that have the ability to control type 2 diabetes. Various extracts of leaves, roots, barks, and fruits from different medicinal plants, herbs, and other plants have been studied to exhibit inhibitory activity against AG (Kawada et al. 2006). Thiocyclitol, a 13-membrane ring compound commonly known as 13-MRT, has been used against diabetes because it is capable of inactivating maltase and sucrase enzymes, respectively (Oe and Ozaki 2008). Therefore, vegetables rich in AGI activities could be helpful in controlling type 2 diabetes, since vegetables are rich in fiber and/or have higher nitrate content; they are stated to be able to reduce both cholesterol and blood pressure.

## 8.10 Plant Metabolism and Secondary Metabolites

Metabolites have various functions such as fuel, structure, signaling, stimulating, and inhibitory roles on enzymes, catalytic activity (usually in the form of enzyme cofactors), and defensive activity, and they often interact with other organisms. Plant metabolites are classified into two: primary and secondary metabolites. Primary metabolites contain carbohydrates, fatty acids, amino acids, and nucleic acids and are commonly used for cell maintenance (Kliebenstein and Osbourn 2012), whereas secondary metabolites are essential for normal growth, development, and plant protection. Secondary metabolites have various important functions, such as being used in pharmaceuticals, food additives, flavorings, and other industrial applications. Plants have been shown to use these secondary metabolites for protection due to their antibiotic, antifungal, and antiviral activity, which has a potential to defend plants from pathogenic infections (Kossel 1891). Secondary metabolites are categorized on the basis of their chemical structure (e.g., rings containing sugar), their composition (containing nitrogen or not), their solubility in different solvents, or their pathway by which they are synthesized (e.g., phenylpropanoid producing tannins). All these secondary metabolites (phenolics, terpenes, steroids, alkaloids, and flavonoids) are having biological significance (Bourgaud et al. 2001).

## 8.11 Importance and Main Role of Secondary Metabolites

Primary metabolites have a broad variety of defensive functions against microorganisms such as viruses, bacteria, fungi, and herbivores (arthropods and vertebrates). Secondary metabolites also perform special roles, including acting as competitive weapons against (1) other bacteria, fungi, amoebae, plants, insects, and other species; (2) metal transporting agents; (3) symbiosis agents between microbes and plants, nematodes, insects, and higher animals; (4) sex hormones; and (5) differentiation agents. Although antibiotics are not needed for sporulation, some secondary metabolites (including antibiotics) stimulate spore formation and slow down or stimulate germination, (6) protect dormant or initiated spore from amoebae intake, or (7) clean up the immediate environment of competing microorganisms during germination (Demain and Fang 2000).

## 8.12 Antioxidant Potential of Plant Phenols

Antioxidant significantly inhibits oxidation of oxidizing substrates when present at low concentrations (Halliwell 2007). Plants function as a source of exogenous (dietary) antioxidants. It has been estimated that two-thirds of plant species are medicinally important and also have significant antioxidant potential (Krishnaiah et al. 2011). Antioxidants may be synthesized under in vivo conditions (e.g., reduced glutathione (GSH), superoxide dismutase (SOD), etc.) or taken as supplementary dietary antioxidants (Halliwell 2007; Sies 1997). The antioxidant potential of plants is gradually attenuated by accelerated oxidative stress as one of the causative agents for the development of various diseases, such as neurodegenerative and cardiovascular diseases. In addition, supplementation of exogenous antioxidants or enhancement of the functions of endogenous antioxidant system to combat the detrimental effects of oxidative stress (Alam et al. 2013). Polyphenols are distinguished by one or more aromatic rings having one or more hydroxyl groups. The antioxidant potential of phenolic compounds depends on the presence of free hydroxyls and the conjugation of side chains to aromatic rings (Moran et al. 1997). In addition to their individual ability, they also associate with other physiological antioxidants, including ascorbate or tocopherol, and synergistically improve and enhance their biological effects (Croft 1998). It has been shown under experimental conditions that the antioxidant capacity of plant phenolics depends on various factors, such as electron donation, power reduction, and metal ion chelating capability (Rice-Evans et al. 1997).

## 8.13 Classification of Secondary Metabolites

Approximately 20% of plant species have been studied to contain alkaloids such as terpenoid indole alkaloids, tropane alkaloids, and purine alkaloids (Ziegler and Facchini 2008). Terpenoids are another class of secondary metabolites containing more than 40,000 monoterpene, sesquiterpene, and diterpene (Aharoni et al. 2005) compounds with substantial antioxidant activity under in vitro conditions. Plant polyphenols are classified into five main groups, such as phenolic acids, flavonoids, lignans, stilbenes, and tannins (Duthie et al. 2000; Myburgh 2014; Blokhina et al. 2003). Flavonoids and phenolic acids are one of the main groups of plant phenols, biosynthetically derived from acetate as well as shikimate acid pathways, some of which are derived from phenylalanine or tyrosine shikimate pathways (Dewick 2009) (Table 8.2).

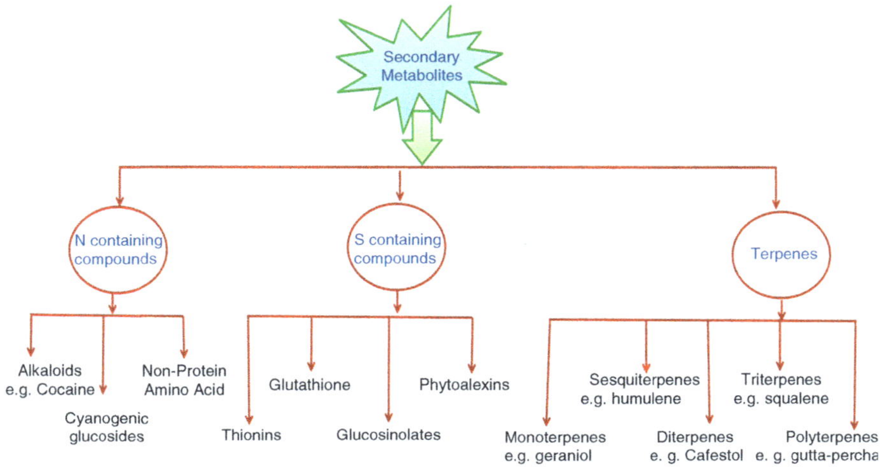


## 8.14 Role of Secondary Metabolites in Thyroid Disease

Growth, development, reproductive functions, and metabolism are the normal physiological functions maintained by the endocrine system. Similar hormones are produced by different endocrine glands: thyroid hormones (T3, T4), ovaries (estrogen, progesterone), testis (testosterone), and adrenal glands (catecholamines, mineralocorticoids, glucocorticoids, androgens). The natural growth and production of body thyroid hormones is very significant. Thyroid hormones are the main regulators of the basal metabolic rate (Oppenheimer et al. 1987).

Among endocrine disorders, thyroid disorder is very common, and different environmental conditions and substances may interfere with the biosynthesis and metabolism of thyroid hormones (Yen 2001; Mondal et al. 2016; Boas et al. 2009; Pearce and Braverman 2009; Zoeller 2010) and nutritional factors, especially when the diet is deficient in iodine (Divi and Doerge 1996). Moudgal et al. (1958) researched the antithyroid activity of flavonoids, such as inhibition of thyroid biosynthesis and reduction of iodine uptake (Gaitan 1996) and, in addition, the antithyroid and goitrogenic effects of pearl millet, a staple food eaten by local and

**Table 8.2** Classification of polyphenolic compounds and their natural sources

| S. no. | Classification of polyphenols | Food sources | Compounds | References |
|---|---|---|---|---|
| 1. | **Flavonoids** (flavonols, flavones, flavanones, isoflavones, anthocyanins) | Black tea, onion, wine, walnuts, apple, shallots, green tea, blueberries, almonds, oranges, chocolate, spinach, pepper, garlic, citrus, soybean, cabbage, beans, potatoes, grapes, red wine, pomegranate, cherries, rice | Kaempferol<br>Quercetin<br>Isorhamnetin<br>Myricetin<br>Apigenin<br>Luteolin<br>Wogonin<br>Tangeretin<br>Pelargonidin<br>Cyanidin<br>Delphinidin<br>Malvidin<br>Petunidin | Basheer and Kerem (2015)<br>Corcoran et al. (2012)<br>Bingham (2006)<br>Andrés-Lacueva et al. (2010) |
| 2. | **Phenolic acids** (hydroxybenzoic acids, hydroxycinnamic acids) | Green, black tea, raspberries, strawberries, pomegranate, mango, blackberries, grapes, wine, nuts, cherries, blueberries, apple, kiwis, coffee, cereals, red wine | Gallic acid<br>Vanillic acids<br>Protocatechuic acid<br>Curcumin<br>*p*-Coumaric acid<br>Caffeic acid<br>Ferulic acid<br>Sinapinic acid | Basheer and Kerem (2015)<br>Fraga et al. (2019)<br>Liu et al. (2019) |
| 3. | **Non-flavonoids** (stilbenes, lignans, tannin) | Red wine, grape fruits, cereals, flaxseed, linseed, algae, pumpkin, potato, vegetables, fruits, leguminous plant, grapes | Resveratrol<br>Syringaresinol<br>Ecoisolariciresinol<br>Secoisolariciresinol<br>Matairesinol<br>Medioresinol<br>Sesamin<br>Pinoresinol<br>Lariciresinol<br>Gallotannins (gallic acid)<br>Ellagitannins (ellagic acid, punicalin, punicalagin) | Basheer and Kerem (2015)<br>Wiciński et al. (2020)<br>Kawabata et al. (2019) |

poor people in various African and Asian countries. Pearl millet produces glycosyl flavones, apigenin and luteolin. Flavonoids reduce both the organification and the release of thyroid hormones.

Reactive oxygen species are formed at the active site by compound-mediated oxidation of the phenolic suicide substrate, and inactivation occurs by covalent binding of these radicals to the catalytic amino acid radical(s) of compound II (Divi and Doerge 1994). Myricetin and naringin demonstrate a noncompetitive inhibition of tyrosine iodination with respect to iodine ion and linear mixed-type suppression with respect to hydrogen peroxide. Myricetin and naringin interact with

TPO compounds I and II, but do not bind to enzymatic iodinating organisms or native TPO.

Biochanin A is considered to be a substituent substrate for iodination by means of a competitive bond between tyrosine and the alternative substrate (biochanin A) for the enzymatic iodinate species (EOI). This results in the complete blockade of tyrosine iodination due to the higher affinity of biochanin A (Divi and Doerge 1994). Biochanin A possesses an inhibitory effect on the development of thyroid peroxidase, which can cause TSH to increase, contributing to the growth of thyroid gland and goiter, particularly when taken at higher concentrations. Divi et al. (1997) reported that soy products (genistein and daidzein) are capable of inhibiting thyroperoxidase tyrosine iodination activity. Soya intake leads to a sharp rise in TSH levels, with a remarkable link between the basal levels of daidzein and thyrotropin (Hampl et al. 2008). It has been studied that catechins possess the ability to inhibit the activity of thyroperoxidase as well as decrease the levels of T3 and T4 in serum while increasing the amount of TSH level (Chandra and De 2010). In addition to this, green as well as black tea extract alters physiology, architectures such as thyroid gland enlargement, hypertrophy and/or thyroid follicle hyperplasia, and inhibition of thyroid peroxidase activity and 50-deiodinase type I, elevation in the thyroid function $Na^{+}$-$K^{+}$-ATPase, with decreased serum T3, T4, and TSH levels. In addition, treatment with genistein and daidzein lowers the overall T3 and T4 and raises the TSH level in orchiectomized middle-aged rats (Šošić-Jurjević et al. 2010).

## 8.15 Plant Polyphenols and Hepatitis

The liver is the largest and key gland that is strongly engaged in a range of metabolic activities, such as energy output, glucose storage (glycogen) and its metabolism, and the use of carbohydrate for cholesterol synthesis. The liver plays an important role in transforming excess fatty acids into ketone bodies, which serve as a source of energy during fasting and starvation. It also helps to preserve homeostasis (Chiang 2014). Hepatitis is a common liver disease caused by various strains of hepatitis A, B, and C viruses. The most popular forms are hepatitis B (HBV) and hepatitis C viruses (HCV). HBV is a significant cause of liver cirrhosis and hepatocellular carcinoma (Tomé-Carneiro et al. 2013).

Various phytochemicals, such as wogonin and polyphenols (geranin), have both in vitro and in vivo antihepatic activity (Zhang et al. 2016; Xiong et al. 2015). These polyphenolic compounds prevent entry of the virus, inhibit viral antigen secretion, and inhibit DNA replication (Parvez et al. 2016). Green tea also contains several useful flavonoids, such as EGCG, which is well-known to inhibit the entry of HBV into hepatocytes by inducing clathrin-dependent endocytosis of sodium taurocholate by co-transporting polypeptide from plasma membrane and protein degradation, as well as inhibiting clathrin-mediated transferrin endocytosis (Huang et al. 2014). While epicatechins are present in green tea, they inhibit the replication of the viral genome via cyclooxygenase-2 and also reduce viral inflammation (Lin et al. 2013). These flavonoids inhibit HCV entry into hepatoma cell lines and primary human hepatocytes (Calland et al. 2015; Ciesek et al. 2011). In addition to this, delphinidin, which is present in blue-purple flowers and berries, induces bulging of the viral envelope in order to prevent HCV from binding to the cell surface (Calland et al. 2015). Delphinidin acts directly on viral particles and prevents the entry of HCV. Delphinidin is likely to prevent the docking of the virus to the cell surface and to function at an early stage of entry. In addition, Calland et al. (2015) reported that delphinidin is active against all HCV genotypes and shows its impact on virion itself. Delphinidin prevents both HCVpp and HCVcc infections and acts on the viral particle. In addition, apigenin inhibits HCV replication by lowering the mature miRNA122 level, as miRNA122 acts as a positive liver-specific miRNA for replication control (Shibata et al. 2014). Microarray analyzes the expression levels of mature miRNAs, such as miR122 and miR103, which is decreased by apigenin (Ohno et al. 2013). Shibata et al. (2014) reported that inhibition of extracellular signal-regulated kinase (ERK) activity was due to decreased phosphorylation of the RNA-binding protein trans-activation response (TRBP) (Paroo et al. 2009). The levels of the mature miR122 are increased by the overexpression of the TRBP (SD). It is evident that the reduction in miR122 maturation also depended on the activity of TRBP inhibited by apigenin (Ohno et al. 2013). In addition, quercetin has been reported to have antiviral activity against various viruses. Quercetin substantially reduces the replication of the viral genome, the formation of infectious HCV particles, and the specific infectivity of newly developed viral particles (Rojas et al. 2016). In addition to quercetin, HCV is decreased by inhibiting NS3 protease activity (Bachmetov et al. 2012). It prevents HCV replication of the genome and reduces HCV-specific infectivity by influencing the morphogenesis of infectious particles. Quercetin is the active substance responsible for inhibiting the activity of NS3 protease and eventually reducing the development of HCV (Bachmetov et al. 2012).

## 8.16 Secondary Metabolites in the Prevention of Hepatorenal Toxicity

The liver and kidney are two main organs involved in different metabolic processes, such as glucose metabolism, drug transformation, and xenobiotic excretion (Abdel-Daim et al. 2014). The liver and kidney are also extremely exposed to different toxic chemicals (Abdel-Daim et al. 2013; Al-Sayed and Abdel-Daim 2014). Oxidative stress is known to be a key factor in renal failure, liver damage, atherosclerosis, inflammation, and carcinogenesis (Abdel-Daim et al. 2014; Abdel-Daim and Ghazy 2015). Oxidative stress has also been experimentally associated with liver damage and hepatic fibrosis. Continuous hepatocellular damage delays the mechanism of regeneration of damaged tissues and overpowers the protective capacity of the liver. ROS and lipid peroxidation products are produced in wounded hepatocytes that stimulate the transformation of hepatic stellate cells (HSCs) into fibrogenic myofibroblast-like cells and form collagen mass in the liver (Li et al. 2012; Shin et al. 2010). The danger of cellular injury can also be avoided by the use of antioxidants such as dietary polyphenols due to antioxidant potential and numerous other health-promoting effects.

Natural antioxidants, including dietary polyphenols, can also prevent the risk of cell injury (Han et al. 2007). The hepatoprotective potential of polyphenols was assessed by examining the activity of various liver-specific enzymes such as alanine aminotransferase (ALT), aspartate aminotransferase (AST), alkaline phosphatase (ALP), lactate dehydrogenase (LDH), bilirubin, cholesterol, and protein in control serum and treated mice. The nephroprotective effect of plant polyphenols has been estimated by analyzing various functional renal enzymes such as uric acid, urea, and creatinine levels in the circulatory serum.

Proanthocyanidins are natural antioxidants found in a large range of plants, such as fruits and vegetables (Bagchi et al. 2000). These proanthocyanidins have a wide range of pharmacological and therapeutic activity against oxidative stress. Proanthocyanidin derived from grapes is used as a dietary supplement due to its numerous health benefits, such as hepatoprotective, cardioprotective, anti-fibrogenic, and chemopreventive activities (Li et al. 2012). Proanthocyanidins are complex polymers of polyhydroxy flavan-3-ol constitutive groups, the most common of which are (+) -catechin and (−)-epicatechin in the case of procyanidin form, or (+) gallocatechin, and (−)-epigallocatechin, in the case of prodelphinide form (Hellström et al. 2007). Earlier studies indicated that proanthocyanidins have a higher potential to defend tissues/cells against free radical and lipid peroxidation attacks than vitamins C and E and β-carotene (Bagchi et al. 2000). Grape seed proanthocyanidin extract has a higher protective activity against thioacetamide-mediated hepatic fibrosis in the liver (Li et al. 2012), while GSPE also protects the liver against the fibrogenic activity of dimethylnitrosamine in rats (Shin et al. 2010). In addition, GSPE has an important protective role against acetaminophen-mediated liver and kidney damage by decreasing oxidative stress and ALT activity and inhibiting apoptosis and necrotic cell death (Bagchi et al. 2000). GSPE also protects against cyclosporine A- and cisplatin-induced nephropathy and restores renal function. This nephroprotective potential of GSPE has been explained due to its antioxidant activity, restores the tubular damage, and increases the process of regeneration (Ulusoy et al. 2012) (Table 8.3).

## 8.17 Secondary Metabolites in Tuberculosis and Their Potency Against Tuberculosis

*Mycobacterium tuberculosis* (Mtb), a highly infectious bacterium, causes tuberculosis (TB) and affects one-third of the world's population. Treatment against tuberculosis began 73 years ago when streptomycin was discovered, numerous medications were put on the market, but the disease remains at the top of causes of death worldwide. Tuberculosis infection occurs by swallowing bacteria through alveolar macrophages; bacilli bypass and multiply continuously by preventing phagosome-lysosome fusion. More and more macrophages and other immune cells are then transported to the site of infection, leading to the creation of an organized

**Table 8.3** Diverse group of phytotherapeutics having bioactive potential against metabolic disorders

| Sr. no. | Phytotherapeutic | Parameters | Metabolic disorders | Properties | References |
|---|---|---|---|---|---|
| 1. | Green tea (*Camellia sinensis*) | Inhibit α-amylase and α-glucosidase activity<br>Inhibiting fat digestion<br>Triglyceride level↓ | Hypertension<br>Diabetes<br>Hypercholesterolemia<br>Obesity<br>Hypertension<br>Osteoporosis | Antioxidant<br>Anti-inflammatory<br>Anticancer<br>Cholesterol lowering cardiovascular properties<br>Antidiabetic properties<br>Anti-obesity | Naveed et al. (2018)<br>Bedrood et al. (2018) |
| 2. | Blueberries (*Cyanococcus*) | LDL cholesterol↓<br>Triglycerides↓<br>Adiponectin↓<br>HDL cholesterol↑ | Cardiovascular disorders<br>Hypertension<br>Diabetes<br>Obesity | Antioxidant<br>Anti-inflammatory | Kalea et al. (2009)<br>Martineau et al. (2006) |
| 3. | Pomegranate (*Punica granatum*) | Insulin sensitivity↑<br>Inhibition of α-glucosidase<br>Cholesterol↓<br>Reducing inflammation | Blood pressure<br>Cardiovascular diseases<br>Hyperlipidemia<br>Diabetes | Anti-inflammatory<br>Antioxidant<br>Hypoglycemic<br>Anticancer<br>Hypolipidemic antimicrobial activity | Aprotosoaie et al. (2016) |
| 4. | Curcumin (*Curcuma longa*) | Glucose level↓<br>NF-κB activity↓<br>Reduction in BMI | Hyperlipidemia<br>Diabetes<br>Cardiovascular risk hyperinsulinemia<br>Obesity<br>Hypertriglyceridemia | Anti-inflammatory<br>Anticancer<br>Hypoglycemic antioxidant<br>Antiviral<br>Antimicrobial activities | Gupta et al. (2013)<br>Milani et al. (2017)<br>de Melo et al. (2018) |
| 5. | Olive oil (*Olea europaea*) | Improving blood sugar<br>Blood pressure control<br>Reducing LDL oxidation<br>Reduce body weight | Cardiovascular diseases<br>Blood pressure<br>Obesity<br>Dyslipidemia | Anti-inflammatory<br>Antioxidant<br>Antimicrobial activities | Visioli and Bernardini (2011)<br>Estruch et al. (2006) |

cellular architecture known as granuloma (Barry et al. 2009). It has been stated that if HIV patients are infected with tuberculosis, the death rate of these patients has been estimated to be one in every three HIV patients (World Health Organization 2016). There are various types of drugs available for tuberculosis treatment as described below.

## 8.17.1 First Line Drugs

The first-line medications are used for the treatment of new patients. These patients are not expected to have any resistance to any of the TB medications, e.g., (rifampin (RIF), isoniazid (INH), ethambutol (EMB), streptomycin, and pyrazinamide (PZA).

### 8.17.1.1 Isoniazid

Isoniazid (INH) was developed in 1952 and has been used to treat tuberculosis as a particular antituberculosis medication and one of the most effective drugs (Bernstein et al. 1952). INH is not active under anaerobic conditions. It is only active against rising bacilli tubers. INH joins mycobacterial cells by passive diffusion (Bardou et al. 1998). *Mycobacterium tuberculosis* is particularly susceptible to isoniazid. Hepatotoxicity and neurotoxicity are the side effects of isoniazid.

### 8.17.1.2 Rifampicin

Rifampicin (RIF) is an antituberculosis drug that was introduced in 1972. It binds the β-subunit to RNA polymerase (rpoB) (Ramaswamy and Musser 1998). RNA polymerase is the key transcription enzyme responsible for the expression of mycobacterial genes and eventually inhibits the activity of bacterial transcription and thus destroys the organism. Rifampicin is active against vigorously growing and non-growing bacilli (Mitchison 1979). RIF has very few adverse reactions. It can cause gastrointestinal discomfort. Hepatotoxicity occurs less often than with isoniazid.

### 8.17.1.3 Ethambutol

Ethambutol (EMB) is chemically referred to as dextro-2, 2′-(ethylenediimino)-di-1-butanol. It is an effective first-line and anti-mycobacterial medication used in the treatment of tuberculosis. It plays a key role in drug-resistant TB chemotherapy (American Thoracic Society 2003). EMB improves the effects of other medications such as aminoglycosides, quinolones, and rifamycin. Nausea, vomiting, stomach pain, color blindness, swelling of lips or eyes, loss of appetite, headache, rash, itching, breathlessness, swelling of the face, dizziness, blurred vision, and numbness or tingling of the fingers or toes are typical side effects of ethambutol (Jnawali and Ryoo 2013).

### 8.17.1.4 Pyrazinamide

Pyrazinamide (PZA) is a first-line drug used to treat tuberculosis. It is a short-term chemotherapy drug used in the treatment of MDR-TB (Mitchison 1985). It is an effective drug because it inhibits semi-dormant bacilli living in acidic environments

(Mitchison 1985). Pyrazinamide is an active agent against *M. tuberculosis* at 5.5 acid pH (Konno et al. 1967). Hypersensitivity and gastrointestinal upset are common side effects of pyrazinamide.

#### 8.17.1.5 Streptomycin

Streptomycin (SM) was the first drug used in the treatment of TB in 1948 (British Medical Research Council 1948). It is an antibiotic aminocyclitol glycoside that destroys vigorously developing bacilli tubers with a minimum inhibitory concentration (MIC) of 2–4 μg/mL and is ineffective against nongrowing bacilli (Mitchison 1985). It inhibits protein synthesis and interferes with translation rereading as it binds to 16S rRNA (British Medical Research Council 1948; Gale 1981). The most common side effects of streptomycin are ototoxicity, nephrotoxicity, vestibular impairment, hearing loss and renal toxicity.

### 8.17.2 Second-Line Medications

#### 8.17.2.1 Fluoroquinolone

Fluoroquinolones (FQs) are commonly used for the treatment of bacterial infections, gastrointestinal and urinary tract infections, sexually transmitted diseases, and chronic osteomyelitis (Spies et al. 2008). They are used as second-line medications for the treatment of tuberculosis. Levofloxacin, moxifloxacin, ciprofloxacin, and ofloxacin are FQs medications. They exhibit antimicrobial and antibiotic properties and have excellent in vitro and in vivo anti-TB activity (Bartlett et al. 2000; Wang et al. 2006). But they also display numerous side effects such as rash, dizziness, headache, and gastrointestinal intolerance.

#### 8.17.2.2 Aminoglycosides (Kanamycin, Amikacin, and Capreomycin)

Drugs used in the treatment of multidrug-resistant tuberculosis are aminoglycosides (KAN), amikacin (AMK), and cyclic polypeptide capreomycin (CAP). All these drugs demonstrate their effect at the stage of protein translation. AMK and KAN have a high degree of cross-resistance between them (Alangaden et al. 1995; Jugheli et al. 2009; Maus et al. 2005). CAP is a possible candidate to replace AMK or KAN and is structurally dissimilar to aminoglycosides (Johansen et al. 2006; World Health Organization (WHO) 2008). When CAP resistance develops among MDR-TB cases, the risk of treatment failure and mortality increases (Migliori et al. 2008). Changes in 16S rRNA are due to AMK/KAN and CAP drug resistance (Alangaden et al. 1995; Maus et al. 2005; Johansen et al. 2006; Via et al. 2010). The side effects of these medications are renal toxicity and hearing loss.

#### 8.17.2.3 Ethionamide and Prothionamide Ethionamide

Since 1956, ethionamide (ETH, 2-ethylisonicotinamide) has been used as an antituberculosis medication and is a derivative of isonicotinic acid. Ethionamide and prothionamide PTH (2-ethyl-4-pyridinecarbothioamide) are active agents of EtaA/EthA (a monooxygenase) (DeBarber et al. 2000) and inhibit targets such as isoniazid (INH) (Banerjee et al. 1994). Ethionamide undergoes many changes until it enters the bacterial cell. Flavin monooxygenase oxidizes the sulfate group and is then converted to 2-ethyl-4-aminopyridine. The intermediate products formed between ethionamide and 2-ethyl-4-aminopyridine tend to be toxic to mycobacteria, but the structure of these intermediate products is unknown. These intermediate products can be highly unstable compounds. The side effects of ethionamide are stomach issues.

#### 8.17.2.4 *P*-Aminosalicylic Acid

*P*-Aminosalicylic acid (PAS) was used to treat TB in amalgamation with isoniazid and streptomycin. It was the first antibiotic to demonstrate anti-TB activity (Zhang et al. 1992). Once rifampicin and other potent drugs have been discovered, their use as a first-line drug has been discontinued. While PAS benefits are minimal and very toxic, PAS is still used for the treatment of XDR TB. Gastrointestinal disorders are the most common side effects associated with PAS.

#### 8.17.2.5 Cycloserine

Cycloserine (CS) is used as an antibiotic in the treatment of TB. Since its action mechanism is still not quite clear, it is thought that cycloserine prevents the TB bacteria from producing peptidoglycans. These peptidoglycans are needed for the development of bacterial cell walls. This results in the weakening of the cell wall of the bacteria, which destroys the bacteria. Compared to other drugs, cycloserine has high gastric tolerance and lacks cross-resistance to other compounds. The biggest downside of this drug is that it induces psychological side effects. Cycloserine is the basis of care for MDR and XDR TB (Caminero 2006).

| Sr. no. | First-line drugs | Adverse effects | Second-line drugs | Adverse effects | References |
|---|---|---|---|---|---|
| 1. | **Isoniazid** | Hepatotoxicity and neurotoxicity | **Fluoroquinolones** | Gastrointestinal intolerance, rashes, dizziness, and headache | Bardou et al. (1998), Wang et al. (2006) |
| 2. | **Rifampicin** | Gastrointestinal upset and hepatotoxicity | **Aminoglycosides (kanamycin, amikacin, and capreomycin)** | Renal toxicity, hearing loss | Mitchison (1979), Zaunbrecher et al. (2009) |
| 3. | **Ethambutol** | Dizziness; blurred vision; color blindness; nausea; vomiting; stomach pain; loss of appetite; headache; rash; itching; breathlessness; swelling of the face, lips, or | **Ethionamide/ prothionamide ethionamide** | Gastrointestinal side effects | Jnawali and Ryoo (2013) |

(continued)

| Sr. no. | First-line drugs | Adverse effects | Second-line drugs | Adverse effects | References |
|---|---|---|---|---|---|
| | | eyes; numbness or tingling in the fingers or toes | | | |
| 4. | **Pyrazinamide** | Hypersensitivity reactions and gastrointestinal upset | ***p*-Aminosalicylic acid** | Gastrointestinal disturbances | Zhang and Mitchison (2003), Leung et al. (2010) |
| 5. | **Streptomycin** | Ototoxicity and nephrotoxicity, vestibular dysfunction, auditory damage, renal toxicity | **Cycloserine** | Adverse psychiatric effects | British Medical Research Council (1948), Gale (1981) |

## 8.18 Therapeutic Use of Phytopolyphenols

In view of these details, some alternate therapeutic methods are required to cure this life-threatening disease. At present, more attention is paid to plant polyphenols due to their limited side effects, which is why polyphenol compounds could be a good option for combating and reducing the percentage of tuberculosis. *Mycobacterium tuberculosis* H37Rv (ATCC 27294) is a commonly used target strain for the evaluation of anti-TB efficacy of any medication or other therapeutic agents (Pauli et al. 2005).

Various secondary metabolites such as green tea polyphenol (9-epigallocatechin-3-gallate) have been found to inhibit Mtb survival in human macrophages (Anand et al. 2006). The downregulation of the host molecule tryptophan aspartate containing coat protein (TACO) expression of the epigallocatechin-3-gallate gene was followed by inhibition of Mtb survival within macrophages as assessed by flow cytometry and colony counts. Anand et al. (2006) studied that pre-treatment with EGCG inhibited mycobacterial entry by 18% as analyzed by fluorescence-activated cell sorting (FACS) for the detection of percent of lipoarabinomannan (LAM) fluorescence (indicating the number of bacteria present) using monoclonal anti-LAM antibody. At 12 o'clock, survival was reduced to just 26%. In the case of post-treatment EGCG, survival at 48 h was reduced to 55%. However, the colony forming unit (CFU) of the related THP-1 macrophages experiment did not suggest survival at 12 h of infection. LAM is a cell wall antigen of mycobacteria and can be found in macrophages even after the bacteria have partially degraded or are transformed into lysosomes (Anand et al. 2006). It is evident from the above data that the EGCG is actually hindering entry of *M. tuberculosis* and prevents the survival of the pathogen. In addition, dihydro-β-agarofuran sesquiterpenes (15-007-acetoxy-6β, 9β-dibezoyloxydihydro-β-agarofuran) isolated from the leaves of *Celastrus vulcanicola* Donn. show anti-TB activity against the strain MDR in which the MIC values are 0.0062 mg/mL, naphthoquinones, plumbagin, and its dimers maritinone and 3.3′-biplumbagin from *Diospyros anisandra* S.F.Blake, plumericin isolated from *Plumeria bicolor* Ruiz & Pav MIC values 0.0015–0.002 mg/mL and MBC (minimum bactericidal concentration) values 0.003–0.004 mg/mL and, 3′-biplumbagin from *Diospyros anisandra*. In addition, a compound called 7-methyljuglone was isolated from the roots of *Euclea natalensis* (Lall et al. 2005). This compound demonstrated activity against intracellular strain Mtb Erdman within macrophages J774.1 at a concentration of 0.57 μg mL plus 1. Minimum inhibitory concentrations of 7-methyljuglone were documented to be approximately four times higher against drug-resistant isolates. In addition, *Chamaedorea tepejilote Liebm.* (Arecaceae) and *Lantana hispida* Kunth are medicinal plants from which ursolic acid (UA) and oleanolic acid (OA) are extracted (used in traditional Mexican medicine). These two compounds are used to treat respiratory problems such as colds, cough, bronchitis, and pneumonia (Jiménez-Arellanes et al. 2013). Jiménez-Arellanes et al. (2013) stated that these two compounds had intracellular anti-TB activity in the infected Mtb macrophage cell line J774A. Mtb-infected BALB/c mice have been used to test the anti-TB activity of UA and OA. Compared to the control animals, UA- and OA-treated animals showed higher expressions of IFN-Δ and TNF-5-007 in the lungs (Jiménez-Arellanes et al. 2013). Jiménez-Arellanes et al. (2013) reported that the UA and OA combination was found to be very successful in destroying intracellular Mtb.

## References

Abdel-Daim MM, Ghazy EW (2015) Effects of Nigella sativa oil and ascorbic acid against oxytetracycline-induced hepato-renal toxicity in rabbits. Iran J Basic Med Sci 18(3):221

Abdel-Daim MM, Abuzead SM, Halawa SM (2013) Protective role of Spirulina platensis against acute deltamethrin-induced toxicity in rats. PLoS One 8(9):e72991

Abdel-Daim MM, Abd Eldaim MA, Mahmoud MM (2014) Trigonella foenum-graecum protection against deltamethrin-induced toxic effects on haematological, biochemical, and oxidative stress parameters in rats. Can J Physiol Pharmacol 92(8):679–685

Aharoni A, Jongsma MA, Bouwmeester HJ (2005) Volatile science? Metabolic engineering of terpenoids in plants. Trends Plant Sci 10(12):594–602

Akilen R, Tsiami A, Devendra D, Robinson N (2012) Cinnamon in glycaemic control: systematic review and meta analysis. Clin Nutr 31(5):609–615

Alam MN, Bristi NJ, Rafiquzzaman M (2013) Review on in vivo and in vitro methods evaluation of antioxidant activity. Saudi Pharm J 21(2):143–152

Alangaden GJ, Manavathu EK, Vakulenko SB, Zvonok NM, Lerner SA (1995) Characterization of fluoroquinolone-resistant mutant strains of Mycobacterium tuberculosis selected in the laboratory and isolated from patients. Antimicrob Agents Chemother 39(8):1700–1703

Alić B, Gurbeta L, Badnjević A, Badnjević-Čengić A, Malenica M, Dujić T, Čaušević A, Bego T (2017) Classification of metabolic syndrome patients using implemented expert system. In: CMBEBIH 2017. Springer, Singapore, pp 601–607

Almroth BC, Sturve J, Berglund Å, Förlin L (2005) Oxidative damage in eelpout (Zoarces viviparus), measured as protein carbonyls and TBARS, as biomarkers. Aquat Toxicol 73 (2):171–180

Al-Sayed E, Abdel-Daim MM (2014) Protective role of Cupressuflavone from Cupressus macrocarpa against carbon tetrachloride-induced hepato- and nephrotoxicity in mice. Planta Med:80

American Thoracic Society (2003) CDC; Infectious Diseases Society of America: treatment of tuberculosis. MMWR 52(11):1–77

Anand PK, Kaul D, Sharma M (2006) Green tea polyphenol inhibits Mycobacterium tuberculosis survival within human macrophages. Int J Biochem Cell Biol 38(4):600–609

Andrés-Lacueva C, Medina-Remon A, Llorach R, Urpi-Sarda M, Khan N, Chiva-Blanch G, Zamora-Ros R, Rotches-Ribalta M, Lamuela-Raventos RM (2010) Phenolic compounds: chemistry and occurrence in fruits and vegetables. In: Fruit and vegetable phytochemicals: chemistry, nutritional value, and stability. Wiley-Blackwell, Oxford

Andújar I, Recio MC, Giner RM, Ríos JL (2012) Cocoa polyphenols and their potential benefits for human health. Oxidative Med Cell Longev 2012

Aprotosoaie AC, Luca SV, Miron A (2016) Flavor chemistry of cocoa and cocoa products—an overview. Compr Rev Food Sci Food Saf 15(1):73–91

Azab AE, Albasha MO (2018) Hepatoprotective effect of some medicinal plants and herbs against hepatic disorders induced by hepatotoxic agents. J Biotechnol Bioeng 2(1):8–23

Bachmetov L, Gal-Tanamy M, Shapira A, Vorobeychik M, Giterman-Galam T, Sathiyamoorthy P, Golan-Goldhirsh A, Benhar I, Tur-Kaspa R, Zemel R (2012) Suppression of hepatitis C virus by the flavonoid quercetin is mediated by inhibition of NS3 protease activity. J Viral Hepat 19(2): e81–e88

Bagchi D, Bagchi M, Stohs SJ, Das DK, Ray SD, Kuszynski CA, Joshi SS, Pruess HG (2000) Free radicals and grape seed proanthocyanidin extract: importance in human health and disease prevention. Toxicology 148(2–3):187–197

Baldwa VS, Bhandari CM, Pangaria A, Goyal RK (1977) Clinical trial in patients with diabetes mellitus of an insulin-like compound obtained from plant source. Ups J Med Sci 82(1):39–41

Banerjee A, Dubnau E, Quemard A, Balasubramanian V, Um KS, Wilson T, Collins D, De Lisle G, Jacobs WR (1994) inhA, a gene encoding a target for isoniazid and ethionamide in Mycobacterium tuberculosis. Science 263(5144):227–230

Bardou F, Raynaud C, Ramos C, Laneelle MA, Lanfelle G (1998) Mechanism of isoniazid uptake in Mycobacterium tuberculosis. Microbiology 144(9):2539–2544

Barry CE, Boshoff HI, Dartois V, Dick T, Ehrt S, Flynn J, Schnappinger D, Wilkinson RJ, Young D (2009) The spectrum of latent tuberculosis: rethinking the biology and intervention strategies. Nat Rev Microbiol 7(12):845–855

Bartlett JG, Dowell SF, Mandell LA, File TM Jr, Musher DM, Fine MJ (2000) Practice guidelines for the management of community-acquired pneumonia in adults. Clin Infect Dis 31 (2):347–382

Basheer L, Kerem Z (2015) Interactions between CYP3A4 and dietary polyphenols. Oxid Med Cell Longev

Basu A, Sanchez K, Leyva MJ, Wu M, Betts NM, Aston CE, Lyons TJ (2010) Green tea supplementation affects body weight, lipids, and lipid peroxidation in obese subjects with metabolic syndrome. J Am Coll Nutr 29(1):31–40

Bedrood Z, Rameshrad M, Hosseinzadeh H (2018) Toxicological effects of Camellia sinensis (green tea): a review. Phytother Res 32(7):1163–1180

Belcaro G, Ledda A, Hu S, Cesarone MR, Feragalli B, Dugall M (2013) Greenselect phytosome for borderline metabolic syndrome. Evid Based Complement Alternat Med

Bernstein J, Lott WA, Steinberg BA, Yale HL (1952) Chemotherapy of experimental tuberculosis: V. Isonicotinic acid hydrazide (nydrazid) and related compounds. Am Rev Tuberc 65 (4):357–364

Bingham M (2006) The metabolism of polyphenols by the human gut microbiota. Gastrointest Microbiol:155–168

Bischoff HBAG (1994) Pharmacology of alpha-glucosidase inhibition. Eur J Clin Investig 24:3–10

Blokhina O, Virolainen E, Fagerstedt KV (2003) Antioxidants, oxidative damage and oxygen deprivation stress: a review. Ann Bot 91(2):179–194
Boas M, Main KM, Feldt-Rasmussen U (2009) Environmental chemicals and thyroid function: an update. Curr Opin Endocrinol Diabetes Obes 16(5):385–391
Bourgaud F, Gravot A, Milesi S, Gontier E (2001) Production of plant secondary metabolites: a historical perspective. Plant Sci 161(5):839–851
Brennan ML, Wu W, Fu X, Shen Z, Song W, Frost H, Vadseth C, Narine L, Lenkiewicz E, Borchers MT, Lusis AJ (2002) A tale of two controversies defining both the role of peroxidases in nitrotyrosine formation in vivo using eosinophil peroxidase and myeloperoxidase-deficient mice, and the nature of peroxidase-generated reactive nitrogen species. J Biol Chem 277 (20):17415–17427
Cabrera C, Artacho R, Giménez R (2006) Beneficial effects of green tea—a review. J Am Coll Nutr 25(2):79–99
Calland N, Sahuc ME, Belouzard S, Pène V, Bonnafous P, Mesalam AA, Deloison G, Descamps V, Sahpaz S, Wychowski C, Lambert O (2015) Polyphenols inhibit hepatitis C virus entry by a new mechanism of action. J Virol 89(19):10053–10063
Caminero JA (2006) Treatment of multidrug-resistant tuberculosis: evidence and controversies. Int J Tuberc Lung Dis 10(8):829–837
Carbonel AF, Calió ML, Santos MA, Bertoncini CA, Sasso GDS, Simões RS, Simões MJ, Soares JM Jr (2015) Soybean isoflavones attenuate the expression of genes related to endometrial cancer risk. Climacteric 18(3):389–398
Casazza K, Dulin-Keita A, Gower BA, Fernandez JR (2009) Differential influence of diet and physical activity on components of metabolic syndrome in a multiethnic sample of children. J Am Diet Assoc 109(2):236–244
Chandra AK, De N (2010) Goitrogenic/antithyroidal potential of green tea extract in relation to catechin in rats. Food Chem Toxicol 48(8–9):2304–2311
Chaturvedi P (2012) Antidiabetic potentials of Momordica charantia: multiple mechanisms behind the effects. J Med Food 15(2):101–107
Chen L, Sun P, Wang T, Chen K, Jia Q, Wang H, Li Y (2012) Diverse mechanisms of antidiabetic effects of the different procyanidin oligomer types of two different cinnamon species on db/db mice. J Agric Food Chem 60(36):9144–9150
Cheng DM, Kuhn P, Poulev A, Rojo LE, Lila MA, Raskin I (2012) In vivo and in vitro antidiabetic effects of aqueous cinnamon extract and cinnamon polyphenol-enhanced food matrix. Food Chem 135(4):2994–3002
Chiang J (2014) Liver physiology: metabolism and detoxification
Chu C, Deng J, Man Y, Qu Y (2017) Green tea extracts epigallocatechin-3-gallate for different treatments. BioMed Res Int 2017
Ciesek S, von Hahn T, Colpitts CC, Schang LM, Friesland M, Steinmann J, Manns MP, Ott M, Wedemeyer H, Meuleman P, Pietschmann T (2011) The green tea polyphenol, epigallocatechin-3-gallate, inhibits hepatitis C virus entry. Hepatology 54(6):1947–1955
Corcoran RB, Ebi H, Turke AB, Coffee EM, Nishino M, Cogdill AP, Brown RD, Della Pelle P, Dias-Santagata D, Hung KE, Flaherty KT (2012) EGFR-mediated reactivation of MAPK signaling contributes to insensitivity of BRAF-mutant colorectal cancers to RAF inhibition with vemurafenib. Cancer Discov 2(3):227–235
Council MR (1948) Streptomycin treatment of pulmonary tuberculosis. Br Med J 2:769–782
Croft KD (1998) The chemistry and biological effects of flavonoids and phenolic acids. Ann N Y Acad Sci 854(1):435–442
Dasgupta A, Klein K (2014) Antioxidants in food, vitamins and supplements: prevention and treatment of disease. Academic Press
Davies GJ, Gloster TM, Henrissat B (2005) Recent structural insights into the expanding world of carbohydrate-active enzymes. Curr Opin Struct Biol 15(6):637–645

de Melo ISV, dos Santos AF, Bueno NB (2018) Curcumin or combined curcuminoids are effective in lowering the fasting blood glucose concentrations of individuals with dysglycemia: systematic review and meta-analysis of randomized controlled trials. Pharmacol Res 128:137–144

DeBarber AE, Mdluli K, Bosman M, Bekker LG, Barry CE (2000) Ethionamide activation and sensitivity in multidrug-resistant Mycobacterium tuberculosis. Proc Natl Acad Sci 97 (17):9677–9682

Del Rio D, Rodriguez-Mateos A, Spencer JP, Tognolini M, Borges G, Crozier A (2013) Dietary (poly) phenolics in human health: structures, bioavailability, and evidence of protective effects against chronic diseases. Antioxid Redox Signal 18(14):1818–1892

Demain AL, Fang A (2000) The natural functions of secondary metabolites. In: History of modern biotechnology I. Springer, Berlin, pp 1–39

Dewick PM (2009) The shikimate pathway: aromatic amino acids and phenylpropanoids. Med Nat Prod 137:86

Diepvens K, Westerterp KR, Westerterp-Plantenga MS (2007) Obesity and thermogenesis related to the consumption of caffeine, ephedrine, capsaicin, and green tea. Am J Physiol Regul Integr Comp Physiol

Divi RL, Doerge DR (1994) Mechanism-based inactivation of lactoperoxidase and thyroid peroxidase by resorcinol derivatives. Biochemistry 33(32):9668–9674

Divi RL, Doerge DR (1996) Inhibition of thyroid peroxidase by dietary flavonoids. Chem Res Toxicol 9(1):16–23

Divi RL, Chang HC, Doerge DR (1997) Anti-thyroid isoflavones from soybean: isolation, characterization, and mechanisms of action. Biochem Pharmacol 54(10):1087–1096

dos Reis SA, da Conceição LL, e Dias MM, Siqueira NP, Rosa DD, de Oliveira LL, da Matta SLP, Peluzio MDCG (2019) Kefir reduces the incidence of pre-neoplastic lesions in an animal model for colorectal cancer. J Funct Foods 53:1–6

Duthie GG, Duthie SJ, Kyle JA (2000) Plant polyphenols in cancer and heart disease: implications as nutritional antioxidants. Nutr Res Rev 13(1):79–106

el Bilbeisi AH, Hosseini S, Djafarian K (2017) Dietary patterns and metabolic syndrome among type 2 diabetes patients in Gaza Strip, Palestine. Ethiop J Health Sci 27(3):227–238

Estruch R, Martínez-González MA, Corella D, Salas-Salvadó J, Ruiz-Gutiérrez V, Covas MI, Fiol M, Gómez-Gracia E, López-Sabater MC, Vinyoles E, Arós F (2006) Effects of a Mediterranean-style diet on cardiovascular risk factors: a randomized trial. Ann Intern Med 145(1):1–11

Evans JL, Bahng M (2014) Non-pharmaceutical intervention options for type 2 diabetes: diets and dietary supplements (botanicals, antioxidants, and minerals). Endotext. MDText. com, Inc., South Dartmouth, MA

Finkel T, Holbrook NJ (2000) Oxidants, oxidative stress and the biology of ageing. Nature 408 (6809):239–247

Fraga CG, Galleano M, Verstraeten SV, Oteiza PI (2010) Basic biochemical mechanisms behind the health benefits of polyphenols. Mol Asp Med 31(6):435–445

Fraga CG, Croft KD, Kennedy DO, Tomás-Barberán FA (2019) The effects of polyphenols and other bioactives on human health. Food Funct 10(2):514–528

Fridlyand LE, Philipson LH (2006) Oxidative reactive species in cell injury: mechanisms in diabetes mellitus and therapeutic approaches. Ann N Y Acad Sci 1066(1):136–151

Fujiki H, Sueoka E, Watanabe T, Suganuma M (2015) Primary cancer prevention by green tea, and tertiary cancer prevention by the combination of green tea catechins and anticancer compounds. J Cancer Prevent 20(1):1

Fuller S, Stephens JM (2015) Diosgenin, 4-hydroxyisoleucine, and fiber from fenugreek: mechanisms of actions and potential effects on metabolic syndrome. Adv Nutr 6(2):189–197

Gaitan E (1996) Flavonoids and the thyroid. Nutrition 12(2):127–129

Gale EF (1981) Molecular basis of antibiotic action

Grover JK, Yadav SP (2004) Pharmacological actions and potential uses of Momordica charantia: a review. J Ethnopharmacol 93(1):123–132

Guo H, Ling W (2015) The update of anthocyanins on obesity and type 2 diabetes: experimental evidence and clinical perspectives. Rev Endocr Metab Disord 16(1):1–13
Gupta SC, Patchva S, Aggarwal BB (2013) Therapeutic roles of curcumin: lessons learned from clinical trials. AAPS J 15(1):195–218
Haffner SM, Valdez RA, Hazuda HP, Mitchell BD, Morales PA, Stern MP (1992) Prospective analysis of the insulin-resistance syndrome (syndrome X). Diabetes 41(6):715–722
Halliwell B (2007) Biochemistry of oxidative stress. Biochem Soc Trans 35:1147–1150
Hampl R, Ostatnikova D, Celec PETER, Putz Z, Lapcík O, Matucha PETR (2008) Short-term effect soy consumption on thyroid hormone levels and correlation with phytoestrogen level in healthy subjects. Endocr Regul *42*(2):53
Han X, Shen T, Lou H (2007) Dietary polyphenols and their biological significance. Int J Mol Sci 8 (9):950–988
Hanhineva K, Törrönen R, Bondia-Pons I, Pekkinen J, Kolehmainen M, Mykkänen H, Poutanen K (2010) Impact of dietary polyphenols on carbohydrate metabolism. Int J Mol Sci 11 (4):1365–1402
Hannan JMA, Ali L, Rokeya B, Khaleque J, Akhter M, Flatt PR, Abdel-Wahab YHA (2007) Soluble dietary fibre fraction of Trigonella foenum-graecum (fenugreek) seed improves glucose homeostasis in animal models of type 1 and type 2 diabetes by delaying carbohydrate digestion and absorption, and enhancing insulin action. Br J Nutr 97(3):514–521
Hellström J, Sinkkonen J, Karonen M, Mattila P (2007) Isolation and structure elucidation of procyanidin oligomers from Saskatoon berries (Amelanchier alnifolia). J Agric Food Chem 55 (1):157–164
Hertog MG, Feskens EJ, Kromhout D, Hollman PCH, Katan MB (1993) Dietary antioxidant flavonoids and risk of coronary heart disease: the Zutphen Elderly Study. Lancet 342 (8878):1007–1011
Ho GTT, Kase ET, Wangensteen H, Barsett H (2017) Effect of phenolic compounds from elderflowers on glucose-and fatty acid uptake in human myotubes and HepG2-cells. Molecules 22(1):90
Hu FB (2002) Dietary pattern analysis: a new direction in nutritional epidemiology. Curr Opin Lipidol 13(1):3–9
Hu FB (2013) Resolved: there is sufficient scientific evidence that decreasing sugar-sweetened beverage consumption will reduce the prevalence of obesity and obesity-related diseases. Obes Rev 14(8):606–619
Huang HC, Tao MH, Hung TM, Chen JC, Lin ZJ, Huang C (2014) (−)-Epigallocatechin-3-gallate inhibits entry of hepatitis B virus into hepatocytes. Antivir Res 111:100–111
Hurrell R, Egli I (2010) Iron bioavailability and dietary reference values. Am J Clin Nutr 91 (5):1461S–1467S
Jetté L, Harvey L, Eugeni K, Levens N (2009) 4-Hydroxyisoleucine: a plant-derived treatment for metabolic syndrome. Curr Opin Investig Drugs 10(4):353–358
Jiang B, Ji M, Liu W, Chen L, Cai Z, Zhao Y, Bi X (2016) Antidiabetic activities of a cucurbitane-type triterpenoid compound from Momordica charantia in alloxan-induced diabetic mice. Mol Med Rep 14(5):4865–4872
Jiménez-Arellanes A, Luna-Herrera J, Cornejo-Garrido J, López-García S, Castro-Mussot ME, Meckes-Fischer M, Mata-Espinosa D, Marquina B, Torres J, Hernández-Pando R (2013) Ursolic and oleanolic acids as antimicrobial and immunomodulatory compounds for tuberculosis treatment. BMC Complement Alternat Med 13(1):258
Jnawali HN, Ryoo S (2013) First-and second-line drugs and drug resistance. In: Tuberculosis-current issues in diagnosis and management, pp 163–180
Johansen SK, Maus CE, Plikaytis BB, Douthwaite S (2006) Capreomycin binds across the ribosomal subunit interface using tlyA-encoded 2′-O-methylations in 16S and 23S rRNAs. Mol Cell 23(2):173–182
Joseph B, Jini D (2013) Antidiabetic effects of Momordica charantia (bitter melon) and its medicinal potency. Asian Pac J Trop Dis 3(2):93–102
Jugheli L, Bzekalava N, de Rijk P, Fissette K, Portaels F, Rigouts L (2009) High level of cross-resistance between kanamycin, amikacin, and capreomycin among Mycobacterium tuberculosis

isolates from Georgia and a close relation with mutations in the rrs gene. Antimicrob Agents Chemother 53(12):5064–5068

Kajimoto T, Node M (2009) Inhibitors against glycosidases as medicines. Curr Top Med Chem 9 (1):13–33

Kalea AZ, Clark K, Schuschke DA, Klimis-Zacas DJ (2009) Vascular reactivity is affected by dietary consumption of wild blueberries in the Sprague-Dawley rat. J Med Food 12(1):21–28

Kalailingam P, Kannaian B, Tamilmani E, Kaliaperumal R (2014) Efficacy of natural diosgenin on cardiovascular risk, insulin secretion, and beta cells in streptozotocin (STZ)-induced diabetic rats. Phytomedicine 21(10):1154–1161

Kanikarla-Marie P, Jain SK (2016) Hyperketonemia and ketosis increase the risk of complications in type 1 diabetes. Free Radic Biol Med 95:268–277

Kant AK (2004) Dietary patterns and health outcomes. J Am Diet Assoc 104(4):615–635

Kaplan NM (1989) The deadly quartet: upper-body obesity, glucose intolerance, hypertriglyceridemia, and hypertension. Arch Intern Med 149(7):1514–1520

Kausar S, Wang F, Cui H (2018) The role of mitochondria in reactive oxygen species generation and its implications for neurodegenerative diseases. Cells 7(12):274

Kawabata K, Yoshioka Y, Terao J (2019) Role of intestinal microbiota in the bioavailability and physiological functions of dietary polyphenols. Molecules 24(2):370

Kawada Y, Miura M, Gomyo T (2006) Inhibitory effect of vegetables, fruits and herbs on α-glucosidase in an immobilized enzyme assay system. Food Sci Technol Res 12(4):275–277

Kim Y, Keogh JB, Clifton PM (2016) Polyphenols and glycemic control. Nutrients 8(1):17

Kliebenstein DJ, Osbourn A (2012) Making new molecules–evolution of pathways for novel metabolites in plants. Curr Opin Plant Biol 15(4):415–423

Konno K, Feldmann FM, McDermott W (1967) Pyrazinamide susceptibility and amidase activity of tubercle bacilli. Am Rev Respir Dis 95(3):461–469

Kossel A (1891) Ueber die chemische Zusammensetzung der Zelle. Du Bois-Reymond's Archiv/ Arch Anat Physiol Physiol Abt:181–186

Kovacevic Z, McGivan JD (1983) Mitochondrial metabolism of glutamine and glutamate and its physiological significance. Physiol Rev 63(2):547–605

Krishnaiah D, Sarbatly R, Nithyanandam R (2011) A review of the antioxidant potential of medicinal plant species. Food Bioprod Process 89(3):217–233

Lall N, Meyer JM, Wang Y, Bapela NB, Van Rensburg CEJ, Fourie B, Franzblau SG (2005) Characterization of intracellular activity of antitubercular constituents the roots of Euclea natalensis. Pharm Biol 43(4):353–357

Landgrave-Gómez J, Mercado-Gómez O, Guevara-Guzmán R (2015) Epigenetic mechanisms in neurological and neurodegenerative diseases. Front Cell Neurosci 9:58

Leung ELH, Fiscus RR, Tung JW, Tin VPC, Cheng LC, Sihoe ADL, Fink LM, Ma Y, Wong MP (2010) Non-small cell lung cancer cells expressing CD44 are enriched for stem cell-like properties. PLoS One 5(11):14062

Li J, Li J, Li S, He B, Mi Y, Cao H, Zhang C, Li L (2012) Ameliorative effect of grape seed proanthocyanidin extract on thioacetamide-induced mouse hepatic fibrosis. Toxicol Lett 213 (3):353–360

Li Y, Yao J, Han C, Yang J, Chaudhry MT, Wang S, Liu H, Yin Y (2016) Quercetin, inflammation and immunity. Nutrients 8(3):167

Lin YT, Wu YH, Tseng CK, Lin CK, Chen WC, Hsu YC, Lee JC (2013) Green tea phenolic epicatechins inhibit hepatitis C virus replication via cycloxygenase-2 and attenuate virus-induced inflammation. PLoS One *8*(1):e54466

Liu W, Zhai Y, Heng X, Che FY, Chen W, Sun D, Zhai G (2016) Oral bioavailability of curcumin: problems and advancements. J Drug Target 24(8):694–702

Liu K, Luo M, Wei S (2019) The bioprotective effects of polyphenols on metabolic syndrome against oxidative stress: evidences and perspectives. Oxidative Med Cell Longev 2019

Lu Z, Jia Q, Wang R, Wu X, Wu Y, Huang C, Li Y (2011) Hypoglycemic activities of A-and B-type procyanidin oligomer-rich extracts from different Cinnamon barks. Phytomedicine 18 (4):298–302

Lusis AJ (2000) Atherosclerosis. Nature 407(6801):233–241

Mandel SA, Amit T, Weinreb O, Reznichenko L, Youdim MB (2008) Simultaneous manipulation of multiple brain targets by green tea catechins: a potential neuroprotective strategy for Alzheimer and Parkinson diseases. CNS Neurosci Ther 14(4):352–365

Martín-Peláez S, Covas MI, Fitó M, Kušar A, Pravst I (2013) Health effects of olive oil polyphenols: recent advances and possibilities for the use of health claims. Mol Nutr Food Res 57(5):760–771

Martineau LC, Couture A, Spoor D, Benhaddou-Andaloussi A, Harris C, Meddah B, Leduc C, Burt A, Vuong T, Le PM, Prentki M (2006) Anti-diabetic properties of the Canadian lowbush blueberry Vaccinium angustifolium Ait. Phytomedicine 13(9-10):612–623

Maus CE, Plikaytis BB, Shinnick TM (2005) Molecular analysis of cross-resistance to capreomycin, kanamycin, amikacin, and viomycin in Mycobacterium tuberculosis. Antimicrob Agents Chemother 49(8):3192–3197

Migliori GB, Lange C, Centis R, Sotgiu G, Mütterlein R, Hoffmann H, Kliiman K, De Iaco G, Lauria FN, Richardson MD, Spanevello ANTONIO (2008) Resistance to second-line injectables and treatment outcomes in multidrug-resistant and extensively drug-resistant tuberculosis cases. Eur Respir J 31(6):1155–1159

Milani A, Basirnejad M, Shahbazi S, Bolhassani A (2017) Carotenoids: biochemistry, pharmacology and treatment. Br J Pharmacol 174(11):1290–1324

Mitchison DA (1979) Basic mechanisms of chemotherapy. Chest 76(6):771–780

Mitchison DA (1985) The action of antituberculosis drugs in short-course chemotherapy. Tubercle 66(3):219–225

Mondal S, Raja K, Schweizer U, Mugesh G (2016) Chemistry and biology in the biosynthesis and action of thyroid hormones. Angew Chem Int Ed 55(27):7606–7630

Moran JF, Klucas RV, Grayer RJ, Abian J, Becana M (1997) Complexes of iron with phenolic compounds from soybean nodules and other legume tissues: prooxidant and antioxidant properties. Free Radic Biol Med 22(5):861–870

Moudgal NR, Raghupathy E, Sarma PS (1958) Studies on goitrogenic agents in food: III. Goitrogenic action of some glycosides isolated from edible nuts. J Nutr 66(2):291–303

Mozaffarian D, Hao T, Rimm EB, Willett WC, Hu FB (2011) Changes in diet and lifestyle and long-term weight gain in women and men. N Engl J Med 364(25):2392–2404

Murakami A, Ohnishi K (2012) Target molecules of food phytochemicals: food science bound for the next dimension. Food Funct 3(5):462–476

Myburgh KH (2014) Polyphenol supplementation: benefits for exercise performance or oxidative stress? Sports Med 44(1):57–70

Nagao T, Hase T, Tokimitsu I (2007) A green tea extract high in catechins reduces body fat and cardiovascular risks in humans. Obesity 15(6):1473–1483

Napoli C, Ignarro LJ (2009) Nitric oxide and pathogenic mechanisms involved in the development of vascular diseases. Arch Pharm Res 32(8):1103–1108

Naseri R, Farzaei F, Haratipour P, Nabavi SF, Habtemariam S, Farzaei MH, Khodarahmi R, Tewari D, Momtaz S (2018) Anthocyanins in the management of metabolic syndrome: a pharmacological and biopharmaceutical review. Front Pharmacol 9:1310

Naveed M, BiBi J, Kamboh AA, Suheryani I, Kakar I, Fazlani SA, FangFang X, Yunjuan L, Kakar MU, Abd El-Hack ME, Noreldin AE (2018) Pharmacological values and therapeutic properties of black tea (Camellia sinensis): a comprehensive overview. Biomed Pharmacother 100:521–531

Newby PK, Tucker KL (2004) Empirically derived eating patterns using factor or cluster analysis: a review. Nutr Rev 62(5):177–203

Nikolić A, Nikolić D, Stanimirović V (2007) Metabolički sindrom X ili sindrom insulinske rezistencije. Vojnosanit Pregl 64(1):45–51

Oe H, Ozaki S (2008) Hypoglycemic effect of 13-membered ring thiocyclitol, a novel α-glucosidase inhibitor from Kothala-himbutu (Salacia reticulata). Biosci Biotechnol Biochem 72(7):1962–1964

Ohno M, Shibata C, Kishikawa T, Yoshikawa T, Takata A, Kojima K, Akanuma M, Kang YJ, Yoshida H, Otsuka M, Koike K (2013) The flavonoid apigenin improves glucose tolerance through inhibition of microRNA maturation in miRNA103 transgenic mice. Sci Rep 3:2553

Olokoba AB, Obateru OA, Olokoba LB (2012) Type 2 diabetes mellitus: a review of current trends. Oman Med J 27(4):269

Oppenheimer JH, Schwartz HL, Mariash CN, Kinlaw WB, Wong NCW, Freake HC (1987) Advances in our understanding of thyroid hormone action at the cellular level. Endocr Rev 8 (3):288–308

Paroo Z, Ye X, Chen S, Liu Q (2009) Phosphorylation of the human microRNA-generating complex mediates MAPK/Erk signaling. Cell 139(1):112–122

Parvez MK, Arbab AH, Al-Dosari MS, Al-Rehaily AJ (2016) Antiviral natural products against chronic hepatitis B: recent developments. Curr Pharm Des 22(3):286–293

Pauli GF, Case RJ, Inui T, Wang Y, Cho S, Fischer NH, Franzblau SG (2005) New perspectives on natural products in TB drug research. Life Sci 78(5):485–494

Pearce EN, Braverman LE (2009) Environmental pollutants and the thyroid. Best Pract Res Clin Endocrinol Metab 23(6):801–813

Pierce JP, Faerber S, Wright FA, Rock CL, Newman V, Flatt SW, Kealey S, Jones VE, Caan BJ, Gold EB, Haan M (2002) A randomized trial of the effect of a plant-based dietary pattern on additional breast cancer events and survival: the Women's Healthy Eating and Living (WHEL) Study. Control Clin Trials 23(6):728–756

Prasain JK, Carlson SH, Wyss JM (2010) Flavonoids and age-related disease: risk, benefits and critical windows. Maturitas 66(2):163–171

Ramaswamy S, Musser JM (1998) Molecular genetic basis of antimicrobial agent resistance in Mycobacterium tuberculosis: 1998 update. Tuber Lung Dis 79(1):3–29

Reilly JJ, Armstrong J, Dorosty AR, Emmett PM, Ness A, Rogers I, Steer C, Sherriff A (2005) Early life risk factors for obesity in childhood: cohort study. BMJ 330(7504):1357

Rice-Evans C, Miller N, Paganga G (1997) Antioxidant properties of phenolic compounds. Trends Plant Sci 2(4):152–159

Rodríguez G, Moreno LA (2006) Is dietary intake able to explain differences in body fatness in children and adolescents? Nutr Metab Cardiovasc Dis 16(4):294–301

Rojas Á, Del Campo JA, Clement S, Lemasson M, García-Valdecasas M, Gil-Gómez A, Ranchal I, Bartosch B, Bautista JD, Rosenberg AR, Negro F (2016) Effect of quercetin on hepatitis C virus life cycle: from viral to host targets. Sci Rep 6(1):1–9

Rose WC (1933) The role of proteins in metabolism

Saeidnia S, Abdollahi M (2013) Antioxidants: friends or foe in prevention or treatment of cancer: the debate of the century. Toxicol Appl Pharmacol 271(1):49–63

Sauvaire Y, Baissac Y, Leconte O, Petit P, Ribes G (1996) Steroid saponins from fenugreek and some of their biological properties. In: Saponins used in food and agriculture. Springer, Boston, pp 37–46

Sheng X, Zhang Y, Gong Z, Huang C, Zang YQ (2008) Improved insulin resistance and lipid metabolism by cinnamon extract through activation of peroxisome proliferator-activated receptors. PPAR Res 2008

Shibata C, Ohno M, Otsuka M, Kishikawa T, Goto K, Muroyama R, Kato N, Yoshikawa T, Takata A, Koike K (2014) The flavonoid apigenin inhibits hepatitis C virus replication by decreasing mature microRNA122 levels. Virology 462:42–48

Shibib BA, Khan LA, Rahman R (1993) Hypoglycaemic activity of Coccinia indica and Momordica charantia in diabetic rats: depression of the hepatic gluconeogenic enzymes glucose-6-phosphatase and fructose-1, 6-bisphosphatase and elevation of both liver and red-cell shunt enzyme glucose-6-phosphate dehydrogenase. Biochem J 292(1):267–270

Shih CC, Shlau MT, Lin CH, Wu JB (2014) Momordica charantia ameliorates insulin resistance and dyslipidemia with altered hepatic glucose production and fatty acid synthesis and AMPK phosphorylation in high-fat-fed mice. Phytother Res 28(3):363–371

Shiina Y, Funabashi N, Lee K, Murayama T, Nakamura K, Wakatsuki Y, Daimon M, Komuro I (2009) Acute effect of oral flavonoid-rich dark chocolate intake on coronary circulation, as compared with non-flavonoid white chocolate, by transthoracic Doppler echocardiography in healthy adults. Int J Cardiol 131(3):424–429

Shin MO, Yoon S, Moon JO (2010) The proanthocyanidins inhibit dimethylnitrosamine-induced liver damage in rats. Arch Pharm Res 33(1):167–173

Shixian Q, VanCrey B, Shi J, Kakuda Y, Jiang Y (2006) Green tea extract thermogenesis-induced weight loss by epigallocatechin gallate inhibition of catechol-O-methyltransferase. J Med Food 9(4):451–458

Sies H (1997) Oxidative stress: oxidants and antioxidants. Exp Physiol Transl Integr 82(2):291–295

Šošić-Jurjević B, Filipović B, Ajdžanović V, Savin S, Nestorović N, Milošević V, Sekulić M (2010) Suppressive effects of genistein and daidzein on pituitary–thyroid axis in orchiectomized middle-aged rats. Exp Biol Med 235(5):590–598

Spies FS, Da Silva PEA, Ribeiro MO, Rossetti ML, Zaha A (2008) Identification of mutations related to streptomycin resistance in clinical isolates of Mycobacterium tuberculosis and possible involvement of efflux mechanism. Antimicrob Agents Chemother 52(8):2947–2949

Srichamroen A, Thomson AB, Field CJ, Basu TK (2009) In vitro intestinal glucose uptake is inhibited by galactomannan from Canadian fenugreek seed (Trigonella foenum graecum L) in genetically lean and obese rats. Nutr Res 29(1):49–54

Sroka Z, Cisowski W (2003) Hydrogen peroxide scavenging, antioxidant and anti-radical activity of some phenolic acids. Food Chem Toxicol 41(6):753–758

Stutz AE, Wrodnigg TM (2011) Imino sugars and glycosyl hydrolases. Adv Carbohydr Chem Biochem 66:187–298

Suliburska J, Bogdanski P, Szulinska M, Stepien M, Pupek-Musialik D, Jablecka A (2012) Effects of green tea supplementation on elements, total antioxidants, lipids, and glucose values in the serum of obese patients. Biol Trace Elem Res 149(3):315–322

Tan MJ, Ye JM, Turner N, Hohnen-Behrens C, Ke CQ, Tang CP, Chen T, Weiss HC, Gesing ER, Rowland A, James DE (2008) Antidiabetic activities of triterpenoids isolated from bitter melon associated with activation of the AMPK pathway. Chem Biol 15(3):263–273

Teiten MH, Dicato M, Diederich M (2014) Hybrid curcumin compounds: a new strategy for cancer treatment. Molecules 19(12):20839–20863

Tomé-Carneiro J, Gonzálvez M, Larrosa M, Yáñez-Gascón MJ, García-Almagro FJ, Ruiz-Ros JA, Tomás-Barberán FA, García-Conesa MT, Espín JC (2013) Resveratrol in primary and secondary prevention of cardiovascular disease: a dietary and clinical perspective. Ann N Y Acad Sci 1290(1):37–51

Tsai CH, Chen ECF, Tsay HS, Huang CJ (2012) Wild bitter gourd improves metabolic syndrome: a preliminary dietary supplementation trial. Nutr J 11(1):4

Uemura T, Hirai S, Mizoguchi N, Goto T, Lee JY, Taketani K, Nakano Y, Shono J, Hoshino S, Tsuge N, Narukami T (2010) Diosgenin present in fenugreek improves glucose metabolism by promoting adipocyte differentiation and inhibiting inflammation in adipose tissues. Mol Nutr Food Res 54(11):1596–1608

Ulusoy S, Ozkan G, Yucesan FB, ERSÖZ Ş, Orem A, Alkanat M, Yuluğ E, Kaynar K, Al S (2012) Anti-apoptotic and anti-oxidant effects of grape seed proanthocyanidin extract in preventing cyclosporine A-induced nephropathy. Nephrology 17(4):372–379

Vendrame S, Daugherty A, Kristo AS, Riso P, Klimis-Zacas D (2013) Wild blueberry (Vaccinium angustifolium) consumption improves inflammatory status in the obese Zucker rat model of the metabolic syndrome. J Nutr Biochem 24(8):1508–1512

Verspohl EJ, Bauer K, Neddermann E (2005) Antidiabetic effect of Cinnamomum cassia and Cinnamomum zeylanicum in vivo and in vitro. Phytother Res 19(3):203–206

Via LE, Cho SN, Hwang S, Bang H, Park SK, Kang HS, Jeon D, Min SY, Oh T, Kim Y, Kim YM (2010) Polymorphisms associated with resistance and cross-resistance to aminoglycosides and capreomycin in Mycobacterium tuberculosis isolates from South Korean patients with drug-resistant tuberculosis. J Clin Microbiol 48(2):402–411

Vinayagam R, Xu B (2015) Antidiabetic properties of dietary flavonoids: a cellular mechanism review. Nutr Metab 12(1):60

Virgili F, Marino M (2008) Regulation of cellular signals from nutritional molecules: a specific role for phytochemicals, beyond antioxidant activity. Free Radic Biol Med 45(9):1205–1216

Visioli F, Bernardini E (2011) Extra virgin olive oil's polyphenols: biological activities. Curr Pharm Des 17(8):786–804

Vocadlo DJ, Davies GJ (2008) Mechanistic insights into glycosidase chemistry. Curr Opin Chem Biol 12(5):539–555

Wang H, Nair MG, Strasburg GM, Chang YC, Booren AM, Gray JI, DeWitt DL (1999) Antioxidant and antiinflammatory activities of anthocyanins and their aglycon, cyanidin, from tart cherries. J Nat Prod 62(2):294–296

Wang JY, Hsueh PR, Jan IS, Lee LN, Liaw YS, Yang PC, Luh KT (2006) Empirical treatment with a fluoroquinolone delays the treatment for tuberculosis and is associated with a poor prognosis in endemic areas. Thorax 61(10):903–908

Weiss R, Bremer AA, Lustig RH (2013) What is metabolic syndrome, and why are children getting it? Ann N Y Acad Sci 1281(1):123

Wiciński M, Gębalski J, Mazurek E, Podhorecka M, Śniegocki M, Szychta P, Sawicka E, Malinowski B (2020) The influence of polyphenol compounds on human gastrointestinal tract microbiota. Nutrients 12(2):350

Wood LG, Fitzgerald DA, Gibson PC, Cooper DM, Garg ML (2000) Lipid peroxidation as determined by plasma isoprostanes is related to disease severity in mild asthma. Lipids 35 (9):967

World Health Organization (2008) Policy guidance on drug-susceptibility testing (DST) of second-line antituberculosis drugs (No. WHO/HTM/TB/2008.392). World Health Organization

World Health Organization (2016) World malaria report 2015. World Health Organization

Xiao JB, Hogger P (2015) Dietary polyphenols and type 2 diabetes: current insights and future perspectives. Curr Med Chem 22(1):23–38

Xiong W, Ma X, Wu Y, Chen Y, Zeng L, Liu J, Sun W, Wang D, Hu Y (2015) Determine the structure of phosphorylated modification of icariin and its antiviral activity against duck hepatitis virus A. BMC Vet Res 11(1):1–9

Yang YS, Su YF, Yang HW, Lee YH, Chou JI, Ueng KC (2014) Lipid-lowering effects of curcumin in patients with metabolic syndrome: a randomized, double-blind, placebo-controlled trial. Phytother Res 28(12):1770–1777

Yang X, Belosay A, Hartman JA, Song H, Zhang Y, Wang W, Doerge DR, Helferich WG (2015) Dietary soy isoflavones increase metastasis to lungs in an experimental model of breast cancer with bone micro-tumors. Clin Exp Metastasis 32(4):323–333

Yang L, Ling W, Du Z, Chen Y, Li D, Deng S, Liu Z, Yang L (2017) Effects of anthocyanins on cardiometabolic health: a systematic review and meta-analysis of randomized controlled trials. Adv Nutr 8(5):684–693

Yen PM (2001) Physiological and molecular basis of thyroid hormone action. Physiol Rev 81 (3):1097–1142

Zaunbrecher MA, Sikes RD Jr, Metchock B, Shinnick TM, Posey JE (2009) Over-expression of the chromosomally encoded aminoglycoside acetyltransferase eis confers kanamycin resistance in Mycobacterium tuberculosis. Proc Natl Acad Sci U S A 106(47):20004–20009

Zhang Y, Mitchison D (2003) The curious characteristics of pyrazinamide: a review. Int J Tuberc Lung Dis 7(1):6–21

Zhang Y, Heym B, Allen B, Young D, Cole S (1992) The catalase—peroxidase gene and isoniazid resistance of Mycobacterium tuberculosis. Nature 358(6387):591–593

Zhang YB, Zhan LQ, Li GQ, Wang F, Wang Y, Li YL, Ye WC, Wang GC (2016) Dimeric matrine-type alkaloids from the roots of Sophora flavescens and their anti-hepatitis B virus activities. J Org Chem 81(15):6273–6280

Ziegler J, Facchini PJ (2008) Alkaloid biosynthesis: metabolism and trafficking. Annu Rev Plant Biol 59:735–769

Zoeller TR (2010) Environmental chemicals targeting thyroid. Hormones 9(1):28–40

# Metal Complexes of Plant Secondary Metabolites with Therapeutic Potential

9

Valentina Uivarosi, Alexandra-Cristina Munteanu, Mihaela Badea, and Rodica Olar

**Abstract**

Among the vast and diverse assortment of organic compounds produced by plants, secondary metabolites are extremely attractive to the pharmaceutical industry as leads for the development of drug candidates. A very broad classification of plant secondary metabolites includes four major classes: terpenoids, phenolic compounds, alkaloids, and sulfur-containing compounds. Furthermore, metal complexation can complement the biological activities of plant secondary metabolites and serve as structural scaffolds for the design of bioactive compounds. Distinct spectroscopic signatures, versatile spatial arrangements around the metal center, adjustable ligand exchange kinetics, and fine-tuned redox properties are some of the unique features of metal complexes. Among the secondary metabolites, especially phenolic compounds and some alkaloids possess excellent chelating properties. This chapter will report recent trends and achievements related to the numerous biological activities of the metal complexes of plant secondary metabolites. Due to their structural diversity, these metal complexes exhibit a wide range of pharmacological activities, including antioxidant, antitumor, antimicrobial, anti-inflammatory, antidiabetic, antineurodegenerative, immunostimulant, estrogenic, etc. Furthermore, aspects regarding the binding of these complexes to nucleic acids and other nontraditional targets will be discussed.

V. Uivarosi (✉) · A.-C. Munteanu
Department of General and Inorganic Chemistry, Faculty of Pharmacy, Carol Davila University of Medicine and Pharmacy, Bucharest, Romania
e-mail: valentina.uivarosi@umfcd.ro; alexandra.ticea@umfcd.ro

M. Badea · R. Olar
Department of Inorganic Chemistry, Faculty of Chemistry, University of Bucharest, Bucharest, Romania
e-mail: mihaela.badea@chimie.unibuc.ro; rodica.olar@chimie.unibuc.ro


A. K. Sharma, A. Sharma (eds.), *Plant Secondary Metabolites*,
https://doi.org/10.1007/978-981-16-4779-6_9

**Keywords**

Plant secondary metabolites · Metal complexation · Bioactive compounds · Pharmacological activities · Binding

## 9.1 Introduction

Metabolites are intermediates and products of metabolic reactions (Croteau et al. 2000). The term metabolite usually refers to small molecules. Plants produce a large diversity of such small organic compounds. Some, involved in growth, development, and reproduction of the organism, are called primary metabolites (Bennett and Wallsgrove 1994). The great majority of them, however, do not appear to be directly involved in growth- and development-related processes inside the plant. These substances are generally referred to as secondary metabolites. Their functions in plants include defense against herbivores, plant pathogens, or other plants, protection against UV light, oxidation and other physical stressors, synthesis of compounds to attract pollinating and seed-dispersing animals, and role in communication between plants and symbiotic microorganisms (Theis and Lerdau 2003). If absence of primary metabolites usually results in immediate death, loss of secondary metabolites causes long-term impairment of the interactions between the plant and its environment, esthetics, or its ability to reproduce (Gurib-Fakim 2006). Note that primary and secondary metabolites cannot be readily classified on the basis of their chemical structures, precursors, or biosynthetic origins and their classification can vary widely.

As secondary metabolites are known to play defensive roles in plants, they can be induced by infection or wounding. Alkaloids are synthesized by plants to either kill or retard the development of the herbivores and pathogens (War et al. 2012). Glucosinolates play a significant role in the response to external stimuli, in connection with plant defense against insects and pathogens (Singh 2017). Tannins have diverse functions in plants, including defense against herbivores and pathogens, regulating dispersal in fruits and seeds, and as modulators of nutrient cycling and plant tolerance to abiotic stress (Constabel et al. 2014). Several flavonoids can function as shields against UV-B irradiation (Mierziak et al. 2014). An interesting strategy developed by plants for their protection refers to the synthesis of highly active "prodrug" phytochemicals that are activated upon damage. Glycosides account for numerous such molecules.

Interestingly, numerous medicinal plants contain broad-spectrum phytochemicals which display weak or moderate bioactivity. It has been suggested that in order to overcome the problem of pathogen resistance, plants use combinations of pleiotropic multi-targeted phytochemical complexes (Efferth and Koch 2011). Therefore, the interplay between diverse bioactive molecules can result in increased bioactivity.

Humans use secondary metabolites as medicines (most polyphenolic compounds), flavorings (e.g., terpenes), and recreational drugs (e.g., morphine, caffeine). When used for medicinal purpose, natural compounds usually display good biological activity, yet the generally poor selectivity and oral bioavailability impede their further clinical use. One of the main causes of poor oral bioavailability is represented by the presence of free hydroxyl groups in the structure of numerous secondary metabolites, which can be readily conjugated in the human body and further eliminated (Hussain et al. 2019). Increased lipophilicity of secondary metabolites, achieved, for example, by metal chelation, results in improved intestinal absorption and biotransformation (Bone and Mills 2013). Their ability to target multiple cellular components/pathways may also result in lack of selectivity. Several plant secondary metabolites possess the ability to chelate metal ions; therefore, one strategy, employed to overcome these shortages, has been metal chelation. Structural features which render metal binding possible include the presence of vicinal hydroxyl groups, keto-hydroxy moiety, or diketo motif.

In drug design, the multitude of ligands and binding geometries allow for fine tailoring of the local environment around the metal center (Ndagi et al. 2017; Liu et al. 2020b). Moreover, addition of a phytochemical ligand mitigates the response of living cells to the metal ion and allows for a more complex mode of action.

## 9.2 Classification

Although wide variations exist in the literature, plant secondary metabolites can be classified on the basis of chemical structure, composition (containing nitrogen or not), or the pathway by which they are synthesized (e.g., phenylpropanoid, which produces tannins). Four large molecule families are generally considered: terpenes, nitrogen-containing compounds, phenolics, and sulfur-containing compounds.

### 9.2.1 Terpenes

Terpenes are among the most widespread and diverse groups of natural products. Terpenes consist of a group of hydrocarbon-based natural products; their chemical structures are based on multiple condensations of isoprene units. Terpenes are classified by the number of carbons in monoterpenes (C10), sesquiterpenes (C15), diterpenes (C20), triterpenes (C30), tetraterpenes (C40), and polyterpenes (Fig. 9.1). The most comprehensive studies regarding the biological activity of terpenes are related to the prevention and treatment of cancer. Taxol derivatives (paclitaxel and docetaxel) are among the widely used drugs in cancer chemotherapy. Other important therapeutic uses of terpenoids include antioxidant, antimicrobial, antifungal, antiviral, hypoglycemic, anti-inflammatory, antiparasitic, immunomodulatory, and as skin permeation enhancers (Brahmkshatriya and Brahmkshatriya 2013). Generally, natural terpenoids do not possess the ability to chelate metal ions.

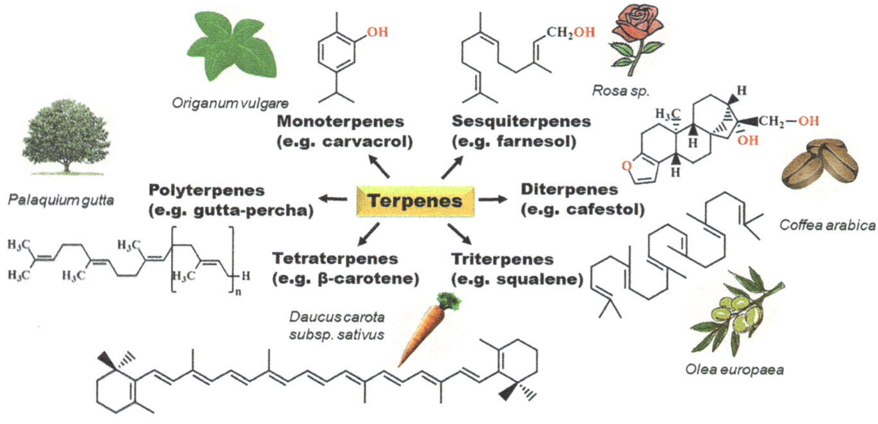


**Fig. 9.1** Classification of natural terpenes; representatives of each subclass, which can be isolated from various plants (shown in pink oval shapes), are given in parentheses, alongside with their chemical structure

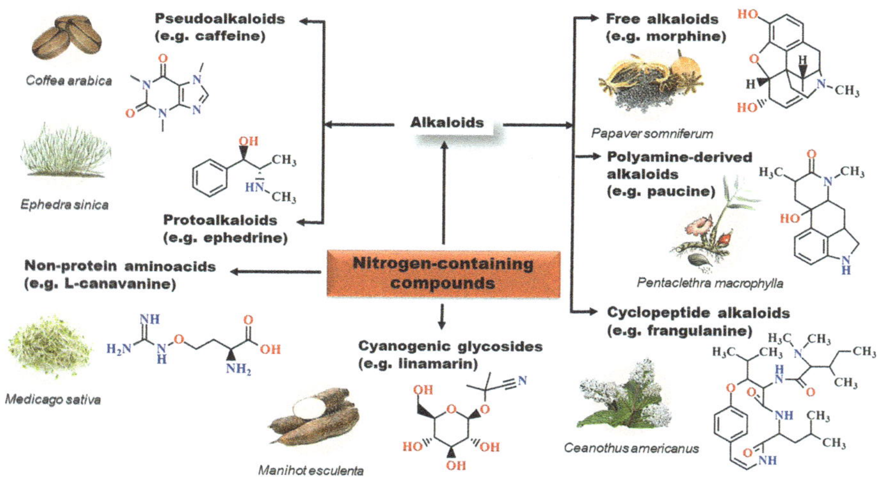


**Fig. 9.2** Classification of natural nitrogen-containing compounds; representatives of each subclass, which can be isolated from various plants (shown in pink oval shapes), are given in parentheses, alongside with their chemical structure (Katerova et al. 2012)

## 9.2.2 Nitrogen-Containing Compounds

Although most N-containing compounds possess intrinsic ability to bind metal ions, only a relatively small number of studies (see below) report on the biological activity of their coordination complexes. Most such studies involve naturally occurring *alkaloids* as ligands (Fig. 9.2). Their biological activities extend over a large range, including antibacterial, antiviral, analgesic, antinociceptive, sympathomimetic, stimulant of the central nervous system, etc. (Waller and Nowacki 1978).

### 9.2.3 Phenolics

Based on their structure, they can be divided into flavonoids and non-flavonoids (Fig. 9.3).

*Flavonoids* more than 8000 different representatives are known thus far. Flavonoids comprise a diverse set of compounds and perform a wide range of functions. Flavonoids consist of various groups of plant metabolites which include chalcones, aurones, flavanones, isoflavonoids, flavones, flavonols, flavanones, flavanonols, catechins, anthocyanins, etc. (Fig. 9.4). Most flavonoids are good chelating agents, and therefore a large body of studies regarding the biological activities of flavonoid metal complexes exist in the literature. Among the wide range of biological activities, flavonoids possess antioxidant, anticancer, antimicrobial, and antiviral properties (Falcone Ferreyra et al. 2012).

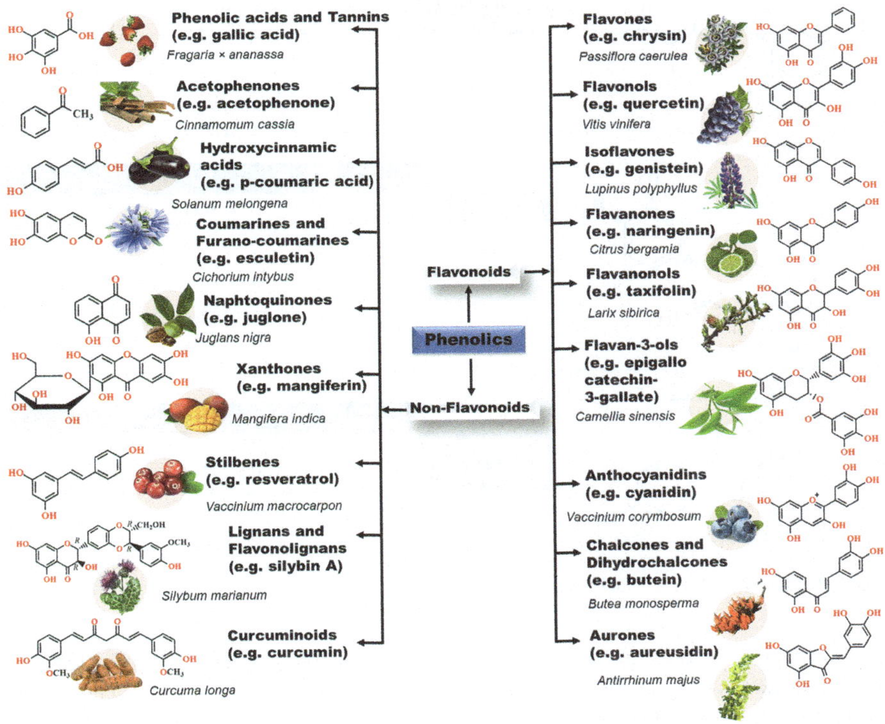


**Fig. 9.3** Classification of natural phenolics; representatives of each subclass, which can be isolated from various plants (shown in pink oval shapes), are given in parentheses, alongside with their chemical structure

**Flavonols**

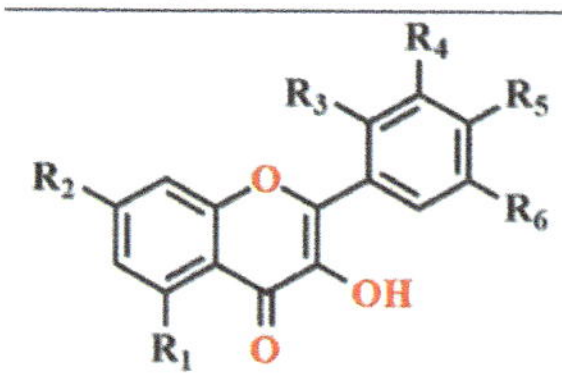


| | R1 | R2 | R3 | R4 | R5 | R6 |
|---|---|---|---|---|---|---|
| Quercetin | OH | OH | H | OH | OH | H |
| Morin | OH | OH | OH | H | OH | H |
| Kaempferol | OH | OH | H | H | OH | H |
| Fisetin | H | OH | H | OH | OH | H |
| Galangin | OH | OH | H | H | H | H |
| Myricetin | OH | OH | H | OH | OH | OH |

**Flavones**

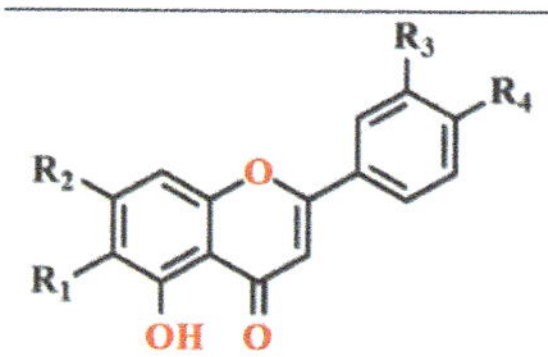


| | R1 | R2 | R3 | R4 |
|---|---|---|---|---|
| Primuletin | H | H | H | H |
| Chrysin | H | OH | H | H |
| Apigenin | H | OH | H | OH |
| Baicalein | OH | OH | H | H |
| Luteolin | H | OH | OH | OH |
| Diosmetin | H | OH | OH | $OCH_3$ |

**Flavanone**

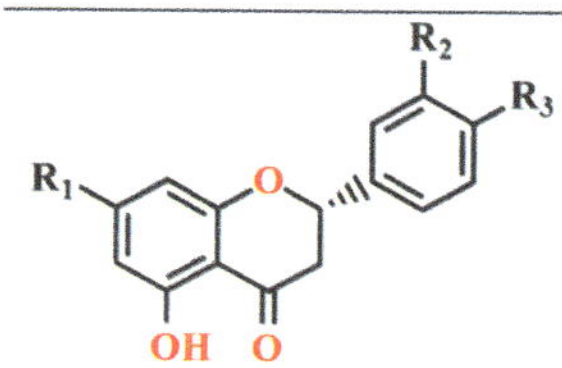


| | R1 | R2 | R3 |
|---|---|---|---|
| Naringenin | OH | H | OH |
| Hesperetin | OH | OH | $OCH_3$ |

**Flavanonols**

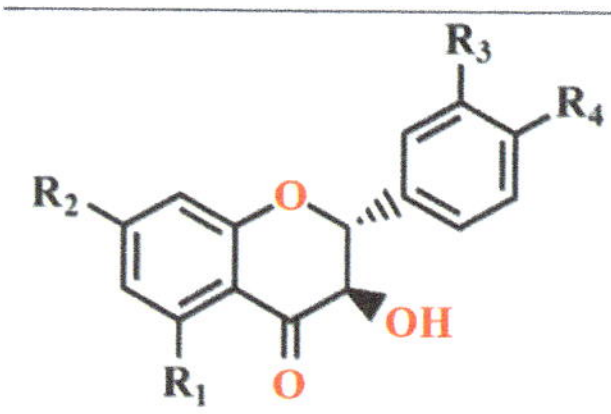


| | R1 | R2 | R3 | R4 |
|---|---|---|---|---|
| Taxifolin | OH | OH | OH | OH |

**Isoflavones**

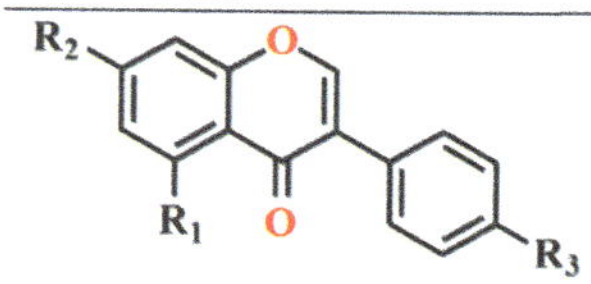


| | R1 | R2 | R3 |
|---|---|---|---|
| Genistein | OH | OH | OH |
| Daidzein | H | OH | OH |

**Fig. 9.4** Representatives of natural flavonoids

*Coumarins* belong to a widespread family of plant metabolites called the benzopyranones (Fig. 9.3). In plants, these compounds can occur in seed coats, fruits, flowers, roots, leaves, and stems, although in general the greatest concentration are found in fruits and flowers (Stefanachi et al. 2018).

*Stilbenes* are present in bryophytes, pteridophyte, gymnosperms, and angiosperms. Resveratrol, for instance, can be found in *Vitis species* (grapes), red wine, and other plant species. Stilbenoids exert various biological activities ranging from cardio- and neuroprotection, hypoglycemic properties, anti-inflammatory, anti-cancer, etc. (Akinwumi et al. 2018).

*Curcumin* is a secondary metabolite, intensely studied for its antioxidant, anti-cancer, antimicrobial, and anti-inflammatory activities (Pröhl et al. 2016; Dar et al. 2017). It has the ability to form biologically active metal complexes (see below).

### 9.2.4 Sulfur-Containing Compounds

S-containing compounds play important roles as defense molecules in the Asteraceae, Alliaceae, and Brassicaceae families. Studies regarding biological activities have been related to their antimicrobial, anticancer, anti-inflammatory, neuroprotective, etc. properties (Abdalla and Mühling 2019). To the extent of our knowledge, no metal complexes with proven structure and biological activity have been reported for sulfur-containing compounds (Fig. 9.5).

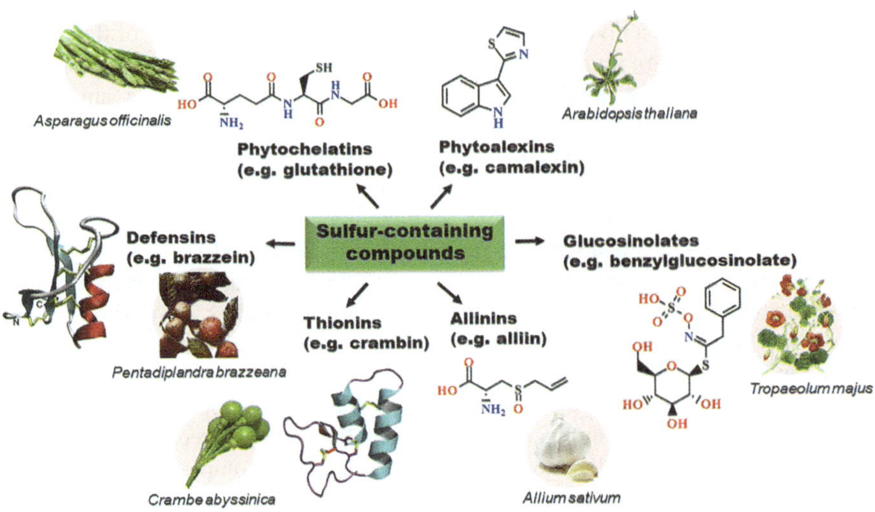


**Fig. 9.5** Classification of natural sulfur-containing compounds; representatives of each subclass, which can be isolated from various plants (shown in pink oval shapes), are given in parentheses, alongside with their chemical structure. (The structure of brazzein was reproduced from van der Weerden and Anderson (2013), with permission. The structure of crambin was reproduced from Westermann and Craik (2010), with permission)

## 9.3 Antioxidant Activity

Antioxidants are active compounds that fight against reactive oxygen species (ROS), reactive nitrogen species (RNS), and free radicals that play significant roles in various diseases, including cancer, neurodegenerative diseases (e.g., Alzheimer's disease), inflammatory disorders (arteriosclerosis, rheumatoid arthritis, some autoimmune diseases), aging, and diabetes. Phenolic compounds exert antioxidative functions either through electron donation (Thomas 2000), hydrogen atom transfer (Musialik et al. 2009), or metal chelation (Dimitrić Marković et al. 2011). Hydrogen atom transfer has been correlated to the bond dissociation enthalpy (BDE) of the compound, and electron donating ability to the ionization potential (IP). Low BDE or IP translates to high hydrogen atom transferring or electron donating ability (Chen et al. 2009).

There is an extensive body of evidence in the literature regarding the antioxidant properties of secondary metabolites, which depend largely on the number and position of the hydroxyl groups available in their structure. With respect to the antioxidant activity of their metal complexes, factors like O–H bond dissociation energies, ionization potential, redox behavior, and steric hindrance have been found to play crucial roles. Whenever possible, certain aspects in relation to structure–activity relationship have been discussed throughout this subchapter.

*Flavonoids* are among the major antioxidant constituents of our diet. Flavonoids have been shown to react with electron-deficient radicals such as peroxyls or $DPPH^{\cdot}$, this activity being correlated with the acidity of the phenolic hydroxyl groups in their structure and the stability of the resulted radical (Musialik et al. 2009). The redox potential has also been inversely correlated to the antioxidant activity of flavonoids. Several flavonols (e.g., myricetin, quercetin), flavones (e.g., luteolin), and flavan-3-ols (e.g., catechin) and flavonoid glycosides (e.g., rutin) have higher antioxidant activity correlated to lower oxidation potentials (Zhang et al. 2011). The existing literature has called attention to a number of factors that favor the antioxidant activity of flavonoids, namely, the presence of a catechol moiety (3′,4′-dihydroxy) on the B ring (Yasarawan et al. 2013), a double bond conjugated to a 4-carbonyl group, and hydroxyl groups in C3 and C5 (see Fig. 9.6; Porfírio et al. 2014).

Since quercetin presents the most simple chemical structure for which all of the abovementioned requirements are met, a plethora of studies have explored the antioxidant activity of its metal chelates. It should be noted that all molar ratios will be given as ligand: metal ion.

For instance, a series of metal complexes of flavonol quercetin (1:2), flavonoid glycoside rutin (3:2), flavonol galangin (1:1), and flavan-3-ol catechin (1:1) with Cu(II), Fe(II), Zn(II), and Al(III) have been studied. Quercetin complexes were found to be the most potent antioxidants, with Al(III) complexes being more active among the series of complexes bearing the same flavonoid ligand. Based on the fact that Al(III) has vacant p-orbitals that are good π acceptors, the improved antioxidant activity of its complexes can be explained by the strong stabilization effect of the aluminum ion on the flavonoid phenoxyl radicals (de Souza and De Giovani 2004). Several 1:1 morin, quercetin, and fisetin (flavonols) complexes with Fe(II), formed at

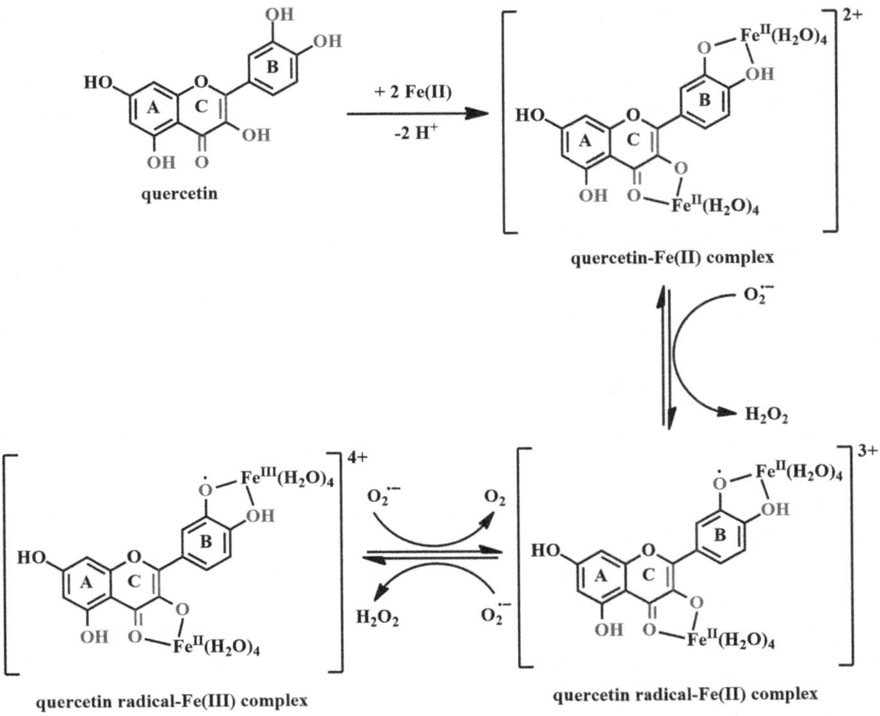


**Fig. 9.6** Mechanisms of superoxide scavenging by flavonoid complexes. The dismutation of superoxide ion by flavonoid complexes could be explained by the redox cycling of the metal and the ligand: the metal complex can be oxidized to a Fe(II)–semiquinone complex by superoxide radical, which is oxidized by another superoxide radical to form the Fe(III)–semiquinone complex. (The scheme was reproduced from De Souza et al. (2003), with permission). Factors that favor the antioxidant activity of flavonoids and their metal complexes have been colored in purple in the structure of quercetin: the catechol moiety (3′,4′-dihydroxy) on the B ring (most reactive redox group)—promotes electron displacement, which in turn will increase the stability of the radical; a C2=C3 double bond conjugated to a 4-carbonyl group—increases the electron displacement from the B ring; hydroxyl groups in C3 and C5—allows for electron displacement from the 4-oxo group (Porfírio et al. 2014)

physiological pH, showed a significant increase in their antioxidant capacity. The most effective antioxidant was morin–Fe(II), yet the most dramatic increase in the antioxidant activity was registered for the quercetin–Fe(II) complex. In contrast, the 1:1 catechin (flavan-3-ol) and chrysin (flavone) complex formation did not influence their antioxidant capacity in a significant manner (Porfírio et al. 2014). Complexes of morin, quercetin, and primuletin with Cu(II) (2:1) and Fe(III) (3:1) have been studied in comparison with the free flavonoids against stable (DPPH$^{\cdot}$, Tempo$^{\cdot}$) and more reactive ($OH^{\cdot}$ and $O_2^{\cdot -}$) free radicals. It was found that the radical scavenging ability varies in the order quercetin> morin> primuletin, and the same trend is valid for the metal complexes, with quercetin complexes being the most active. Cu(II) complexes were generally more active than their Fe(III) correspondents, with the exception of quercetin (Jabeen et al. 2017a).

Notably, the reaction of metal chelates with free radicals (especially superoxide) is generally facilitated when the metal coordination sites are either free or occupied by an easily displaceable ligand (such as water molecules). Therefore, complexes in which the flavonoid/metal ion molar ratio is 1:1 or 2:1 will be more efficient scavengers in comparison to 3:1 metal complexes (Graf et al. 1984). The 2:1 complexes of flavonoids (catechin, quercetin, rutin, fisetin, luteolin, taxifolin, kaempferol) with Cu(II), Fe(II), or Fe(III) were more effective than the parent compounds in scavenging superoxide radicals generated by the xanthine oxidase/ hypoxanthine system in isolated rat hepatocytes. Cu(II) and Fe(III) complexes have been reported to possess higher activity than the corresponding Fe(II) chelates, with quercetin and rutin complexes being the most active compounds of the series. Interestingly, the differences in scavenging efficacy in favor of the Cu (II) complexes were remarkable for all flavonoid complexes, except quercetin and fisetin (Moridani et al. 2003). Similarly, a 2:1 quercetin–Fe(II) complex was also shown to possess superior free radical (DPPH$^{\cdot}$) scavenging activity compared to free quercetin (Raza et al. 2016).

Coordination (generally at the 4-keto-3-hydroxyl or "maltol-like" site when available) increases the antioxidant activities of the free flavonoids, which can be correlated to the lower oxidation potentials of the complexes compared to those of the free flavonoids. A comprehensive study reported the syntheses of 1:2, 2:3, and 1: 1 complexes of quercetin, rutin, and 3-hydroxyflavone with Cu(II) and Fe(II). Metal coordination resulted in lower oxidation potentials of the complexes compared to those of the free flavonoids; metal complexes were also more efficient in terms of scavenging superoxide radicals, with the two quercetin (coordinated via the maltol moiety) complexes being the most active compound of the series (De Souza et al. 2003).

The coordination complexes of quercetin with Cu(II), Zn(II), Ni(II), Co(II), and Fe(II) (2:1) were tested as antioxidant agents in zebrafish (*Danio rerio*) in relation to their potential use against Alzheimer's disease. The Fe(II) complex presented the highest antioxidant activity (da Silva et al. 2020). Similarly, the ability of 2:1 quercetin complexes with Mn(II), Co(II), Ni(II), Cu(II), Zn(II), and Pb(II) to act as superoxide radical scavengers has been determined to be higher than that of quercetin (Zhou et al. 2001b). The 3:1 quercetin complexes with trivalent lanthanide metal ions (La, Nd, Eu, Gd, Tb, Dy, Tm, and Y) have displayed higher activity than free quercetin, yet were only marginally more potent than the 2:1 quercetin complexes with divalent transitional metal ions (Zhou et al. 2001a).

Interestingly, the radical scavenging activity of the flavonols myricetin, quercetin, fisetin, and flavone luteolin has been greatly improved by complexation with Fe(III) in the presence of solid substrates. In the resulting biofunctional metal–phenolic networks or films, the ligand/metal ratio was reported to be of approximately 3:1. Flavonol/Fe(III) films proved to be twice more potent than the luteolin/Fe(III) network when the %DPPH consumption was calculated (□80% vs. <40%), presumably due to the absence of the 3-OH group. In addition to the higher antioxidant efficacy and increased water solubility, the scavenging activity of the quercetin/Fe (III) films was preserved over at least three cycles (Bertleff-Zieschang et al. 2017).

Increasing the number of metal centers in the complexes has been employed as another successful strategy to improve the free radical scavenging activity of the free flavonoids. 1:2 complexes of quercetin with Cu(II) (Bukhari et al. 2009), Mg (II) (Ghosh et al. 2015), and Co(II) (Birjees Bukhari et al. 2008) have also been reported to possess higher free radical scavenging activity compared to the free flavonoid. In addition, one study investigated the protective effects of a 1: 2 quercetin–Ge(IV) complex against $OH^{\cdot}$ radicals on human erythrocytes. The complex was found to effectively reduce or eliminate the oxidative damage inflicted by ROS to erythrocytes, and subsequently to induce the partial recovery of topographic and nanostructural properties of cell membranes (Li et al. 2012).

Mixed-ligand complexes of Ni(II) and Cu(II) with quercetin (Q) or naringenin (nar) and heterocyclic imine (2,2′,6′,2″-terpyridine = terpy or 2,2′-bipyiridine = bpy) ligands have been found to exert high radical scavenger activities. The complexes have been assigned the general formulae [Ni(narH)(terpy)Cl], [Cu (narH)(terpy)Cl], and [Cu(QH1)(bpy)($NO_3$)], the last one being the most active antioxidant agent from this group (Alper et al. 2019). Complexes of Cu(II) with curcumin and quercetin and bpy as ancillary ligand were found to act as more potent antioxidant agents compared to the free flavonoids in the DPPH test. In accordance with the higher antioxidant potential of quercetin in comparison with curcumin, the scavenging activity of the quercetin complex was found to be more potent than that of the curcumin complex (Halevas et al. 2020).

Several studies have been concerned with the mechanisms involved in the antioxidant activity of metal complexes of quercetin and other flavonoids. A 2:1 quercetin–Cr(III) complex displayed higher antioxidant activity compared to quercetin. The mechanism of action regarding the antioxidant activity of quercetin mainly involves hydrogen atom transfer in an oxidative process. Since Cr(III) chelation induces a decrease of the ionization potential, the complex is thought to exert antioxidant activity through a combined mechanism, involving both hydrogen atom transfer and electron donation (Chen et al. 2009).

For 1:2 complexes of quercetin and rutin with Cu(II) and 1:2 complexes of quercetin and rutin with Zn(II), the antioxidant capacity was determined using the ABTS method. Notably, rutin (quercetin-3-rutinoside or sophorin) is a flavonoid glycoside, in which the glycosidic bond is formed between quercetin and the disaccharide rutinoside. The rutin–Zn complex was the most active compound; the activities of the rutin–Cu(II) and quercetin–Cu(II) complexes are only marginally higher in comparison with free quercetin, while quercetin–Zn displayed ten times lower antioxidant capacity than the ligand (Bratu et al. 2014). Further studies have been employed for a 1:2 rutin–Zn(II) complex, which has proven superior superoxide and free radical scavenging activities in comparison to the free ligand (Ikeda et al. 2015). In addition, 1:2 rutin–Cr(III) (Panhwar and Memon 2014), 2:1 rutin–Ni (II) (Raza et al. 2017), and 2:1 rutin–VO(IV) (Roy et al. 2015a) complexes have shown slightly improved antioxidant activity compared to free rutin; the vanadyl complex has been reported to be essentially nontoxic at 20 ppm in BALB/c mice. A

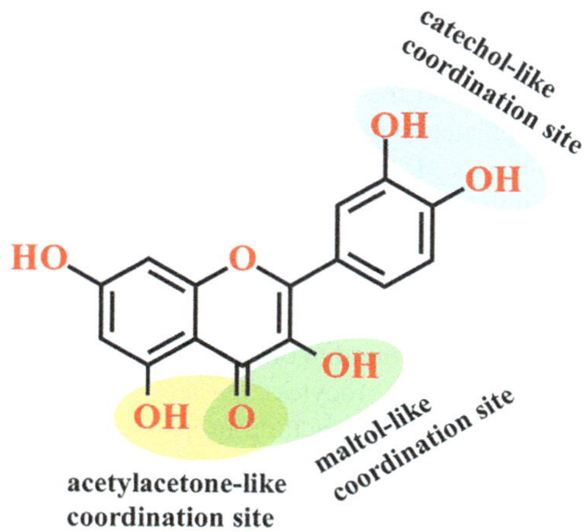


**Fig. 9.7** Possible coordination modes of flavonoids; the structure of quercetin is given as an example

common assay used for these complexes was the DPPH test. The mechanism of the reaction with DPPH˙ radicals can explain the higher activities displayed by the metal chelates. Both rutin and its complexes react with DPPH˙ by means of H transfer from one of the hydroxyl groups in the B ring, via homolytic dissociation of the O–H bond. This reaction results in the formation of an intermediate semiquinone radical, which can be stabilized via electron delocalization over the conjugated system of either free rutin or its complex. In the free flavonoid, the intramolecular hydrogen bond between the vicinal hydroxyl groups in ring B hinders the detachment of H by DPPH. In contrast, this hydrogen bond is lost upon coordination (at the catechol site) to the metal center, which renders the complex to display a more potent antioxidant activity (Panhwar and Memon 2014). Moreover, this phenomenon is a result of the electron withdrawing effect of the positively charged metal ion, which facilitates H abstraction by the DPPH˙ radical (Altun and Şuözer 2017).

In contrast, Sb(III) chelation in a 2:1 quercetin–Sb(III) complex (Tong et al. 2016) and Tb(III) chelation in an incompletely characterized quercetin–Tb(III) complex proved unfavorable in terms of antioxidant activity in comparison to the free ligand (Ezzati Nazhad Dolatabadi et al. 2014). Similarly, a 1:1 quercetin–Cd (II) (Ravichandran et al. 2014) and a Sn(II) complex (Dehghan and Khoshkam 2012) displayed lower antioxidant activities than the parent flavonoid; the lower antioxidant activity has been explained by the fact that chelation of these metal ions increases the oxidation potential of the complexes compared to the free flavonoid. Similarly, complexes of morin (at the 3,4 site) (Panhwar and Memon 2012) and rutin (at the catechol site) (Panhwar and Memon 2013) with Sn(II) resulted in a decrease of the antioxidant activity. Noteworthy, the radical scavenging activity of kaempferol has also been shown to decrease dramatically upon coordination to Sn (II) at the 3,4-site, correlated to an increase in the oxidation potential of the complex (Yang et al. 2020).

Structurally similar to quercetin, fisetin (Fig. 9.7) lacks the 5-OH group and prefers a "maltol-like" coordination. A 1:1 fisetin complex with Cu(II) acted as a more efficient photoprotector of tryptophan against photogenerated ROS compared to the free flavonoid. One of the photo-oxidizable amino acid residues, tryptophan, has been used as a model for the analysis of possible protection effects in biological environments. Moreover, the complex displayed effective quenching of photodynamically generated molecular singlet oxygen ($O_2(^1\Delta_g)$). The in vitro interaction of the complex, however, with $O_2(^1\Delta_g)$ is a partially chemical (irreversible) process, while fisetin interacts with singlet oxygen species via a physical (reversible) process, a desirable feature for a ROS scavenger (Muñoz et al. 2020). On the contrary, chrysin has been shown to act as a "sacrifice" scavenger, interacting with $O_2(^1\Delta_g)$ via a chemical (irreversible) process, while for its Cu(II) complex, only a small chemical contribution has been reported. Thus, for chrysin, the $O_2(^1\Delta_g)$ scavenging ability was greatly improved with complexation (Muñoz et al. 2016).

Flavonols lacking free ortho-catechol structure in the B ring (e.g., morin, galangin, kaempferol) form relatively unstable radicals, which results in low superoxide and $OH^{\cdot}$ scavenging activity. Interestingly, metal coordination in the 2:1 morin–oxovanadium(IV) complex (at the 3,4-site) improved the antiradical activity of morin with respect to $OH^{\cdot}$ and $O_2^{\cdot-}$ radicals, yet the free flavonoid was more effective as a peroxyl ($ROO^{\cdot}$) scavenger (Naso et al. 2013). The 1:1 kaempferol complex with Zn(II) efficiently scavenged β-carotene radical cation (short-lived, but strong oxidizing agent) and displayed 16 times higher radical scavenging activity against $DPPH^{\cdot}$ radicals compared to the free ligand. Notably, β-carotene radical cation scavenging mechanism involves electron transfer; in contrast, for $DPPH^{\cdot}$, hydrogen atom transfer is prevalent. Therefore, both mechanisms are presumed to be involved in the radical scavenging activity of the 1:1 kaempferol–Zn(II) complex, depending on the reaction partner (Xu et al. 2018). 2:1 complexes of taxifolin or dihydroquercetin (flavanonol) with Zn(II), Cu(II), and Ca(II) have been shown to alter the lipid peroxidation level in blood plasma and catalase activity, the Zn (II) complex being the most potent antioxidant of the series (Stolpovskaya et al. 2017).

Flavones, lacking the 3-OH group on the C ring, most commonly bind metal ions at the 4-C=O, 5-OH site ("acetylacetone-like" coordination). A 1:1 Apigenin (coordinated at the 4-C=O, 5-OH site) complex with VO(IV) exerted only moderate antioxidant activity, in terms of superoxide dismutase (SOD)-mimetic and $ROO^{\cdot}$, $DPPH^{\cdot}$, and $OH^{\cdot}$ radical scavenging activities, yet higher than the free ligand. Both apigenin and its VO(IV) complex have been reported to cause a significant dose-dependent increase in ROS accumulation in mitochondria. Although not yet completely elucidated, it was suggested that the mechanism of cancer cell killing is, at least in part, oxidative stress-related, more effective in the HeLa cell line (Martínez Medina et al. 2017).

Moreover, the free radical scavenging activity of luteolin was found to increase significantly upon coordination (2:1 molar ratio) to VO(IV) (Roy et al. 2015b) and Mn(II) (Dong et al. 2017). The $DPPH^{\cdot}$ radical scavenging activity was tested for both complexes, and the results favor the VO(IV) complex, albeit the difference in efficacy is rather small. In contrast, a 1:1 luteolin–VO(IV) complex, in which coordination occurred via the two *cis*-deprotonated hydroxyl groups in the B ring, "catechol-like" coordination (Naso et al. 2016d), appears to be a far less potent DPPH$^{\cdot}$ radical scavenger than the corresponding 2:1, "acetylacetone-like" coordinated complex. Therefore, the "acetylacetone-like" coordination in metal flavone complexes is more favorable with regard to free radical scavenging activity as compared to "catechol-like" coordination.

Another flavone, chrysin (5,7-dihydroxyflavone), and its 2:1 VO(IV) complex were not remarkably active against $O_2^{\cdot-}$ and DPPH$^{\cdot}$ radicals. However, they were efficient as scavengers for ABTS$^+$ and OH$^{\cdot}$ radicals, with chrysin–VO(IV) being more active than the free flavone (Naso et al. 2010). A 2:1 chrysin–Cu(II) complex was found to practically be inactive against DPPH$^{\cdot}$ radicals. The involvement of the 5-OH group in coordination, after deprotonation, proved to disfavor free radical scavenging efficacy, since the phenolate form of chrysin is more likely to take part in electron transfer reactions and, hence, improve the antioxidant activity (Selvaraj et al. 2011). On the other hand, when chrysin is bound to the metal ion in a monodentate fashion, via 7-OH, the activity of the resulting complex is improved to a great extent. A 1:1 chrysin (monodentate, 7-OH)–organogermanium (IV) complex displayed stronger OH$^{\cdot}$ scavenging effects than free chrysin and inhibited ROS-dependent oxidative damage in normal rat hepatocyte BRL cells. Pretreatment with the organogermanium complex in the presence of $H_2O_2$ was found to alleviate oxidative damage of the cell membrane and mitigate the recovery of protein function (Jiang et al. 2013).

Mixed ligand complexes of chrysin and Cu(II), where the ancillary ligand is either 2,2′-bipyridine, 1,10-phenanthroline, 4,7-dimethyl-1,10-phenanthroline, or bathophenanthroline, have also been studied. According to results provided by the ABTS method, bathophenanthroline was the most active, and all complexes displayed stronger antioxidant activity than the parent flavonoid (Mutlu Gençkal 2020).

A trinuclear oxo-centered Cr(III) complex of primuletin (5-hydroxyflavone) has been found to scavenge DPPH$^{\cdot}$ radicals more potently than free primuletin. 5-Hydroxyflavone possesses the most simple structure for a flavone, with a single –OH group, which binds Cr(III) in a deprotonated manner. Therefore, the superior antiradical activity of the complex can only be explained by the fact that Cr(III) is a more efficient electron donor than the H atom. The only possible mechanism underlying the free scavenging effects of the primuletin–Cr(III) complex is sequential proton loss electron transfer (SPLET; Alexiou et al. 2015). In vivo experiments have shown that the Fe(III)–primuletin complex can display antioxidant activity by decreasing ROS accumulation and enhancing SOD and catalase activities (Jabeen et al. 2017a).

The complexes of Ce(IV) with naringenin (flavanone), chrysin (flavone), and daidzein (isoflavone) were assayed using a DPPH radical scavenging method in different aqueous solutions of DMSO (50–80% v/v) at physiological pH. The radical scavenging activity of the complexes was found to be correlated to several parameters including the polarity parameter of the media and the hydrogen bond donating acidity parameter. Notably, the isoflavone–Ce(IV) complex was found to be the most potent antioxidant in this assay (Jabbari and Gharib 2012).

The mixed-ligand complex of Mg(II) with hesperidin (hesperetin-7-rutinoside), coordinated at the 4-C=O, 5-OH site, and 1,10′-phenanthroline (1:2:1), was found to display a much improved scavenging activity against superoxide radical than free hesperidin and vitamin C (more than tenfold higher). Notably, a lower oxidation potential of coordinated hesperidin (+0.58 V vs. $Ag/Ag^{+}$) compared to that of the free flavonoid (+0.66 V vs. $Ag/Ag^{+}$) has been correlated to the higher activity displayed by the complex (Oliveira et al. 2013b).

Genistein (isoflavone) complexation with Cu(II), subsequently reduced to Cu(I), in methanol/chloroform solutions, in a 1:1 molar ratio, increases the scavenging activity of the free flavonoid against the $DPPH^{\cdot}$ radical by more than a factor of 4. However, only a minor increase in the activity has been reported for β-carotene radical cation upon metal binding; taking into account that the difference in the oxidation potentials for genistein and the genistein–copper complex was measured to be very small, −0.016 V, electron transfer is presumably not an important component of the antioxidant effects. The increased activity of the complex is, therefore, most likely correlated to kinetic factors, resulting from the extension of the conjugated bonding system of genistein (Yang et al. 2017).

Upon coordination, the oxidation potential of the flavonoid/semiquinone redox pair decreases. In consequence, the flavonoid–metal complexes are more readily oxidized by the superoxide ions than the corresponding free flavonoids (De Souza et al. 2003). Moreover, the additional radical scavenging metal center can provide superoxide dismutase-mimetic properties (Uivarosi and Munteanu 2017). Pulse radiolysis studies have shown that the metal complexes of natural phenolic compounds with Mn(II), Fe(II), or Cu(II) are able to undergo electron transfer reactions and exert SOD-mimetic activity (Mahal et al. 2005). This concept has been further illustrated for a 1:1 rutin–Cu(II) complex, which was proven efficient for the inhibition of superoxide production and microsomal lipid peroxidation (Afanas'Ev et al. 1998). Under in vitro and ex vivo conditions, it was found that the Cu(II) complex was a highly efficient free radical scavenger, more potent than the Fe(III) complex. In support of this, an argument has been made that the Cu(II)–rutin complex has acquired an additional superoxide-dismuting copper center. Interestingly, Fe(III)–rutin has been found to promote the spontaneous production of oxygen radicals by macrophages, possibly due to the reduction to the prooxidant Fe (II) species (Afanas'eva et al. 2001). Similar results have been reported for rutin, taxifolin, (−)-epicatechin, and luteolin complexes with Fe(II), Fe(III), and Cu(II). The superoxide scavenging activities of the metal complexes were higher than those of the corresponding free flavonoids, with Cu(II) complexes being the most potent antioxidants of the series (Kostyuk et al. 2004). Metal complexation improves the antioxidant activity to a greater extent when less active flavonoids are used as ligands (e.g., chrysin; Naso et al. 2010; Sulpizio et al. 2016).

When coordination of the metal ion takes place at the catechol site in the B ring, however, the resulted complex generally displays lower antioxidant activity (Porfírio et al. 2014; Yang et al. 2014), correlated to an increase in the oxidation potential of the complexes with lower antioxidant strength (Yang et al. 2020). For flavonoid glycosides, e.g., diosmin (diosmetin 7-O-rutinoside), coordination to VO(IV) ions via *cis*-deprotonated hydroxyl groups of the rutinose disaccharide does not have a remarkable influence on the ROO$^{\cdot}$, DPPH$^{\cdot}$, and OH$^{\cdot}$ radical scavenging activity of the free ligand. This coordination mode does not influence to any large extent the electron delocalization from the metal center to the flavonoid's aromatic system and therefore stability of the intermediary semiquinone radical (Naso et al. 2016a). Similarly, the 1:1 hesperidin–VO(IV) complex, formed via coordination at the *cis*-deprotonated OH groups of the rutinose disaccharide, improved the SOD-mimetic activity of the free ligand, but not the radical (DPPH$^{\cdot}$, ABTS) scavenging activity of the complex (Etcheverry et al. 2008).

*Curcumin* (Fig. 9.3). Several drawbacks (i.e., hydrolytic instability involving the diketo moiety in the intracellular milieu and the poor bioavailability) prompted towards the synthesis of numerous derivatives, including metal complexes, which have been shown to generally improve and/or broaden its therapeutic potential (Banerjee and Chakravarty 2015).

The antioxidant, with emphasis on the SOD-mimetic activity of Cu(II)–curcumin complexes, for instance, has been found to improve up to tenfold when metal complexes possess increased structural flexibility (Barik et al. 2007). A mononuclear 1:1 curcumin complex with Cu(II) with a flexible distorted orthorhombic geometry displayed highly efficient (at nM concentrations) SOD- mimicking activity (Barik et al. 2005). The SOD-mimetic activity of the 1:1 complex was superior to that of a more rigid, square planar 2:1 curcumin–Cu(II) complex (Barik et al. 2007) and to that of free curcumin. Notably, the complex was found to be fully regenerated after the reaction with $O_2^{\cdot-}$, which argues for its catalytic activity in the neutralization reaction with superoxide radicals. However, the rate constant characterizing the reaction of the 1:1 complex with free radicals (DPPH$^{\cdot}$) was ten times lower relative to the free ligand (Barik et al. 2005). Correlated to the antioxidant activity, the complex was found to be very effective in protecting irradiated lymphocytes against radiation-induced suppression of the enzymatic activities of SOD, catalase, and glutathione peroxidase. Compared to curcumin, the Cu(II) complex showed superior ROS scavenging activity and more effective protection against radiation-induced protein damage and lipid peroxidation in splenic lymphocytes (Kunwar et al. 2007).

Further efforts to increase the pharmacological profile of curcumin derivatives included the synthesis of a series of vanadyl complexes. The ternary curcumin–VO(IV)–(2,2′-bipyridine) complex has proved as highly efficient $O_2^{\cdot-}$ scavenger in yeast cells (Halevas et al. 2019). In contrast, curcumin–VO(IV)–acetylacetone only displayed slightly improved $O_2^{\cdot-}$ scavenging activity compared to free curcumin, yet it was found to be less active than the binary 2:1 curcumin–VO(IV) complex (Adam et al. 2017). In turn, 3:1 curcumin complexes with Ga(III) and In(III), however, proved to be more active antioxidants (as measured by Trolox equivalent antioxidant capacity assay) than binary 2:1 curcumin–VO(IV) complex (Mohammadi et al. 2005).

A 1:1 curcumin–Zn(II) complex has been administered per os in a rat model of acute alcoholism. The complex was found to prevent in a dose-dependent manner the ethanol-induced elevation of serum malondialdehyde and suppression of glutathione peroxidase and SOD activities (Yu et al. 2014). Similar activity has been reported for the same complex in mouse testis (Lu et al. 2015).

Coordination to Mn(II) in a 1:1 molar ratio, on the other hand, was shown to decrease the DPPH˙ radical scavenging activity of curcumin. This observation is theoretically justified by means of lower bond dissociation enthalpy and higher HOMO and LUMO energy values of the free ligand compared to the values calculated for the corresponding Mn(II) complex (Gorgannezhad et al. 2016).

Several *phenolic acids and tannins* have been used as ligands in the synthesis of antioxidant metal complexes. Binary complexes of gallic acid (Fig. 9.3) with Fe(III), Cr(III), and Cu(II) and the corresponding ternary complexes with the nonnatural amino acid L-norleucine as ancillary ligand display free radical scavenging activities. All Fe(III) and Cu(II) complexes showed good to excellent antioxidant activity, with the 3:1 gallic acid–Fe(III) complex being the most active compound of the series (Fazary et al. 2011). A Zn(II) ternary complex comprising gallic acid and quercetin as ligands was reported to act as a more efficient DPPH˙ scavenger than the free ligands, the Mn(II) corresponding complex and ascorbic acid (Yesufu et al. 2020). Other metal complexes with non-flavonoidic polyphenolic plant secondary metabolites with improved antioxidant activity include a 2:1 complex of the flavolignan silibinin with VO(IV) (Naso et al. 2011), a 1:1 chlorogenic acid–VO (IV) complex (Naso et al. 2014).

*Resveratrol* (Fig. 9.3) is a polyphenolic plant secondary metabolite for which the antioxidant activity is well-documented (Banez et al. 2020). The chemical structure of the natural compound is not optimal for metal binding, nor for the formation of stable coordination complexes. One study, however, reports on the free radical scavenging activity of an organogermanium sesquioxide 1:1 complex with resveratrol. The complex has proven to act as a more efficient DPPH˙ and OH˙ radicals scavenger than free resveratrol (Yao et al. 2012).

Only a small number of studies report on the antioxidant activity of *N-containing compounds*. For instance, mixed-ligand complexes of theophylline, nicotinamide, and thiocyanate with Co (II), Ni (II), and Cu (II) ions (2:2:2:1) displayed higher than the free ligands, although notwithstanding, DPPH˙ scavenging activities (Altun and Şuözer 2017). A 4:1 caffeine–selenium complex has also been proved active in vitro against DPPH˙, OH˙, and $O_2^{\cdot -}$ radicals (Altalhi et al. 2019).

## 9.4 Cytotoxic Activity for Cancer Therapy

Cancer is a leading cause of death worldwide, deeply impacting the lives of millions. Currently, cancer treatment involves surgery, radiotherapy, chemotherapy, and/or immunotherapy. Although some breakthroughs in biological targeted approaches have occurred (e.g., immunological therapy), an unmet clinical need remains for a large section of the patient population, meaning novel wide-spectrum chemotherapy agents are still required. The costs involved in treatment (experts postulate that the current escalation of costs is unsustainable) and drug discovery are immense. The ongoing task of searching for more effective therapeutic strategies needs to take into consideration more efficient and cost-effective drugs. Several classes of natural compounds and their metal complexes have been taken into consideration as novel anticancer agents due to their promiscuous affinity towards a plethora of cellular targets and pathways. Among these, cyclin-dependent kinases, DNA topoisomerases I and II and actin polymerization, androgen receptor signaling, tumor-suppressor protein p53, and nuclear factor-kappa B (NF-κB) pathways can be listed. Consequently, plant secondary metabolites and their metal complexes modulate a series of cancer-related processes, such as cellular proliferation and differentiation, apoptosis, necrosis, metastasis, angiogenesis, and reversal of multidrug resistance (Uivarosi et al. 2019a).

Among *flavonols*, quercetin, kaempferol, and morin have been used as ligands in metal complexes with anticancer activity. Correlated to the upregulation of p53, downregulation of Bcl2, and inhibition of the mTOR/Akt pathway, a 2:1 quercetin–Ru(II) complex has been found to induce apoptosis in colorectal adenocarcinoma HT-29 cells. In vivo, in rat models of colon cancer, the complex was shown to increase CAT, SOD, and glutathione levels and to possess antiangiogenic activity (Roy et al. 2018). Heteroleptic Ru(II)–quercetin complexes with 2,2′-bipyridine as co-ligand displayed strong antiproliferative activity towards breast adenocarcinoma (MCF-7 cell line; IC50 = 0.39 μM, 72 h). The corresponding complex bearing 1,10-phenanthroline as co-ligand was practically inactive, with IC50 > 100 μM (Zahirović et al. 2017). Quercetin-conjugated tetrakis (dimethylsulfoxide)-dichlororuthenium(II) displayed cytotoxic activity ($IC_{50}$ = 10 μM, 48 h) on non-small cell lung cancer cell line (A549), but not on normal cell lines. The mechanism underlying the reported cytotoxic effects of the complex involves intracellular ROS generation and DNA fragmentation (Lakshmi et al. 2019) (Fig. 9.8).

Quercetin exerts electrostatic interactions with DNA, while its bulkier complexes commonly bind DNA via major or minor groove interactions and/or intercalation (Uivarosi and Munteanu 2017). The cytotoxic effects of a 1:1 quercetin–Zn (II) complex against three tumor cell lines (HepG2, SMMC7721, and A549) have been correlated to its ability to act as a DNA intercalator. The Zn(II) complex has been shown to induce typical morphologic changes of apoptosis in HepG2 cells, namely, nuclear shrinkage, chromatin condensation, or DNA fragmentation (Tan et al. 2009a). In a dose-dependent manner, the complex was shown to act as an antimetastatic agent by inhibiting the motility and invasiveness of BFTC-905 (human bladder cancer) cells. The Zn(II) complex has been found to interact with key factors involved in the neoplastic progression of cancer cells, like phosphoprotein kinase B (p-Akt) and membrane type-1 matrix metalloproteinase (MT1-MMP) (Lee and Tuyet 2019). An increase of the DNA binding affinities has been observed for quercetin complexes with Cu(II) (Ni et al. 2007), Mn(II) (Jun et al. 2007), Tb(III), Eu(III) (Li et al. 2009), Fe(II) (Raza et al. 2016), $Cu^{II}$–$Sn_2^{IV}$–quercetin and $Zn^{II}$–$Sn_2^{IV}$–quercetin complexes (Tabassum et al. 2013), and valine–quercetin diorganotin(IV) (Parveen et al. 2016). Correlated with its affinity to bind GC-rich DNA sequences, a Ni(II)–quercetin complex displayed moderate antiproliferative activity against human hepatocellular liver carcinoma (HepG2), human hepatoma

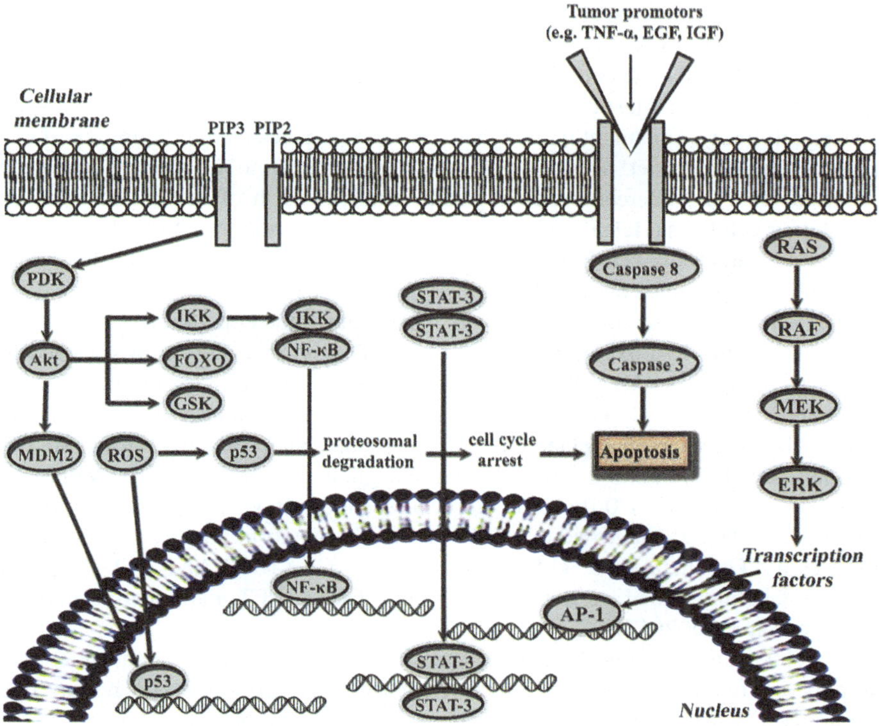


**Fig. 9.8** Schematic diagram of several molecular targets and downstream signaling pathways of plant secondary metabolites and their metal complexes. (Reproduced from Uivarosi et al. (2019b), with permission)

(SMMC7721), and human lung adenocarcinoma A549 cell lines. The 2:1 quercetin–Ni(II) complex was shown to induce apoptotic events in HepG2 cells, namely, downregulation of survivin (via intercalation into its GC-rich core promoter region) and Bcl2, as well as upregulation of p53 (Tan et al. 2010). Similar effects have been reported for the corresponding 2:1 quercetin complexes with Cu(II) (Tan et al. 2009b) and Mn(II) (Tan et al. 2011). A quercetin–lanthanum complex has been shown to possess antiproliferative effects on human cervical carcinoma cells. The antiproliferative activity was shown to be correlated with dose-dependent pro-oxidative effects and the ability of the complex to cleave DNA (Durgo et al. 2011). DNA interaction has also been demonstrated in vitro for 3:1 quercetin complexes with lanthanide(III) metal ions. Additionally, the complexes have been shown to exert $O_2^{\cdot-}$ radical scavenging activities and display superior antiproliferative activity relative to the free ligand (Zhou et al. 2001a). Slightly improved cytotoxic effects have been reported for 2:1 quercetin complexes with the divalent transition metal ions Cu(II) and Zn(II) (Zhou et al. 2001b).

Compared to the free flavonol, the 2:1 quercetin–VO(IV) complex does not improve the activity on UMR106 osteoblast proliferation. However, at smaller doses (up to 40 μM), the complex was found to stimulate type I collagen production and to promote osteoblast differentiation (on both MC3T3E1 normal and UMR106 cancer cell lines). Treatment with higher concentrations resulted in inhibition of cell proliferation (Ferrer et al. 2006). Similarly, weak cytotoxic activity has been reported for a 2:1 quercetin–Ge (IV) complex, after 72 h of incubation time, on four tumor cell lines (Hela (human uterine cervix cancer cell line), SPC-A-1 (human lung adenocarcinoma cell line), EC9706 (human esophageal cancer cell line), and PC-3 (human prostate cancer cell line)) (Zhai et al. 2012).

The structural resemblance between morin and quercetin (with only a small difference in the hydroxylation pattern on the B ring; *meta*-morin as opposed to *ortho*-quercetin) is striking. Several studies, however, report on the poor activity of heteroleptic Ru(II)–polypiridyl complexes with morin as an ancillary ligand on MDAMB-435S (human melanoma), FaDU (human pharynx carcinoma), MCF-7 (human ductal carcinoma), and U87 (human glioblastoma) cell lines (Munteanu et al. 2020) or on SW620 (human colorectal adenocarcinoma, metastatic), HepG2, MCF-7, and HeLa (Zahirović et al. 2017). However, morin has been successfully used as a $^{68}$Ga chelator for kidney cancer cells' (CAKI-1, CAKI-2, ACHN, and 786-O) labeling aimed at the development of new PET radiopharmaceuticals (Sentkowska et al. 2017).

A 1:1 kaempferol–zinc(II) complex exerted two times higher antiproliferative activity on human esophageal cancer (EC9706) cell line as compared to free kaempferol. The complex induced morphologic changes typical for apoptotic cells, correlated to increased levels of intracellular calcium ions (Tu et al. 2016). Mixed-ligand metal complexes of kaempferol and 1,10-phenanthroline/2,2′-bipyridine with Cu(II) and Zn(II) have been shown to possess higher antiproliferative activity as compared to free kaempferol on human breast carcinoma cells (MDA-MB-231). The anticancer activity is presumed to be linked to the ability of the complexes to act as strong DNA intercalators, as well as to induce DNA cleavage (Wang et al. 2014). A 1:1 kaempferol–Ru(II) complex displayed moderate, yet selective, cytotoxic effects on the non-small cell lung cancer cell line, A549, relative to normal cells. The complex was shown to induce membrane rupture and DNA damage in the cancerous cells (Thangavel et al. 2018). It should be stated that the kaempferol–Ru(II) complex displayed the highest, albeit still moderate, antiproliferative activity in the above-presented series of kaempferol–metal complexes.

Metal binding to either coordination site of the *flavone* luteolin (Fig. 9.4) leads to increased anticancer properties relative to the free ligand. A "catechol-like" coordination has been reported in $[VO(lut)(H_2O)_2]Na \cdot 3H_2O$. The luteolin–VO(IV) complex displayed stronger cytotoxic activity towards MDAMB231 breast cancer cells ($IC_{50}=17$ μM, 24 h), as compared to the activity on A549 lung cancer cells ($IC_{50}=60$ μM, 24 h). Treatment with VO(lut) generated cytoplasmic and nuclear membrane damages, as well as oxidative stress-related processes, which were found to cause cancer cells to undergo mitotic arrest (Naso et al. 2016d). In a follow-up study, the VO(IV) complex induced a decrease in cellular invasion, migration, and adhesion in the MDAMB231 human breast cancer cell line. In addition, VO(lut) displayed potent cytotoxic activity towards CT26 colon cancer cells ($IC_{50}$=0.9 μM, 24 h) and inhibited tumor growth and liver metastasis in a xenograft model of colon cancer (Naso et al. 2016c). A 2:1 luteolin–VO(IV) complex (in which an "acetylacetone-like" coordination mode is proposed) induced DNA fragmentation and apoptosis in human colon carcinoma HT-29 cells via activation of p53, caspase-3, and Akt expression. Furthermore, the complex has been shown to alter mitochondria functions and decrease the angiogenic factor VEGF levels in HT-29 cells (Roy et al. 2015b).

As compared to luteolin, apigenin (Fig. 9.4) lacks the 3′-OH group in the C ring; therefore, the "acetylacetone-like" bidentate coordination renders the most stable metal complexes. A 1:1 apigenin–VO(IV) complex exerted good anticancer activity on lung A549 ($IC_{50}$=2.2 μM, 72 h) and cervix HeLa ($IC_{50}$=9.7 μM, 72 h) cancer cell lines via intracellular ROS generation (Martínez Medina et al. 2017). Relative to the corresponding homoleptic 2:1 chrysin and genistein chelates with Cu(II), apigenin–Cu(II) was slightly less active against 518A2 melanoma, HCT-116 colon, KB-V1/Vbl cervix, and MCF-7/Topo breast carcinoma cells. All complexes induced G2/M cell cycle arrest on 518A2 melanoma cells used as a model of metastasis and reduced cell migration in wound healing assays (Spoerlein et al. 2013). In both of the abovementioned cases, metal coordination led to a drastic increase in the anticancer/antimetastatic activities.

Two 8-nitrogen saturated heterocycle substituted derivatives of chrysin (5,7-dihydroxyflavone), flavopiridol and P-276-00, have been shown to disrupt cell cycle progression and trigger apoptosis by strong inhibition of regulatory cyclin-dependent kinases (CDKs; Senderowicz 1999). Phase II clinical trials involving flavopiridol against acute myeloid leukemia are ongoing (Deep et al. 2018), and P-276-00 has been studied in phase II clinical trials for several types of cancer, including advanced refractory neoplasms and multiple myeloma (Jain et al. 2012). Therefore, chrysin is considered a lead structure for further development of anticancer drug candidates. In the search for such new molecules, among numerous other chrysin derivatives, several metal complexes have been studied. The cytotoxic effects of $[VO(chrysin)_2EtOH]_2$ have been studied on MG-63 osteosarcoma (Leon et al. 2013; León et al. 2016b) and HT-29 colon (León et al. 2015) human cell lines. Treatment with the complex in the MG-63 cell line induced apoptosis, disruption of the mitochondrial membrane potential, caspase-3 upregulation, and DNA fragmentation. Its mechanism of action involves ROS generation and alterations in the GSH/GSSG ratio (Leon et al. 2013) and activation/inactivation of key protein kinases, i.e., FAK, AKT1, ERK1, and ERK2 (León et al. 2016b). In a follow-up study, the complex has been tested in vitro on a 3D multicellular spheroid model and in vivo in a xenograft murine model of osteosarcoma. Treatment with $[VO(chrysin)_2EtOH]_2$ for 11 days resulted in significant reduction of the tumor volume and suppression of tumor growth (León et al. 2016a).

A chrysin–organogermanium(IV) complex triggered cell cycle arrest and induced apoptosis as a result of impaired mitochondrial function, in Colo205 colorectal cancer cells. In addition, the complex was shown to inhibit new vessel formation, which indicates potential antimetastatic activity (Yang et al. 2015). Another organometallic compound, a chrysin–organotin complex, triggered apoptosis and nuclear fragmentation in MCF-7 breast cancer cells. The anticancer activity of the complex was linked to caspase-3 activation and ROS generation (Xuan et al. 2016).

In agreement with a large body of studies in the literature (Notaro and Gasser 2017; Poynton et al. 2017), Ru(II) coordination has proven to improve the anticancer activity of free chrysin. A 2:1 chrysin–Ru(II) complex was found to suppress cellular growth and induce cell cycle arrest and apoptosis in MCF-7 breast cancer cells. In vivo studies revealed that the complex reduced hyperplastic lesions in the mammary tissues of rats used as breast cancer models, by triggering apoptosis and hampering with the angiogenic pathway (Roy et al. 2019). Further attempts to improve chrysin's activity involved the synthesis of heteroleptic complexes of Ru (II) bearing chrysin as an ancillary ligand. The antiproliferative activity of Ru (II) complexes with trithiacyclononane and chrysin proved unsatisfactory against PC-3 prostate, MG-63 osteosarcoma, and MCF-7 and MDA-MB-231 breast adenocarcinoma cells, with $IC_{50}$ values > 100 μM (Marques et al. 2019). Ru(II)–polypyridyl complexes with chrysin (bearing 2,2′-dipyridyl, 1,10-phenanthroline, or bathophenanthroline as co-ligands) have been tested on various cancer cell lines (breast, cervical, colorectal, hepatic, pharynx, glioblastoma, melanoma). $IC_{50}$ values (ranging from 15.5 to >50 μM) indicate moderate activity, yet much improved relative to the free ligand (Zahirović et al. 2017; Munteanu et al. 2020).

Primuletin (5-hydroxyflavone; Fig. 9.4) displays very weak anticancer activity (IC50 > 100 μM). However, its activity has been improved upon coordination with either lanthanide ions (Munteanu et al. 2016), Al(III), Ga(III), and In(III) (Munteanu et al. 2018), or upon binding to a Ru(II)–bis(bathophenanthroline) precursor (Munteanu et al. 2020). Interestingly, primuletin seems to offer an optimal scaffold for efficient interactions of the resulting complexes with macromolecules (DNA and transport proteins).

Up to date, few studies have been concerned with the anticancer activity of metal complexes bearing isoflavones as ligands. Coordination of genistein to Cu (II) proved favorable for the anticancer/antimetastatic activity. The 2:1 genistein–Cu(II) was more cytotoxic than the corresponding complexes of apigenin and chrysin against 518A2 melanoma, HCT-116 colon, KB-V1/Vbl cervix, and MCF-7/Topo breast carcinoma cells. Additionally, in 518A2 melanoma cells, the complex was shown to interfere with key factors that regulate cell–cell adhesion, namely, induced actin cytoskeleton remodeling and stimulated cadherin–catenin complex formation, while downregulating the expression of the matrix metalloproteinases MMP-2 and MMP-9. These findings indicate a potential antimetastatic effect of the complex (Spoerlein et al. 2013). A Ru–polypyridyl complex bearing genistein as co-ligand, $[Ru(DIP)_2(gen)](PF_6)$, was found to display cytotoxic activity comparable to that of cisplatin and doxorubicin. Notably, the genistein complex displayed stronger activity as compared to the corresponding primuletin, chrysin, and morin chelates. Interestingly, [Ru(DIP)2(gen)](PF6) showed strong activity towards MDA-MB-435S ($IC_{50}$ = 2.64 μM; 48 h) cells, which are commonly used for the study of metastasis; in contrast, genistein was not cytotoxic ($IC_{50}$ >100 μM). Additionally, the complex was found to be taken up, via a passive transportation mechanism, more efficiently by MDA-MB-435S cells in comparison to MCF-7 breast cancer cells. The effects of the complex on cell proliferation have been linked to its effects on cell metabolism, since [Ru(DIP)$_2$(gen)](PF$_6$) was shown to inhibit mitochondrial respiration and interfere with the cytosolic glycolytic pathway (Munteanu et al. 2020).

Several studies have been concerned with the design of new anticancer metal (mainly Cu(II)) complexes bearing naringenin and hesperetin (*flavanones*; Fig. 9.4) as ligands. Homoleptic 2:1 complexes of naringenin/hesperetin with Cu(II) have been tested in vitro against HepG2 hepatocellular, SGC-7901 gastric, and HeLa cervical carcinoma cell lines. The hesperetin complex was more active. Relative to the free ligand, the hesperetin complex displayed higher activity against all tested cell lines, yet naringenin–Cu(II) has only shown improved cytotoxic activity in HepG2 cells (Tan et al. 2009c). Moreover, a polynuclear complex of naringenin with oxovanadyl (IV) displayed cytotoxic activity in the MDAMB231 breast cancer cell line, correlated to ROS generation, cell membrane damage, DNA fragmentation, and decreased mitochondrial membrane potential. Additionally, the complex induced cell cycle arrest and caspase-3/7 activation (Islas et al. 2015).

Heteroleptic Cu(II) complexes with 1,10-phenanthroline and naringenin/hesperetin displayed cytotoxic effects against A549 lung carcinoma cells via a mitochondria-independent apoptotic mechanism. The naringenin complex was found to be less active than the corresponding hesperetin chelate ($IC_{50}$ = 16.42 μM vs. 5.82 μM; 24 h) (Tamayo et al. 2016). Another Cu(II) heteroleptic complex of naringenin, bearing 2,2′,6′,2″-terpyridine as co-ligand, displayed strong activity against MCF-7 breast adenocarcinoma ($IC_{50}$ <0.8 μM; 48 h). The complex was shown to induce apoptosis through a caspase-3/7-dependent pathway. Noteworthy, the corresponding Ni(II) complex was less active against the same cell line ($IC_{50}$ = 22.8 μM; 48 h; Alper et al. 2019).

The 5-OH and 4-C=O groups in the structure of rutin (*flavonoid glycoside*) have been reported to be involved in coordination with Cu(II) (Roy et al. 2016) and Zn(II) (Ikeda et al. 2015). The 2:1 rutin–Cu(II) complex exerted weak cell cytotoxicity against HeLa cells in vitro ($IC_{50}$ = 35 μM; 72 h vs. rutin: $IC_{50}$ > 150 μM; 72 h; Roy et al. 2016). A 1:2 binuclear rutin–Zn(II) complex (in which coordination occurs at both the 4,5- and 3′,4′-sites) was shown to possess weak cytotoxic activity against several leukemia, multiple myeloma, and melanoma cell lines in vitro ($IC_{50}$ > 90 μM; 24 h). However, in vivo studies have shown rutin–Zn(II) to induce apoptosis and DNA fragmentation and to show synergistic antitumor activity when administered in combination with paclitaxel. Some undesirable side effects of chemotherapy, i.e., anemia and myelosuppression, have also been ameliorated as a result of the simultaneous rutin-Zn(II) and paclitaxel administration. Furthermore, the complex was well tolerated (practically noncytotoxic) against normal cells or when administered to BALB/c mice (Ikeda et al. 2015).

A 1:1 diosmin–VO(IV) complex (in which coordination occurs via two *cis* -OH groups in the sugar moiety of diosmin) was shown to improve the cytotoxic activity of the free flavonoid against A549 lung and T47D, SKBR3, and MDAMB231 breast cancer cell lines. Albeit not yet elucidated, in the breast cancer cells, the mechanism underlying the anticancer activity does not appear to involve apoptotic cell death, ROS generation, or caspase-3/7 activation (Naso et al. 2016b).

Furthermore, VO(IV) complexes with the *flavonolignan* silibinin have been reported to display antiproliferative activity (Naso et al. 2011; Leon et al. 2014, 2015). Cancer cell death was correlated with the ability of the complex to generate intracellular ROS (Naso et al. 2011). The complexes induce caspase-3-dependent apoptosis and $G_2$/M cell cycle arrest, on MG-63 osteosarcoma (Leon et al. 2014) and HT-29 colon cancer cells (León et al. 2015).

The anticancer activity of *curcumin* (Fig. 9.3) and its metal complexes has been intensively studied. Cu(II), Zn(II), VO(IV), Pt(II), Pd(II), and Ru(II) chelation has been reported to improve the bioactivity of the free ligand and expand the range of potential therapeutic targets. An increase of the anticancer activity of curcumin has been reported upon VO(IV) (Thompson et al. 2004; Mohammadi et al. 2005), Ga (III), and In(III) (Mohammadi et al. 2005) coordination in L1210 mouse lymphoma cells. Moreover, a polyvinylpyrrolidone-based solid dispersion of Zn(II)–curcumin was shown to severely impair tumor growth in mouse models of hepatocellular carcinoma. The mechanism underlying the antitumor activity is related to the modulation of gut microbiome-mediated Zn homeostasis (Wu et al. 2019). Moreover, complexation of Zn(II) with curcumin has been shown to increase the water solubility of the free ligand and its cellular uptake and bioavailability (Xing et al. 2014). Small interfering RNA (siRNA) (Mahmoodi Chalbatani et al. 2019)-loaded pH-sensitive nanospheres of Zn(II)–curcumin inhibited the cellular proliferation in vitro in human bladder cancer cells and in vivo in xenograft murine models. Interestingly, the acidic pH in the cancer cells causes the acid-labile Zn(II)–O bond to break and therefore, leads to intracellular release of curcumin. Moreover, these nanospheres protect siRNA from enzymatic degradation and allow its efficient delivery inside the cancer cells (Xing et al. 2014).

However, metal coordination in three binuclear curcumin–M(II) (where M = Cu, Zn, Cd) hydroxo complexes has proved unfavorable in regard to the cytotoxic activity of the free ligand against a panel of cancer cell lines. The lower cytotoxicity of the metal complexes can be due to pre-exhaustion of curcumin redox potential as a result of reduction of $NO_3^-$ ions (present in the reaction milieu during synthesis) to $NO_2$ and $O_2^{\cdot-}$ (Khalil et al. 2014). Half-sandwich organometallic rhodium and ruthenium complexes of curcumin have been tested in vitro against a multidrug-resistant human colonic adenocarcinoma cell line. The complexes displayed similar or lower anticancer activity compared to that of curcumin, correlated with their poor stability upon interactions with serum proteins and cell culture medium components (Mészáros et al. 2019).

A ternary Pd(II)–curcumin–bipyridine complex displayed moderate cytotoxicity against MCF-7, HeLa, and A549 cells. However, mechanistic studies revealed that the complex triggered cell cycle arrest in the S phase, disrupted mitochondrial membrane potential, and induced apoptosis via a ROS-dependent pathway (Li et al. 2018). A Ru(II)–arene complex, (p-Cymene)Ru(curcuminato)chloro, was found to display moderate cytotoxic activity against HCT116 colorectal ($IC_{50}$ =

13.98 μM), MCF-7 breast (19.58 μM) and A2780 ovarian (23.38 μM), U-87 glioblastoma (29.36 μM), and A549 lung (62.33 μM) cancer cell lines (Caruso et al. 2012). A series of arene–Ru(II) complexes of curcumin was shown to inhibit proteasomes in cultured HCT116 colon cancer cells and consequently to induce apoptosis. [($\eta^6$-Benzene)Ru(curcuminato)Cl] was the most active compound of the series, as compared to the corresponding complexes, in which the arene moiety was *p-i*$PrC_6H_4Me$ or $C_6Me_6$ (Bonfili et al. 2012).

Numerous curcumin complexes (see below) have been reported to display significant photochemotherapeutic potential for applications in photodynamic therapy (PDT). A Pt(II) complex with curcumin, [PtCl(curc)(DMSO)], showed high cytotoxicity against MCF-7 breast and SW620 colorectal cancer cell lines and improved selectivity as compared to free curcumin and cisplatin. Upon irradiation with visible light (note that light promotes Pt–O bond breakage), the activity of both the free ligand and its Pt(II) complex increased. However, the photosensitizing effect was reported to be more pronounced for free curcumin as compared with its Pt (II) complex (Censi et al. 2019).

Furthermore, $[Pt(NH_3)_2(cur)](NO_3)$, where Hcur is curcumin, has been reported to display similar photocytotoxic activity to that of free curcumin. The complex was found to release curcumin inside HaCaT skin keratinocytes, BT474 and T47D breast epithelial carcinoma, and Hep3B hepatocarcinoma cells upon irradiation with visible light. As a result, photocytotoxic effects ($IC_{50}$ values of 12–18 μM), explained by a mechanism involving ROS-mediated apoptotic cell death, have been detected (Mitra et al. 2017). Upon inclusion in polymeric nanoparticles, the cellular uptake in and the anticancer activity against A549 non-small lung cancer cells of $[Pt(NH_3)_2(cur)]$ $(NO_3)$ was shown to improve significantly. In vivo, these nanoparticles have displayed enhanced antimetastatic activity and reduced side effects as compared to the free Pt(II) complex (Chen et al. 2019).

A heteroleptic Co(III) complex of curcumin bearing a tetradentate phenolate-based co-ligand was designed as to be activated by visible light irradiation and release curcumin inside HeLa cervical and MCF-7 breast cancer cells. Poor cellular uptake, however, has been correlated to the lack of activity for the Co(III)–curcumin complex ($IC_{50}$ >100 μM). Nonetheless, replacement of the natural compound with mitocurcumin (a synthetic dicationic bis-(triphenylphosphonium) derivative of curcumin) has resulted in great increase ($IC_{50}$ = 3.9 μM) of the photocytotoxic activity (Garai et al. 2016).

Improved visible light-induced cytotoxicity in cancer cells has also been reported for a dinuclear Fe(III)–curcumin heteroleptic complex bearing acetylacetonate (acac) as an ancillary ligand. The complex displayed phototoxic activity at low-micromolar $IC_{50}$ values in HeLa ($IC_{50}$ = 3.1 μM) and MCF-7 ($IC_{50}$ = 4.9 μM) cancer cells. The complex triggers cell apoptosis via light-induced ROS generation (Sarkar et al. 2016).

The therapeutic potential of oxovanadium(IV) complexes of curcumin has also been exploited in the search for new photosensitizing drugs for PDT. VO (IV) complexes bearing curcumin and bis-(2-pyridylmethyl)amine derivatives, for instance, displayed photoactivated plasmid DNA cleavage and anticancer activity (albeit similar to that of free curcumin) in HeLa cells (Balaji et al. 2013).

Porphyrin-based organic dyes currently approved for PDT possess a number of attributes, namely, mitochondrial localization (to avoid mutations in the nuclear DNA), red light activation (for higher tissue penetration), and the ability to induce apoptotic cell death (Prasad et al. 2014). Therefore, the design of new PDT agents has tried to retain and improve these features while adding new. A ternary VO (IV) complex of curcumin and a dipicolylamine (dpa)-based N,N,N-donor ancillary ligand with pendant di-iodinated boron-dipyrromethene (BODIPY) moiety has been designed for mitochondria-targeted photocytotoxic activity. The complex displayed low-micromolar photocytotoxicity ($IC_{50}$ values of 2–6 μM) in HeLa and MCF-7 cancer cells, via a mechanism involving oxidative stress-dependent apoptotic cell death (Bhattacharyya et al. 2017). Mitochondrial localization and intracellular ROS generation have also been reported for VO(IV)–curcumin complexes bearing either a naphthalimide (**1**) (Prasad et al. 2014) or a (acridinyl)dipyridophenazine derivative (**2**) (Banerjee et al. 2014). The (acridinyl)dipyridophenazine complex has been shown to be more active against HaCaT keratinocyte cells ((**1**): $IC_{50}$ = 6.3 μM; (**2**): 0.18 μM) and in MCF-7 breast cancer cells ($IC_{50}$ = 5.4 μM).

However, for a VO(IV) complex of curcumin with 2-(2′-pyridyl)-1,10-phenanthroline, both cytosolic and nuclear localization have been observed. The complex displayed high selectivity (light vs. dark and cancer vs. normal cells) and photocytotoxicity against HeLa ($IC_{50}$ = 3.4 μM) and MCF-7 ($IC_{50}$ = 4.5 μM) cancer cells, by triggering apoptotic pathways as a result of light-assisted intracellular ROS generation (Banerjee et al. 2015). A similar mechanism of action has been reported for heteroleptic complexes of VO(IV) with curcumin and ferrocenyl-terpyridine derivatives. The visible light-induced cytotoxicity in HeLa and Hep G2 cancer cells (while the complexes remained essentially nontoxic in dark) was correlated to a high cellular uptake due to the lipophilic character of the ferrocenyl moiety (Balaji et al. 2014).

Further studies identified the potential of Cu(II) and Zn(II) curcumin complexes as potential photosensitizers. Cu(II) complexes of curcumin and N-ferrocenylmethyl-appended-L-amino acids, for instance, displayed high photocytotoxicity in HeLa ($IC_{50}$ = 2.9–4.2 μM) and MCF-7 ($IC_{50}$ = 12.2–15.7 μM) cancer cells, accompanied by low dark toxicity. As for other curcumin-based metal complexes designed for PDT applications, the complexes were found to trigger apoptosis and to prefer cytosolic localization (Goswami et al. 2013). In addition, Cu(II) and Zn(II) complexes bearing curcumin and acridinyl-dipyridophenazine derivatives displayed high cytotoxic activity upon irradiation with visible light ($IC_{50}$ = 0.3–4.5 μM) in HeLa, MCF-7 and HepG2 cancer cells. Further mechanistic studies revealed that cellular apoptosis is induced as a result of intracellular photo-generation of singlet oxygen and hydroxyl radicals (Mukherjee et al. 2019).

*Non-flavonoidic* plant secondary metabolites have also been used in the design of new anticancer agents. Zn(II) chelation has been shown to increase the antiangiogenic effects of ellagic acid, by exerting inhibitory effects on MMP-2 activity, matrix-induced tube formation, and cell migration on HUVEC umbilical vein endothelial cells. The abovementioned effects make good arguments for a potential antimetastatic activity of the complex (Huang et al. 2011). Binary and ternary Fe(III), Cr(III), and Cu(II) complexes of gallic acid and the nonnatural amino acid L-norleucine were evaluated against Hep-2 laryngeal, Daoy medulloblastoma, MCF-7 breast, and WiDr colon cancer cell lines. The Cu(II) complexes have been shown to be more cytotoxic than the moderately active corresponding Fe(III) and weakly active Cr(III) complexes, correlated with the antioxidant activity (Fazary et al. 2011).

Several complexes with gallic acid have been designed for theranostic applications. A coordinatively unsaturated Fe(III)–gallic acid complex has been developed in combination with upconversion luminescence nanoparticles for MRI monitorization and therapy of tumors. Fe(III) ions are released from the nanoformulations under acidic conditions in the tumor microenvironment. In vivo studies in mice bearing LS180 colorectal tumors highlighted a remarkable ability of these nanocarriers to target cancer cells and allow for MRI monitorization of tumors. Avid binding affinity towards unsaturated transferrin proteins in serum was shown to be a possible reason for the high targeting ability towards solid tumors (Zhang et al. 2019). In order to ensure selective delivery into the tumor cells and minimize side effects, a Fe(III)–gallic acid complex has been encapsulated into pH-sensitive nanocarriers. The Fe(III)–gallic acid complex strongly absorbs light within the near-infrared (NIR) range, which can penetrate into deep tissues, and subsequently convert the NIR light into heat for photothermal therapy (PTT; thermal ablation of cancer cells) applications. Metabolic breakdown of the nanoparticles occurs under neutral physiological conditions, but not in the weak acidic tumor microenvironments. Therefore, the pH-sensitive Fe(III)–gallic acid nanoparticles were shown to accumulate at the tumor site while being easily metabolized in other tissues. Further in vivo studies, in a mouse model of breast cancer, indicated the potential use of these Fe(III)–gallic acid-loaded nanoparticles for photoacoustic-imaging-guided PTT (Zeng et al. 2016).

As opposed to the classic anticancer agents which exert their activity by inducing apoptosis, tributyltin(IV)–ferulic acid has been shown to induce autophagic cell death in HCT116, HT-29, and Caco-2 colon cancer cells. Cell death was accompanied by G2/M cell cycle arrest, increased membrane permeabilization and emergence of autophagic vacuoles, and increase in markers of autophagy (Pellerito et al. 2020). Another synthetic derivative of a phenolic acid, a 1:1 VO(IV)–chlorogenic acid complex, displayed moderate selectivity (cancer vs. normal cells) and cytotoxicity against SKBR3 breast cancer cell line. The cytotoxic effects are assumed to occur via caspase-independent apoptotic cell death (Naso et al. 2014). Apoptotic cell death has also been reported for a 2:1 *o*-carvacrotinate–Cu (II) complex which also showed moderate cytotoxic activity against A2780 ($IC_{50}$ = 54 μM) ovarian carcinoma cells (Sutradhar et al. 2019).

Very few *alkaloid derivatives* of natural origin have been used as ligands in metal complexes tested for anticancer activity. Two Ca(II) complexes containing monodeprotonated caffeate ligands (Maciejewska et al. 2009). Zn(II) and Cd

(II) halide complexes with caffeine and antipyrine cause higher levels of DNA damage and necrotic death in MCF-7 breast cancer cells while inducing low levels of DNA damage and apoptotic events in DPSC mesenchymal stem cells. Cd (II) complexes are more active than the corresponding Zn(II) chelates (Rukk et al. 2019).

## 9.5 Antimicrobial Activity

### 9.5.1 Antibacterial and Antifungal Activity

Natural compounds (especially those produced by microorganisms) have led over the years to the discovery of highly effective drugs against pathogens (e.g., penicillins, glycopeptides, tetracyclines). Yet, bacterial resistance is currently expanding at an increasing and alarming pace. Therefore, there is an urgent need for the development of new, more efficient, therapeutical strategies. At physiological pH, the outer cell walls of all microbes are negatively charged (the major component of bacterial cell wall is represented by a class of negatively charged phospholipids—phosphatidylethanolamines) (Uivarosi et al. 2019a). Hence, new drugs (including metal complexes) are synthesized as to possess a positively charged component, designed as to target the oppositely charged biological structures, such as the cell walls of some microorganisms.

Gram-positive (G+) bacteria have a fairly porous and simple cell wall structure that generally allows exogenous molecules to pierce their outer layer. Gram-negative (G−) bacteria, however, have more complex cell wall structures, with an outer layer that is less permeable. Infections caused by Gram-negative bacteria are usually more difficult to treat, and further understanding regarding the mechanisms of envelope permeability and the drug influx/efflux systems is necessary.

Metal complexes of flavonoids display moderate to high activity on G+ bacteria, yet the activity is generally not retained for G− strains. A Mn(II) complex of the flavonol quercetin and a Cd(II) complex of the semi-oxidized form of the ligand displayed bactericidal effects against the Gram-positive *Bacillus cereus*. Notably, no activity has been reported for the G− bacteria *Escherichia coli*, *Klebsiella pneumoniae*, or *Pseudomonas aeruginosa* (Bravo and Anacona 2001). Modest activity has also been reported for heteroleptic Ru(II) complexes of quercetin/morin/chrysin against the abovementioned G− bacteria (Zahirović et al. 2017), for a quercetin–Fe(II) complex (Raza et al. 2016) and for Ni(II) and Zn(II) complexes of quercetin (Kalinowska et al. 2016) against *E. coli* strains. However, the complexes bearing 2,2′-bipyridine as ancillary ligand showed moderate activity towards the Gram-negative *Acinetobacter baumannii* (Zahirović et al. 2017).

Curcumin and its homoleptic Cr(III), Mn(II), Fe(III), Ni(II), Cu(II), and Zn (II) complexes have not displayed antimicrobial activity against Gram-positive, Gram-negative, or fungal strains. However, the Co(II)–curcumin complex was moderately active against *Bacillus subtilis* (G+), *Staphylococcus aureus* (G+), and *P. aeruginosa* (G−) (Refat 2013). A Ga(III)–curcumin showed no significant antibacterial activity against *S. aureus* and *E. coli* (Jahangoshaei et al. 2015). Binary and ternary VO(IV) complexes of acetylacetone and curcumin have good inhibitory activity against all tested bacterial strains (G− : *E. coli*, *P. aeruginosa*, G+ : *S. aureus*, *Bacillus megaterium*), with minimal inhibitory concentrations (MICs) in the range of 20–30 μM. The binary complex of curcumin displayed higher antimicrobial activity relative to the ternary compound and free curcumin (Adam et al. 2017).

A Cu(II)–*O*-carvacrotinate complex was found to be more active on yeasts (*Candida albicans* and *Saccharomyces cerevisiae*) rather than bacteria (G− : *E. coli*, *P. aeruginosa*, G+ : *S. aureus*, *Enterococcus faecalis*) (Sutradhar et al. 2019). Tannic acid has been shown to form in situ complexes with metal ions (Zn (II), Zr(IV), Ag(I)) and chitosan from jute fabrics. The resulted complexes show moderate activity on *S. aureus* bacteria and *C. albicans* (Higazy et al. 2010).

Ca(II), Mn(II), Cu(II), Zn(II), and Cd(II) complexes of ferulic acid have been tested against *E. coli*, *B. subtilis*, *P. aeruginosa*, *S. aureus*, *Proteus vulgaris*, and *C. albicans*. All complexes showed increased activity relative to the free ligand. Zn (II) and Cd(II) complexes displayed the highest activity against all tested microorganisms, with 82–99% growth inhibition. Metal coordination increases the lipophilicity of the complexes relative to ferulic acid and its sodium salt. Thus, the lipophilic character of the complexes allows them to interact more efficiently with the phospholipid bilayers of the cell membranes and increases their uptake into bacterial cells (Kalinowska et al. 2014).

A Zn(II) complex of *p*-coumaric acid displayed strong growth inhibitory effects against *B. subtilis*, *C. albicans*, *Proteus vulgaris*, and *S. aureus*. In contrast, the antibacterial activity against *P. aeruginosa* was weak (Kalinowska et al. 2013). The Zn(II) complex was also shown to be the most active compound in comparison with its Co(II), Ni(II), and Cu(II) complexes of *p*-coumaric acid when tested against *E. faecalis*, *C. albicans*, *E. coli*, *P. aeruginosa*, and *S. aureus* (Koc et al. 2016).

Free caffeine is essentially non-active against *E. coli*, *P. aeruginosa*, and *S. aureus*. However, its mixed-ligand Cd(II), Ni(II), Cu(II), and Zn(II) complexes bearing caffeine and an azo dye co-ligand displayed antibacterial activity; the Cd (II) and Zn(II) complexes have been shown to be the most active compounds of the series (Bouhdada et al. 2019a). Complexes of the general formula $[M(caffeine)_4](PF_6)_2$, where M = Fe(II), Co(II), Mn(II), Cd(II), Zn(II), Cu(II), Ni(II), have been evaluated against a panel of G+/− bacteria. The Fe(II), Co(II), Mn(II), Zn(II), and Cu(II) complexes show no growth inhibitory effects when tested against *E. coli*, *S. aureus*, *K. pneumoniae*, *K. oxytoca*, and *P. putida*. The Ni(II) complex showed moderate activity against *S. aureus*, while the Cd(II) complex was moderately active on all tested microorganisms except *E. coli* (Hamdani and Amane 2019).

Zn(II) coordination was shown to increase the activity of another alkaloid, nicotine, and broaden its spectrum of activity (Zaidi et al. 2012). Out of a series of Co(II), Ni(II), and Cu(II) mixed-ligand complexes bearing nicotinamide, theophylline, and thiocyanate, the Co(II) complex was the only active compound. The antimicrobial activity was tested against the G− bacteria *E. coli*, *Salmonella typhimurium*, *Listeria monocytogenes*, and the fungus *C. albicans* (Altun and Şuözer

2017). However, Sc(III) and Nb(V) binary complexes of theophylline displayed higher in vitro antibacterial activity relative to the free ligand (tested on *E. coli*, *P. aeruginosa*, *B. subtilis*, and *S. aureus*) and lower antifungal activity (tested on *Aspergillus flavus* and *C. albicans*) (El-Habeeb and Refat 2018). Ternary transition metal complexes (Cd(II), Ni(II), Mn(II), Fe(II), Cu(II), and Zn(II)) with theophylline and oxalate as co-ligands have been tested against a panel of five G+/− bacterial strains and three fungi. The Cd(II) complex was the only active compound of the series on all tested microorganisms. The Zn(II) complex displayed similar activity compared to the corresponding Cd(II) compound; however, it was not active against *E. coli* and *S. saprophyticus*. All of the other complexes are practically nontoxic towards the tested microorganisms (Bouhdada et al. 2019b). In a comparative study involving a series of heteroleptic arene–Sn(IV) complexes of theophylline and theobromine, theophylline was shown to act as a more suitable scaffold for the antimicrobial activity (Jain et al. 2013).

A detailed structure–activity relationship analysis would be problematic, given that studies have used different bacterial strains and working conditions/assays and results are given as either growth inhibition % or as the diameter of the inhibition zone. However, metal coordination increases the lipophilicity of the free ligands and hence, the antibacterial activity. Furthermore, Zn(II) and Cd(II) complexes generally display higher activity as compared to other transition metals.

The *antiviral activity* of metal complexes of plant secondary metabolites has been less studied. Zn(II) and Cu(II) complexes of taxifolin displayed in vitro and in vivo activity against the influenza virus A/Aichi/2/68 (H3N2). The Cu(II) complex was shown to possess direct virucidal activity against A/Duck/Potsdam (H5N2) viral strain (Trofimova et al. 2015). In addition, Fe(III) coordination to caffeic acid has been reported to increase by over 100-fold the activity of the free ligand. The complex was highly active in the early stages of infection against the herpes simplex viruses HSV1 and HSV2, and showed moderate activity against vaccinia virus and a VSV–Ebola pseudotyped virus (Langland et al. 2018).

## 9.6 Antidiabetic Activity

Type 2 diabetes mellitus is associated with insulin resistance and impaired insulin secretion, which cause rampant glucose production and decreased transport and use of glucose in peripheral tissues. The antidiabetic activity may come as a result of inhibition of various type 2 diabetes-related enzymes or processes, e.g., in vivo experiments for metal complexes of plant secondary metabolites have correlated their antidiabetic activity with ROS scavenging effects and antidyslipidemic activity (that is because insulin modulates the function of adipocytes and insulin resistance is associated with an aberrant behavior of the adipose tissue).

Vanadium salts have been shown to possess insulin-mimetic properties, having the ability to lower blood glucose, increase glucose oxidation, as well as modulate adipocyte function (Shechter and Shisheva 1993). In order to improve their activity and pharmacokinetic profiles, numerous coordination complexes have been synthesized and tested for antidiabetic activity. Bis(maltolato)oxovanadium (IV) complex (BMOV), for instance, has been identified as a promising drug candidate. On the basis of structural similarity between flavonoids and the maltol ligand in BMOV, the hypoglycemic activity of bis(quercetinato)oxovanadium(IV), BQOV, was tested in streptozotocin-induced diabetic Balb/c mice. The results of the study, namely, strong hypoglycemic effects and negligible nephrotoxicity, were promising for the potential use of metal chelates of flavonoids as insulin-mimetics (Shukla et al. 2006). Other VO(IV) complexes of free glycosylated flavonoids have been tested for hypoglycemic activities in alloxan-induced diabetic rats. As a result, kaempferol-3-neohesperidoside has been identified as a suitable ligand for enhancing the insulin-mimicking activity of free VO(IV) (Cazarolli et al. 2006).

Cu(II) and Fe(III) complexes of quercetin, morin, and primuletin have also been shown in vivo to lower blood glucose levels and to antagonize alloxan-induced loss of body mass (Jabeen et al. 2017b). Diabetes mellitus was found to be associated with Zn deficiency. In order to correct this diabetes-induced status, Zn supplementation through Zn(II) complexes has been assessed as a viable option for the treatment of diabetes. Additionally, (Al-Ali et al. 2016) Zn(II) complexes were shown to possess insulin-mimetic properties . Several Zn(II) complexes have also been studied as insulin-mimetics. A Zn(II)–morin complex, for instance, was found to display antidiabetic, antidyslipidemic, and antioxidant activities in high-fat diet-fed streptozotocin-induced type 2 diabetic rats (Sendrayaperumal et al. 2014). A Zn (II)–diosmin complex, studied in the same type of diabetic rat model, was shown to lower blood glucose levels and improve insulin sensitivity via enhanced protein metabolism and regulation of the muscle and liver glycogen levels (Gopalakrishnan and Subramanian 2016). Moreover, a curcumin–Zn(II) complex was proven as an efficient hypoglycemic agent and insulin-mimetic in streptozotocin-induced diabetic rats. Notably, Zn(II)–curcumin significantly reduced glycosylated hemoglobin and lipid profile parameters, and showed negligible nephrotoxicity (Al-Ali et al. 2016).

## 9.7 Anti-inflammatory Activity

Long-term therapy with nonsteroidal anti-inflammatory drugs (NSAIDs) puts the patients at risk of gastrointestinal and cardiovascular side effects. Hence, further development of new anti-inflammatory drugs, with safer profiles for the long-term use, is still necessary. Metal complexes of some secondary metabolites have been synthesized as to interact with pro-inflammatory protein (e.g., cytokines, chemokines) or chemical mediators (e.g., histamine, serotonin, eicosanoids), inhibit the expression of inflammation-related genes, or exhibit antioxidant and prooxidant effects.

The anti-inflammatory activity of rutin was shown to increase upon complexation with Fe(III) and Cu(II). The Cu(II)–rutin complex displayed higher anti-inflammatory activity relative to the Fe(III) complex and free rutin, possibly due to the addition of the copper center responsible for the SOD-mimetic activity of the complex. Since inflammatory processes have been correlated with ROS overproduction, the antioxidant activity of the complexes is probably linked to their anti-inflammatory effects (Li et al. 2011). All five binary complexes of La(III), Ho (III), Yb(III), Lu(III), and Y(III) of luteolin showed inhibitory effects on carrageenan-induced paw edema in mice, albeit weaker than those displayed by free luteolin or dexamethasone (Afanas'eva et al. 2001). A curcumin–Au(I) complex displayed antiarthritic properties, evaluated by measuring its ability to inhibit paw edema in mice (Sharma et al. 1987). Aryl–Hg(II) complexes with theophylline showed some anti-inflammatory activity. Although the presence of the theophylline moiety in the complexes enables the drug to interact more efficiently with the lipid bilayers in the cell membrane, the presence of both –Cl and –$NO_2$ groups in the co-ligand proved to have a more dramatic effect on activity (Marwaha et al. 1995). Furthermore, analgesic activity has been reported for a $Cu_2(fen)_4(caf)_2$ (where fen = fenoprofenate anion; caf = caffeine) complex in mice, which has been correlated to the ability of the complex to decrease the inflammation process (Gumilar et al. 2012).

## 9.8 Anti-Alzheimer Activity

Alzheimer's disease (AD) is a chronic, progressive neurodegenerative disorder, characterized by cognitive impairment, memory loss, dementia, and the behavioral disturbances associated with it (Shaik et al. 2016). AD is triggered by the excess of amyloid-β peptide (Aβ) aggregates in the brain. Aβ self-aggregates into oligomers, fibrils, and senile plaques, and thus formed Aβ aggregates can act as modulators of mitochondrial activity and proteasome function and activators of inflammatory processes.

Aβ is involved in the hyperphosphorylation of protein tau, which plays a role in axonal transport regulation. Tau hyperphosphorylation leads to the formation of neurofibrillary tangles and toxic soluble tau. In AD, the levels of acetylcholine (ACh) are decreased in several brain areas associated with learning, memory, emotional responses, and behavior. Current therapeutic strategies for AD involve the development of new inhibitors of Aβ aggregation, tau hyperphosphorylation, and enzyme (acetylcholinesterase (AChE), monoamineoxidase (MAO) A and B) inhibition.

Cu(II), Zn(II), Ni(II), Co(II), and Fe(II) complexes of quercetin (da Silva et al. 2020) and *cis*-[Ru(hesperidin)(1,10-phenanthroline)$_2$]($PF_6$) showed in vitro AChE inhibition (Oliveira et al. 2013a). Inhibition of tau aggregation has been reported for ternary curcumin-conjugated Ru(II) complexes bearing either bipyridine or phenanthroline (Liu et al. 2020a). More pronounced than for the free ligand, inhibitory effects on Aβ aggregation have been reported for Cu(II) and Zn(II)–curcumin complexes (Banerjee 2014). A neutral, lipophilic, heteroleptic Ga(III) Schiff base–curcumin complex has been designed as to traverse the blood–brain barrier and bind to Aβ plaques (Lange et al. 2016).

## 9.9 Other Activities

*Obesity* is a metabolic disease usually associated with hyperlipidemia, insulin resistance, type 2 diabetes mellitus, hypertension, cardiovascular diseases, etc. A Zn(II)–diosmin complex displayed hypolipidemic effect in high-fat diet-fed streptozotocin-induced type 2 diabetes rats. This activity was evidenced from the restoration in the levels of lipids, lipoprotein cholesterol, and fatty acids to roughly normal levels upon oral administration of the complex (Gopalakrishnan and Subramanian 2016). A rutin–Ca(II) complex presented antihyperlipidemic activity and protective effects against organ pathological changes in rats with experimental hyperlipidemia (Zhang et al. 2016).

A Zn(II)–curcumin complex exerted *gastro-protective* effects against several rodent models of gastric ulcer (Mei et al. 2009, 2011, 2012, 2013). The complex prevented pylorus ligation-induced lesions in rats by inhibiting NF-κB activation and the secretion of pro-inflammatory cytokines (Mei et al. 2009). Zn(II)–curcumin prevented formation of ulcer lesions induced by ethanol, significantly inhibiting the expression of TNF-α and interleukin 6 (Mei et al. 2012) and MMP-9 (Mei et al. 2013).

Depression and anxiety are chronic and disabling mental disorders, characterized by a decrease in an individual's ability to experience interest or pleasure (Uivarosi et al. 2019a). Most antidepressant drugs modulate the levels of the neurotransmitters serotonin [5-hydroxytryptamine (5-HT)], dopamine (DA), and noradrenaline (NA). In vivo experiments in rodent models of depression, show that of a Zn(II)–curcumin complex display *antidepressant* activity by indirectly releasing 5-HT (Mei et al. 2011).

Metal complexes of plant secondary metabolites have been used as biomaterials for bone tissue engineering. Mixed-ligand copper(II) complexes of quercetin have been shown to stimulate *osteogenesis* and *angiogenesis*, with the chelates bearing 1,10-phenanthroline as co-ligand exerting the highest activity (Vimalraj et al. 2018). Zn(II)-chelated morin (1:2) has been shown to promote osteoblast differentiation in MG63 human osteoblast-like cells and C3H10T1/2 mouse mesenchymal stem cells (MSCs; Vimalraj et al. 2019). Interestingly, Sr(II)–morin complexation has been achieved with morin incorporated into octadecylphosphonic acid Langmuir monolayers; the resulting films were deposited on Ti surfaces. The final coatings exhibited high surface free energy and increased polarity, which allowed for further serum protein adsorption and stimulation of osteoblast growth and differentiation (Cruz et al. 2019). Binary and ternary (when the co-ligand is 1,10-phenanthroline) Cu(II) complexes with silibinin and Cu–silibinin complexes were found to promote osteoblast differentiation by stimulating calcium deposition, alkaline phosphatase (ALP) activity, and the expression of osteoblastic markers (Rajalakshmi et al. 2018). Out of a series of mononuclear Eu(III), Gd(III), or Lu(III) complexes of curcumin, only the Eu(III) complex promoted osteoblast differentiation in MG63 cells (Mawani and Orvig 2014).

Furthermore, a quercetin–Gd(III) complex exerted superior contrast properties when used as positive contrast enhancer for magnetic resonance imaging, in comparison to a commercially available contrast agent (gadopentetate dimeglumine). The quercetin–Gd(III) complex was nontoxic towards both normal and cancer cells (Muthurajan et al. 2015). Furthermore, orally administered tannic acid labeled with $^{99m}$Tc has been proposed for stomach ulcer radioimaging (Ibrahim et al. 2013).

## 9.10 Conclusions

Small molecules of plant origin have been and continue to be important leads for drug development. Among the wide variety of synthetic derivatives of secondary metabolites of plant origin, metal complexation generally aims at improving and/or broadening the activity spectrum of the corresponding free compound and on improving its biopharmaceutical profile. With few exceptions, these attempts have proven successful, whether regarding the decrease in side effects or the increase in cellular uptake and bioavailability (see, for instance, the case of curcumin), activity, and/or selectivity as compared to the naturally occurring compounds. Awareness should be raised towards the mixed-ligand complexes, which have been shown, in numerous cases, to display much improved activity.

There are, without any doubt, an impressive number of studies with reference to the biological activity of the metal complexes of plant secondary metabolites available in the literature. Metal coordination generally results in enhanced in vitro/in vivo activity; however, up to date, there is no clinical evidence to support this. Further studies could be addressed to a better understanding of their mechanisms of action and their biotransformation in the animal body.

**Acknowledgement** This research was supported by the grant PED 383/2020 of Ministry of Research, the Executive Unit for Financing Higher Education, Research, Development and Innovation (UEFISCDI).

## References

Abdalla MA, Mühling KH (2019) Plant-derived sulfur containing natural products produced as a response to biotic and abiotic stresses: a review of their structural diversity and medicinal importance. J Appl Bot Food Qual 92:204–215. https://doi.org/10.5073/JABFQ.2019.092.029

Adam MSS, Youssef MM, Aboelghar MF et al (2017) Synthesis and characterization of binary and ternary oxovanadium complexes of N,N′-(2-pyridyl)thiourea and curcumin: Catalytic oxidation potential, antibacterial, antimicrobial, antioxidant and DNA interaction studies. Appl Organomet Chem 31:1–10. https://doi.org/10.1002/aoc.3650

Afanas'Ev IB, Ostrachovich EA, Korkina LG (1998) Effect of rutin and its copper complex on superoxide formation and lipid peroxidation in rat liver microsomes. FEBS Lett 425:256–258. https://doi.org/10.1016/S0014-5793(98)00244-0

Afanas'eva IB, Ostrakhovitch EA, Mikhal'chik EV et al (2001) Enhancement of antioxidant and anti-inflammatory activities of bioflavonoid rutin by complexation with transition metals. Biochem Pharmacol 61:677–684. https://doi.org/10.1016/S0006-2952(01)00526-3

Akinwumi BC, Bordun KAM, Anderson HD (2018) Biological activities of stilbenoids. Int J Mol Sci 19:1–25. https://doi.org/10.3390/ijms19030792

Al-Ali K, Fatah HSA, El-Badry YAM (2016) Dual effect of curcumin-zinc complex in controlling diabetes mellitus in experimentally induced diabetic rats. Biol Pharm Bull 39:1774–1780. https://doi.org/10.1248/bpb.b16-00137

Alexiou ADP, Decandio CC, Almeida SDN et al (2015) A trinuclear oxo-chromium(III) complex containing the natural flavonoid primuletin: synthesis, characterization, and antiradical properties. Molecules 20:6310–6318. https://doi.org/10.3390/molecules20046310

Alper P, Erkisa M, Genckal HM et al (2019) Synthesis, characterization, anticancer and antioxidant activity of new nickel(II) and copper(II) flavonoid complexes. J Mol Struct 1196:783–792. https://doi.org/10.1016/j.molstruc.2019.07.009

Altalhi T, Kobeasy MI, Gobouri AA et al (2019) Synthesis of selenium(0) and zinc(II) biomolecules in nano-structured forms as new antioxidant agents: chemical and biological studies. Russ J Gen Chem 89:800–805. https://doi.org/10.1134/S1070363219040261

Altun Ö, Şuözer M (2017) Synthesis, spectral analysis, stability constants, antioxidant and biological activities of Co (II), Ni (II) and Cu (II) mixed ligand complexes of nicotinamide, theophylline and thiocyanate. J Mol Struct 1149:307–314. https://doi.org/10.1016/j.molstruc.2017.07.069

Balaji B, Somyajit K, Banik B et al (2013) Photoactivated DNA cleavage and anticancer activity of oxovanadium(IV) complexes of curcumin. Inorganica Chim Acta 400:142–150. https://doi.org/10.1016/j.ica.2013.02.025

Balaji B, Balakrishnan B, Perumalla S et al (2014) Photoactivated cytotoxicity of ferrocenyl-terpyridine oxovanadium(IV) complexes of curcuminoids. Eur J Med Chem 85:458–467. https://doi.org/10.1016/j.ejmech.2014.07.098

Banerjee R (2014) Inhibitory effect of curcumin-Cu(II) and curcumin-Zn(II) complexes on amyloid-beta peptide fibrillation. Bioinorg Chem Appl 2014:325873., 8 pages. https://doi.org/10.1155/2014/325873

Banerjee S, Chakravarty AR (2015) Metal complexes of curcumin for cellular imaging, targeting, and photoinduced anticancer activity. Acc Chem Res 48:2075–2083. https://doi.org/10.1021/acs.accounts.5b00127

Banerjee S, Prasad P, Khan I et al (2014) Mitochondria targeting photocytotoxic oxidovanadium (IV) complexes of curcumin and (acridinyl)dipyridophenazine in visible light. Zeitschrift fur Anorg und Allg Chemie 640:1195–1204. https://doi.org/10.1002/zaac.201300569

Banerjee S, Dixit A, Karande AA, Chakravarty AR (2015) Remarkable selectivity and photo-cytotoxicity of an oxidovanadium(IV) complex of curcumin in visible light. Eur J Inorg Chem 2015:447–457. https://doi.org/10.1002/ejic.201402884

Banez MJ, Geluz MI, Chandra A et al (2020) A systemic review on the antioxidant and anti-inflammatory effects of resveratrol, curcumin, and dietary nitric oxide supplementation on human cardiovascular health. Nutr Res 78:11–26. https://doi.org/10.1016/j.nutres.2020.03.002

Barik A, Mishra B, Shen L et al (2005) Evaluation of a new copper(II)-curcumin complex as superoxide dismutase mimic and its free radical reactions. Free Radic Biol Med 39:811–822. https://doi.org/10.1016/j.freeradbiomed.2005.05.005

Barik A, Mishra B, Kunwar A et al (2007) Comparative study of copper(II)-curcumin complexes as superoxide dismutase mimics and free radical scavengers. Eur J Med Chem 42:431–439. https://doi.org/10.1016/j.ejmech.2006.11.012

Bennett RN, Wallsgrove RM (1994) Secondary metabolites in plant defence mechanisms. New Phytol 127:617–633. https://doi.org/10.1111/j.1469-8137.1994.tb02968.x

Bertleff-Zieschang N, Rahim MA, Ju Y et al (2017) Biofunctional metal-phenolic films from dietary flavonoids. Chem Commun 53:1068–1071. https://doi.org/10.1039/c6cc08607a

Bhattacharyya U, Kumar B, Garai A et al (2017) Curcumin "drug" stabilized in oxidovanadium (IV)-BODIPY conjugates for mitochondria-targeted photocytotoxicity. Inorg Chem 56:12457–12468. https://doi.org/10.1021/acs.inorgchem.7b01924

Birjees Bukhari S, Memon S, Mahroof Tahir M, Bhanger MI (2008) Synthesis, characterization and investigation of antioxidant activity of cobalt-quercetin complex. J Mol Struct 892:39–46. https://doi.org/10.1016/j.molstruc.2008.04.050

Bone K, Mills S (2013) 2 - Principles of herbal pharmacology. In: Principles and practice of phytotherapy, 2nd edn. Churchill Livingstone, Saint Louis, pp 17–82

Bonfili L, Pettinari R, Cuccioloni M et al (2012) Arene-RuII complexes of curcumin exert antitumor activity via proteasome inhibition and apoptosis induction. ChemMedChem 7:2010–2020. https://doi.org/10.1002/cmdc.201200341

Bouhdada M, EL Amane M, El Hamzaoui N (2019a) Synthesis, spectroscopic studies, X-ray powder diffraction data and antibacterial activity of mixed transition metal complexes with sulfonate azo dye, sulfamate and caffeine ligands. Inorg Chem Commun 101:32–39. https://doi.org/10.1016/j.inoche.2019.01.005

Bouhdada M, EL Amane M, Mohammed BB, Yamni K (2019b) Synthesis, spectroscopic studies, X-ray powder diffraction data and biological activity of mixed transition metal complexes with oxalato and theophylline ligands. J Mol Struct 1177:391–400. https://doi.org/10.1016/j.molstruc.2018.09.047

Brahmkshatriya PP, Brahmkshatriya PS (2013) Terpenes: chemistry, biological role, and therapeutic applications. In: Ramawat KG, Mérillon J-M (eds) Natural products: phytochemistry, botany and metabolism of alkaloids, phenolics and terpenes. Springer, Berlin, pp 2665–2691

Bratu MM, Birghila S, Miresan H et al (2014) Biological activities of Zn(II) and Cu(II) complexes with quercetin and rutin: antioxidant properties and UV-protection capacity. Rev Chim 65:544–549

Bravo A, Anacona JR (2001) Metal complexes of the flavonoid quercetin: antibacterial properties. Transit Met Chem 26:20–23

Bukhari SB, Memon S, Mahroof-Tahir M, Bhanger MI (2009) Synthesis, characterization and antioxidant activity copper-quercetin complex. Spectrochim Acta A Mol Biomol Spectrosc 71: 1901–1906. https://doi.org/10.1016/j.saa.2008.07.030

Caruso F, Rossi M, Benson A et al (2012) Ruthenium-arene complexes of curcumin: X-ray and density functional theory structure, synthesis, and spectroscopic characterization, in vitro antitumor activity, and DNA docking studies of (p-cymene)Ru(curcuminato)chloro. J Med Chem 55:1072–1081. https://doi.org/10.1021/jm200912j

Cazarolli LH, Zanatta L, Jorge AP et al (2006) Follow-up studies on glycosylated flavonoids and their complexes with vanadium: their anti-hyperglycemic potential role in diabetes. Chem Biol Interact 163:177–191. https://doi.org/10.1016/j.cbi.2006.07.010

Censi V, Caballero AB, Pérez-Hernández M et al (2019) DNA-binding and in vitro cytotoxic activity of platinum(II) complexes of curcumin and caffeine. J Inorg Biochem 198:110749. https://doi.org/10.1016/j.jinorgbio.2019.110749

Chen W, Sun S, Cao W et al (2009) Antioxidant property of quercetin-Cr(III) complex: the role of Cr(III) ion. J Mol Struct 918:194–197. https://doi.org/10.1016/j.molstruc.2008.08.008

Chen Y, Chen C, Zhang X et al (2019) Platinum complexes of curcumin delivered by dual-responsive polymeric nanoparticles improve chemotherapeutic efficacy based on the enhanced anti-metastasis activity and reduce side effects. Acta Pharm Sin B. https://doi.org/10.1016/j.apsb.2019.10.011

Constabel C, Yoshida K, Walker V (2014) Diverse ecological roles of plant tannins: plant defense and beyond. In: Recent advances in polyphenol research, p 464

Croteau R, Kutchan TM, Lewis NG (2000) Natural products (secondary metabolites). In: Buchanan B, Gruissem W, Jones R (eds) Biochemistry and molecular biology of plants. American Society of Plant Physiologists, Rockville, pp 131–143

Cruz MAE, Soares MPR, Pazin W et al (2019) Interface-driven Sr-morin complexation at Langmuir monolayers for bioactive coating design. Colloids Surf B Biointerf 181:856–863. https://doi.org/10.1016/j.colsurfb.2019.06.039

Dar OA, Malik MA, Shahid-ul-Islam et al (2017) Curcumin and its derivatives—isolation, synthesis, and applications. In: Plant-based natural products, pp 145–174

De Souza RFV, Sussuchi EM, De Giovani WF (2003) Synthesis, electrochemical, spectral, and antioxidant properties of complexes of flavonoids with metal ions. Synth React Inorg Met Chem 33:1125–1144. https://doi.org/10.1081/SIM-120023482

Deep A, Marwaha RK, Marwaha MG et al (2018) Flavopiridol as cyclin dependent kinase (CDK) inhibitor: a review. New J Chem 42:18500–18507. https://doi.org/10.1039/c8nj04306j

Dehghan G, Khoshkam Z (2012) Tin(II)–quercetin complex: synthesis, spectral characterisation and antioxidant activity. Food Chem 131:422–426. https://doi.org/10.1016/j.foodchem.2011.08.074

Dimitrić Marković JM, Marković ZS, Brdarić TP et al (2011) Iron complexes of dietary flavonoids: combined spectroscopic and mechanistic study of their free radical scavenging activity. Food Chem 129:1567–1577. https://doi.org/10.1016/j.foodchem.2011.06.008

Dong H, Yang X, He J et al (2017) Enhanced antioxidant activity, antibacterial activity and hypoglycemic effect of luteolin by complexation with manganese(II) and its inhibition kinetics on xanthine oxidase. RSC Adv 7:53385–53395. https://doi.org/10.1039/c7ra11036g

Durgo K, Ivana H, Ivana Š, Jasna F (2011) Cytotoxic and genotoxic effects of the quercetin/lanthanum complex on human cervical carcinoma cells in vitro. Arch Ind Hyg Toxicol 62:221

Efferth T, Koch E (2011) Complex interactions between phytochemicals. The multi-target therapeutic concept of phytotherapy. Curr Drug Targets 12:122–132. https://doi.org/10.2174/138945011793591626

El-Habeeb AA, Refat MS (2018) Three new complexes of theophylline drug with Sc(III), Nb(V), and W(VI) ions: spectroscopic, thermal stability, and antimicrobial studies. Russ J Gen Chem 88:2170–2176. https://doi.org/10.1134/S1070363218100237

Etcheverry SB, Ferrer EG, Naso L et al (2008) Antioxidant effects of the VO(IV) hesperidin complex and its role in cancer chemoprevention. J Biol Inorg Chem 13:435–447. https://doi.org/10.1007/s00775-007-0332-9

Ezzati Nazhad Dolatabadi J, Mokhtarzadeh A, Ghareghoran SM, Dehghan G (2014) Synthesis, characterization and antioxidant property of quercetin-Tb(III) complex. Adv Pharm Bull 4:101–104. https://doi.org/10.5681/apb.2014.016

Falcone Ferreyra ML, Rius SP, Casati P (2012) Flavonoids: biosynthesis, biological functions, and biotechnological applications. Front Plant Sci 3:222. https://doi.org/10.3389/fpls.2012.00222

Fazary AE, Hernowo E, Angkawijaya AE et al (2011) Complex formation between ferric(III), chromium(III), and cupric(II) metal ions and (O,N) and (O,O) donor ligands with biological relevance in aqueous solution. J Solution Chem 40:1965–1986. https://doi.org/10.1007/s10953-011-9768-1

Ferrer EG, Salinas MV, Correa MJ et al (2006) Synthesis, characterization, antitumoral and osteogenic activities of quercetin vanadyl(IV) complexes. J Biol Inorg Chem 11:791–801. https://doi.org/10.1007/s00775-006-0122-9

Garai A, Pant I, Banerjee S et al (2016) Photorelease and cellular delivery of mitocurcumin from its cytotoxic cobalt(III) complex in visible light. Inorg Chem 55:6027–6035. https://doi.org/10.1021/acs.inorgchem.6b00554

Ghosh N, Chakraborty T, Mallick S et al (2015) Synthesis, characterization and study of antioxidant activity of quercetin—magnesium complex. Spectrochim Acta Part A Mol Biomol Spectrosc 151:807–813. https://doi.org/10.1016/j.saa.2015.07.050

Gopalakrishnan V, Subramanian SP (2016) Biochemical evaluation of antidyslipidemic properties of a new zinc-diosmin complex studied in high fat diet fed-low dose streptozotocin induced experimental type 2 diabetes in rats. Der Pharm Lett 8:105–113

Gorgannezhad L, Dehghan G, Ebrahimipour SY et al (2016) Complex of manganese (II) with curcumin: spectroscopic characterization, DFT study, model-based analysis and antiradical activity. J Mol Struct 1109:139–145. https://doi.org/10.1016/j.molstruc.2015.12.051

Goswami TK, Gadadhar S, Gole B et al (2013) Photocytotoxicity of copper(II) complexes of curcumin and N-ferrocenylmethyl-L-amino acids. Eur J Med Chem 63:800–810. https://doi.org/10.1016/j.ejmech.2013.03.026

Graf E, Mahoney JR, Bryant RG, Eaton JW (1984) Iron-catalyzed hydroxyl radical formation. J Biol Chem 259:3620–3624

Gumilar F, Agotegaray M, Bras C et al (2012) Anti-nociceptive activity and toxicity evaluation of Cu(II)-fenoprofenate complexes in mice. Eur J Pharmacol 675:32–39. https://doi.org/10.1016/j.ejphar.2011.11.049

Gurib-Fakim A (2006) Medicinal plants: traditions of yesterday and drugs of tomorrow. Mol Aspects Med 27, 1:–93. https://doi.org/10.1016/j.mam.2005.07.008

Halevas E, Papadopoulos TA, Swanson CH et al (2019) In-depth synthetic, physicochemical and in vitro biological investigation of a new ternary V(IV) antioxidant material based on curcumin. J Inorg Biochem 191:94–111. https://doi.org/10.1016/j.jinorgbio.2018.10.010

Halevas E, Pekou A, Papi R et al (2020) Synthesis, physicochemical characterization and in vitro biological evaluation of two novel Cu(II) complexes based on natural products curcumin and quercetin. J Inorg Biochem 208:111083. https://doi.org/10.1016/j.jinorgbio.2020.111083

Hamdani HEL, Amane MEL (2019) Preparation, spectral, antimicrobial properties and anticancer molecular docking studies of new metal complexes [M(caffeine)4](PF6)2; M = Fe(II), Co(II), Mn(II), Cd(II), Zn(II), Cu(II), Ni(II). J Mol Struct 1184:262–270. https://doi.org/10.1016/j.molstruc.2019.02.049

Higazy A, Hashem M, ElShafei A et al (2010) Development of anti-microbial jute fabrics via in situ formation of cellulose-tannic acid-metal ion complex. Carbohydr Polym 79:890–897. https://doi.org/10.1016/j.carbpol.2009.10.019

Huang ST, Yang RC, Wu HT et al (2011) Zinc-chelation contributes to the anti-angiogenic effect of ellagic acid on inhibiting MMP-2 activity, cell migration and tube formation. PLoS One 6: e18986. https://doi.org/10.1371/journal.pone.0018986

Hussain MB, Hassan S, Waheed M et al (2019) Bioavailability and metabolic pathway of phenolic compounds. In: Soto-Hernández M, García-Mateos R, Palma-Tenango M (eds) Plant physiological aspects of phenolic compounds. InTechOpen, Rijeka. Ch. 5

Ibrahim IT, El-Tawoosy M, Talaat HM (2013) Labeling of tannic acid with technetium-99m for diagnosis of stomach ulcer. Radiochemistry 55:423–427. https://doi.org/10.1134/S1066362213040140

Ikeda NEA, Novak EM, Maria DA et al (2015) Synthesis, characterization and biological evaluation of Rutin-zinc(II) flavonoid -metal complex. Chem Biol Interact 239:184–191. https://doi.org/10.1016/j.cbi.2015.06.011

Islas MS, Naso LG, Lezama L et al (2015) Insights into the mechanisms underlying the antitumor activity of an oxidovanadium(IV) compound with the antioxidant naringenin. Albumin binding studies. J Inorg Biochem 149:12–24. https://doi.org/10.1016/j.jinorgbio.2015.04.011

Jabbari M, Gharib F (2012) Solvent dependence on antioxidant activity of some water-insoluble flavonoids and their cerium(IV) complexes. J Mol Liq 168:36–41. https://doi.org/10.1016/j.molliq.2012.02.001

Jabeen E, Janjua NK, Ahmed S et al (2017a) Radical scavenging propensity of Cu2 +, Fe3 + complexes of flavonoids and in-vivo radical scavenging by Fe3 +-primuletin. Spectrochim Acta Part A Mol Biomol Spectrosc 171:432–438. https://doi.org/10.1016/j.saa.2016.08.035

Jabeen E, Janjua NK, Ahmed S et al (2017b) DFT predictions, synthesis, stoichiometric structures and anti-diabetic activity of Cu (II) and Fe (III) complexes of quercetin, morin, and primuletin. J Mol Struct 1150:459–468. https://doi.org/10.1016/j.molstruc.2017.09.003

Jahangoshaei P, Hassani L, Mohammadi F et al (2015) Investigating the effect of gallium curcumin and gallium diacetylcurcumin complexes on the structure, function and oxidative stability of the peroxidase enzyme and their anticancer and antibacterial activities. J Biol Inorg Chem 20:1135–1146. https://doi.org/10.1007/s00775-015-1295-x

Jain SK, Bharate SB, Vishwakarma RA (2012) Cyclin-dependent kinase inhibition by flavoalkaloids. Mini Rev Med Chem 12:632–649

Jain R, Singh R, Kaushik NK (2013) Activities of some novel organotin (IV): purine base complexes. J Chem 2013. 568195, 12 pages

Jiang J, Yao S, Cai HH et al (2013) Synthesis and synergetic effects of chrysin-organogermanium (IV) complex as potential anti-oxidant. Bioorganic Med Chem Lett 23:5727–5732. https://doi.org/10.1016/j.bmcl.2013.07.073

Jun T, Bochu W, Liancai Z (2007) Hydrolytic cleavage of DNA by quercetin zinc(II) complex. Bioorg Med Chem Lett 17:1197–1199. https://doi.org/10.1016/j.bmcl.2006.12.023

Kalinowska M, Mazur L, Piekut J et al (2013) Synthesis, crystal structure, spectroscopic properties, and antimicrobial studies of a zinc(II) complex of p-coumaric acid. J Coord Chem 66:334–344. https://doi.org/10.1080/00958972.2012.756480

Kalinowska M, Piekut J, Bruss A et al (2014) Spectroscopic (FT-IR, FT-Raman, 1H, 13C NMR, UV/VIS), thermogravimetric and antimicrobial studies of Ca(II), Mn(II), Cu(II), Zn(II) and Cd (II) complexes of ferulic acid. Spectrochim Acta Part A Mol Biomol Spectrosc 122:631–638. https://doi.org/10.1016/j.saa.2013.11.089

Kalinowska M, Świderski G, Matejczyk M, Lewandowski W (2016) Spectroscopic, thermogravimetric and biological studies of Na(I), Ni(II) and Zn(II) complexes of quercetin. J Therm Anal Calorim 126:141–148. https://doi.org/10.1007/s10973-016-5362-5

Katerova Z, Todorova D, Tasheva K, Sergiev I (2012) Influence of ultraviolet radiation on plant secondary metabolite production. Genet Plant Physiol 2:113–144

Khalil MI, Al-Zahem AM, Qunaibit MM (2014) Synthesis, characterization, and antitumor activity of binuclear curcumin-metal(II) hydroxo complexes. Med Chem Res 23:1683–1689. https://doi.org/10.1007/s00044-013-0727-9

Koc S, Köse D, Avci E, Köse K (2016) Synthesis and thermal characterization of p-coumaric acid complexes of CoII, NiII, CuII and ZnII metal cations and biological applications. Hittite J Sci Eng 3:15–22. https://doi.org/10.17350/HJSE19030000027

Kostyuk VA, Potapovich AI, Strigunova EN, et al (2004) Experimental evidence that flavonoid metal complexes may act as mimics of superoxide dismutase. Arch Biochem Biophys 428:204–208. https://doi.org/10.1016/j.abb.2004.06.008

Kunwar A, Narang H, Indira Priyadarsini K et al (2007) Delayed activation of PKCδ and NFκB and higher radioprotection in splenic lymphocytes by copper (II)-curcumin (1:1) Complex as compared to curcumin. J Cell Biochem 102:1214–1224. https://doi.org/10.1002/jcb.21348

Lakshmi BA, Bae JY, An JH, Kim S (2019) Facile design and spectroscopic characterization of novel bio-inspired Quercetin-conjugated tetrakis (dimethylsulfoxide)dichlororuthenium (II) complex for enhanced anticancer properties. Inorganica Chim Acta 495:118989. https://doi.org/10.1016/j.ica.2019.118989

Lange JL, Hayne DJ, Roselt P et al (2016) A gallium(III) Schiff base-curcumin complex that binds to amyloid-β plaques. J Inorg Biochem 162:274–279. https://doi.org/10.1016/j.jinorgbio.2016.02.029

Langland J, Jacobs B, Wagner CE et al (2018) Antiviral activity of metal chelates of caffeic acid and similar compounds towards herpes simplex, VSV-Ebola pseudotyped and vaccinia viruses. Antiviral Res 160:143–150. https://doi.org/10.1016/j.antiviral.2018.10.021

Lee YH, Tuyet PT (2019) Synthesis and biological evaluation of quercetin–zinc (II) complex for anti-cancer and anti-metastasis of human bladder cancer cells. Vitr Cell Dev Biol Anim 55:395–404. https://doi.org/10.1007/s11626-019-00363-2

Leon IE, Di Virgilio AL, Porro V et al (2013) Antitumor properties of a vanadyl(IV) complex with the flavonoid chrysin [VO(chrysin)2EtOH]2 in a human osteosarcoma model: the role of oxidative stress and apoptosis. Dalt Trans 42:11868–11880. https://doi.org/10.1039/c3dt50524c

Leon IE, Porro V, Di Virgilio AL et al (2014) Antiproliferative and apoptosis-inducing activity of an oxidovanadium(IV) complex with the flavonoid silibinin against osteosarcoma cells. J Biol Inorg Chem 19:59–74. https://doi.org/10.1007/s00775-013-1061-x

León IE, Cadavid-Vargas JF, Tiscornia I et al (2015) Oxidovanadium(IV) complexes with chrysin and silibinin: anticancer activity and mechanisms of action in a human colon adenocarcinoma model. J Biol Inorg Chem 20:1175–1191. https://doi.org/10.1007/s00775-015-1298-7

León IE, Cadavid-Vargas JF, Resasco A et al (2016a) In vitro and in vivo antitumor effects of the VO-chrysin complex on a new three-dimensional osteosarcoma spheroids model and a

xenograft tumor in mice. J Biol Inorg Chem 21:1009–1020. https://doi.org/10.1007/s00775-016-1397-0
León IE, Díez P, Etcheverry SB, Fuentes M (2016b) Deciphering the effect of an oxovanadium (IV) complex with the flavonoid chrysin (VOChrys) on intracellular cell signalling pathways in an osteosarcoma cell line. Metallomics 8:739–749. https://doi.org/10.1039/c6mt00045b
Li J, Kang J, Lu J et al (2009) Determination of calf thymus DNA using resonance light-scattering quenching method based on the terbium (Ш) (Tb3+)/europium (Ш) (Eu3+)–quercetin system. J Lumin 129:906–911. https://doi.org/10.1016/j.jlumin.2009.03.015
Li J, Wang L, Bai H (2011) Synthesis, characterization , and anti-inflammatory activities of rare earth metal complexes of luteolin. Med Chem Res Chem Res 20:88–92. https://doi.org/10.1007/s00044-009-9289-2
Li S, Xie W, Cai H et al (2012) Hydroxyl radical scavenging mechanism of human erythrocytes by quercetin—germanium (IV) complex. Eur J Pharm Sci 47:28–34. https://doi.org/10.1016/j.ejps.2012.04.019
Li Y, Gu Z, Zhang C et al (2018) Synthesis, characterization and ROS-mediated antitumor effects of palladium(II) complexes of curcuminoids. Eur J Med Chem 144:662–671. https://doi.org/10.1016/j.ejmech.2017.12.027
Liu W, Hu X, Zhou L et al (2020a) Orientation-inspired perspective on molecular inhibitor of tau aggregation by curcumin conjugated with ruthenium(II) complex scaffold. J Phys Chem B 124: 2343–2353. https://doi.org/10.1021/acs.jpcb.9b11705
Liu X-W, Tang Y-C, Liu N-Y et al (2020b) Topo I inhibition, DNA photocleavage, molecular docking and cytotoxicities of two new phenanthroline-based ruthenium complexes. Appl Organomet Chem 34:e5312. https://doi.org/10.1002/aoc.5312
Lu WP, Mei XT, Wang Y et al (2015) Zn(II)-curcumin protects against oxidative stress, deleterious changes in sperm parameters and histological alterations in a male mouse model of cyclophosphamide-induced reproductive damage. Environ Toxicol Pharmacol 39:515–524. https://doi.org/10.1016/j.etap.2014.12.014
Maciejewska G, Zierkiewicz W, Adach A et al (2009) Atypical calcium coordination number: hysicochemical study, cytotoxicity, DFT calculations and in silico pharmacokinetic characteristics of calcium caffeates. J Inorg Biochem 103:1189–1195. https://doi.org/10.1016/j.jinorgbio.2009.05.010
Mahal HS, Kapoor S, Satpati AK, Mukherjee T (2005) Radical scavenging and catalytic activity of metal-phenolic complexes. J Phys Chem B 109:24197–24202. https://doi.org/10.1021/jp0549430
Mahmoodi Chalbatani G, Dana H, Gharagouzloo E et al (2019) Small interfering RNAs (siRNAs) in cancer therapy: a nano-based approach. Int J Nanomed 14:3111–3128. https://doi.org/10.2147/IJN.S200253
Marques J, Silva AMS, Marques MPM, Braga SS (2019) Ruthenium(II) trithiacyclononane complexes of 7,3′,4′-trihydroxyflavone, chrysin and tectochrysin: synthesis, characterisation, and cytotoxic evaluation. Inorganica Chim Acta 488:71–79. https://doi.org/10.1016/j.ica.2019.01.003
Martínez Medina JJ, Naso LG, Pérez AL et al (2017) Apigenin oxidovanadium(IV) cation interactions. Synthesis, spectral, bovine serum albumin binding, antioxidant and anticancer studies. J Photochem Photobiol A Chem 344:84–100. https://doi.org/10.1016/j.jphotochem.2017.05.007
Marwaha SS, Kaur J, Sodhi GS (1995) Structure determination and anti-inflammatory activity of some purine complexes. Met Based Drugs 2:13–17. https://doi.org/10.1155/mbd.1995.13
Mawani Y, Orvig C (2014) Improved separation of the curcuminoids, syntheses of their rare earth complexes, and studies of potential antiosteoporotic activity. J Inorg Biochem 132:52–58. https://doi.org/10.1016/j.jinorgbio.2013.12.004
Mei X, Luo X, Xu S et al (2009) Gastroprotective effects of a new zinc(II)-curcumin complex against pylorus-ligature-induced gastric ulcer in rats. Chem Biol Interact 181:316–321. https://doi.org/10.1016/j.cbi.2009.06.022

Mei X, Xu D, Xu S et al (2011) Gastroprotective and antidepressant effects of a new zinc(II)-curcumin complex in rodent models of gastric ulcer and depression induced by stresses. Pharmacol Biochem Behav 99:66–74. https://doi.org/10.1016/j.pbb.2011.04.002

Mei X, Xu D, Xu S et al (2012) Novel role of Zn(II)-curcumin in enhancing cell proliferation and adjusting proinflammatory cytokine-mediated oxidative damage of ethanol-induced acute gastric ulcers. Chem Biol Interact 197:31–39. https://doi.org/10.1016/j.cbi.2012.03.006

Mei XT, Xu DH, Xu SK et al (2013) Zinc(II)-curcumin accelerates the healing of acetic acid-induced chronic gastric ulcers in rats by decreasing oxidative stress and downregulation of matrix metalloproteinase-9. Food Chem Toxicol 60:448–454. https://doi.org/10.1016/j.fct.2013.07.075

Mészáros JP, Poljarevic JM, Gál GT et al (2019) Comparative solution and structural studies of half-sandwich rhodium and ruthenium complexes bearing curcumin and acetylacetone. J Inorg Biochem 195:91–100. https://doi.org/10.1016/j.jinorgbio.2019.02.015

Mierziak J, Kostyn K, Kulma A (2014) Flavonoids as important molecules of plant interactions with the environment. Molecules 19:16240–16265. https://doi.org/10.3390/molecules191016240

Mitra K, Gautam S, Kondaiah P, Chakravarty AR (2017) Platinum(II) complexes of curcumin showing photocytotoxicity in visible light. Eur J Inorg Chem 2017:1753–1763. https://doi.org/10.1002/ejic.201601078

Mohammadi K, Thompson KH, Patrick BO et al (2005) Synthesis and characterization of dual function vanadyl, gallium and indium curcumin complexes for medicinal applications. J Inorg Biochem 99:2217–2225. https://doi.org/10.1016/j.jinorgbio.2005.08.001

Moridani MY, Pourahmad J, Bui H et al (2003) Dietary flavonoid iron complexes as cytoprotective superoxide radical scavengers. Free Radic Biol Med 34:243–253. https://doi.org/10.1016/S0891-5849(02)01241-8

Mukherjee N, Raghavan A, Podder S et al (2019) Photocytotoxic activity of copper(II) and zinc (II) complexes of curcumin and (acridinyl)dipyridophenazine. ChemistrySelect 4:9647–9658. https://doi.org/10.1002/slct.201902281

Muñoz VA, Ferrari GV, Montaña MP et al (2016) Effect of Cu2+-complexation on the scavenging ability of chrysin towards photogenerated singlet molecular oxygen (O2(1Δg)). Possible biological implications. J Photochem Photobiol B Biol 162:597–603. https://doi.org/10.1016/j.jphotobiol.2016.07.027

Muñoz VA, Dimarco Palencia FCD, Sancho MI et al (2020) Experimental and theoretical study of the stability of the complex fisetin–Cu(II) and a comparative study of free ligand and complex interaction with molecular singlet oxygen. Photochem Photobiol. https://doi.org/10.1111/php.13213

Munteanu A-C, Badea M, Olar R et al (2016) Synthesis and structural investigation of new bio-relevant complexes of lanthanides with 5-hydroxyflavone: DNA binding and protein interaction studies. Molecules 21:1737

Munteanu A-C, Badea M, Olar R et al (2018) Cytotoxicity studies, DNA interaction and protein binding of new Al (III), Ga (III) and In (III) complexes with 5-hydroxyflavone. Appl Organomet Chem 32:e4579. https://doi.org/10.1002/aoc.4579

Munteanu A-C, Notaro A, Jakubaszek M et al (2020) Synthesis, characterization, cytotoxic activity, and metabolic studies of ruthenium(II) polypyridyl complexes containing flavonoid ligands. Inorg Chem 59:4424–4434. https://doi.org/10.1021/acs.inorgchem.9b03562

Musialik M, Kuzmicz R, Pawlowski TS, Litwinienko G (2009) Acidity of hydroxyl groups: an overlooked influence on antiradical properties of flavonoids. J Org Chem 74:2699–2709. https://doi.org/10.1021/jo802716v

Muthurajan T, Rammanohar P, Rajendran NP et al (2015) Evaluation of a quercetin-gadolinium complex as an efficient positive contrast enhancer for magnetic resonance imaging. RSC Adv 5:86967–86979. https://doi.org/10.1039/c5ra16405b

Mutlu Gençkal H (2020) New heteroleptic Cu(II) complexes of chrysin with 2,2'–bipyridine and substituted 1,10–phenanthrolines: synthesis, characterization, thermal stability and antioxidant activity. J Mol Struct 1209:127917. https://doi.org/10.1016/j.molstruc.2020.127917

Naso L, Ferrer EG, Lezama L et al (2010) Role of oxidative stress in the antitumoral action of a new vanadyl(IV) complex with the flavonoid chrysin in two osteoblast cell lines: relationship with the radical scavenger activity. J Biol Inorg Chem 15:889–902. https://doi.org/10.1007/s00775-010-0652-z

Naso LG, Ferrer EG, Butenko N et al (2011) Antioxidant, DNA cleavage, and cellular effects of silibinin and a new oxovanadium(IV)/silibinin complex. J Biol Inorg Chem 16:653–668. https://doi.org/10.1007/s00775-011-0769-8

Naso LG, Lezama L, Rojo T et al (2013) Biological evaluation of morin and its new oxovanadium (IV) complex as antioxidant and specific anti-cancer agents. Chem Biol Interact 206:289–301. https://doi.org/10.1016/j.cbi.2013.10.006

Naso LG, Valcarcel M, Roura-Ferrer M et al (2014) Promising antioxidant and anticancer (human breast cancer) oxidovanadium(IV) complex of chlorogenic acid. Synthesis, characterization and spectroscopic examination on the transport mechanism with bovine serum albumin. J Inorg Biochem 135:86–99. https://doi.org/10.1016/j.jinorgbio.2014.02.013

Naso L, Martínez VR, Lezama L et al (2016a) Antioxidant, anticancer activities and mechanistic studies of the flavone glycoside diosmin and its oxidovanadium(IV) complex. Interactions with bovine serum albumin. Bioorganic Med Chem 24:4108–4119. https://doi.org/10.1016/j.bmc.2016.06.053

Naso L, Martínez VR, Lezama L et al (2016b) Antioxidant , anticancer activities and mechanistic studies of the flavone glycoside diosmin and its oxidovanadium ( IV ) complex. Interactions with bovine serum albumin. Bioorg Med Chem 24:4108–4119. https://doi.org/10.1016/j.bmc.2016.06.053

Naso LG, Badiola I, Clavijo JM, et al (2016c) Inhibition of the metastatic progression of breast and colorectal cancer in vitro and in vivo in murine model by the oxidovanadium(IV) complex with luteolin. Bioorg Med Chem 24:6004–6011. doi: 10.1016/j.bmc.2016.09.058

Naso LG, Lezama L, Valcarcel M et al (2016d) Bovine serum albumin binding, antioxidant and anticancer properties of an oxidovanadium(IV) complex with luteolin. J Inorg Biochem 157:80–93. https://doi.org/10.1016/j.jinorgbio.2016.01.021

Ndagi U, Mhlongo N, Soliman ME (2017) Metal complexes in cancer therapy—an update from drug design perspective. Drug Des Devel Ther 11:599–616. https://doi.org/10.2147/DDDT.S119488

Ni Y, Du S, Kokot S (2007) Interaction between quercetin-copper(II) complex and DNA with the use of the Neutral Red dye fluorophor probe. Anal Chim Acta 584:19–27. https://doi.org/10.1016/j.aca.2006.11.006

Notaro A, Gasser G (2017) Monomeric and dimeric coordinatively saturated and substitutionally inert Ru(II) polypyridyl complexes as anticancer drug candidates. Chem Soc Rev 46:7317–7337. https://doi.org/10.1039/C7CS00356K

Oliveira RMM, De Souza Daniel JF, Carlos RM (2013a) Synthesis, spectroscopic characterization and biological activity of cis-[Ru(hesperidin)(1,10′-phenanthroline)2](PF6) complex. J Mol Struct 1031:269–274. https://doi.org/10.1016/j.molstruc.2012.09.066

Oliveira RMM, De Souza Daniel JF, De Aguiar I et al (2013b) Structural effects on the hesperidin properties obtained by chelation to magnesium complexes. J Inorg Biochem 129:35–42. https://doi.org/10.1016/j.jinorgbio.2013.08.005

Panhwar QK, Memon S (2012) Synthesis, spectral characterization and antioxidant activity of Tin (II)-morin complex. Pakistan J Anal Environ Chem 13:159–168

Panhwar QK, Memon S (2013) Synthesis, characterization and antioxidant study of Tin(II)-rutin complex: exploration of tin packaging hazards. Inorganica Chim Acta 407:252–260. https://doi.org/10.1016/j.ica.2013.08.001

Panhwar QK, Memon S (2014) Synthesis, characterisation, and antioxidant study of Cr(III)-rutin complex. Chem Pap 68:614–623. https://doi.org/10.2478/s11696-013-0494-6

Parveen S, Tabassum S, Arjmand F (2016) Human topoisomerase I mediated cytotoxicity profile of l-valine-quercetin diorganotin(IV) antitumor drug entities. J Organomet Chem 823:23–33. https://doi.org/10.1016/j.jorganchem.2016.09.015

Pellerito C, Emanuele S, Ferrante F et al (2020) Tributyltin(IV) ferulate, a novel synthetic ferulic acid derivative, induces autophagic cell death in colon cancer cells: from chemical synthesis to biochemical effects. J Inorg Biochem 205:110999. https://doi.org/10.1016/j.jinorgbio.2020.110999

Porfírio DA, Ferreira RDQ, Malagutti AR, Valle EMA (2014) Electrochemical study of the increased antioxidant capacity of flavonoids through complexation with iron(II) ions. Electrochim Acta 141:33–38. https://doi.org/10.1016/j.electacta.2014.07.046

Poynton FE, Bright SA, Blasco S et al (2017) The development of ruthenium(II) polypyridyl complexes and conjugates for in vitro cellular and in vivo applications. Chem Soc Rev 46:7706–7756. https://doi.org/10.1039/C7CS00680B

Prasad P, Pant I, Khan I et al (2014) Mitochondria-targeted photoinduced anticancer activity of oxidovanadium(IV) complexes of curcumin in visible light. Eur J Inorg Chem 2014:2420–2431. https://doi.org/10.1002/ejic.201402001

Pröhl M, Schubert US, Weigand W, Gottschaldt M (2016) Metal complexes of curcumin and curcumin derivatives for molecular imaging and anticancer therapy. Coord Chem Rev 307:32–41. https://doi.org/10.1016/j.ccr.2015.09.001

Rajalakshmi S, Vimalraj S, Saravanan S et al (2018) Synthesis and characterization of silibinin/phenanthroline/neocuproine copper(II) complexes for augmenting bone tissue regeneration: an in vitro analysis. J Biol Inorg Chem 23:753–762. https://doi.org/10.1007/s00775-018-1566-4

Ravichandran R, Rajendran M, Devapiriam D (2014) Antioxidant study of quercetin and their metal complex and determination of stability constant by spectrophotometry method. Food Chem 146: 472–478. https://doi.org/10.1016/j.foodchem.2013.09.080

Raza A, Xu X, Xia L et al (2016) Quercetin-iron complex: synthesis, characterization, antioxidant, DNA binding, DNA cleavage, and antibacterial activity studies. J Fluoresc 26:2023–2031. https://doi.org/10.1007/s10895-016-1896-y

Raza A, Bano S, Xu X et al (2017) Rutin–nickel complex: synthesis, characterization, antioxidant, DNA binding, and DNA cleavage activities. Biol Trace Elem Res 178:160–169. https://doi.org/10.1007/s12011-016-0909-7

Refat MS (2013) Synthesis and characterization of ligational behavior of curcumin drug towards some transition metal ions: chelation effect on their thermal stability and biological activity. Spectrochim Acta Part A Mol Biomol Spectrosc 105:326–337. https://doi.org/10.1016/j.saa.2012.12.041

Roy S, Majumdar S, Singh AK et al (2015a) Synthesis, characterization, antioxidant status, and toxicity study of vanadium-rutin complex in Balb/c Mice. Biol Trace Elem Res 166:183–200. https://doi.org/10.1007/s12011-015-0270-2

Roy S, Mallick S, Chakraborty T et al (2015b) Synthesis, characterisation and antioxidant activity of luteolin–vanadium(II) complex. Food Chem 173:1172–1178. https://doi.org/10.1016/j.foodchem.2014.10.141

Roy AS, Tripathy DR, Samanta S et al (2016) DNA damaging, cell cytotoxicity and serum albumin binding efficacy of the rutin-Cu(II) complex. Mol Biosyst 12:1687–1701. https://doi.org/10.1039/c6mb00161k

Roy S, Das R, Ghosh B, Chakraborty T (2018) Deciphering the biochemical and molecular mechanism underlying the in-vitro and in-vivo chemotherapeutic efficacy of ruthenium quercetin complex in colon cancer. Mol Carcinog 57:700–721. https://doi.org/10.1002/mc.22792

Roy S, Sil A, Chakraborty T (2019) Potentiating apoptosis and modulation of p53, Bcl2, and Bax by a novel chrysin ruthenium complex for effective chemotherapeutic efficacy against breast cancer. J Cell Physiol 234:4888–4909. https://doi.org/10.1002/jcp.27287

Rukk NS, Kuzmina LG, Shamsiev RS et al (2019) Zinc(II) and cadmium(II) halide complexes with caffeine: synthesis, X-ray crystal structure, cytotoxicity and genotoxicity studies. Inorganica Chim Acta 487:184–200. https://doi.org/10.1016/j.ica.2018.11.036

Sarkar T, Butcher RJ, Banerjee S et al (2016) Visible light-induced cytotoxicity of a dinuclear iron (III) complex of curcumin with low-micromolar IC50 value in cancer cells dedicated to professor Animesh Chakravorty on the occasion of his 80th birthday. Inorganica Chim Acta 439:8–17. https://doi.org/10.1016/j.ica.2015.09.026

Selvaraj S, Krishnaswamy S, Devashya V et al (2011) Investigations on membrane perturbation by chrysin and its copper complex using self-assembled lipid bilayers. Langmuir 27:13374–13382. https://doi.org/10.1021/la2029356

Senderowicz AM (1999) Flavopiridol: the first cyclin-dependent kinase inhibitor in human clinical trials. Invest New Drugs 17:313–320

Sendrayaperumal V, Iyyam Pillai S, Subramanian S (2014) Design, synthesis and characterization of zinc-morin, a metal flavonol complex and evaluation of its antidiabetic potential in HFD-STZ induced type 2 diabetes in rats. Chem Biol Interact 219:9–17. https://doi.org/10.1016/j.cbi.2014.05.003

Sentkowska A, Kilian K, Cheda Ł (2017) Ga (III) complex with morin for kidney cancer cell labelling. Appl Organomet Chem 31:e3882. https://doi.org/10.1002/aoc.3882

Shaik JB, Palaka BK, Penumala M et al (2016) Synthesis, pharmacological assessment, molecular modeling and in silico studies of fused tricyclic coumarin derivatives as a new family of multifunctional anti-Alzheimer agents. Eur J Med Chem 107:219–232. https://doi.org/10.1016/j.ejmech.2015.10.046

Sharma KK, Chandra S, Basu DK (1987) Synthesis and antiarthritic study of a new orally active diferuloyl methane (curcumin) gold complex. Inorganica Chim Acta 135:47–48. https://doi.org/10.1016/S0020-1693(00)83260-0

Shechter Y, Shisheva A (1993) Vanadium salts and the future treatment of diabetes. Endeavour 17:27–31. https://doi.org/10.1016/0160-9327(93)90008-q

Shukla R, Barve V, Padhye S, Bhonde R (2006) Reduction of oxidative stress induced vanadium toxicity by complexing with a flavonoid, quercetin: a pragmatic therapeutic approach for diabetes. BioMetals 19:685–693. https://doi.org/10.1007/s10534-006-9005-3

da Silva WMB, de Oliveira PS, Alves DR et al (2020) Synthesis of quercetin-metal complexes, in vitro and in silico anticholinesterase and antioxidant evaluation, and in vivo toxicological and anxiolitic activities. Neurotox Res 37:893–903. https://doi.org/10.1007/s12640-019-00142-7

Singh A (2017) Glucosinolates. In: Mérillon J-M, Ramawat KG (eds) Glucosinolates and plant defense. Springer International Publishing, Cham, pp 237–246

de Souza RFV, De Giovani WF (2004) Antioxidant properties of complexes of flavonoids with metal ions. Redox Rep 9:97–104. https://doi.org/10.1179/135100004225003897

Spoerlein C, Mahal K, Schmidt H, Schobert R (2013) Effects of chrysin, apigenin, genistein and their homoleptic copper(II) complexes on the growth and metastatic potential of cancer cells. J Inorg Biochem 127:107–115. https://doi.org/10.1016/j.jinorgbio.2013.07.038

Stefanachi A, Leonetti F, Pisani L et al (2018) Coumarin: a natural, privileged and versatile scaffold for bioactive compounds. Molecules 23:250. https://doi.org/10.3390/molecules23020250

Stolpovskaya EV, Trofimova NN, Babkin VA (2017) Evaluation of antioxidant activity of dihydroquercetin complexes with biogenic metal ions. Russ J Bioorganic Chem 43:742–746. https://doi.org/10.1134/S1068162017070160

Sulpizio C, Müller STR, Zhang Q et al (2016) Synthesis, characterization, and antioxidant activity of Zn2+ and Cu2+ coordinated polyhydroxychalcone complexes. Monatsh Chem 147:1871–1881. https://doi.org/10.1007/s00706-016-1822-7

Sutradhar M, Fernandes AR, Paradinha F et al (2019) A new Cu(II)-O-carvacrotinate complex: synthesis, characterization and biological activity. J Inorg Biochem 190:31–37. https://doi.org/10.1016/j.jinorgbio.2018.09.018

Tabassum S, Zaki M, Afzal M, Arjmand F (2013) New modulated design and synthesis of quercetin-Cu(II)/Zn(II)-Sn2(IV) scaffold as anticancer agents: in vitro DNA binding profile, DNA cleavage pathway and Topo-I activity. Dalton Trans 42:10029–10041. https://doi.org/10.1039/c3dt50646k

Tamayo LV, Gouvea LR, Sousa AC et al (2016) Copper(II) complexes with naringenin and hesperetin: cytotoxic activity against A 549 human lung adenocarcinoma cells and investigation on the mode of action. BioMetals 29:39–52. https://doi.org/10.1007/s10534-015-9894-0

Tan J, Wang B, Zhu L (2009a) DNA binding, cytotoxicity, apoptotic inducing activity, and molecular modeling study of quercetin zinc(II) complex. Bioorganic Med Chem 17:614–620. https://doi.org/10.1016/j.bmc.2008.11.063

Tan J, Wang B, Zhu L (2009b) DNA binding and oxidative DNA damage induced by a quercetin copper(II) complex: potential mechanism of its antitumor properties. J Biol Inorg Chem 14:727–739. https://doi.org/10.1007/s00775-009-0486-8

Tan M, Zhu J, Pan Y et al (2009c) Synthesis, cytotoxic activity, and DNA binding properties of copper (II) complexes with hesperetin, naringenin, and apigenin. Bioinorg Chem Appl 2009: 347872. https://doi.org/10.1155/2009/347872

Tan J, Zhu L, Wang B (2010) From GC-rich DNA binding to the repression of survivin gene for quercetin nickel (II) complex: implications for cancer therapy. BioMetals 23:1075–1084. https://doi.org/10.1007/s10534-010-9353-x

Tan J, Zhu L, Wang B (2011) GC (guanine-cytosine)-selective DNA-binding and antitumor activity of a quercetin-manganese (II) complex. Chem Biodivers 8:1550–1559. https://doi.org/10.1002/cbdv.201000313

Thangavel P, Viswanath B, Kim S (2018) Synthesis and characterization of kaempferol-based ruthenium (II) complex: a facile approach for superior anticancer application. Mater Sci Eng C 89:87–94. https://doi.org/10.1016/j.msec.2018.03.020

Theis N, Lerdau M (2003) The evolution of function in plant secondary metabolites. Int J Plant Sci 164:S93–S102. https://doi.org/10.1086/374190

Thomas MJ (2000) The role of free radicals and antioxidants. Nutrition 16:716–718. https://doi.org/10.1016/S0899-9007(00)00343-9

Thompson KH, Böhmerle K, Polishchuk E et al (2004) Complementary inhibition of synoviocyte, smooth muscle cell or mouse lymphoma cell proliferation by a vanadyl curcumin complex compared to curcumin alone. J Inorg Biochem 98:2063–2070. https://doi.org/10.1016/j.jinorgbio.2004.09.011

Tong L, Zhang J, Liu Q, Zhai G (2016) Sb(III)–quercetin complex: synthesis, characterization and antioxidant activity. Life Sci J 13:127–133. https://doi.org/10.7537/marslsj130716.15. Keywords

Trofimova NN, Babkin VA, Kiselev OI (2015) Complex compounds of zinc and copper(II) ions with dihydroquercetin and their antiviral activity. Russ Chem Bull 64:1430–1436. https://doi.org/10.1007/s11172-015-1027-y

Tu LY, Pi J, Jin H et al (2016) Synthesis, characterization and anticancer activity of kaempferol-zinc (II) complex. Bioorganic Med Chem Lett 26:2730–2734. https://doi.org/10.1016/j.bmcl.2016.03.091

Uivarosi V, Munteanu A (2017) Flavonoid complexes as promising anticancer metallodrugs. In: Justino GC (ed) Flavonoids - from biosynthesis to human health. InTech, Rijeka

Uivarosi V, Munteanu A-C, Niţulescu GM (2019a) Chapter 2—An overview of synthetic and semisynthetic flavonoid derivatives and analogues: perspectives in drug discovery. In: Atta-ur-Rahman BT-S (ed) Studies in natural products chemistry. Elsevier, Amsterdam, pp 29–84

Uivarosi V, Munteanu AC, Sharma A, Singh Tuli H (2019b) Metal complexation and patent studies of flavonoid. In: Singh Tuli H (ed) Current aspects of flavonoids: their role in cancer treatment. Springer, Singapore, pp 39–89

Vimalraj S, Rajalakshmi S, Raj Preeth D et al (2018) Mixed-ligand copper(II) complex of quercetin regulate osteogenesis and angiogenesis. Mater Sci Eng C 83:187–194. https://doi.org/10.1016/j.msec.2017.09.005

Vimalraj S, Rajalakshmi S, Saravanan S et al (2019) Zinc chelated morin promotes osteoblast differentiation over its uncomplexed counterpart. Process Biochem 82:167–172. https://doi.org/10.1016/j.procbio.2019.04.008

Waller GR, Nowacki EK (1978) The role of alkaloids in plants. In: Waller GR, Nowacki EK (eds) Alkaloid biology and metabolism in plants. Springer, Boston, MA, pp 143–181

Wang Q, Huang M, Huang Y et al (2014) Synthesis, characterization, DNA interaction, and antitumor activities of mixed-ligand metal complexes of kaempferol and 1,10-phenanthroline/2,2'-bipyridine. Med Chem Res Chem Res 23:2659–2666. https://doi.org/10.1007/s00044-013-0863-2

War AR, Paulraj MG, Ahmad T et al (2012) Mechanisms of plant defense against insect herbivores. Plant Signal Behav 7:1306–1320. https://doi.org/10.4161/psb.21663

van der Weerden NL, Anderson MA (2013) Plant defensins: common fold, multiple functions. Fungal Biol Rev 26:121–131. https://doi.org/10.1016/j.fbr.2012.08.004

Westermann J-C, Craik DJ (2010) 5.09—Plant peptide toxins from nonmarine environments. In: Liu H-W, LBT-CNPII M (eds) Comprehensive natural products II. Elsevier, Oxford, pp 257–285

Wu R, Mei X, Ye Y et al (2019) Zn(II)-curcumin solid dispersion impairs hepatocellular carcinoma growth and enhances chemotherapy by modulating gut microbiota-mediated zinc homeostasis. Pharmacol Res 150:104454. https://doi.org/10.1016/j.phrs.2019.104454

Xing Z-H, Wei J-H, Cheang T-Y et al (2014) Bifunctional pH-sensitive Zn(II)–curcumin nanoparticles/siRNA effectively inhibit growth of human bladder cancer cells in vitro and in vivo. J Mater Chem B 2:2714–2724. https://doi.org/10.1039/C3TB21625J

Xu Y, Qian LL, Yang J et al (2018) Kaempferol binding to zinc(II), efficient radical scavenging through increased phenol acidity. J Phys Chem B 122:10108–10117. https://doi.org/10.1021/acs.jpcb.8b08284

Xuan HZ, Zhang JH, Wang YH et al (2016) Anti-tumor activity evaluation of novel chrysin-organotin compound in MCF-7 cells. Bioorganic Med Chem Lett 26:570–574. https://doi.org/10.1016/j.bmcl.2015.11.072

Yang AH, Shi XY, Li X et al (2014) Spectroscopic and electrochemical studies on the evaluation of the radical scavenging activities of luteolin by chelating iron. RSC Adv 4:25227–25233. https://doi.org/10.1039/c4ra01396d

Yang F, Gong L, Jin H et al (2015) Chrysin-organogermanium (IV) complex induced Colo205 cell apoptosis-associated mitochondrial function and anti-angiogenesis. Scanning 37:246–257. https://doi.org/10.1002/sca.21205

Yang J, Xu Y, Liu HY et al (2017) Genistein binding to copper (II)-solvent dependence and effects on radical scavenging. Molecules 22:1757. https://doi.org/10.3390/molecules22101757

Yang ZY, Qian LL, Xu Y et al (2020) Kinetic studies on radical scavenging activity of kaempferol decreased by Sn(II) binding. Molecules 25:1975. https://doi.org/10.3390/molecules25081975

Yao S, Jiang J, Yang P, Cai J (2012) Synthesis, characterization and antioxidant activity of a novel organogermanium sesquioxide with resveratrol. Bull Korean Chem Soc 33:1121–1122. https://doi.org/10.5012/bkcs.2012.33.4.1121

Yasarawan N, Thipyapong K, Sirichai S, Ruangpornvisuti V (2013) Fundamental insights into conformational stability and orbital interactions of antioxidant (+)-catechin species and complexation of (+)-catechin with zinc(II) and oxovanadium(IV). J Mol Struct 1047:344–357. https://doi.org/10.1016/j.molstruc.2013.05.038

Yesufu HB, Musa SA, Mohammed GT, Fugu MB (2020) Synthesis, characterization and antioxidant properties of metal (II) complexes of gallic acid and quercetin. Int J Res Innov Appl Sci 5: 226–231

Yu C, Mei XT, Zheng YP, Xu DH (2014) Zn(II)-curcumin protects against hemorheological alterations, oxidative stress and liver injury in a rat model of acute alcoholism. Environ Toxicol Pharmacol 37:729–737. https://doi.org/10.1016/j.etap.2014.02.011

Zahirović A, Kahrović E, Cindrić M et al (2017) Heteroleptic ruthenium bioflavonoid complexes: from synthesis to in vitro biological activity. J Coord Chem 70:4030–4053. https://doi.org/10.1080/00958972.2017.1409893

Zaidi M, Wattoo F, Hamid M et al (2012) Antibacterial activities of nicotine and its zinc complex. African J Microbiol Res 6:5134–5137. https://doi.org/10.5897/AJMR11.1209

Zeng JF, Cheng M, Wang Y et al (2016) pH-responsive Fe(III)-gallic acid nanoparticles for in vivo photoacoustic-imaging-guided photothermal therapy. Adv Healthc Mater 5:772–780. https://doi.org/10.1002/adhm.201500898

Zhai G, Zhu W, Duan Y et al (2012) Synthesis, characterization and antitumor activity of the germanium-quercetin complex. Main Gr Met Chem 35:103–109. https://doi.org/10.1515/mgmc-2012-0026

Zhang D, Chu L, Liu Y et al (2011) Analysis of the antioxidant capacities of flavonoids under different spectrophotometric assays using cyclic voltammetry and density functional theory. J Agric Food Chem 59:10277–10285. https://doi.org/10.1021/jf201773q

Zhang J, Cui C, Zhang H et al (2016) Synthesis, characterization and antihyperlipidemic of rutin-calcium (II) complex. Life Sci J 13:13–21. https://doi.org/10.7537/marslsj130816.03.Keywords

Zhang P, Hou Y, Zeng J et al (2019) Coordinatively unsaturated Fe3+ based activatable probes for enhanced MRI and therapy of tumors. Angew Chem Int Ed 58:11088–11096. https://doi.org/10.1002/anie.201904880

Zhou J, Wang L, Wang J, Tang N (2001a) Synthesis, characterization, antioxidative and antitumor activities of solid quercetin rare earth(III) complexes. J Inorg Biochem 83:41–48. https://doi.org/10.1016/S0162-0134(00)00128-8

Zhou J, Wang L, Wang J, Tang N (2001b) Antioxidative and anti-tumour activities of solid quercetin metal (II) complexes. Transit Met Chem 26:57–63

# Phytoalexins: Implications in Plant Defense and Human Health 10

Indu Sharma, Abhinay Thakur, Aditi Sharma, Narayan Singh, Rahul Kumar, and Ashutosh Sharma

**Abstract**

The sessile habit of plants led to the evolution of plant secondary metabolites that were aiding to the plant defense. One of such groups of secondary metabolites is phytoalexins, which are low molecular weight compounds of antimicrobial nature and are produced by plants in response to the attacking pathogens. Phytoalexins are diverse in their chemical nature, but have a function in common, i.e., plant defense. With the climate change and development of pure line crop varieties, plant diseases are continuously challenging the efforts toward enhancement of crop productivity. The property of phytoalexins to accumulate at infection sites and inhibit the growth of the pathogenic microbes makes them potential antimicrobial agents, which may be exploited for the disease resistance. Since many genes encoding the enzymes involved in their biosynthetic pathway are now well studied, the expression of these genes can be manipulated to engineer future crops for better resistance toward plant pathogens. Besides their key role in plant defense, phytoalexins are also helpful in promoting human health. Some of them are known to possess various bioactive properties such as antioxidant, anticancer, antidiabetic, antiparasitic, cardioprotective, neuroprotective, and

I. Sharma
Department of Botany, SBBS University, Jalandhar, Punjab, India

A. Thakur
Department of Zoology, DAV College, Jalandhar, Punjab, India

A. Sharma
Faculty of Agricultural Sciences, DAV University, Jalandhar, Punjab, India

Collage of Horticulture and Forestry, Thunag, Mandi, Dr. Y. S. Parmar University of Horticulture and Forestry, Nanuni, Solan, India

N. Singh · R. Kumar · A. Sharma (✉)
Faculty of Agricultural Sciences, DAV University, Jalandhar, Punjab, India



A. K. Sharma, A. Sharma (eds.), *Plant Secondary Metabolites*,
https://doi.org/10.1007/978-981-16-4779-6_10

growth-stimulating. The present chapter, therefore, highlights their diverse chemical nature, regulation of their biosynthesis, distribution, accumulation, role in plant defense, and potential in promoting human health.

**Keywords**

Shikimic acid pathway · Flavonoids · Germ tube · Biosynthetic pathway · Phenylalanine ammonia lyase · Plant disease · Transgenic plants

## 10.1 Introduction

Plants face a serious threat from the plant pathogens that are mainly of microbial origin. The plants, therefore, evolved several mechanisms for defending themselves from the invading pathogens that include production of PR proteins (Sharma et al. 2021), production of thick cuticular wax and several other kinds of structural defenses (Agrios 2005), etc. Besides this, plants also produce some antimicrobial substances upon pathogen invasion that helps in defending them against the pathogens.

The phytoalexins (Greek words: *phyton*–plant and *alexos*—defend) play a role in plant defense and are not present in healthy plants (Chripkova et al. 2016). These are referred to as species-specific substances, since the synthesis of phytoalexins is stimulated by elicitors which are the specific substances. After some hours or days of the pathogen attack, the phytoalexins are produced in small amounts and trigger the defense response in the affected plants. It was found that *Solanum tuberosum* inoculated with an incompatible race of *Phytophthora infestans* induced the synthesis of a putative defense-related metabolite, which resulted into the resistance toward that provided resistance to a compatible race of the pathogen (Müller et al. 1939). In 1940, these induced putative antimicrobials were termed as phytoalexins, whose production was found to be triggered in response to *P. infestans* attack in potato (Müller and Börger 1940; Müller 1940), and therefore, the term phytoalexins was first time used for describing such antimicrobial substances.

The phytoalexins have been described thereafter as a diverse group of plant-based antimicrobial compounds with low molecular weights, which can be synthesized/accumulated in plants in response to their exposure to the pathogenic microorganisms (Paxton 1980, 1981; VanEtten et al. 1994). Another somewhat similar group of plant-based antimicrobials has been referred to as phytoanticipins (VanEtten et al. 1994). The phytoalexins and phytoanticipins are distinguished on the basis of their production and response toward the invading pathogens. Although their production may seem arbitrary, this distinction is based on fundamental differences in the responses of the host plants to the microbial invasions. Phytoalexin play a role in plant disease resistance, when a pathogen induces an active response (that is stimulated upon exposure of plant pathogen); whereas a phytoanticipin plays a role in plant disease resistance response only through passive interaction (even without the pathogen exposure).

Phytoalexins are low molecular weight, inducible secondary metabolites, which possess antimicrobial potential (Bailey and Mansfield 1982; Chripkova et al. 2016;

Pedras et al. 2007). After exposure of plants to various biological factors such as bacteria, viruses, fungi, etc.; physical factors such as heat shock, UV radiation, injury, etc.; or chemical factors such as heavy metals, the phytoalexins have been produced through de novo synthesis. The first phytoalexin was isolated from pea (*Pisum sativum*) after infection by *Ascochyta pisi*, and this phytoalexin was chemically characterized as (+)-pisatin (Cruickshank and Perrin 1960).

## 10.2 Chemically Diverse Nature of Phytoalexins

On the basis of plant species synthesizing phytoalexins in response to biotic or abiotic stresses, different kinds of phytoalexins are produced, belonging to structurally diverse chemical classes. For example, copper chloride-induced abiotic stress in *Erucastrum canariense* Webb & Berthel. (family: Brassicaceae) stimulated the phytoalexin production which was analyzed as phytoalexin erucalexin (Pedras and To 2017). Glyceollins phytoalexins derived from soybeans under abiotic/biotic stress have been categorized into three classes, viz., glyceollin I (GI), glyceollin II (GII), and glyceollin III (GIII), which are synthesized de novo from soy isoflavone daidzein (Pham et al. 2019).

Phytoalexins include various kinds of species-specific substances which may be included in various chemical classes like alkaloids, terpenoids, flavonoids, isoflavonoids, phenolics, indole compounds, etc. (Chripkova et al. 2016). Phytoalexins are, therefore, a structurally diverse group of antimicrobial compounds that may be produced in response to a particular plant pathogen. Various chemical classes of these phytoalexins (Bizuneh 2020) have been reported to possess antimicrobial properties against different phytopathogens (Table 10.1).

Biosynthesis of these phytoalexins involves about 32 different classes of biosynthetic pathways in different host plants (Zhang et al. 2010). The structural composition of phytoalexin depends on the plant species/family and pathogen/abiotic stress that induce their accumulation. Different plant families, e.g., Solanaceae, Poaceae, Fabaceae, Euphorbiaceae, Linaceae, Orchidaceae, Chenopodiaceae, Asteraceae, Moraceae, Rutaceae, Rosaceae, Piperaceae, Amaryllidaceae, Apiaceae, Convolvulaceae, etc., are known to produce various antimicrobial compounds (Tiku 2020). More than 350 phytoalexins have been chemically characterized from different plant families (Darvill and Albersheim 1984). Among these plant families, the biosynthetic pathways, genes involved, and metabolic products are also highly diverse (Jeandet et al. 2014; Ahmed and Kovinich 2020). For example, members of the family Solanaceae are known to synthesize terpenoid phytoalexins like rishitin, and the phenylpropanoid phytoalexins like chlorogenic acid and caffeic acid (Henfling et al. 1980; Ohnishi et al. 1994). Further, scopoletin and capsidiol, which are also biosynthesized in *N. tabacum* (family: Solanaceae) are induced after the infection by *Botrytis cinerea*, *Phytophthora nicotianae*, and *Phytophthora palmivora* (El Oirdi et al. 2010; Lee et al. 2017). However, several diterpenoids like oryzalexins, momilactones, zealexins, kauralexins, deoxyanthocyanidins, apigeninidin, and sakuranetin are produced by the members of Poaceae (Poloni

**Table 10.1** Chemical diversity and antimicrobial action of different phytoalexins

| Chemical nature | Phytoalexins | Pathogen infection | Reference(s) |
|---|---|---|---|
| Flavonoids | Flavones: Apigenin, luteolin, and baicalein<br>Flavonols: Kaempferol, quercetin, myricetin<br>Flavanones: Betagarin, naringenin, hesperitin, and liquiritigenin<br>Flavanonols: Dihydrokaempferol, dihydromyricetin, and dihydroquercetin | *Colletotrichum lupine, Colletotrichum trifolii, Colletotrichum sublineolum*, and *Cochliobolus heterostrophus* | Ahuja et al. (2012), Schnippenkoetter et al. (2017) |
| Isoflavonoids | Isoflavones: Daidzein, formononetin, and genistein, wighteone, betavulgarin<br>Isoflavanone: Kievitone, vestitone<br>Pterocarpans: Medicarpin, maackiain, pisatin, phaseollin, and glyceollins<br>Isoflavans: Vestitol<br>Coumestans: Coumestrol | *Nectria haematococca, Colletotrichum trifolii, Phoma medicaginis, Macrophomina phaseolina, Sclerotinia sclerotiorum* and *Phytophthora sojae, Aspergillus niger, Aspergillus oryzae, Aspergillus awamori, Aspergillus sojae and Rhizopus oligosporus, Fusarium solani* f. sp. *glycines* | Kumar and Pandey (2013) |
| Stilbenoids | Stilbenes and stilbene oligomers, dihydrostilbenes, resveratrol | *Aspergillus caelatus, A. flavus, A. parasiticus and A. niger, Rhizopus oligosporus, Plasmopara viticola, Erysiphe necator,* and *Botrytis cinerea* | Jeandet et al. (2019) |
| Coumarins | Furanocoumarins, Marmesin, and psoralen, their coumarin precursor, umbelliferone, and the methoxylated Psoralen derivatives, xanthotoxin, bergapten, and isopimpinellin | *Phytophthora citrophthora, Phytophthora sojae, Botrytis cinerea, Alternaria alternata,* and *Sclerotinia sclerotiorum* | Dugrand-Judek et al. (2015) |
| Phenolics | Xanthotoxin, 6-methoxymellein, *p*-hydroxybenzoic acid, protocatechuic acid, caffeic acid, and Sinapic acid | *Colletotrichum sublineolum, Rhizoctonia solani* | Ahuja et al. (2012) |
| Anthocyanidin | Apigeninidin, cyanidin, luteolinidin, proanthocyanidins | *Helminthosporium maydis, Colletotrichum* | Smeriglio et al. (2016) |

(continued)

**Table 10.1** (continued)

| Chemical nature | Phytoalexins | Pathogen infection | Reference(s) |
|---|---|---|---|
| | | *graminicola*, *Fusarium oxysporum*, *Gibberella zeae*, *Gliocladium roseum*, *Alternaria solani*, and *Phytophthora infestans* | |
| Sesquiterpenoids | Furanosesquiterpene: Ipomeamarone Eudesmane-type sesquiterpenes: 7-hydroxycostal and 7-hydroxycostol Eremophilane-type sesquiterpenes: Capsidiol and debneyol Vetispirane-, nor-eudesmane-, and seco-edusmane-type sesquiterpenes: Solavetivone, lubimin, Phytuberin, phytuberol, rishitin, and glutinosone Sesquiterpene lactones: Cichoralexin, Lettucenin A, and costunolide Cad inane-type sesquiterpenes: Hemigossypol, 2,7-dihydroxycadalene, 7-hydroxycalamenene, and mansonones | *Ceratocystis fimbriata*, *Phytophthora infestans*, *Helminthosporium carbonum*, *Cladosporium cucumerinum*, *Phytophthora infestans*, and *potato virus X* *Tobacco mosaic virus* | Li et al. (2015), Lee et al. (2017) |
| Monoterpenoids | Piquerol A and piquerinol, cupressotropolone A and B, *p*-thujaplicin, tropolone | *Piqueria trinervia*, *Phoma macdonaldii*, *Fusarium moniliforme*, *Helminthosporium pedicel/atum*, and *Paecilomyces elegans*, *Diplodia pinea* f. sp. *cupressi*, *Verticillium dahliae*, and *Alternaria alternata* | Jassbi et al. (2017) |
| Diterpenoids | Momilactones A and B, oryzalexins A–F, oryzalexin S, Phytocassanes A–E and ent-10-oxodepressin, Oryzalides, oryzalic acids, and oryzadiones | *Magnaporthe oryzae* *Aspergillus flavus* *Xanthomonas oryzae* pv. *oryzae* | Inoue et al. (2013), Schmelz et al. (2014) |

(continued)

**Table 10.1** (continued)

| Chemical nature | Phytoalexins | Pathogen infection | Reference(s) |
|---|---|---|---|
| Indole phytoalexins | Brassinin, cyclobrassinin and 1-methoxybrassinin, Camalexin, indole-3-carboxylic acid, Hydroxyindole 4-hydroxy-indole-3-carbonyl Nitriles | *Pseudomonas syringae*, *Alternaria brassicicola*, and *Botrytis cinerea* and by some abiotic stresses, such as $AgNO_3$ and amino acid starvation | Stahl et al. (2016), Pastorczyk et al. (2020) |

and Schirawski 2014). Meyer et al. (2017) reported kauralexin production that led to enhanced resistance against fungal pathogen *Cercospora zeina* by RNA-Seq analysis of resistant and susceptible subtropical maize (family: Poaceae).

## 10.3 Biosynthesis of Phytoalexins

Phytoalexins (unlike phytoanticipins) are not present in the plants all the time, but are synthesized in plant upon pathogen attack or because of abiotic stresses, from the precursor molecules already available in the plants. This "on demand" synthesis of phytoalexins is mainly a result of de novo synthesis (Ahuja, et al. 2012) of the enzymes, thereby leading to the production of specific phytoalexins. It has been observed that majority of these phytoalexin precursors are directed from any of the three basic metabolic routes or a combination their off (Shukla et al. 2019), viz., the condensation of a cinnamyl-CoA and three acetyl-CoA units (many phytoalexins from the family Leguminosae), phenylpropanoic-polymalonic acid pathway (many flavonoid phytoalexins like isoflavones, pterocarpans, isoflavonoids, coumestans, aryl benzofurans, etc. as well as the stilbene phytoalexins and their derivatives like dihydrophenanthrenes), and the shikimate pathway (like betagarin that accumulates in *Beta vulgaris* following the infection by *Cercospora beticola*). A detailed classical account of the structures, chemical classification, biosynthesis, and degradation is available elsewhere (Dixon et al. 1983).

Biosynthesis of many phenylpropanoid phytoalexins have been reported from the plants belonging to the families Leguminosae, Solanaceae, Convolvulaceae, Umbelliferae, and Gramineae (Shukla et al. 2019). Glyceollins, medicarpin, pisatin, phaseollin, and maackiain are synthesized by members of the family Leguminosae and are known to be the best studied phytoalexins in terms of biosynthesis, enzymology, and molecular biology (Kumar and Pandey 2013). However, a lot of information on the biochemistry and molecular biology has been accumulated by the research on model plant species discussed in detail under a separate section of the present chapter. The model plant *Arabidopsis thaliana* from Brassicaceae family produces the indole alkaloid camalexin, hydroxyindole 4-hydroxy-indole-3-carbonyl, indole-3-carboxylic acid nitriles, etc. (Pastorczyk et al. 2020). Various indole phytoalexins such as camalexin and brassinin are known to show antifungal action against *Alternaria brassicicola* and *Albugo candida* (Pedras et al. 2009).

The flavonoid biosynthesis is one of the most extensively studied areas of secondary metabolites that are antimicrobials, and the identification and characterization of these compounds has been done by the different molecular techniques (Miranda et al. 2012; Baskar et al. 2018). Schnippenkoetter et al. (2017) have recently reported the antifungal properties of several flavonoids against different species of *Colletotrichum* and *Cochliobolus*. Since the production/accumulation of phytoalexin provides an active chemical defense to the plants against fungi and other pathogens, several attempts have been made to understand the mode of action of different phytoalexins. The mode of action of different phytoalexins is found to be diverse like their chemical nature, involving the inhibition of mycelial growth, inhibition of fungal enzymes, disruption of plasma membrane, disorganization of cellular contents, cytoplasmic granulation, etc. (Arruda et al. 2016).

## 10.4 Distribution in Plant Families

Phytoalexins such as isoflavones, sesquiterpenoids, diterpenes, and sulfurous indole phytoalexins have been reported respectively in the families Fabaceae, Solanaceae, Poaceae, and Brassicaceae (Chripkova et al. 2016; Gross 1993; Pedras and Ahiahonu 2005). The members of family Brassicaceae are referred to as cruciferous plants (rapeseed-mustard, broccoli, turnip, etc.), which produce various types of sulfur-containing and/or indole-containing phytoalexins or indole-sulfur phytoalexins in response to pathogen attack (Klein and Sattely 2017). Brassinin has been known as the parent cruciferous phytoalexin which is derived from the glucosinolates.

Various plant species of family Poaceae have been reported to produce chiefly the diterpenoid phytoalexins, i.e., kauralexins, zealexins (acid sesquiterpenoids), etc. Zealexins are diterpenoid phytoalexins formed in corn in response to microbial attack (*Aspergillus flavus*, *A. sojae*, *Colletotrichum sublineolum*, *Cochliobolus heterostrophus*, *Fusarium graminearum*, *Rhizopus microsporus*, *Ostrinia nubilalis*, *Ustilago maydis*) (Arruda et al. 2016; Huffaker et al. 2011). However, in rice leaves, sakuranetin (5,4′-dihydroxy-7-methoxyflavanone), a flavonoid phytoalexin, has been produced (Kodama et al. 1992), whereas in sorghum 3-deoxyanthocyanidin phytoalexins [viz., apigeninidin {2-(4-hydroxyphenyl)benzopyrylium chloride} and luteolinidin {2-(3,4-dihydroxyphenyl)-chromenylium 5,7-diol}] have been produced (Wharton and Nicholson 2000). Recently, Kariya et al. (2020) revealed that diterpenoid phytoalexins such as phytocassanes, momilactones, and oryzalexins have also been produced in rice. The phytocassanes (A and D) and momilactones (A and B) have been induced through UV light irradiations in rice leaves. Further, structure determination of these compounds revealed that an isomer of momilactone A possesses an abietane skeleton, whereas another isomer was di-dehydrogenated phytocassane A, which were called as oryzalactone and phytocassane G, respectively.

The spot blotch disease is caused by *Bipolaris sorokiniana* in family Poaceae, and when leaves of wheat were inoculated by *B. sorokiniana,* production of cinnamic acid amides (*N*-cinnamoyl-9-hydroxy-8-oxotryptamine, *N*-cinnamoyl-8-oxotryptamine) and *p*-coumaric acid amides (agmatine, hydroxyagmatine, hydroxyputrescine, hydroxydehydroagmatine) was induced (Ube et al. 2019). Out of these, two cinnamic acid amides were reported to be accumulated under *Fusarium graminearum* pathogen infection, $CuCl_2$ stress, and jasmonic acid treatments, revealing these amides to be phytoalexins. Moreover, 25 phenylamides have been observed to be accumulated after *B. sorokiniana* infection in wheat leaves such as hydroxycinnamic acid amides of serotonin, tryptamine, putrescine, and agmatine. Therefore, in wheat (family: Poaceae), two types of phenylamides have been found to be accumulated, viz., (1) *p*-coumaric acid amides (putrescine- and agmatine-related amines) and (2) cinnamic acid amides (indole amines).

Stilbenoids (phenylpropanoid compounds) have been reported to be accumulated in pine (family: Pinaceae), grape (family: Vitaceae), and peanut (family: Fabaceae) in response to various biotic/abiotic stresses (Ahuja et al. 2012; Yang et al. 2018). The stilbenoids have been identified as phytoalexins that protect various plants from microbial infections. Stilbenoid prenyltransferases (encoded by genes *AhR4DT-1* and *AhR3′DT-1*) have been reported to be a key step in the diversification of non-prenylated stilbenoids (peanut phytoalexins) into prenylated stilbenoids in peanut (Yang et al. 2018). More than 45 prenylated stilbenoids and its derivatives have been reported from peanut (family: Fabaceae). These prenylated stilbenoids in peanut can be categorized into two groups, viz., (1) derivative at the C-4 position and (2) derivative at the C-3′ position.

## 10.5 Differential Mechanisms of Action in Phytopathogens with Different Lifestyles

Since phytoalexins have been distributed among diverse chemical classes, their mechanisms of action is also different. Moreover, the mechanism of a particular phytoalexin may also differ against different types of pathogens.

Sesquiterpenoid phytoalexins from potato, i.e., lubimin and rishitin, are known to be associated with the resistance toward both *Phytophthora infestans* (late blight pathogen) and *Alternaria solani* (early blight pathogen). Gene silencing transgenic potato (*sesquiterpene cyclase* gene silenced using RNAi) deficient in phytoalexins (lubimin and rishitin) showed reduced post-invasive resistance to the avirulent *P. infestans* isolate. However, *A. solani* was found to penetrate leaf epidermal cells of the transgenic that led to the production of severe disease symptoms. Therefore, it was concluded that the antimicrobial phytoalexins differ in their mechanism to induce disease resistance based on the nature of the pathogen. The sesquiterpene cyclase-mediated phytoalexins may participate in pre-invasive resistance to necrotrophic pathogens like *A. solani* and post-invasive resistance toward hemi-biotrophic pathogens like *P. infestans* (Yoshioka et al. 2019).

## 10.6 De Novo Biosynthesis/Accumulation of Phytoalexins and Its Regulation

Phytoalexins are synthesized de novo (Ahuja, et al. 2012), against a number of biotic and abiotic stresses. Various elicitors of biotic origin, chemical elicitors, and a combination of these two have been reported to enhance the production of phytoalexins. In an experiment to study the effect of a biotic elicitor (autoclaved spore suspensions of *Verticillium dahliae*) were used to study the accumulation of the phytoalexin solavetivone, in the callus cultures of *Solanum melongena* (eggplant), an increased was observed both in dose- and time-dependent manner. The highest accumulation was observed in the susceptible eggplant cultivar *Long Purple*, 72 h after the elicitation with maximum dose of the *V. dahliae* spore suspensions (i.e., 2.0 mL) used (Kiran et al. 2017). In another experiment, they found that arachidonic acid was quite effective in stimulating the formation of phytoalexins (i.e., solavetivone and lubimin) in the callus suspension culture in the eggplant variety Long Purple, susceptible to wilting caused by *V. dahliae*. Under the in vitro conditions, solavetivone was formed in the tissue phase, whereas lubimin was formed in the filtrate phase (Kıran and Ellialtıoğlu 2019).

However, the mechanism of action of the elicitation of biotic or abiotic origin could be wide apart. It was found that the combination of a biotic elicitor (wall glucan elicitor from the *Phytophthora sojae*, a phytopathogen) has an additive effect on the elicitation of glyceollin I (Farrell et al. 2017). Among the abiotic elicitors, attempts have also been made to test the efficacy of nanomaterial elicitors for the enhanced production of phytoalexins. Ag (silver) nanoparticles induced the biosynthesis of phytoalexins, i.e., camalexin and related compounds (hydroxycamalexin O-hexoside and hydroxycamalexin malonyl-hexoside) in *Arabidopsis thaliana* (Kruszka et al. 2020).

Further, the transcriptional regulation of phytoalexin genes has also been reported from various plants; however, our current understanding on the transcriptional regulation is mainly developed from the studies in rice. Activation of *OsDCL1a* (a rice DICER-like ribonuclease) in *OsDCL1a* activation mutants suppresses the pathogen-inducible host defense response and negatively regulates diterpenoid phytoalexin production via the downregulation of diterpenoid phytoalexin biosynthetic genes that are weakly induced during pathogen infection than the wild type (Salvador-Guirao et al. 2019).

Various transcription factors have also been reported to be involved in the regulation of phytoalexin biosynthesis. In an experiment to find out the mechanism of regulation of the expression of rice diterpenoid phytoalexins, momilactones, it was found that OsTGAP1 (a chitin oligosaccharide elicitor-inducible basic leucine zipper transcription factor) was a key regulator. The knockout mutant for *OsTGAP1* had almost nil expression of the five clustered genes for production of momilactones upon elicitation, whereas the overexpression of *OsTGAP1* was leading to enhanced expression of these clustered genes, thereby leading to hyperaccumulation of momilactones upon elicitation. It was also recorded that the increased expression of *OsTGAP1* leads to the upregulation of *OsDXS3* (a rice MEP pathway gene), consequently leading to inductive production of momilactones, the diterpenoid phytoalexins produced in rice (Okada et al. 2009). However, interestingly, the overexpression of another bZIP transcription factor (basic leucine zipper TF), OsbZIP79, was found to suppress the production of diterpenoid phytoalexin in rice (Miyamoto et al. 2015). Another transcription factor, DPF of another class (a basic helix-loop-helix TF), was also found to play a key role in the biosynthesis of diterpenoid phytoalexins (momilactones and phytocassanes) in rice (Yamamura et al. 2015). Expression of all the diterpenoid phytoalexins biosynthetic genes and accumulation of both momilactones and phytocassanes were found increased its overexpressing lines, and were found decreased its knockdown lines.

It was found that a WRKY transcription factor of maize, ZmWRKY79, is involved in the accumulation of terpenoid phytoalexins. Its overexpression in maize protoplasts enhanced the expression of genes involved in maize terpenoid phytoalexin biosynthesis. ZmWRKY79 was found to play the role of a potential master regulator of the terpenoid phytoalexins through the involvement in phytohormone metabolism/signaling and ROS scavenging. The transient overexpression of *ZmWRKY79* in tobacco was able to provide resistance against *Rhizoctonia solani* infection via reducing ROS production (Fu et al. 2018). Since the promoter of *ZmWRKY79* contained a lot of G-boxes (the target cis-elements of MYC transcription factors), it was anticipated that maize MYC transcription factors might be involved in the regulation of *ZmWRKY79* gene, and thereby could have a role in JA-mediated phytoalexin regulation. Further, the overexpression of a MYC transcription factor gene, i.e., *ZmMYC2*, in *Arabidopsis* enhanced resistance toward *Botrytis cinerea* linking ZmMYC2 with a possible role in regulating JA-mediated defense responses (Fu et al. 2020).

Further, an ERF2-like transcription factor was found to positively regulate the production of a sesquiterpenoid phytoalexin, capsidiol, via direct transactivation of a capsidiol biosynthetic gene, EAS12, that cis accumulated in *Nicotiana attenuata* plants in response to *Alternaria alternata* (Song et al. 2019). Moreover, a NAC family transcription factor, $GmNAC_{42\text{-}1}$, was also recorded to be an essential positive regulator of glyceollin biosynthesis as the overexpression of $GmNAC_{42\text{-}1}$ in hairy roots was capable of increasing glyceollin synthesis by more than tenfold upon the elicitation (Jahan et al. 2019).

Moreover, PevD1 (a fungal effector from *Verticillium dahlia*) is capable of inducing HR (hypersensitive responses) like necrosis and SAR (systemic acquired resistance) by drastically inducing the expression of *Nbnrp1* (an asparagine-rich protein, i.e., NRP of *Nicotiana benthamiana*), thereby inducing resistance response. *Nbnrp1*-RNAi (gene silencing) lines showed reduced PevD1-induced immune responses by the modulation of differential expression genes (DEGs). These DEGs were found to be related to the sesquiterpenoid and triterpenoid biosynthesis, flavone and flavonol biosynthesis as well as the phenylpropanoid biosynthesis pathways, etc. (Liang et al. 2020).

## 10.7 Phytoalexins from Two Model Plants

In the recent years, our knowledge on plant defense and disease resistance are primarily based on the experimentation on model plant species. In this regard, *Arabidopsis* and rice have been two excellent model plants for plant defense at molecular level; however, the recent studies have also been using some alternative model plants. In the subsequent sections, we have tried to compile the information available on the phytoalexins of these two model plants available till date, to the best of our knowledge. The chemical structures of the major phytoalexins from these two model plants have been shown in Fig. 10.1.

### 10.7.1 Arabidopsis

*Arabidopsis thaliana* is a model plant of family Brassicaceae. In this plant, tryptophan-derived metabolites like an indolic phytoalexin, i.e., camalexin and indolic glucosinolates, indole-3-carboxylic acid derivatives, indole-3-carbonyl nitriles, etc., have been reported as end products of microbe-induced metabolic pathways (Pastorczyk et al. 2020); however, the major phytoalexin found in *A. thaliana* is camalexin (Kruszka et al. 2020), which has been extensively studied since its discovery.

Tsuji et al. (1992) reported the accumulation of a sulfur-containing indole derivative phytoalexin, i.e., camalexin (3-thiazol-2′-yl-indole), upon the inoculation of leaves of *A. thaliana* with the wheat pathogen, *Pseudomonas syringae* pv *syringae*, leading to a hypersensitive reaction (Fig. 10.1). Thereafter, five different

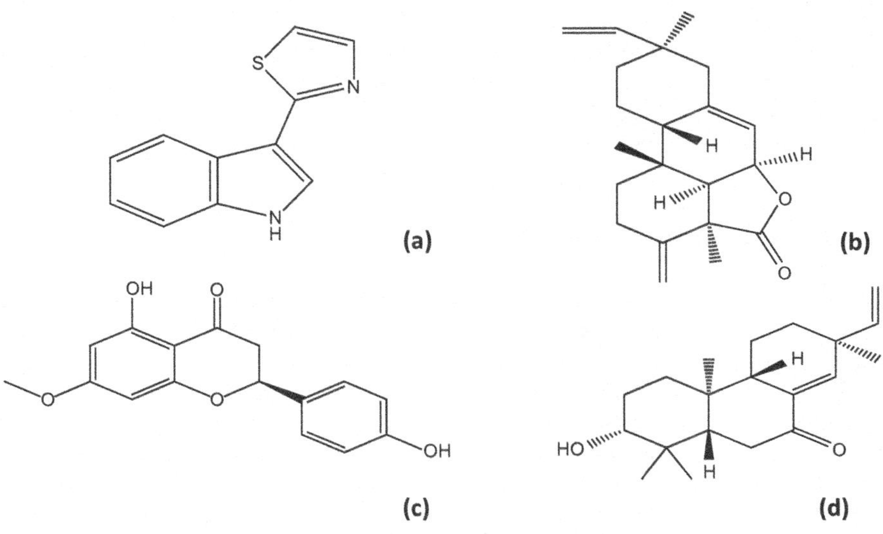


**Fig. 10.1** Phytoalexins from *Arabidopsis thaliana* and *Oryza sativa*. (**a**) Camalexin (from *Arabidopsis*), (**b**) momilactone A (from rice), (**c**) sakuranetin (from rice), and (**d**) oryzalexin A (from rice)

types of phytoalexin-deficient mutants have been identified in *Arabidopsis* (pad1-1, pad2-1, pad3-1, pad4-1, and pad5-1) which could produce the reduced amounts of camalexin upon the inoculation with bacterial pathogens (Glazebrook and Ausubel 1994; Glazebrook et al. 1997). Any such deficiency in phytoalexin production may lead to the enhanced susceptibility toward some phytopathogens. Thomma et al. (1999) reported that deficiency in camalexin production enhances the susceptibility of *A. thaliana* to its fungal pathogen *Alternaria brassicicola*. It has been found that silver nanoparticles are also able to induced the biosynthesis of camalexin and the related compounds (hydroxycamalexin O-hexoside and hydroxycamalexin malonyl-hexoside) in *A. thaliana* (Kruszka et al. 2020).

### 10.7.2 Rice

Rice is a well-studied model plant crop plant, belonging to the family Poaceae. The phytoalexins produced in rice plants have been classified in two chemical classes (Kariya et al. 2020), viz., diterpenes (momilactones A and B, oryzalexins A–F, oryzalexin S, phytocassanes A–E, and ent-10-oxodepressin) and flavanones (sakuranetin) (Cartwright et al. 1981, Kodama et al. 1992). Besides these, phenylamides are also now considered as the new class of rice phytoalexins that includes N-cinnamoyltryptamine and N-benzoyltryptamine (Morimoto et al. 2018).

Sakuranetin, the flavonoid phytoalexin in rice, is not only induced by plant pathogens but also by ultraviolet (UV) irradiation, $CuCl_2$ stress, and/or jasmonic acid (JA) treatment(s) (Shimizu et al. 2012). The phytotoxin coronatine also induces production of the two rice phytoalexins (viz., sakuranetin and momilactone A) in rice leaves. This coronatine-induced sakuranetin production was also found to be under the control of kinetin and ascorbic acid (AsA), as in case of JA, suggesting that both may interact at the same active site(s) (Tamogami and Kodama 2000).

In an experiment to determine the types/amounts of the diterpenoid phytoalexins that were analyzed in rice germplasm (wild and cultivated), it was found that there was a variation in amount at cultivar level. Further, two new diterpenoids, viz., oryzalactone and phytocassane G, were also recorded (Kariya et al. 2020).

## 10.8 Role in Promoting Human Health

Phytoalexins have been evolved as plant defense molecules, which protect these sessile organisms against a variety of pathogens. This has led to the perception that they might have some inherent bioactive, antimicrobial, or immune-modulatory properties which could have a wide range of implications in human health too. Therefore, in the recent past, they have been looked at for any such properties. Consequently, the scientific evidences pertaining a wide array of medicinal-/health-promoting properties, viz. antioxidant, antitumor, anticarcinogenic, antidiabetic, antiparasitic, cardioprotective, neuroprotective, growth-stimulating properties etc., started to emerge in the scientific literature. The list of phytoalexins reported to possess human health-promoting properties is shown in Table 10.2. The antiproliferative activity of 1-methoxybrassinin, a cruciferous phytoalexin, has been reported in leukemic T-Jurkat cells which was revealed through significant change in the fraction of cells having a sub-G0/G1 DNA content (Pilátová et al. 2005). This fraction of cells was significantly increased which is a marker of cell cycle arrest/cell death and induction of apoptosis. Further, induction of apoptosis was observed and confirmed through annexin V staining too.

A natural pentacyclic triterpenoid phytoalexins from olives have been referred to as maslinic acid which is having a huge potential in promoting human health. Maslinic acid has been reported to possess antidiabetic, antiparasitic, antitumor, cardioprotective, growth-stimulating, and neuroprotective agent (Pavel et al. 2016). The benzylamide derivative of maslinic acid, i.e., benzyl (2α, 3β) 2,3-diacetoxy-olean-12-en-28-amide, has also been reported to possess cytotoxic potential against A375 human malignant melanoma and B164A5 murine melanoma cell lines. Besides, antimicrobial/antibacterial/antifungal activity of this benzylamide derivative has been analyzed against *Enterococcus faecalis*, *Streptococcus pneumoniae*, *Escherichia coli*, *Bacillus cereus*, *Streptococcus pyogenes*, *Yersinia enterocolitica*, *Klebsiella pneumoniae*, *Proteus mirabilis*, *Staphylococcus aureus*, *Pseudomonas aeruginosa*, and *Candida albicans*. Benzyl (2α, 3β) 2,3-diacetoxy-olean-12-en-28-amide has been reported to be most effective against infections with *Staphylococcus aureus* and *Streptococcus pyogenes*. Various fungal pathogens and abiotic factors such as UV radiations or heat shock in soy plants trigger the formation of glyceollins phytoalexins (Bamji and Corbitt 2017).

In vitro studies on tumor cell lines and in vivo studies on animal models have revealed the anti-estrogenic activity of glyceollins which has been demonstrated through suppression of estrogen-responsive tumors by affecting estrogen receptors. Further analysis of lipid and glucose metabolism emphasized the osteoinductive, antioxidant, antimicrobial, and potential neuroprotective effect on central nervous system activity of glyceollins in a dose-dependent manner. Glyceollins phytoalexins have also been reported to play a crucial role in stimulation of signaling pathways of protein kinase, lipid kinase, and estrogen receptor (Pham et al. 2019). The glyceollins have been reported to possess antibacterial, antioxidative, antifungal, antinematode, anti-estrogenic, antiproliferative, anti-inflammatory, and anticholesterolemic potential. Due to numerous health-promoting benefits, glyceollins have been recommended as therapeutic/nutraceutical/dietary supplements.

The stimulation of Nrf2 {nuclear factor (erythroid-derived 2)-like 2}/HO-1(heme oxygenase 1) signaling pathway by the glyceollins phytoalexins derived from soybean improved the cognitive function of mice revealing its neuroprotective role against glutamate-induced damage (Seo et al. 2018). Glyceollins phytoalexins have also been reported to possess Nrf2-dependent memory-enhancing potential in the mice challenged with scopolamine. In human colorectal carcinoma HCT116 cells, indole phytoalexin derivative K-453 [(±)-*trans*-1,2-dimethoxy-2′-(3,5-bis-trifluoromethylphenylamino) spiro {indoline-3,5′ [4′,5′] dihydrothiazol}] induced mitochondrial-mediated apoptosis (Tischlerova et al. 2017). This apoptosis was

**Table 10.2** Phytoalexins reported to possess human health-promoting properties

| Sr. no. | Phytoalexin | Origin of phytoalexins (plant(s)) | Chemical class | Structure | Property(ies) | Reference(s) |
|---|---|---|---|---|---|---|
| 1 | Maslinic acid | Olives | Pentacyclic triterpenoid | 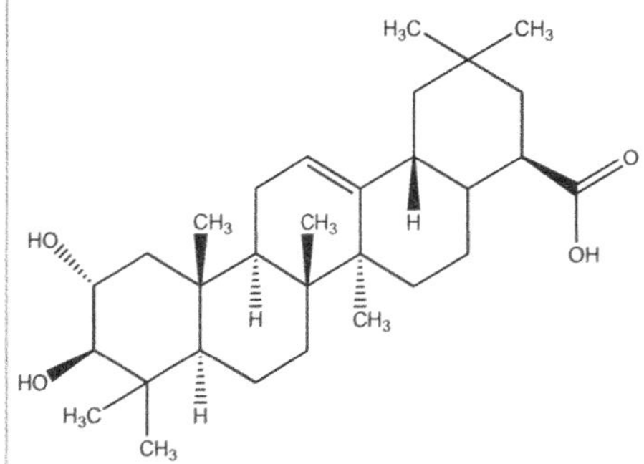 | Antimicrobial, cytotoxic against cancer cells | Pavel et al. (2016) |
| 2 | Resveratrol | Many plants belonging to different families (like grapes raspberries, blueberries, mulberries, scots pine, eastern white pine, and knotweed | Stilbenoid (a type of natural phenol) | 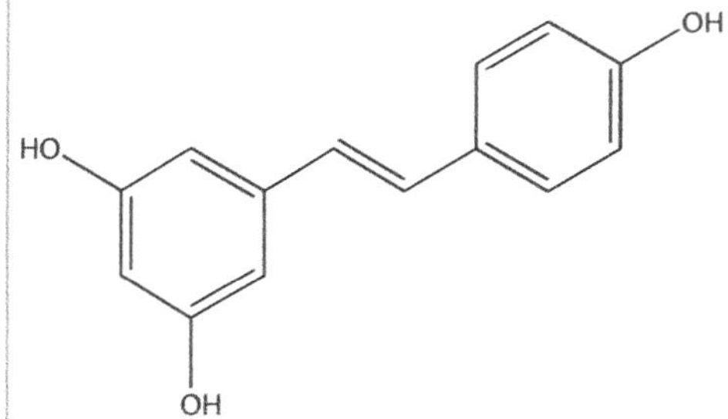 | Antioxidant, anti-inflammatory, antiproliferative, proapoptotic, and anti-angiogenic properties and useful in prostate cancer prevention | Jasiński et al. (2013) |

| | | | | | | |
|---|---|---|---|---|---|---|
| 3 | 1-Methoxybrassinin | Crucifers | Indole compound | 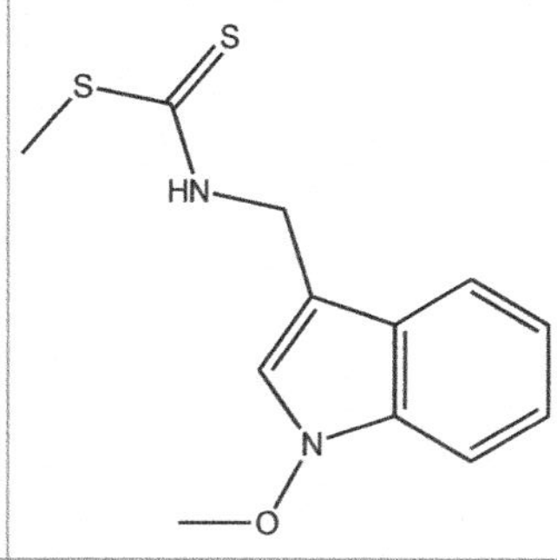 | Cell cycle arrest, induction of apoptosis, and antiproliferative activity | Pilátová et al. (2005) |
| 4 | Glyceollins (glyceollins I, II, and III) | Soybean | Pterocarpans (derivatives of isoflavonoids) | 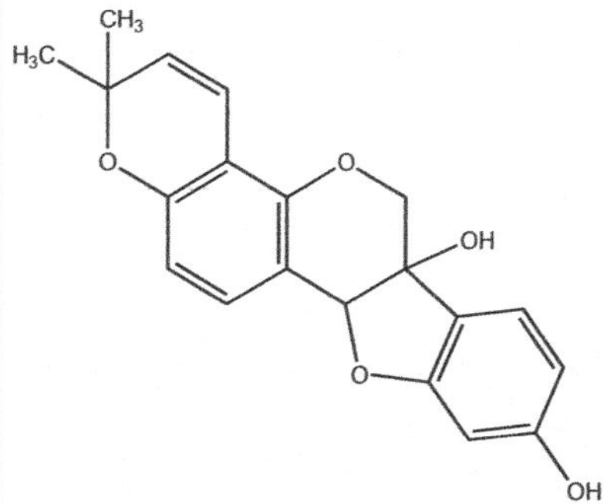 | Anti-estrogenic, antitumor, osteoinductive, antioxidant, antimicrobial, and potential central nervous system activity, improvement of cognitive function, activation of Nrf2/HO-1 signaling pathway, neuroprotective potential | Bamji and Corbitt (2017), Seo et al. (2018), Pham et al. (2019), Tischlerova et al. (2017) |

(continued)

**Table 10.2** (continued)

| Sr. no. | Phytoalexin | Origin of phytoalexins (plant(s)) | Chemical class | Structure | Property(ies) | Reference(s) |
|---|---|---|---|---|---|---|
| | | | | 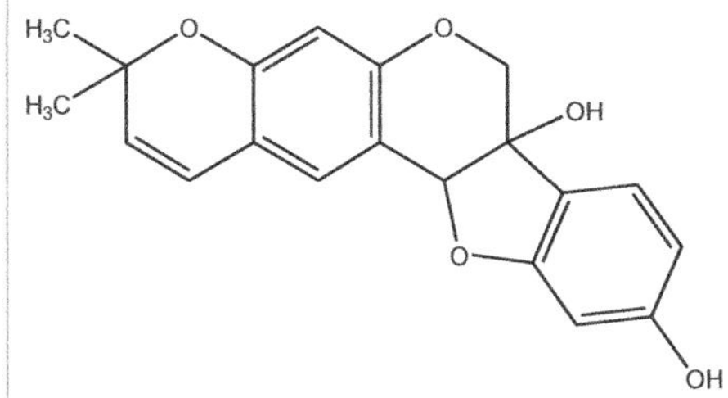 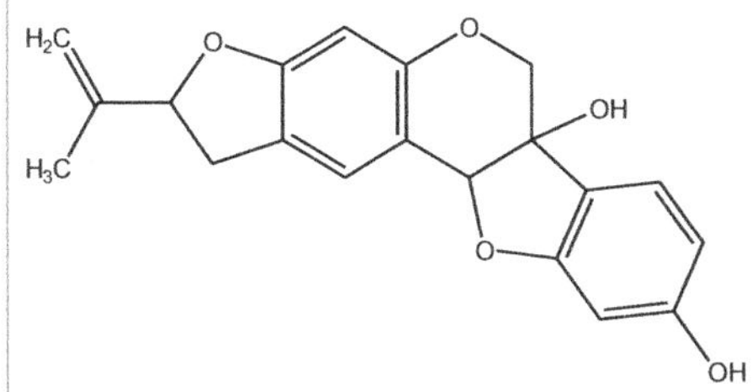 | | |
| 5 | Indole phytoalexin derivative K-453 | Crucifers | Indole compound | 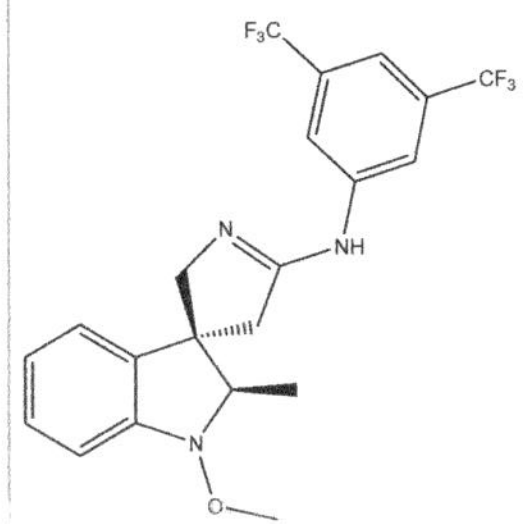 | Induced mitochondrial-mediated apoptosis in human colorectal carcinoma HCT116 cells | Tischlerova et al. (2017) |

| | | | | | | |
|---|---|---|---|---|---|---|
| 6 | Camalexin | *Arabidopsis* (crucifer) | Indole compound | 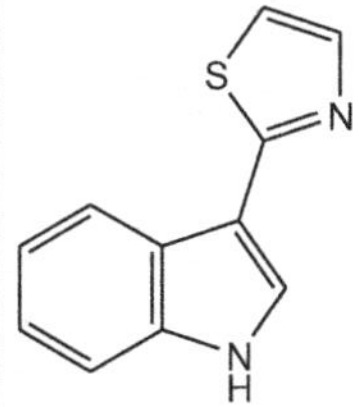 | Cytotoxic to prostate cancer cells | Smith et al. (2013) |
| 7 | Gossypol | Cotton (Malvaceae) | Terpenoid aldehyde formed by sesquiterpene dimerization | 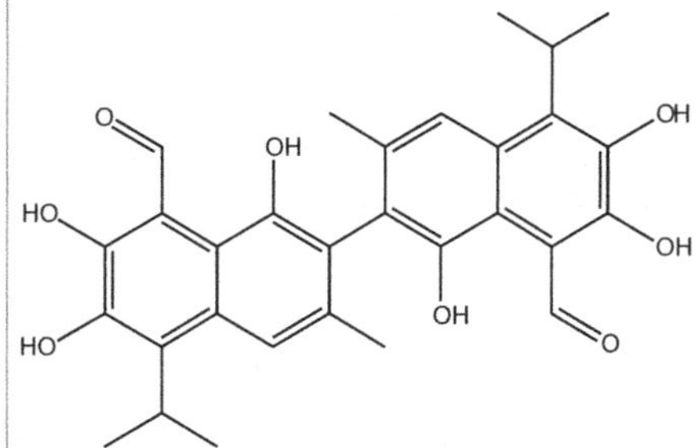 | Useful against malignant mesothelioma | Benvenuto et al. (2018) |
| 8 | Sakuranetin | Rice (Poaceae) | Methoxylated flavanone | 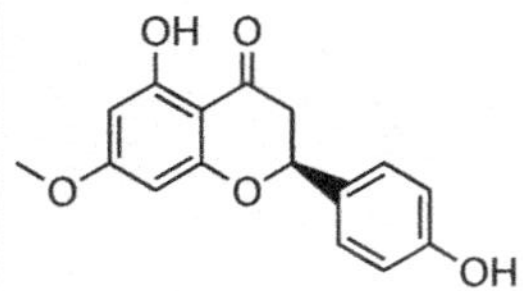 | Antiproliferative, antiviral, antioxidant, antimicrobial, anti-inflammatory, antiparasitic, antimutagenic, and anti-allergic properties | Stompor (2020) |

reported to be associated with the cleavage of PARP, activation of caspase-3 and caspase-9, release of cytochrome-*c*, stimulation of phosphorylation of p38 MAPK, altered concentrations of Bcl-2 family members, and loss of mitochondrial membrane potential as well as downregulation of NF-κB1 and RelA proteins resulting in the loss of their anti-apoptotic activity.

Resveratrol is a plant stilbenoid (a type of natural polyphenol) phytoalexin that is known to possess a number of beneficial health effects like antioxidant, anti-inflammatory, antiproliferative, proapoptotic, and anti-angiogenic. It has also been reported to have cardioprotective, neuroprotective, and immunomodulatory properties (Jasiński et al. 2013).

The phytoalexins from model plants like *O. sativa* and *A. thaliana* have been tested for their potential therapeutic benefits. In an experiment to evaluate the cytotoxic effect of camalexin (an indole-type phytoalexin) on prostate cancer cells, it was found that the more aggressive prostate cancer cells expressing higher levels of ROS (reactive oxygen species) are more sensitive to camalexin than the less aggressive prostate cancer cells (Smith et al. 2013). Besides camalexin, many other indole-type phytoalexins have been found to possess anticancer properties (Chripkova et al. 2016), whereas a rice methoxylated flavanone phytoalexin, i.e., sakuranetin is known to possess many antiproliferative, antiviral, antimicrobial, antioxidant, anti-inflammatory, antimutagenic, antiparasitic, and anti-allergic properties (Stompor 2020).

However, some phytoalexins have also been found to be toxic to humans. Gossypol, a sesquiterpenoid phytoalexin produced in cotton (Tian et al. 2016) against its pathogens and herbivores, besides being useful against the malignant mesothelioma (Benvenuto et al. 2018), is generally considered as a toxic metabolite (Sunilkumar et al. 2006). It may cause heart- and liver-related problems in humans (Rathore et al. 2020). Therefore, careful evaluation of toxicity/dosage is of paramount importance before their addition into foodstuffs in the form of nutraceuticals.

## 10.9 Phytoalexins as Food Preservatives

Since phytoalexins act as antimicrobial substances in plant defense, their possible role in food preservations has also been anticipated. They have the potential to be utilized as sustainable natural food preservation substances considering that they are abundantly present in plant kingdom. However, relatively few attempts have been made to study their role in food preservation. The prospective role as food preservatives has been discussed by Ejike et al. (2013). The increased accumulation and shelf life of phytoalexins in processed foods are currently major limiting factors in their applications in food industry. Moreover, the detoxification of phytoalexins by some microorganisms may affect their potential to act as natural preservatives (Jeandet 2015).

## 10.10 Transgenic Expression of Phytoalexins in Plants

Some key genes involved in the phytoalexin metabolism have been identified to promote disease resistance by their transgenic expression in other plants. One of the first genes tried for the above purpose was *stilbene synthase* gene, *STS* (coding for the enzyme catalyzing the last step of the phenylalanine/polymalonate pathway) in grapevine, resulting in the enhanced disease resistance (Jeandet et al. 2002). The transgenic expression of a fruit-specific *VqSTS6* gene from a wild relative, viz., *Vitis quinquangularis*, in grape plants also led to the enhanced resistance toward *Uncinula necator* causing powdery mildew (Cheng et al. 2016; Liu et al. 2019). However, such an overexpression of *stilbene synthase* could not be considered as a universal method of induction of resistance to every pathogen. When two genes encoding stilbene synthase from the grapevine were transferred to tomato via *Agrobacterium tumefaciens*-mediated method, it led to the accumulation of the phytoalexin trans-resveratrol that enhanced the resistance of transgenic tomato toward *Phytophthora infestans*, but no significant increase was detected against both *Botrytis cinerea* and *Alternaria solani* (Thomzik et al. 1997), suggesting differential action on the plant pathogens of different kinds. Expression of another *stilbene synthase* gene, *VqSTS36*, increased resistance toward powdery mildew and osmotic stress, but enhanced susceptibility toward *Botrytis cinerea* in both *Arabidopsis* and tomato (Huang et al. 2018).

The transgenic overexpression *CYP71Z18* (a gene involved in maize phytoalexin biosynthesis catalyzing sequential oxidation of β-macrocarpene to form zealexin A1) in rice resulted in accumulation of several new diterpenoids and displayed enhanced resistance against the blast pathogen, *Magnaporthe oryzae* (Shen et al. 2019).

## 10.11 Degradation/Detoxification of Phytoalexins

A warfare is not a one-sided affair and not the acquisition of weapons/counter-weapons. Plant defense using phytoalexins has been studied since the identification of the very first phytoalexin. Recently, the counter-defense strategies used by some phytopathogens have also been reported. It has been recorded that plant pathogens *Colletotrichum dematium* and *C. higginsianum* were able to metabolize and detoxify rapalexin A (an indole isothiocyanate phytoalexin, produced by the members of family Brassicaceae). However, another species of the same genera, *C. lentis*, was not able to detoxify it. Both the fungal pathogens were capable of detoxifying rapalexin A by the addition of the thiol group of L-Cys residue to the isothiocyanate carbon atom (usually catalyzed by glutathione transferases) (Pedras and Thapa 2020). Further, a cytochrome P450 CYP76 family enzyme, sesquiterpenoid phytoalexins hydroxylase, was found to detoxify of the solanaceous sesquiterpenoid phytoalexins, viz., rishitin, lubimin, oxylubimin, and solavetivone *in planta*, as the phytoalexins could have some phytotoxic effects too, besides being responsible for plant defense (toxic to phytopathogens), and need to be detoxified by the plant once the pathogen threat has been eliminated (Camagna et al. 2020).

## 10.12 Conclusion and Future Prospects

Phytoalexins were originally described as low molecular weight, inducible, antimicrobial secondary metabolites of plant with a beneficial role in plant defense. Till date, a wide range of plants belonging to diverse plant families have been found to produce various kinds of phytoalexins. These phytoalexins are belonging to varied chemical classes produced by complex biosynthetic pathways. However, their role in promoting human health has been recently worked out. Some of the beneficial roles include health-promoting effects in humans. A few of them have been reported to have antioxidant, anticancer, cardioprotective, and neuroprotective properties. Due to their antimicrobial properties, their possible role as food preservatives has also been anticipated; however, no such attempt has been yet commercialized due to inconsistency in their amount in elicited plant host. Moreover, due to the diverse chemical nature of different phytoalexins, their mode of action differs considerably. Because they are the products of long biosynthetic routes, scaling up of their production for therapeutic purpose remains a challenge, even after the sufficient information is available about some of the genes encoding enzymes for the phytoalexin biosynthetic pathways. Therefore, more basic research on their genomics and transcriptional regulation is required to optimize their production in bioreactors or *in planta*, at desirable levels for therapeutic use in humans, as food preservatives, and also in plant protection.

**Acknowledgment** Authors are thankful to the Vice-Chancellor of DAV University for the continuous encouragement and infrastructural support.

**Conflicts of Interests** Authors declare that they do not have any conflict of interests.

## References

Agrios GN (2005) Plant pathology, 5th edn. Academic Press, Elsevier. https://doi.org/10.1016/C2009-0-02037-6

Ahmed S, Kovinich N (2020) Regulation of phytoalexin biosynthesis for agriculture and human health. Phytochem Rev 20:483–505. https://doi.org/10.1007/s11101-020-09691-8

Ahuja I, Kissen R, Bones AM (2012) Phytoalexins in defense against pathogens. Trends Plant Sci 17(2):73–90

Arruda RL, Paz ATS, Bara MTF, Côrtes MVDCB, Filippi MCCD, Conceição ECD (2016) An approach on phytoalexins: function, characterization and biosynthesis in plants of the family *Poaceae*. Ciência Rural 46(7):1206–1216

Bailey JA, Mansfield JW (1982) Phytoalexins. Blackie, Glasgow. (No. 632 B3)

Bamji SF, Corbitt C (2017) Glyceollins: soybean phytoalexins that exhibit a wide range of health-promoting effects. J Funct Foods 34:98–105

Baskar V, Venkatesh R, Ramalingam S (2018) Flavonoids (antioxidants systems) in higher plants and their response to stresses, Antioxidants and antioxidant enzymes in higher plants. Springer, Cham, pp 253–268

Benvenuto M, Mattera R, Sticca JI, Rossi P, Cipriani C, Giganti MG, Volpi A, Modesti A, Masuelli L, Bei R (2018) Effect of the BH3 mimetic polyphenol (-)-gossypol (AT-101) on the *in vitro* and *in vivo* growth of malignant mesothelioma. Front Pharmacol 9:1269

Bizuneh GK (2020) The chemical diversity and biological activities of phytoalexins. Adv Tradit Med:1–13

Camagna M, Ojika M, Takemoto D (2020) Detoxification of the solanaceous phytoalexins rishitin, lubimin, oxylubimin and solavetivone via a cytochrome P450 oxygenase. Plant Signal Behav 15(2):1707348

Cartwright DW, Langcake P, Pryce RJ, Leworthy DP, Ride JP (1981) Isolation and characterization of two phytoalexins from rice as momilactones A and B. Phytochemistry 20(3):535–537

Cheng S, Xie X, Xu Y, Zhang C, Wang X, Zhang J, Wang Y (2016) Genetic transformation of a fruit-specific, highly expressed stilbene synthase gene from Chinese wild *Vitis quinquangularis*. Planta 243(4):1041–1053

Chripkova M, Zigo F, Mojzis J (2016) Antiproliferative effect of indole phytoalexins. Molecules 21(12):1626

Cruickshank IA, Perrin DR (1960) Isolation of a phytoalexin from *Pisum sativum* L. Nature 1960(187):799–800

Darvill AG, Albersheim P (1984) Phytoalexins and their elicitors—a defense against microbial infection in plants. Annu Rev Plant Physiol 35(1):243–275

Dixon RA, Dey PM, Lamb CJ (1983) Phytoalexins: enzymology and molecular biology. Adv Enzymol Relat Areas Mol Biol 55(1):69

Dugrand-Judek A, Olry A, Hehn A, Costantino G, Ollitrault P, Froelicher Y, Bourgaud F (2015) The distribution of coumarins and furanocoumarins in citrus species closely matches citrus phylogeny and reflects the organization of biosynthetic pathways. PLoS One 10:e0142757

Ejike CE, Gong M, Udenigwe CC (2013) Phytoalexins from the Poaceae: biosynthesis, function and prospects in food preservation. Food Res Int 52(1):167–177

El Oirdi M, Trapani A, Kamal B (2010) The nature of tobacco resistance against *Botrytis cinerea* depends on the infection structures of the pathogen. Environ Microbiol 12:239–253

Farrell K, Jahan MA, Kovinich N (2017) Distinct mechanisms of biotic and chemical elicitors enable additive elicitation of the anticancer phytoalexin glyceollin I. Molecules 22(8):1261

Fu J, Liu Q, Wang C, Liang J, Liu L, Wang Q (2018) ZmWRKY79 positively regulates maize phytoalexin biosynthetic gene expression and is involved in stress response. J Exp Bot 69(3):497–510

Fu J, Liu L, Liu Q, Shen Q, Wang C, Yang P, Zhu C, Wang Q (2020) ZmMYC2 exhibits diverse functions and enhances JA signaling in transgenic Arabidopsis. Plant Cell Rep 39(2):273–288

Glazebrook J, Ausubel FM (1994) Isolation of phytoalexin-deficient mutants of *Arabidopsis thaliana* and characterization of their interactions with bacterial pathogens. Proc Natl Acad Sci USA 91(19):8955–8959

Glazebrook J, Zook M, Mert F, Kagan I, Rogers EE, Crute IR, Holub EB, Hammerschmidt R, Ausubel FM (1997) Phytoalexin-deficient mutants of Arabidopsis reveal that PAD4 encodes a regulatory factor and that four PAD genes contribute to downy mildew resistance. Genetics 146(1):381–392

Gross D (1993) Phytoalexins of the Brassicaceae/Phytoalexine der Brassicaceae. Zeitschrift für Pflanzenkrankheiten und Pflanzenschutz/J Plant Dis Prot 100:433–442

Henfling J, Bostock R, Kuc J (1980) Effect of abscisic acid on rishitin and lubimin accumulation and resistance to *Phytophthora infestans* and *Cladosporium cucumerinum* in potato tuber tissue slices. Phytopathology 70(11):1074–1078

Huang L, Yin X, Sun X, Yang J, Rahman MZ, Chen Z, Wang X (2018) Expression of a grape *VqSTS36*-increased resistance to powdery mildew and osmotic stress in Arabidopsis but enhanced susceptibility to *Botrytis cinerea* in Arabidopsis and tomato. Int J Mol Sci 19(10):2985

Huffaker A, Kaplan F, Vaughan MM, Dafoe NJ, Ni X, Rocca JR, Alborn HT, Teal PE, Schmelz EA (2011) Novel acidic sesquiterpenoids constitute a dominant class of pathogen-induced phytoalexins in maize. Plant Physiol 156(4):2082–2097

Inoue Y, Sakai M, Yao Q, Tanimoto Y, Toshima H, Hasegawa M (2013) Identification of a novel casbane-type diterpene phytoalexin, ent-10-oxodepressin, from rice leaves. Biosci Biotechnol Biochem 77(4):120891

Jahan MA, Harris B, Lowery M, Coburn K, Infante AM, Percifield RJ, Ammer AG, Kovinich N (2019) The NAC family transcription factor GmNAC42-1 regulates biosynthesis of the anticancer and neuroprotective glyceollins in soybean. BMC Genomics 20(1):1–21

Jasiński M, Jasińska L, Ogrodowczyk M (2013) Resveratrol in prostate diseases-a short review. Central Eur J Urol 66(2):144–149

Jassbi AR, Zare S, Asadollahi M, Schuman MC (2017) Ecological roles and biological activities of specialized metabolites from the genus *Nicotiana*. Chem Rev 117:12227–12280

Jeandet P (2015) Phytoalexins: current progress and future prospects. Molecules 20(2):2770–2774

Jeandet P, Douillet-Breuil AC, Bessis R, Debord S, Sbaghi M, Adrian M (2002) Phytoalexins from the Vitaceae: biosynthesis, phytoalexin gene expression in transgenic plants, antifungal activity, and metabolism. J Agric Food Chem 50(10):2731–2741

Jeandet P, Hébrard C, Deville M-A, Cordelier S, Dorey S, Aziz A, Crouzet J (2014) Deciphering the role of phytoalexins in plant-microorganism interactions and human health. Molecules 19(11):18033–18056

Jeandet P, Clément C, Cordelier S (2019) Regulation of resveratrol biosynthesis in grapevine: new approaches for disease resistance? J Exp Bot 70(2):375–378

Kariya K, Ube N, Ueno M, Teraishi M, Okumoto Y, Mori N, Ueno K, Ishihara A (2020) Natural variation of diterpenoid phytoalexins in cultivated and wild rice species. Phytochemistry 180: 112518

Kıran S, Ellialtıoğlu Ş (2019) The effect of arachidonic acid elicitor on phytoalexin stimulation in eggplant callus suspension culture. Derim 36(2):141–145

Kiran S, Ellialtioğlu ŞŞ, Üstün AS, Mehmetoğlu Ü, Bayraktar H, Dolar FS (2017) Phytoalexin accumulations in the callus culture of two eggplant genotypes by using *Verticillium dahliae* Kleb. Elicitor. Int J Forestry Horticult 3(3):1–8

Klein AP, Sattely ES (2017) Biosynthesis of cabbage phytoalexins from indole glucosinolate. Proc Natl Acad SciUSA 114(8):1910–1915

Kodama O, Miyakawa J, Akatsuka T, Kiyosawa S (1992) Sakuranetin, a flavanone phytoalexin from ultraviolet-irradiated rice leaves. Phytochemistry 31(11):3807–3809

Kruszka D, Sawikowska A, Selvakesavan RK, Krajewski P, Kachlicki P, Franklin G (2020) Silver nanoparticles affect phenolic and phytoalexin composition of *Arabidopsis thaliana*. Sci Total Environ 716:135361

Kumar S, Pandey AK (2013) Chemistry and biological activities of flavonoids: an overview. Sci World J 2013:162750

Lee HA, Kim S, Kim S, Choi D (2017) Expansion of sesquiterpene biosynthetic gene clusters in pepper confers nonhost resistance to the Irish potato famine pathogen. New Phytol 215:1132–1143

Li R, Tee CS, Jiang YL, Jiang XY, Venkatesh PN, Sarojam R, Ye J (2015) A terpenoid phytoalexin plays a role in basal defense of *Nicotiana benthamiana* against *potato virus X*. Sci Rep 5:9682

Liang Y, Li Z, Zhang Y, Meng F, Qiu D, Zeng H, Li G, Yang X (2020) Nbnrp1 mediates *Verticillium dahliae* effector PevD1-triggered defense responses by regulating sesquiterpenoid phytoalexins biosynthesis pathway in *Nicotiana benthamiana*. Gene 768:145280. https://doi.org/10.1016/j.gene.2020.145280

Liu M, Ma F, Wu F, Jiang C, Wang Y (2019) Expression of stilbene synthase *VqSTS6* from wild Chinese *Vitis quinquangularis* in grapevine enhances resveratrol production and powdery mildew resistance. Planta 250(6):1997–2007

Meyer J, Berger DK, Christensen SA, Murray SL (2017) RNA-Seq analysis of resistant and susceptible sub-tropical maize lines reveals a role for kauralexins in resistance to grey leaf spot disease, caused by *Cercospora zeina*. BMC Plant Biol *17*(1):1–20

Miranda L, Maier CS, Stevens JF (2012) Flavonoids. In: eLS. Wiley, Chichester

Miyamoto K, Nishizawa Y, Minami E, Nojiri H, Yamane H, Okada K (2015) Overexpression of the bZIP transcription factor OsbZIP79 suppresses the production of diterpenoid phytoalexin in rice cells. J Plant Physiol 173:19–27

Morimoto N, Ueno K, Teraishi M, Okumoto Y, Mori N, Ishihara A (2018) Induced phenylamide accumulation in response to pathogen infection and hormone treatment in rice (*Oryza sativa*). Biosci Biotechnol Biochem 82:407–416

Müller K (1940) Experimentelle untcrsuchungcn uber die *Phytophthora*-resistem der kartoffel. Arb Biol Reichsasnstalt Landw Forstw Berlin 23:189–231

Müller KO, Börger H (1940) Experimentelle Untersuchungen Über die *Phytophthora infestans*-Resistenz der Kartoffel. Arb. Biol. Reichsanst. Land Forstwirtsch 97:189–231

Müller KO, Meyer G, Klinkowski M (1939) Physiologischgenetische Untersuchungen über die Resistenz der Kartoffel gegenüber *Phytophthora infestans*. Naturwissenschaften 27(46):765–768

Ohnishi M, Morishita H, Iwahashi H, Toda S, Shirataki Y, Kimura M, Kido R (1994) Inhibitory effects of chlorogenic acids on linoleic acid peroxidation and haemolysis. Phytochemistry 36(3):579–583

Okada A, Okada K, Miyamoto K, Koga J, Shibuya N, Nojiri H, Yamane H (2009) OsTGAP1, a bZIP transcription factor, coordinately regulates the inductive production of diterpenoid phytoalexins in rice. J Biol Chem 284(39):26510–26518

Pastorczyk M, Kosaka A, Pis'lewska-Bednarek M, López G, Frerigmann H, Kułak K, Glawischnig E, Molina A, Takano Y, Bednarek P (2020) The role of CYP 71A12 monooxygenase in pathogen-triggered tryptophan metabolism and *Arabidopsis* immunity. New Phytol 225(1):400–412

Pavel IZ, Danciu C, Oprean C, Dehelean CA, Muntean D, Csuk R, Muntean DM (2016) *In vitro* evaluation of the antimicrobial ability and cytotoxicity on two melanoma cell lines of a benzylamide derivative of maslinic acid. Anal Cell Pathol 2016:2787623

Paxton JD (1980) A new working definition of the term "phytoalexin". Plant dis 64(8):734

Paxton JD (1981) Phytoalexins-a working redefinition. J Phytopathol 101(2):106–109

Pedras MSC, Ahiahonu PW (2005) Metabolism and detoxification of phytoalexins and analogs by phytopathogenic fungi. Phytochemistry 66(4):391–411

Pedras MSC, Thapa C (2020) Unveiling fungal detoxification pathways of the cruciferous phytoalexin rapalexin A: sequential L-cysteine conjugation, acetylation and oxidative cyclization mediated by *Colletotrichum* spp. Phytochemistry 169:112188

Pedras MSC, To QH (2017) Defense and signalling metabolites of the crucifer *Erucastrum canariense*: synchronized abiotic induction of phytoalexins and galacto-oxylipins. Phytochemistry 139:18–24

Pedras MSC, Zheng QA, Sarma-Mamillapalle VK (2007) The phytoalexins from Brassicaceae: structure, biological activity, synthesis and biosynthesis. Nat Prod Commun 2(3):319–330

Pedras MSC, Chumala PB, Jin W, Islam MS, Hauck DW (2009) The phytopathogenic fungus *Alternaria brassicicola*: phytotoxin production and phytoalexin elicitation. Phytochemistry 70: 394–402

Pham TH, Lecomte S, Efstathiou T, Ferriere F, Pakdel F (2019) An update on the effects of glyceollins on human health: possible anticancer effects and underlying mechanisms. Nutrients 11(1):79

Pilátová M, Šarišský M, Kutschy P, Miroššay A, Mezencev R, Čurillová Z, Suchý M, Monde K, Mirossay L, Mojžiš J (2005) Cruciferous phytoalexins: antiproliferative effects in T-Jurkat leukemic cells. Leuk Res 29(4):415–421

Poloni A, Schirawski J (2014) Red card for pathogens: phytoalexins in sorghum and maize. Molecules 19(7):9114–9133

Rathore KS, Pandeya D, Campbell LM, Wedegaertner TC, Puckhaber L, Stipanovic RD, Thenell JS, Hague S, Hake K (2020) Ultra-low gossypol cottonseed: selective gene silencing opens up a vast resource of plant-based protein to improve human nutrition. Crit Rev Plant Sci 39(1):1–29

Salvador-Guirao R, Baldrich P, Tomiyama S, Hsing YI, Okada K, San Segundo B (2019). *OsDCL1a* activation impairs phytoalexin biosynthesis and compromises disease resistance in rice. Ann Bot 123(1):79–93

Schmelz EA, Huffaker A, Sims JW, Christensen SA, Lu X, Okada K, Peters RJ (2014) Biosynthesis, elicitation and roles of monocot terpenoid phytoalexins. Plant J 79(4):659–678

Schnippenkoetter W, Lo C, Liu G, Dibley K, Chan WL, White J, Milne R, Zwart A, Kwong E, Keller B, Godwin I, Krattinger SG, Lagudah E (2017) The wheat Lr34 multipathogen resistance gene confers resistance to anthracnose and rust in sorghum. Plant Biotechnol J 15(11):1387–1396

Seo JY, Kim BR, Oh J, Kim JS (2018) Soybean-derived phytoalexins improve cognitive function through activation of Nrf2/HO-1 signaling pathway. Int J Mol Sci 19(1):268

Sharma A, Sharma A, Kumar R, Sharma I, Vats AK (2021) PR proteins: key genes for engineering disease resistance in plants. In: Kumar P, Thakur AK (eds) Crop improvement: biotechnological advances. CRC Press, pp 81–98

Shen Q, Pu Q, Liang J, Mao H, Liu J, Wang Q (2019) CYP71Z18 overexpression confers elevated blast resistance in transgenic rice. Plant Mol Biol 100(6):579–589

Shimizu T, Lin F, Hasegawa M, Okada K, Nojiri H, Yamane H (2012) Purification and identification of naringenin 7-O-methyltransferase, a key enzyme in biosynthesis of flavonoid phytoalexin sakuranetin in rice. J Biol Chem 287(23):19315–19325

Shukla PK, Mishra P, Mishra N (2019) A prospective study on emerging roles of phytoalexins in plant protection. Int J Pharma Biol Sci *10*(3):186–198

Smeriglio A, Barreca D, Bellocco E, Trombetta D (2016) Chemistry, pharmacology and health benefits of anthocyanins. Phytother Res 30:1265–1286

Smith BA, Neal CL, Chetram M, Vo B, Mezencev R, Hinton C, Odero-Marah VA (2013) The phytoalexin camalexin mediates cytotoxicity towards aggressive prostate cancer cells via reactive oxygen species. J Nat Med 67(3):607–618

Song N, Ma L, Wang W, Sun H, Wang L, Baldwin IT, Wu J (2019) An ERF2-like transcription factor regulates production of the defense sesquiterpene capsidiol upon *Alternaria alternata* infection. J Exp Bot 70(20):5895–5908

Stahl E, Bellwon P, Huber S, Schlaeppi K, Bernsdorff F, Vallat-Michel A, Mauch F, Zeier J (2016) Regulatory and functional aspects of indolic metabolism in plant systemic acquired resistance. Mol Plant 9(5):662–681

Stompor M (2020) A review on sources and pharmacological aspects of sakuranetin. Nutrients 12(2):513

Sunilkumar G, Campbell LM, Puckhaber L, Stipanovic RD, Rathore KS (2006) Engineering cottonseed for use in human nutrition by tissue-specific reduction of toxic gossypol. Proc Natl Acad Sci 103(48):18054–18059

Tamogami S, Kodama O (2000) Coronatine elicits phytoalexin production in rice leaves (*Oryza sativa* L.) in the same manner as jasmonic acid. Phytochemistry 54(7):689–694

Thomma BP, Eggermont K, Tierens KFJ, Broekaert WF (1999) Requirement of functional ethyleneinsensitive 2 gene for efficient resistance of Arabidopsis to infection by Botrytis cinerea. Plant Physiol 121(4):1093–1101

Thomzik JE, Stenzel K, Stöcker R, Schreier PH, Hain R, Stahl DJ (1997) Synthesis of a grapevine phytoalexin in transgenic tomatoes (*Lycopersicon esculentum* Mill.) conditions resistance against *Phytophthora infestans*. Physiol Mol Plant Pathol 51(4):265–278

Tian X, Ruan J, Huang J, Fang X, Mao Y, Wang L, Chen X, Yang C (2016) Gossypol: phytoalexin of cotton. Sci China Life Sci 59(2):122–129

Tiku AR (2020) Antimicrobial compounds (phytoanticipins and phytoalexins) and their role in plant defense. In: Mérillon JM, Ramawat K (eds) Co-evolution of secondary metabolites. Reference series in phytochemistry. Springer, Cham, pp 845–868. https://doi.org/10.1007/978-3-319-96397-6_63

Tischlerova V, Kello M, Budovska M, Mojzis J (2017) Indole phytoalexin derivatives induce mitochondrial-mediated apoptosis in human colorectal carcinoma cells. World J Gastroenterol 23(24):4341

Tsuji J, Jackson EP, Gage DA, Hammerschmidt R, Somerville SC (1992) Phytoalexin accumulation in *Arabidopsis thaliana* during the hypersensitive reaction to *Pseudomonas syringae* pv syringae. Plant Physiol 98(4):1304–1309

Ube N, Harada D, Katsuyama Y, Osaki-Oka K, Tonooka T, Ueno K, Taketa S, Ishihara A (2019) Identification of phenylamide phytoalexins and characterization of inducible phenylamide metabolism in wheat. Phytochemistry 167:112098

VanEtten HD, Mansfield JW, Bailey JA, Farmer EE (1994) Two classes of plant antibiotics: phytoalexins versus "phytoanticipins". Plant Cell 6(9):1191–1192

Wharton PS, Nicholson RL (2000) Temporal synthesis and radiolabelling of the sorghum 3-deoxyanthocyanidin phytoalexins and the anthocyanin, cyanidin 3-dimalonyl glucoside. New Phytol 145(3):457–469

Yamamura C, Mizutani E, Okada K, Nakagawa H, Fukushima S, Tanaka A, Maeda S, Kamakura T, Yamane H, Takatsuji H, Mori M (2015) Diterpenoid phytoalexin factor, a bHLH transcription factor, plays a central role in the biosynthesis of diterpenoid phytoalexins in rice. Plant J 84(6):1100–1113

Yang T, Fang L, Sanders S, Jayanthi S, Rajan G, Podicheti R, Thallapuranam SK, Mockaitis K, Medina-Bolivar F (2018) Stilbenoid prenyltransferases define key steps in the diversification of peanut phytoalexins. J Biol Chem 293(1):28–46

Yoshioka M, Adachi A, Sato Y, Doke N, Kondo T, Yoshioka H (2019) RNAi of the sesquiterpene cyclase gene for phytoalexin production impairs pre-and post-invasive resistance to potato blight pathogens. Mol Plant Pathol 20(7):907–922

Zhang P, Dreher K, Karthikeyan A, Chi A, Pujar A, Caspi R, Karp P, Kirkup V, Latendresse M, Lee C, Mueller LA, Muller R, Rhee SY (2010) Creation of a genome-wide metabolic pathway database for *Populus trichocarpa* using a new approach for reconstruction and curation of metabolic pathways for plants. Plant Physiol 153:1479–1491

# Secondary Metabolites from Plants: Role in Plant Diseases and Health Care

11

Rahul Datta, Ashutosh Sharma, and Abhinay Thakur

**Abstract**

Plant secondary metabolites are one of the most diverse and fascinating classes of phytochemicals that accumulate in plant tissues for a variety of functions. These metabolites are believed to play various important roles such as to provide protection to plant against attack by insects, herbivores, and pathogens. Moreover, plants are known for their ability to produce nonenzymatic antioxidants that are capable of neutralizing the reactive oxygen species (ROS)-induced oxidative damage. Curcumin, a polyphenolic compound extracted from the rhizomes of *Curcuma longa* (turmeric), has been shown to have anticancer activity against several types of cancers such as lung, breast, pancreatic, and skin cancers, leukemia, as well as brain tumor. Thus, several pharmaceutical properties, viz., anticancer, antioxidant, antimicrobial, antidiabetic, etc., have been reported from these plant secondary metabolites.

**Keywords**

Plant secondary metabolites · Therapeutic · Health care · Curcumin · Anticancer · Antioxidant

R. Datta
Centre for Agricultural Research and Innovation (CARI), Guru Nanak Dev University, Amritsar, India

A. Sharma
Faculty of Agricultural Sciences, DAV University, Jalandhar, Punjab, India

A. Thakur (✉)
PG Department of Zoology, DAV College, Jalandhar, Punjab, India


A. K. Sharma, A. Sharma (eds.), *Plant Secondary Metabolites*,
https://doi.org/10.1007/978-981-16-4779-6_11

## 11.1 Introduction

Plant secondary metabolites are one of the most diverse and fascinating classes of phytochemicals that accumulate in plant tissues for a variety of functions (Gandhi et al. 2015; Ahmed et al. 2017). These metabolites encompass diverse groups of compounds which have no direct function in important processes, viz., respiration, solute transport, photosynthesis, protein synthesis, nutrient assimilation, etc. (Mera et al. 2019). These metabolites are believed to play various important roles (Fig. 11.1), viz., to provide protection to plant against attack by insect, herbivores, and pathogens (Schwekendiek et al. 2007; Naoumkina et al. 2008), survive from abiotic stresses (temperature, water, radiation, chemicals) (Xu et al. 2008; Schäfer and Wink 2009), act as attractants for pollinators (Kessler and Baldwin 2007; Gonzalez-Teuber and Heil 2009), and act as signal molecules (Xu et al. 2009).

Plant secondary metabolites are low molecular mass natural products produced by plants, fungi, and other organisms with great structural diversity (Dixon 2001; Stevenson et al. 2017). Resistance in plant depends upon the concentration of secondary metabolites present, and their production in plant is very costly for them, resulting in suppression in growth and development (Siemens et al. 2002). Secondary metabolites from the plant have been reported to play a role in defense against herbivores, insects, etc. by acting as deterrence, antifeedant activity, toxicity etc. (Nathan 2013). Their various functions as well as immeasurable pharmacological activities make them suitable candidate for research (Gandhi et al. 2015). Some of the metabolites isolated from the plants have been and are still used to cure several infections and diseases and also used as spices, pesticides, perfumes, toxin, etc. (Wink 2010, 2015). Crop loss due to pathogens is a serious issue throughout the world. Plants are known to respond against pathogens by activating genes, forming

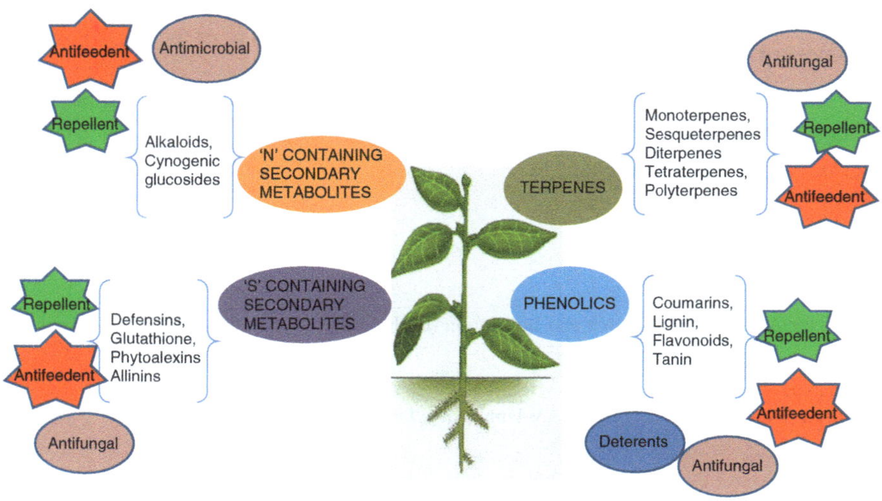


**Fig. 11.1** Plant secondary metabolites and their role in plant defense system

**Table 11.1** Classification of terpenes

| No. of carbon atom | Class | Formula |
|---|---|---|
| 5 | Hemiterpene | $C_5H_8$ |
| 10 | Monoterpene | $C_{10}H_{16}$ |
| 15 | Sesquiterpene | $C_{15}H_{24}$ |
| 20 | Diterpene | $C_{20}H_{32}$ |

pathogen-related proteins, as well as producing several antimicrobial metabolites (Ahuja and Kissen 2012). In response to stresses caused by pathogens, plants produce low molecular mass secondary metabolites with antimicrobial potential known as phytoalexins, a heterogeneous group of compound (Hammerschmidt 1999; Pedras et al. 2011; Schmelz et al. 2011; Huffaker et al. 2011).

During the ancient times, plant-derived products have been used as medicine for cure of common ailments (Crozier et al. 2006). Around 7500 species of plants are being used in ethnomedicines in India (Shankar and Majumdar 1997). Most of the secondary metabolites have pharmacological properties and play an important role in treatment of diseases in both humans and animals and also of great interest to mankind for being used as drugs, oils, waxes, perfumes, flavoring agents, dyes, and many other commercial applications (Kabera et al. 2014; Aniszewski 2007; Park et al. 2001; Ahmed et al. 2017). The pharmaceutical properties of secondary metabolites, viz., anticancerous, antimicrobial, anticancer, anti-inflammatory, antioxidant, etc., play an important role in human health care (Zaynab et al. 2018; Tungmunnithum et al. 2018; Seca and Pinto 2018).

Plant secondary metabolites have been classified into three important types based on their biosynthetic pathway. These are terpenes, phenolics, and alkaloids.

**Terpenes** Terpenes are one of the largest classes of secondary metabolites having five carbon isoterpenoids as their basic structure (Kortbeek et al. 2019). Around 25,000 different types are present in different plants (Chen et al. 2011). Terpenes are classified on the basis of the carbon atom and isoprene unit present in their structure (Table 11.1). The simplest of them is hemiterpene with single isoprene unit and five carbon atoms. If the two isoprene residues are present, then it is called as monoterpene, with three isoprene residues in sesquiterpene, four in diterpenes, etc. (Dewick 2009; Taiz and Zeiger 2010). The synthesis of terpenes depends upon mevalonate and methylerythritol phosphate (MEP) pathways that generate the $C_5$ unit (Singh and Sharma 2015). Most of them are lipophilic and readily interact with biomembranes and membrane protein leading to the increase in the permeability and fluidity of the membranes (Wink 2007, 2015). Terpenes are found in different plant parts and have several important roles, viz., defense, odor, antimicrobial, anti-inflammatory, antiparasitic, insecticidal, plant growth regulators, etc. (Mera et al. 2019; Wink 2015). Steroids, carotenoids, and gibberellic acid belong to terpenes.

**Alkaloids** Alkaloids are extremely heterogenous group of naturally occurring organic compounds having low molecular weight with nitrogen present in their heterocyclic ring (Shields et al. 2008). Most of them are found in higher plant and

**Table 11.2** Classification of plant phenolics

| Sr. no | Number of carbon atom | Basic skeleton | Category |
|---|---|---|---|
| 1 | N | $[C_6–C_3]_n$, $[C_6]_n$ $[C_6–C_3–C_6]_n$ | Lignins, melanins, condensed tannins (flavolans) |
| 2 | 6 | $C_6$ | Simple phenol |
| 3 | 7 | $C_6–C_3$ | Phenolic acids |
| 4 | 8 | $C_6–C_2$ | Phenylacetic acid, hydroxycinnamic acids, phenylpropenes |
| 5 | 9 | $C_6–C_3$ | Coumarins, isocoumarin, chromones |
| 6 | 10 | $C_6–C_4$ | Naphthoquinones |
| 7 | 13 | $C_6–C_1–C_4$ | Xanthones, stilbenes |
| 8 | 14 | $C_6–C_2–C_6$ | Anthraquinones |
| 9 | 15 | $C_6–C_3–C_6$ | Flavonoids |
| 10 | 18 | $[C_6–C_3]_2$ | Lignans, neolignans |
| 11 | 30 | $[C_6–C_3–C_6]$ | Biflavonoids |

mostly accumulate in actively growing parts of plants, bundle sheath, latex vessel, as well as nutritious tissue. Alkaloids are synthesized from amino acids such as phenylalanine, arginine, purines, and putrescine (Ashihara et al. 2008; De Luca and St Pierre 2000). Many of them possess important pharmaceutical properties. Based on their biosynthetic origin, alkaloids are divided into three types, viz., protoalkaloids, true alkaloids, and pseudoalkaloids. Both protoalkaloids and true alkaloids are originated from amino acids, whereas pseudoalkaloids are not derived from them.

**Phenolics** Phenols are the molecules containing one or more hydroxylated benzene ring structures. Synthesis of phenols in plant occurs by phenylpropanoid pathway (Kortbeek et al. 2019). Around 8000 phenolic compounds from the plant has been discovered so far (Kumar and Pandey 2013; Ahmed et al. 2016). Plant phenolics are rich in antioxidants and other bioactive agents which provide benefits for health as well as help in the prevention of many diseases (Tungmunnithum et al. 2018). Phenols are classified into various types on the basis of carbon atom present in their molecule. The simplest molecule is phenol. Table 11.2 shows various types of phenols, viz., acidic phenols, acetophenones, phenylacetic acids, hydroxycinnamic acids, phenylpropenes, coumarins, isocoumarin, chromones, naphthoquinones, xanthones, stilbenes, anthraquinones, flavonoids, lignans, neolignans, and biflavonoids (Kabera et al. 2014). Plant phenolics are complex in their structure; some are soluble in organic solvent while others are not. Phenolics play several important functions, viz., act as deterrents against herbivores and pathogens, attract pollinators, play a role in allelopathy, as well as provide mechanical strength to plant (Alasalvar et al. 2001; Acamovic and Brooker 2005; Edreva et al. 2008). These phytochemical substances have also been possessing several pharmaceutical properties, viz., anticancer, antibacterial, antioxidant, anti-inflammation, cardioprotective, immune system promoting, etc. (Kumar and Pandey 2013; Chen et al. 2015; Meng et al. 2018).

## 11.2 Plant Disease and Resistance

In order to protect themselves against damage, plants have develop a variety of defenses such as constitutive (cell wall, waxy epidermal cuticles, and bark) and inducible (pathogen degrading enzymes, production of toxic chemicals, and deliberate cell suicide) defenses (War et al. 2012). There are two types of pathogens which attack plants. These are biotrophs and necrotrophs. The former attack the host by feeding inside living tissue and keep the host alive, e.g., the powdery mildew fungus *Blumeria graminis* and bacterial rice pathogens *Xanthomonas oryzae*, whereas the latter produce toxins or tissue degrading enzymes that suppress the plant defenses and result in nutrient release. e.g., the gray mold fungus *Botrytis cinerea* and the bacterial soft rot pathogen *Erwinia carotovora* (Freeman and Beattie 2008). Some pathogens that are biotrophic during early stages of infection but become necrotrophic during the latter stages are called hemibiotrophs. e.g., *Magnaporthe grisea*, the causative agent of rice blast disease. Resistance against diseases in plants depends on the activation of coordinated, multicomponent defense mechanisms. One of the most important mechanisms for disease resistance in plants is the increase production of secondary metabolites with high antimicrobial potential, viz., terpenes, tannins, isoflavonoids, coumarins, polyacetylenes, and quinines (Reichling 2018). The synthesis of these metabolites in plants occurs in response to pathogen attack, i.e., perception of microbe-associated molecular pattern (MAMP) (Ahuja and Kissen 2012; Zaynab et al. 2018). Recognition of these molecules resulted in various defense responses, viz., alteration in the cell wall, callose deposition, and production of several defense-related proteins such as glucanases, chitinases, and proteases, resulting in inhibitory effect on pathogens (Van Loon et al. 2006). Those compounds which are produced upon pathogen perception are called as phytoalexins and those which are synthesized in inactive form and converted into active form (toxic form) after pathogen recognition are called as phytoanticipins (Piasecka et al. 2015) (Table 11.3).

## 11.3 Role of Phenolics in Plant Disease Resistance

Phenolics are one of the important plant secondary metabolites which play an important role against pathogens responsible for several types of diseases in plants and include flavonoids, anthocyanins, tannins, lignin, phytoalexins, and coumarins. Plant phenolics are known to act as antifungal, antibacterial, and antiviral compounds (Kumar et al. 2014). Biosynthesis of phenolic compounds increases under stress condition, i.e., pathogen infection. Flavonoid is an important plant phenolic, whose production increases during microbial infection injury, etc.

**Table 11.3** Antimicrobial activity of some secondary plant metabolite (Wink 1988; Rao et al. 2010, Astani and Schnitzler 2014, Yourman and Jeffers 1999; Weidenbörner et al. 1992, Del Río et al. 2003; Friend 1979; Johnson 1976)

| Biological activity | Plant secondary metabolite | | |
|---|---|---|---|
| | Terpenes | Alkaloids | Phenols |
| Antiviral | Linalool, pinene limonene | Lycorine, sparteine | Polyphenols, flavonoids |
| Antibacterial | Citronellal, cineole, essential oils | Berberine | Catechol and coumarin |
| Antifungal | Phellandine, gossypol, limonene, geraniol, citrol, furanocoumarin, rishitin, carvacrol | Solanine, lupanine, | Protocatechuic acid, Chlorogenic acid, tannin, stilbene, benzaldehyde, flavones, oleuropein, nobiletin, genistein |

(Michalak 2006). These compounds play an important role in growth and development and defense against infection by acting as antibacterial and antifungal agents (Reichling 2018). Myricetin and robinetin (flavonols) are among the most active antibacterial agents. Among the flavonoids, isoflavonoids possess antibiotic and antifungal properties and are produced in response to pathogen infection. Being pathogen specific in their toxicity, these toxic molecules inhibit metabolism as well as cellular structure in pathogen (Freeman and Beattie 2008). For example, medicarpin is produced by alfalfa (*Medicago sativa*) during pathogen attack, rishitin is produced by tomatoes and potatoes, and camalexin is produced by *A. thaliana*. Lignin a highly branched heterogenous polymer is the primary component of wood. It is rigid and insoluble and acts as an excellent physical barrier against pathogen (Freeman and Beattie 2008). Furanocoumarins are also produced by several plant species in response to pathogen attack. Several plants are resistant to diseases caused by pathogens due to the presence of more than one antimicrobial compounds in their cells. These compounds are called as phytoanticipins. Tannins important phenolics present in higher concentrations in leaves, fruits, and seeds which provide resistance to young tissues from pathogenic microorganisms. Treatment of wheat with lignin synthesis inhibitors provides resistance against plant pathogen *Puccinia graminis* (Boudet 2000). Resistance in transgenic tomato plant to powdery mildew is due to the increased concentration of 9-hexadecanoic acid in cutin monomers which suppress the germination of powdery mildew spores (Agrios 2005). Maddox et al. (2010) reported that several plant phenolics, viz., phenolic acid, coumarin, stilbene, and flavonoid, showed antibacterial activity against *Xylella fastidiosa*, a pathogenic bacterium that causes diseases in many crop species. Among several plant phenolics, catechol, caffeic acid, and resveratrol exhibited strong anti-*Xylella* activities. Plant phenolics having less complex structure, i.e., catechol and coumarin, possessed both bactericidal and fungicidal activities against pathogens (Cowan 1999). Phenolic phytoalexins can protect plant against pathogens. For example, stilbene phytoalexins produced by Vitaceae (grapes), i.e., resveratrol (3,4′,5-trihydroxy trans-stilbene), play an important role in providing resistance to grapevine from colonization of fungi (Jeandet et al. 2002). Similarly, stilbene also provides resistance to transgenic

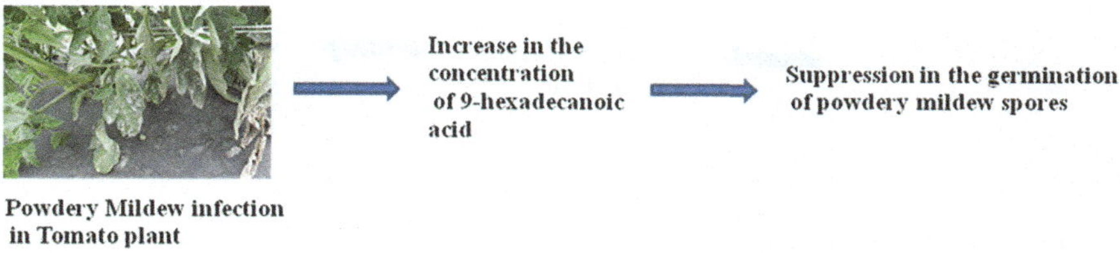


**Fig. 11.2** Antimicrobial activity of plant phenolics

tobacco plant against *Botrytis cinerea* infection (Hain et al. 1993). Among several phenolic metabolites present in plants, polyphenols and flavonoids have been found to possess antiviral properties. Isoflavonoid, a low molecular weight phytoalexin, has been found to show antibiotic and antifungal properties in response to pathogen attack. The best example is medicarpin produced by *Medicago sativa* (alfalfa) and camalexin by *A. thaliana* (Ahuja and Kissen 2012) (Fig. 11.2).

## 11.4 Role of Terpenes in Plant Disease Resistance

Terpenes are one of the largest groups of plant secondary metabolites having 5-C isoterpenoid as their basic structure assembled to form different compounds. On the basis of isoprene unit, terpenes are classified into various types, viz., monoterpenoids (2 units), sesquiterpenoids (3 units), diterpenoids (4 units), and triterpenoids (6 units), possessing variant functional group with addition or subtraction of oxidized methyl group (Perveen and Al-Taweel 2018; Ortiz de Elguea-Culebras et al. 2017; Mera et al. 2019). Essential oils from the plants are rich in monoterpenoids and sesquiterpenoids which are volatile compounds and play an important role in plant defense due to their ability to permeate the skin and show various biological activities (Wojtunik-Kulesza et al. 2019). Various reports are available regarding their biological activities against different organisms. According to Wang et al. (2019) phenolic monoterpenes showed inhibitory effect on plant pathogenic oomycete (*Phytophthora nicotianae*, *Phytophthora capsici*) and hyphomycete (*Alternaria solani*, *Fusarium oxysporum*) fungus. Monoterpenes showed antifungal effect against wood-degrading fungi which is a serious issues nowadays obstructing long-term use of wood (Zhang et al. 2016). For example, mint plant produces monoterpenoids menthol and menthone and pyrethrin produced by *Chrysanthemum* plant (Freeman and Beattie 2008). Gossypol is an important diterpenoid produced by cotton possessing strong antifungal and antibacterial properties (Mellon et al. 2011; Mellon et al. 2014; Przybylski et al. 2009). Saponins are glycosylated triterpenoids having detergent-like properties present in the cell membrane of many plants and are known to inhibit the cell membrane of fungal pathogen (Freeman and Beattie 2008). Oat varieties that contain avenacins (triterpenoid saponins) are resistant to *Gaeumannomyces graminis*, a wheat pathogen. Terpenoids are known to provide defense against various biotic stresses.

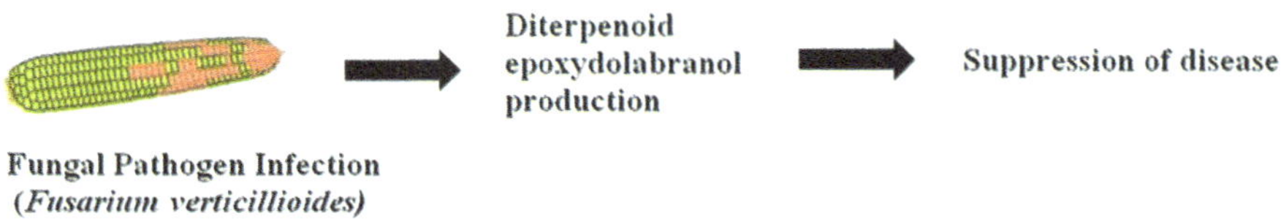


**Fig. 11.3** Antimicrobial activity of terpenes (diterpene)

Diterpenoid epoxydolabranol production is increased rapidly in maize roots during infection of *Fusarium verticillioides* and *Fusarium graminearum*, a fungal pathogen, and it also showed strong antimicrobial activity against both pathogens in vitro studies (Mafu et al. 2018). Several types of diterpenoids, viz., momilactones, oryzalexins, and phytocassanes, are produced by rice plant, and their concentrations are induced in leaves in response to pathogenic fungus *Magnaporthe grisea* (Prisic et al. 2004). Carvacrol has been reported to possess strong antifungal potential against *Pseudomonas aeruginosa*, *Staphylococcus aureus*, and *Candida albicans* (Rao et al. 2010). Essential oils extracted from several plants, viz., *Mentha*, *Cinnamomum*, *Allium*, *Rosmarinus*, and *Origanum*, have been reported to exhibit antiviral and antifungal activity (Seow et al. 2014). Juárez et al. (2015) revealed that essential oils extracted from the endemic Mexican plants *Agastache mexicana* ssp. *olocotziana* and *Porophyllum linaria* showed antifungal activity against several strains of fungi isolated from wheat strains. Several important roles of terpenes have been explored in different species of conifer plant. These plant tissues are enriched with terpenes present as mixture of oleoresin having two main components, i.e., turpentine and rosin, which protect the plant against pathogen damage (Fig. 11.3).

## 11.5 Role of Alkaloids in Plant Disease Resistance

Alkaloids are plant secondary metabolites containing nitrogen atom in their structure (Kabera et al. 2014). These are found in vascular plants, viz., morphine from *Papaver somniferum*, caffeine from *Coffea arabica*, cocaine from *Erythroxylon* sp., nicotine from tobacco, and cannabidiol from *Cannabis sativa* (Freeman and Beattie 2008; Garba and Okeniyi 2012; Kim and Sano 2008; Mazid et al. 2011; Scarpari et al. 2005). They impart defensive role against various pathogens by acting as phytoanticipins and phytoalexins and protect the plants from various diseases (González-Lamothe et al. 2009). Benzylisoquinoline alkaloids, viz., morphine, colchicines, berberine, tubocurarine, etc., play an important role in providing defense to plant against pathogen attack (Ahmed et al. 2017). α-Tomatine, a spirosolane-type alkaloid found in tomato plants, has been reported to exhibit antimicrobial and antifungal activities (Friedman 2002; Chiu and Lin 2008; Ito et al. 2007). Glucosinolates, important plant secondary metabolites containing nitrogen and sulfur, are the most studied plant metabolites and mostly found in family Cruciferae (Arbona and Gómez Cadenas 2016). Glucosinolates are biologically inactive in plants, but during pathogen attack, activation of β-thioglucoside hydrolases or myrosinases initiates glucosinolate hydrolysis and resulted in the production of pungent volatile toxins, viz., isothiocyanates and nitriles, which have toxic effects against microbes (Grubb and Abel 2006; Arbona and Gómez Cadenas 2016). Several alkaloids with biologically active compounds possess antiviral activity. Around 18,000 alkaloids have been reported to show antiviral properties in ancient Chinese herbs (Zaynab et al. 2018). 7-Deoxytrans-dihydronarciclasine, an alkaloid from *Hosta plantaginea*, possesses anti-tobacco mosaic virus (TMV) (Wang et al. 2007). Berberine, an isoquinoline alkaloid found in root and stem of *Berberis* species, has been found to exhibited activity against fungi, bacteria, and viruses (Iwasa et al. 2001;Yi et al. 2007;Domadia et al. 2008). Several quinoline alkaloids (dictamnine, kokusagine, and masculine) isolated from the stem bark of *Teclea afzelii* possess strong antibacterial activity (Lin et al. 2000; Kuete et al. 2008).

## 11.6 Role in Health Care

Secondary metabolites from plants have been long known for their role in health care. Various medicinal plants and their metabolites have been known to show potential against several diseases. Plant phenolics, alkaloids, and terpenoids are rich in several metabolites that can be used in health care. Several pharmaceutical properties, viz. anticancer, antioxidant, antimicrobial, antidiabetic, etc., have been reported from plant secondary metabolites (Jain et al. 2019). Commercial importance of plant secondary metabolites has resulted in more demand and production. Phenolic compounds are one of the largest groups of secondary metabolites and encompass diverse health benefits (Kabera et al. 2014). Various health benefits of phenolics are due to their several pharmaceutical properties, viz. antioxidant, anti-inflammatory, anticarcinogenic, cardioprotective, immune system promoting, antimicrobial, antiseptic, anthelmintic, and other biological properties and skin protection property from UV radiation (Kumar and Pandey 2013; Chen et al. 2015; Działo et al. 2016; Andreu et al. 2018; Meng et al. 2018; Kabera et al. 2014). In many food plants, phenolics are present as phenolic acids such as esters or glycosides along with several natural compounds, viz., alcohols, sterols, flavonoids, glucosides, and hydroxy fatty acids (Dai and Mumper 2010).

## 11.7 Antimicrobial Effect

Antioxidant activities of plant metabolites have been used in research and study long ago. Most of these molecules have been employed as a substitute to artificial ones which lead to carcinogenesis (Carocho et al. 2014). Phenolic compounds and flavonoids are the natural antioxidants which are produced by the plant in response to defense as well as under unfavorable conditions (Tungmunnithum et al. 2018).

Various studies highlighted the antioxidant properties of phenolics and flavonoids from plants (Andreu et al. 2018; Wang et al. 2016; Zahoor et al. 2018).

### 11.7.1 Anticancer

Cancer is the main cause of death worldwide. Several drugs and chemotherapy are used to cure cancer patients worldwide; however, there are several side effects. More than 3000 plants have been reported to possess anticancer property, and many countries around the world are using products from these plants in cancer treatment (Alves-Silva et al. 2017; Tariq et al. 2017). Plant phenolics provide better alternative to these being natural as well as less costly (Tungmunnithum et al. 2018). Curcumin, a polyphenolic compound extracted from the rhizomes of *Curcuma longa* (turmeric), has been reported to show anticancer activity against several types of cancer, viz., lung, breast, pancreatic, and skin cancers, leukemia, as well as brain tumor (Perrone et al. 2015; Pavan et al. 2016; Qadir et al. 2016).

### 11.7.2 Antioxidant Activity

Oxidative stress remains as one of the main causes of the progression and development of several diseases (Kasote et al. 2015). It induces oxidative damage by production of reactive oxygen species (ROS). Antioxidant activities provide defense against these molecules. Plants are known for their ability to produce nonenzymatic antioxidants that are capable of neutralizing this ROS-induced oxidative damage.

## References

Acamovic T, Brooker JD (2005) Biochemistry of plant secondary metabolites and their effects in animals. Proc Nutr Soc 64:403–412

Agrios GN (2005) Chapter six—How plants defend themselves against pathogens. In: Agrios GN (ed) Plant pathology, 5th edn. Academic Press, pp 207–248. ISBN 9780120445653

Ahmed SI, Hayat MQ, Tahir M, Mansoor Q, Ismail M, Keck K, Bates RB (2016) Pharmacologically active flavonoids from the anticancer, antioxidant and antimicrobial extracts of Cassia angustifolia Vahl. BMC Complement Altern Med 16:460

Ahmed E, Arshad M, Khan MZ, Amjad MS, Sadaf HM, Riaz I et al (2017) Secondary metabolites and their multidimensional prospective in plant life. J Pharmacogn Phytochem JPP 205:205–214

Ahuja R, Kissen AM (2012) Bones, phytoalexins in defense against pathogens. Trends Plant Sci 17:73–90

Alasalvar C, Grigor JM, Zhang DL, Quantick PC, Shahidi F (2001) Comparison of volatiles, phenolics, sugars, antioxidant vitamins, and sensory quality of different colored carrot varieties. J Agric Food Chem 49:1410–1416

Andreu L, Nuncio-Jáuregui N, Carbonell-Barrachina AA, Legua P, Hernández F (2018) Antioxidant properties and chemical characterization of Spanish Opuntia ficus-indica Mill. cladodes and fruits. J Sci Food Agric 98(4):1566–1573

Aniszewski T (2007) Alkaloids-secrets of life: alkaloid chemistry, biological significance, applications and ecological role. Elsevier
Arbona V, Gómez Cadenas A (2016) Metabolomics of disease resistance in crops. Cambridge University Press, Cambridge, p 2015
Ashihara H, Sano H, Crozier A (2008) Caffeine and related purine alkaloids: biosynthesis, catabolism, function and genetic engineering. Phytochemistry 69(4):841–856
Astani A, Schnitzler P (2014) Antiviral activity of monoterpenes beta-pinene and limonene against herpes simplex virus in vitro. Iran J Microbiol 6(3):149
Boudet AM (2000) Lignins and lignification: selected issues. Plant Physiol Bioch 38:81–96
Carocho M, Calhelha RC, Queiroz MJR, Bento A, Morales P, Sokovic M, Ferreira IC (2014) Infusions and decoctions of Castanea sativa flowers as effective antitumor and antimicrobial matrices. Ind Crop Prod 62:42–46
Chen F, Tholl D, Bohlmann J, Pichersky E (2011) The family of terpene synthases in plants: a mid-size family of genes for specialized metabolism that is highly diversified throughout the kingdom. Plant J 66(1):212–229
Chen X, Dang TTT, Facchini PJ (2015) Noscapine comes of age. Phytochemistry 111:7–13
Chiu FL, Lin JK (2008) Tomatidine inhibits iNOS and COX-2 through suppression of NF-kappaB and JNK pathways in LPS-stimulated mouse macrophages. FEBS Lett 582(16):2407–2412
Cowan M (1999) Plant products as antimicrobial agents. Clin Microbiol Rev 12:564–582
Crozier A, Yokota T, Jaganath IB, Marks SC, Saltmarsh M, Clifford MN (2006) Secondary metabolites in fruits, vegetables, beverages and other plant based dietary components. In: Plant secondary metabolites: occurrence, structure and role in the human diet, pp 208–302
Dai J, Mumper RJ (2010) Plant phenolics: extraction, analysis and their antioxidant and anticancer properties. Molecules 15(10):7313–7352
De Luca V, St Pierre B (2000) The cell and developmental biology of alkaloid biosynthesis. Trends Plant Sci 5(4):168–173
Del Río JA, Arcas MC, Benavente-García O, Sabater F, Ortuño A (2003) Changes of polymethoxylated flavones levels during development of Citrus aurantium (cv. Sevillano) fruits. Planta Med 64:575–576
Dewick PM (2009) Medicinal natural products: a biosynthetic approach. Wiley. ISBN 9780470741689
Dixon RA (2001) Natural products and plant disease resistance. Nature 411(6839):843–847
Domadia PN, Bhunia A, Sivaraman J, Swarup S, Dasgupta D (2008) Berberine targets assembly of Escherichia coli cell division protein FtsZ. Biochemistry 47(10):3225–3234
Działo M, Mierziak J, Korzun U, Preisner M, Szopa J, Kulma A (2016) The potential of plant phenolics in prevention and therapy of skin disorders. Int J Mol Sci 17(2):160
Edreva A, Velikova V, Tsonev T, Dagnon S, Gürel AL, Aktas L (2008) Stress-protective role of secondary metabolites: diversity of functions and mechanisms. Gen Appl Plant Physiol 34:67–78
Freeman BC, Beattie GA (2008) An overview of plant defenses against pathogens and herbivores. Plant Pathology and Microbiology Publications, p 94
Friedman M (2002) Tomato glycoalkaloids: role in the plant and in the diet. J Agric Food Chem 50 (21):5751–5578
Friend J (1979) Phenolic substances and plant disease. In: Swain T, Harborne J, Van Sumere CF (eds) Biochemistry of plant phenolics. Rec. Adv. Phytochem, vol 12, pp 557–588
Gandhi SG, Mahajan V, Bedi YS (2015) Changing trends in biotechnology of secondary metabolism in medicinal and aromatic plants. Planta 241(2):303–317
Garba S, Okeniyi SO (2012) Antimicrobial activities of total alkaloids extracted from some Nigerian medicinal plants. J Microbiol Antimicrob 4(3):60–63
González-Lamothe R, Mitchell G, Gattuso M, Diarra M, Malouin F, Bouarab K (2009) Plant antimicrobial agents and their effects on plant and human pathogens. Int J Mol Sci 10(8):3400
Gonzalez-Teuber M, Heil M (2009) Nectar chemistry is tailored for both attraction of mutualists and protection from exploiters. Plant Signal Behav 4:809–813

Grubb CD, Abel S (2006) Glucosinolate metabolism and its control. Trends Plant Sci 11:89–100
Hain R, Reif HJ, Krause E et al (1993) Disease resistance results from foreign phytoalexin expression in a novel plant. Nature 361:153–156
Hammerschmidt R (1999) Phytoalexins: what have we learned after 60 years? Annu Rev Phytopathol 37:285–306
Huffaker A et al (2011) Novel acidic sesquiterpenoids constitute a dominant class of pathogen-induced phytoalexins in maize. Plant Physiol 156:2082–2097
Ito S, Ihara T, Tamura H, Tanaka S, Ikeda T, Kajihara H, Dissanayake C, Abdel-Motaal FF, El-Sayed MA (2007) Alpha-Tomatine, the major saponin in tomato, induces programmed cell death mediated by reactive oxygen species in the fungal pathogen Fusarium oxysporum. FEBS Lett 581(17):3217–3222
Iwasa K, Moriyasu M, Yamori T, Turuo T, Lee DU, Wiegrebe W (2001) In vitro cytotoxicity of the protoberberine-type alkaloids. J Nat Prod 64(7):896–898
Jain C, Khatana S, Vijayvergia R (2019) Bioactivity of secondary metabolites of various plants: a review. Int J Pharm Sci Res 10(2):494–498
Jeandet P, Douillet-Breuil AC, Bessis R et al (2002) Phytoalexins from the Vitaceae: biosynthesis, phytoalexin gene expression in transgenic plants, antifungal activity, and metabolism. J Agric Food Chem 50:2731–2741
Johnson PN (1976) Effects of soil phosphate level and shade on plant growth and mycorrhizas. N Z J Bot (New Zealand) 14(4):333–340
Juárez Z, Hernández L, Bach H, Sánchez-Arreola E, Bach H (2015) Antifungal activity of essential oils extracted from Agastache mexicana ssp. Xolocotziana and Porophyllum linaria against post-harvest pathogens. Ind Crop Prod 74:178–182
Kabera JN, Semana E, Mussa AR, He X (2014) Plant secondary metabolites: biosynthesis, classification, function and pharmacological properties. J Pharm Pharmacol 2:377–392
Kasote DM, Katyare SS, Hegde MV, Bae H (2015) Significance of antioxidant potential of plants and its relevance to therapeutic applications. Int J Biol Sci 11(8):982
Kessler D, Baldwin IT (2007) Making sense of nectar scents: the effects of nectar secondary metabolites on floral visitors of Nicotiana attenuata. Plant J 49:840–854
Kim YS, Sano H (2008) Pathogen resistance of transgenic tobacco plants producing caffeine. Phytochemistry 69(4):882–888
Kortbeek RW, van der Gragt M, Bleeker PM (2019) Endogenous plant metabolites against insects. Eur J Plant Pathol 154(1):67–90
Kuete V, Wansi JD, Mbaveng AT, Kana Sop MM, Tadjong AT, Beng VP et al (2008) Antimicrobial activity of the methanolic extract and compounds from Teclea afzelii (Rutaceae). S Afr J Bot 74(4):572–576
Kumar S, Pandey AK (2013) Chemistry and biological activities of flavonoids: an overview. Sci World J 2013:162750
Kumar L, Mahatma MK, Kalariya KA, Bishi SK, Mann A (2014) Plant phenolics: important bio-weapon against pathogens and insect herbivores. Pop Kheti 2(3):149–152
Lin CM, Preston JF 3rd, Wei CI (2000) Antibacterial mechanism of allylisothiocyanate. J Food Prot 63(6):727–734
Maddox CE, Laur LM, Tian L (2010) Antibacterial activity of phenolic compounds against the phytopathogen Xylella fastidiosa. Curr Microbiol 60(1):53–58
Mafu S, Ding Y, Murphy KM, Yaacoobi O, Addison JB, Wang Q, Shen Z, Briggs SP, Bohlmann J, Castro-Falcon G, Hughes CC, Betsiashvili M, Hufaker A, Schmelz EA, Zerbe P (2018) Discovery, biosynthesis and stress-related accumulation of dolabradienederived defenses in maize. Plant Physiol 176(4):2677–2690
Mazid M, Khan TA, Mohammad F (2011) Role of secondary metabolites in defense mechanisms of plants. Biol Med 3(2):232–249
Mellon JE, Zelaya CA, Dowd MK (2011) Inhibitory effects of gossypol-related compounds ongrowth of Aspergillus flavus. Lett Appl Microbiol 52:406–412

Mellon JE, Dowd MK, Beltz SB, Moore GG (2014) Growth inhibitory effects of gossypol and related compounds on fungal cotton root pathogens. Lett Appl Microbiol 59(2):161–168
Meng XH, Liu C, Fan R, Zhu LF, Yang SX, Zhu HT, Wang D, Yang CR, Zhang YJ (2018) Antioxidative flavan-3-ol dimers from the leaves of Camellia fangchengensis. J Agric Food Chem 66:247–254
Mera IFG, Falconí DEG, Córdova VM (2019) Secondary metabolites in plants: main classes, phytochemical analysis and pharmacological activities
Michalak A (2006) Phenolic compounds and their antioxidant activity in plants growing under heavy metal stress. Pol J Environ Stud 15:523–530
Naoumkina MA, He X, Dixon RA (2008) Elicitor-induced transcription factors for metabolic reprogramming of secondary metabolism in Medicago truncatula. BMC Plant Biol 8:132
Nathan SS (2013) Physiological and biochemical effect of neem and other Meliaceae plants secondary metabolites against Lepidopteran insects. Front Physiol 4:359
Ortiz de Elguea-Culebras G, Sánchez-Vioque R, Berruga MI, Herraiz-Peñalver D, Santana-Méridas O (2017) Antifeedant effects of common terpenes from Mediterranean aromatic plants on Leptinotarsa decemlineata. J Soil Sci Plant Nutr 17(2):475–485
Park ES, Moon WS, Song MJ, Kim MN, Chung KH, Yoon JS (2001) Antimicrobial activity of phenol and benzoic acid derivatives. Int Biodeterior Biodegrad 47(4):209–214
Pavan AR, Silva GDBD, Jornada DH, Chiba DE, Fernandes GFDS, Man Chin C, Dos Santos JL (2016) Unraveling the anticancer effect of curcumin and resveratrol. Nutrients 8(11):628
Pedras MSC et al (2011) The phytoalexins from cultivated and wild crucifers: chemistry and biology. Nat Prod Rep 28:1381–1405
Perrone D, Ardito F, Giannatempo G, Dioguardi M, Troiano G, Lo RL, De Lillo A, Laino L, Lo Muzio L (2015) Biological and therapeutic activities, and anticancer properties of curcumin. Exp Ther Med 10(5):1615–1623
Perveen S, Al-Taweel A (eds) (2018) Terpenes and terpenoids. BoD–Books on Demand
Piasecka A, Jedrzejczak-rey N, Bednarek P (2015) Tansley review Secondary metabolites in plant innate immunity: conserved function of divergent chemicals. New Phytol 206:948–964
Prisic S, Xu MM, Wilderman PR, Peters RJ (2004) Rice contains two disparate ent-copalyl diphosphate synthases with distinct metabolic functions. Plant Physiol 136:4228–4236
Przybylski P, Pyta K, Stefańska J, Ratajczak-Sitarz M, Katrusiak A, Huczyński A, Brzezinski B (2009) Synthesis, crystal structures and antibacterial activity studies of aza-derivatives of phytoalexin from cotton plant–gossypol. Eur J Med Chem 44(11):4393–4403
Qadir MI, Naqvi STQ, Muhammad SA, Qadir M, Naqvi ST (2016) Curcumin: a polyphenol with molecular targets for cancer control. APJCP 17(6):2735–2739
Rao A, Zhang Y, Muend S, Rao R (2010) Mechanism of antifungal activity of terpenoid phenols resembles calcium stress and inhibition of the TOR pathway. Antimicrob Agents Chemother 54 (12):5062–5069
Reichling J (2018) Plant–microbe interactions and secondary metabolites with antibacterial, antifungal and antiviral properties. In: Annual plant reviews online, pp 214–347
Scarpari LM, Meinhardt LW, Mazzafera P, Pomella AWV, Schiavinato MA, Cascardo JCM, Pereira GAG (2005) Biochemical changes during the development of witches' broom: the most important disease of cocoa in Brazil caused by *Crinipellis perniciosa*. J Exp Bot 56 (413):865–877
Schäfer H, Wink M (2009) Medicinally important secondary metabolites in recombinant microorganisms or plants: progress in alkaloid biosynthesis. Biotechnol J Healthc Nutr Technol 4(12):1684–1703
Schmelz EA et al (2011) Identity, regulation, and activity of inducible diterpenoid phytoalexins in maize. Proc Natl Acad Sci U S A 108:5455–5460
Schwekendiek A, Spring O, Heyerick A et al (2007) Constitutive expression of a grapevine stilbene synthase gene in transgenic hop (Humulus lupulus L.) yields resveratrol and its derivatives in substantial quantities. J Agric Food Chem 55:7002–7009

Seca AM, Pinto DC (2018) Plant secondary metabolites as anticancer agents: successes in clinical trials and therapeutic application. Int J Mol Sci 19(1):263

Seow YX, Yeo CR, Chung HL, Yuk H-G (2014) Plant essential oils as active antimicrobial agents. Crit Rev Food Sci Nutr 54:625–644

Shankar D, Majumdar B (1997) Beyond the biodiversity convention: the challenge facing the biocultural heritage of India's medicinal plants. In: Bodeker G, Bhat KKS, Burley J, Vantomme P (eds) Medicinal plants for forest conservation and health care. Non-wood forest products, vol 11. FAO, Rome, pp 87–99

Shields VD, Smith KP, Arnold NS, Gordon IM, Shaw TE, Waranch D (2008) The effect of varying alkaloid concentrations on the feeding behavior of gypsy moth larvae, Lymantria dispar (L.) (Lepidoptera: Lymantriidae). Arthropod Plant Interact 2(2):101–107

Siemens DH, Garner SH, Mitchell-Olds T, Callaway RM (2002) Cost of defense in the context of plant competition: Brassica rapa may grow and defend. Ecology 83(2):505–517

Singh B, Sharma RA (2015) Plant terpenes: defense responses, phylogenetic analysis, regulation and clinical applications. 3Biotech 5(2):129–151

Stevenson PC, Nicolson SW, Wright GA (2017) Plant secondary metabolites in nectar: impacts on pollinators and ecological functions. Funct Ecol 31(1):65–75

Taiz L, Zeiger E (2010) Plant physiology, 5th edn. Sinauer Associates Inc., Sunderland, 782 p

Tungmunnithum D, Thongboonyou A, Pholboon A, Yangsabai A (2018) Flavonoids and other phenolic compounds from medicinal plants for pharmaceutical and medical aspects: an overview. Medicines 5(3):93

Van Loon LC, Rep M, Pieterse CMJ (2006) Significance of inducible defense-related proteins in infected plants. Annu Rev Phytopathol 44:135–162

Wang YH, Zhang ZK, Yang FM, Sun QY, He HP, Di YT, Mu SZ, Lu Y, Chang Y, Zheng QT, Ding M, Dong JH, Hao XJ (2007) Benzylphenethylamine alkaloids from Hosta plantaginea with inhibitory activity against tobacco mosaic virus and acetylcholinesterase. J Nat Prod 70:1458–1461

Wang L, Chen C, Su A, Zhang Y, Yuan J, Ju X (2016) Structural characterization of phenolic compounds and antioxidant activity of the phenolic-rich fraction from defatted adlay (Coix lachryma-jobi L. var. ma-yuen Stapf) seed meal. Food Chem 196:509–517

Wang K, Jiang S, Pu T, Fan L, Su F, Ye M (2019) Antifungal activity of phenolic monoterpenes and structure-related compounds against plant pathogenic fungi. Nat Prod Res 33(10):1423–1430

War AR, Paulraj MG, Ahmad T, Buhroo AA, Hussain B, Ignacimuthu S, Sharma HC (2012) Mechanisms of plant defense against insect herbivores. Plant Signal Behav 7(10):1306–1320

Weidenbörner M, Hindorf H, Weltzien HC (1992) An effective treatment of legume seeds with flavonoids and isoflavonoids against storage fungi of the genus Aspergillus. Seed Sci Technol 20:447–463

Wink M (1988) Plant breeding: importance of plant secondary metabolites for protection against pathogens and herbivores. Theor Appl Genet 75(2):225–233

Wink M (2007) Molecular modes of action of cytotoxic alkaloids—from DNA intercalation, spindle poisoning, topoisomerase inhibition to apoptosis and multiple drug resistance. In: Cordell G (ed) The alkaloids, vol 64. Elsevier, Amsterdam, pp 1–48

Wink M (2010) Introduction: biochemistry, physiology and ecological functions of secondary metabolites. In: Annual plant reviews: biochemistry of plant secondary metabolism, vol 40, pp 1–19

Wink M (2015) Modes of action of herbal medicines and plant secondary metabolites. Medicines 2 (3):251–286

Wojtunik-Kulesza KA, Kasprzak K, Oniszczuk T, Oniszczuk A (2019) Natural monoterpenes: much more than only a scent. Chem Biodivers 16(12):e1900434

Xu C, Sullivan JH, Garrett WM et al (2008) Impact of solar ultraviolet-B on the proteome in soybean lines differing in flavonoid contents. Phytochemistry 69:38–48

Xu M, Dong J, Wang H, Huang L (2009) Complementary action of jasmonic acid on salicylic acid in mediating fungal elicitorinduced flavonol glycoside accumulation of Ginkgo biloba cells. Plant Cell Environ 32:960–967

Yi ZB, Yan Y, Liang YZ, Bao Z (2007) Evaluation of the antimicrobial mode of berberine by LC/ESI-MS combined with principal component analysis. J Pharm Biomed Anal 44(1):301–304

Yourman LF, Jeffers SN (1999) Resistance to benzimidazole and dicarboximide fungicides in greenhouse isolates of Botrytis cinerea. Plant Dis 83:569–575

Zahoor M, Zafar R, Rahman NU (2018) Isolation and identification of phenolic antioxidants from Pistacia integerrima gall and their anticholine esterase activities. Heliyon 4(12):e01007

Zaynab M, Fatima M, Abbas S, Sharif Y, Umair M, Zafar MH, Bahadar K (2018) Role of secondary metabolites in plant defense against pathogens. Microb Pathog 124:198–202

Zhang Z, Yang T, Mi N, Wang Y, Li G, Wang L, Xie Y (2016) Antifungal activity of monoterpenes against wood white-rot fungi. Int Biodeterior Biodegradation 106:157–160

# Plant Secondary Metabolites: Their Food and Therapeutic Importance

12

Satish Kumar, Rajni Saini, Priyanka Suthar, Vikas Kumar, and Rakesh Sharma

**Abstract**

Plant secondary metabolites (PSMs) are produced in the form of phytochemicals in various plant parts as a natural defense system against attack of various microorganisms and environmental stresses. The role of these compounds is beyond providing protection, as they are linked to many biochemical pathways inside and outside the plants and possess various well-known therapeutic applications. The extraordinary biological activities of plant secondary metabolites lead to their extensive use as an ingredient in medicines and for therapeutic and other culinary purposes since ages. The minimum effective concentration and effect (positive or negative) of plant secondary metabolites on particular metabolic pathways are the concerns which are still under a trial phase. They occur in very minute quantities within the plant cells, while the purity issues have resulted in manufacturing of their chemical derivatives and their industrial applications as well. Environmental, morphogenetic, and genetic factors and ultimately the processing affect the biosynthesis and the concentration of these PSM present. However, the plants are always in contact with changing conditions of light, water, temperature, pH, insect pest infestation, etc. which may adversely affect the accumulation of secondary metabolites. The present chapter

S. Kumar (✉) · R. Sharma
Department of Food Science and Technology, Dr. Y. S. Parmar University of Horticulture and Forestry, Solan, Himachal Pradesh, India

R. Saini · V. Kumar
Department of Food Science and Technology, Punjab Agricultural University, Ludhiana, Punjab, India

P. Suthar
Food Technology and Nutrition, School of Agriculture, Lovely Professional University, Phagwara, Punjab, India





A. K. Sharma, A. Sharma (eds.), *Plant Secondary Metabolites*,
https://doi.org/10.1007/978-981-16-4779-6_12

has been compiled to give the readers an in-hand information about the plant secondary metabolites with primary objective of their food and functional repute. The overall contents will focus on broad classification of PSMs, various methods adopted for their extraction with their potential advantages and disadvantages, and effect of various methods of food processing on the bioavailability and bioactivity of the PSMs with proposed future research opportunities in their potential therapeutic applications.

**Keywords**

Phytoactive compounds · Extraction · Functional foods · Therapeutic potential · Health Benefits

## 12.1 Introduction

Therapeutic uses of various plants particularly the herbs are significantly based on plant chemistry and have gained significant interest of the researchers particularly in the last decade as food is no more seen as a mere source of energy. For better understanding of the customed use of medicinal plant as essential diet ingredients without the loss of their bioactivity, the need of knowledge behind its chemical composition is of paramount importance. Modern chemistry justifies its role in describing the process and potential role of primary plant metabolites in cell division and growth, storage, respiration, and reproduction, which includes the components of processes such as the citric acid or Krebs cycle, glycolysis, photosynthesis, and other associated pathways. These primary metabolites consist of small molecular structures such as amino acids, tricarboxylic acids, proteins, Krebs cycle intermediates, and polysaccharides. All living cells eventually have similar primary metabolites. However, plants also produce different organic molecules through the metabolic pathways derived from the paths used to manufacture primary metabolites, and these are known as plant secondary metabolites (PSMs) which possess unique carbon skeleton structures. The role of these compounds is, however, not so significant in performing the cell/organism functions, but they play an important part in ensuring the organisms continued existence in the ecosystem. PSMs formation is generally cell, tissue, and organ specific. The same plant genera may often differ on the amount and type of these compounds produced, and this may even vary among the similar species based on several factors including the ecological, geographical, and genetic factors. Protection of plant against different stresses, which can be both biotic (fungi, bacteria, insects, nematodes, or grazing by animals) and abiotic (shading, high temperature and moisture, presence of heavy metals or injury), is significantly the role of these PSMs. Various drugs, fragrances, flavors, dyes, and insecticides are manufactured in the market by using PSMs generally because of their high economic value. On the basis of biosynthesis origin, PSMs are separated into three groups (phenylpropanoids, terpenoids, and polyketides) (Verpoorte and Alfermann 2000). Other than this, alkaloids are the class of PSMs which are nitrogenous organic molecules biologically synthesized from amino acids, e.g., tyrosine, tryptophan, lysine, phenylalanine, and arginine, using enzyme with unique properties (Croteau et al. 2000). Alkaloids are the most important therapeutic agents among the other known compounds. Primary metabolites are ubiquitous and perform various functions related with metabolic systems by actively participating in reproduction and nutrition processes (Croteau et al. 2000). The discrimination between secondary and primary metabolites is sometimes difficult, as some terpenoid compounds have both secondary and primary metabolic roles in the plant cells. Secondary metabolites cover a broad range of compounds which can be highly expressible in stress forming conditions. Pigmentation of cells involves carotenoids and flavonoids in case of flowers and seeds which is the mode of attraction for seed dispersers and pollinators. Therefore, plant reproduction systems also involve these compounds (Winkel-Shirley 2001). Primary metabolites refer to the compounds of proteins, carbohydrates, nucleic acids, lipids, and fats which can be significantly related to the physiology, genetics, and structure, which directly signs their role toward development of the plant. On the contrary, secondary metabolites are most often present in low concentrations as a minor compound and are nonessential to life but are genuinely responsible for the survival fitness of the species under adverse conditions. Biosynthesis pathways are required for production of primary and secondary metabolites, and the synthesis of secondary metabolites is sectionalized at cellular level (Besancon et al. 2008). But these biosynthetic reactions are energy demanding which is derived from the glycolysis, citric acid cycle, electron transport chain, etc. where the oxidation of carbohydrates, amino acids, and fatty acids yields energy-dense molecule called ATP (adenosine triphosphate) which acts as the energy currency to fuel all the reactions. ATP is efficiently recycled via anabolic reactions (reduction reactions) as opposed to catabolic reactions which are oxidation reactions of starting or mother molecules. The chemical reactions are often aided by catalysts which significantly play a role in reducing the activation energy, and one of such is CoA (coenzyme A) which builds ADP (adenosine diphosphate) and pantetheine phosphate (Chapman and Moore 1993). Biosynthetic pathways are commonly undergone via (1) pentose for glycosides, polysaccharides; (2) shikimic acid for tannins, aromatic alkaloids, and phenols; (3) acetate–malonate for alkaloids and phenols; and (4) mevalonic acid for terpenes, steroids, and alkaloids (Burnham et al. 1998). Building blocks derived from 1-deoxylulose 5-phosphate, acetyl coenzyme A, and mevalonic and shikimic acids play a significant role in the biosynthesis of secondary metabolites.

There are various proposed classification systems for the PSMs, and one of such classification systems divides the PSMs into terpenoids, phenolics, alkaloids, and saponins. Glycosides and tannins are considered as special part of them owing to their distinct and specific chemical structure. Albrecht Kossel first ever defined the concept of secondary metabolite, a Nobel Prize winner for medicine and physiology in 1910. The concept of secondary metabolites as an end product in the future was described by Czapek. He explained the secondary modification as the products derived from nitrogen metabolism such as deamination. In the twentieth century, advancement in chromatography techniques resulted in more and more recoveries of these molecules, based on which the phytochemistry discipline has been established. Secondary metabolites possess vital biological effects in response to which they got maximum of medicinal value from ancient or modern communities as well. Their pathogenic protection is due to their antiviral, antibiotic, and antifungal properties; along with this, they also constitute compounds that absorb harmful UV lights that are known to produce serious damage to the leaves of plants. The known forage grasses such as alfalfa or clover extensively pose estrogenic properties and also interact with the fertility of animals.

The classification of secondary metabolites as per the chemical structures is further subdivided into several classes. This chapter includes the nature of plant secondary metabolites, reviewing the main categories of constituents having therapeutic importance. Every section will includes an overview of the class of secondary plant metabolites considering the botanical distribution, structure, and a general discussion about pharmacology, followed by respective examples. Based on the literature information available, the PSMs can be divided into six broad categories including phenolics, terpenes, saponins, alkaloids, lipids, and carbohydrates. Different classes of PSMs have been depicted in Fig. 12.1 along with the subclasses of therapeutic importance. All the classes along with their occurrence, important plant sources, and health apprehensions are discussed hereunder.

## 12.2 Phenolics

Plant secondary metabolites include an array of different compounds and a maximum number of phenolic compounds. The presence of one or more phenolic groups is confirmed as the common characteristic which ranges from simple one aromatic ring to complex polymeric structure. Phenols show their presence in almost every plant, hence a widespread compound among the plant genera which significantly contribute to their taste, color, and flavor. The pharmacological activities of phenols are noteworthy such as anti-inflammatory activity of quercetin is well reported, while silybin is greatly recognized for its antihepatotoxic property. Other examples include phytoestrogenic activity of daidzein and genistein, while naringenin shows insecticidal effects (Goławska et al. 2014). Phenolic compounds are also responsible for the antioxidant property of many plants, especially flavonoids. The classification of phenolics can be done according to their biosynthetic origin or structure. On the basis of structure, they are categorized into two major classes, viz., flavonoids and non-flavonoids, with a large number of compounds under each category.

## 12.3 Flavonoids

These are polyphenolic compounds consisting of 15 carbons, with three carbon bridges connecting two aromatic rings. Being ubiquitous among plants, they are highly concentrated in the epidermis of leaves and the skin of fruits where they significantly play their role as secondary metabolites. The diverse processes which

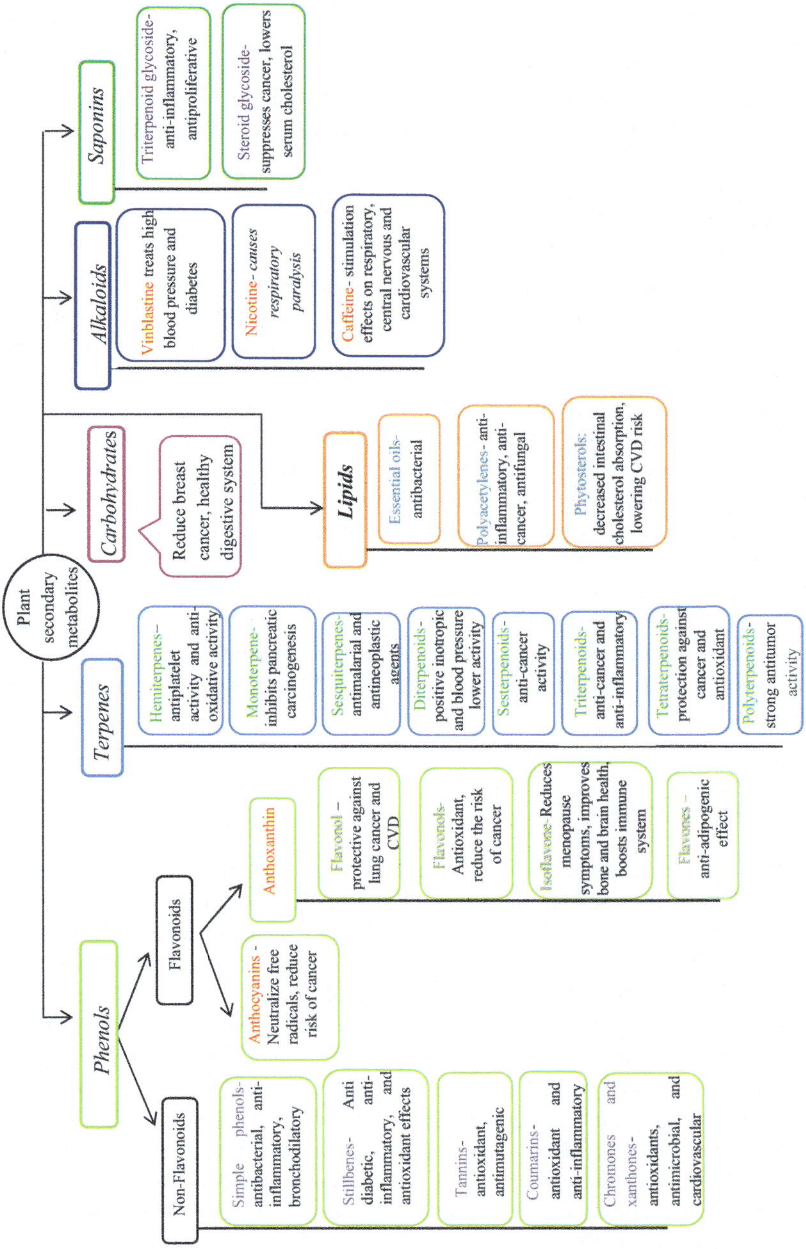


**Fig. 12.1** Classification and therapeutic effects of plant secondary metabolites

involved flavonoids are stimulation of nitrogen-fixing nodules, UV protection, disease resistance, and pigmentation (Pierpoint 2000). Flavonoids are classified further into flavones, flavonols, isoflavones, anthocyanidins, flavan-3-ols, and flavanones. Hydroxyl groups are usually present at positions 4, 5, and 7 of aromatic ring. Sugar-containing flavonoids majorly contain natural glycosides. Availability of hydroxyl and sugar moiety significantly surges flavonoid methyl groups, water solubility, and isopentyl units as substituents make flavonoid lipophilic.

## 12.4 Flavonols

Flavonols are perhaps the most prevalent flavonoid present in the plant kingdom except fungi and algae. The extensive variation in structure and distribution of flavonols has been well documented. Flavonols like quercetin, myricetin, kaempferol, fisetin, and isorhamnetin are commonly found in the form of O-glycosides. Onions, kale, lettuce, tomatoes, apples, grapes, and berries are rich sources of flavonols. Apart from the fresh produce like fruits and vegetables, processed food products like tea and red wine are also very rich sources of flavonols. The intake of flavonols is associated with high antioxidant potential with reduced risk of vascular disease. At position 3 of the C-ring, frequently conjugation occurs, but substitutions can also occur at positions 5, 7, 4, and 3 of the carbon ring. Ample information on levels and amounts of flavonols found in vegetables and fruits is available in the literature which suggests wide variability among various plant sources of same origin. Significant difference in the amounts in seemingly similar produce may be due to varietal differences, and seasonal changes have been recorded.

## 12.5 Flavan-3-ols

This is the complex subclass out of flavonoids which comprises a range of simple monomers (+)-catechin and its isomer (−)-epicatechin, condensed tannins, and the polymeric and oligomeric proanthocyanidins. Unlike planar isoflavones, flavones, flavonols, and anthocyanidins, the presence of saturated C3 in heterolytic C-ring makes flavanones, flavan-3-ols, and proanthocyanidins nonplanar. The chiral centers of flavan-3-ols at C2 and C3 result in four isomers for each level of B-ring hydroxylation out of which (−)-epicatechin and (+)-catechin occur widely in nature and (+)-epicatechin and (−)-catechin are comparatively rare. An extra chiral center is present at C4 of proanthocyanidins (both oligomeric and polymeric), whereas flavanones have only one chiral center at C2. Pairs of enantiomers are not easily detectable in reversed-phase HPLC and so are easily failed to be detected; even though they are difficult to visualize, these chirality differences have a major effect on the 3-D structure of the molecules. But these chiral differences tend to have minimal effect on their redox properties (Unno et al. 2000), whereas significant effect is expected on their binding properties which involves binding or a "lock-and-key" like receptor–ligand, enzyme–substrate, and enzyme–inhibitor interactions. A study by Yang et al. (2015) has shown that humans consuming (−)-epicatechin have excreted some (+)-epicatechin, indicating the possible ring opening and racemization, probably in the GI tract. Further, not only the digestive changes but food processes could also cause this transformation. (+)-Catechin and (−)-epicatechin could potentially form Type B proanthocyanidins when there is oxidative coupling of C4 center of heterocycle and C8 or C6 centers or positions to create oligomers or polymers. Type A proanthocyanidins possess an extra ether bond between C-7 and C-2. Proanthocyanidins could occur as large, complex structures having up to 50 units. Flavan-3-ols form gallocatechins upon hydroxylation and also form esters with gallic acid. Procyanidins are exclusively included in proanthocyanidins that contain (epi)catechin units, and these are the profuse type of proanthocyanidins in plants. However, proanthocyanidins comprising (epi)gallocatechin and (epi)-afzelechin subunits which are known as prodelphinidins and propelargonidins are less common (Balentine et al. 1997). Red wines possess oligomeric procyanidins and prodelphinidins which are extracted from black grape seeds (Auger et al. 2004). Dark chocolates are found to be rich in procyanidins obtained from *Theobroma cacao* roasted seeds (Gu et al. 2006). Green tea (*Camellia sinensis*) contains abundant flavan-3-ols (−)-epicatechin gallate, (−)-epigallocatechin, and (−)-epigallocatechin gallate. As the tea leaves are fermented, its catechin levels plummet, and the principal constituents of black tea are thearubigins and low levels of theaflavins (Del Rio et al. 2004). Theaflavins are often referred to as dimmers as they are obtained from two monomer subunits of flavan-3-ol, but they aren't strict dimers, whereas thearubigin structures are undetermined and described to be flavonoid-derived and are usually cited as tannins despite their inability to convert hides to leather. So, they can be better named or cited as derived polyphenols until well-defined structures are revealed and a relevant nomenclature is applied.

## 12.6 Anthocyanidins

Anthocyanidins, also called as anthocyanins, which are the principal conjugated derivatives, are broadly distributed throughout the kingdom plantae and are the reason behind the attractive purple, red, and blue colors of flowers and fruits. Furthermore, they are also obtained from seeds, stems, roots, and leaves. The main functions of anthocyanins are to protect plants against intense light and UV rays by shading mesophyll cells of leaves and to attract pollinators. The most common occurring anthocyanidins are malvidin, cyanidin, peonidin, petunidin, pelargonidin, and delphinidin. These compounds are found as sugar conjugates in plant tissues and are known as anthocyanins, and also often present in the form of conjugates with organic acids like malic and acetic acids and also with hydroxycinnamates, and even though their conjugation most often occurs at C3, it can also take place at positions 3, 5, and 7. Enzymatic and chemical transformations take place in few products like ports and red wines (matured), and a high number of polyphenols derived from anthocyanins that add to dietary intake of phenols are well established.

## 12.7 Flavanones

Absence of 2,3 double and chiral center at C2 usually characterizes flavanones. The C-ring is adjoined to the B-ring at C2 in the α-configuration. As the flavanone structure is highly reactive, it readily undergoes O-methylation, hydroxylation, and glycosylation reactions. Flavanones are in high concentrations in citrus fruits, and the common flavanone glycoside is found in very high volumes in citrus peel in chemical forms like hesperetin-7-O-rutinoside (hesperidin). Flavanone rutinosides are tasteless compounds. In contrast, flavanone neohesperidoside conjugates possess a bitter taste such as neohesperidin in bitter orange (*Citrus aurantium*) and naringin from grapefruit peel (*Citrus paradisi*). A flavanone usually used as a sweetener in nonalcoholic beers is neohesperidin dihydrochalcone. Citrus flavonoids are reported to have an array of pharmacological effects. They are mostly utilized as potential antioxidants and anti-inflammatory, blood lipid-lowering, and cholesterol-lowering agents either in direct food applications or in the form of therapeutic recommendations by the medical practitioners. Flavanones are also called dihydroflavones.

## 12.8 Isoflavones

Presence of B-ring at C3 position in C2 characterizes isoflavones. They are exclusively found in higher concentrations in leguminous crops like soybean (US Department of Agriculture, Agricultural Research Service 2002). *Genistein, daidzein, and coumestan—coumestrol—usually referred to as phytoestrogens are obtained from lucerne and clovers (Trifolium spp.) and have enough estrogenic activity which could be detrimental to the reproduction of grazing animals such as cows and sheep.* The structural similarity of isoflavonoids with estradiol leads to inhibition of ovulation. So, these animals should be fed with legumes in restricted contents or should be fed with genetically modified legumes with low isoflavonoid concentrations. Isoflavonoids can fight with a number of diseases and are commonly regarded as phytoestrogens because of their estrogenic activity. Genistein has been recently explored for its possible induction of hormonal and metabolic changes, by virtue of which it can influence various disease pathways.

## 12.9 Simple Phenolics

Instead of the ubiquitous aspect of phenolic acid in plants, finding any free phenol is not an easy task. Among plants, free phenol is rare, while among phenols, gallic acid is the most widespread compound which further is composed of gallotannins. The astringent properties of gallic acid are well-known, while various other activities has been confirmed in vitro including antibacterial, antifungal, antiviral, antitumor, anti-inflammatory, antimutagenic, antianaphylactic, bronchodilatory, choleretic, and many more. Retarding the insulin degradation which directly smoothens the muscle relaxation process is carried out by the presence of gallic acid. This group differentiates the phenolic compounds on the basis of their functional groups, which may be carboxylic, hydroxyl, or aldehydic group which further includes vanillin (a phenolic aldehyde), eugenol (a phenolic phenylpropane), and ferulic, caffeic, and salicylic acids. Other than gallic acid, hydroquinone is another widely distributed simple phenol, present in plants in the form of glycoside arbutin. The formation of glycoside is common; coniferin and the other derivatives of phenolic cinnamic acids are lignin precursors which are widely distributed among glycoside (Evans 2009; Hoffmann 2003). The antimicrobial arbutin and anti-inflammatory salicylates are probably the best examples to demonstrate the pharmacological properties of these constituents (Zbigniew et al. 2014). Almost all phenols share one common property of being an antimicrobial agent; phenol itself was used as an antiseptic in surgeries before the invention of other valuable replacements. The pharmacological activities of many plants are contributed by the presence of simple phenols, the most common of which are the diuretic and antimicrobial activities of *Arctostaphylos uva-ursi* (Eric 2002) and rubefacient, stimulant, and analgesic activities of *Capsicum* spp. owing to the presence of capsaicinoids (Spiller et al. 2008). Other examples include the anthelmintic activity of *Dryopteris filix-mas*, the cholagogue activity of *Cynara scolymus*, and the diuretic and anticatarrhal properties of *Solidago virgaurea.*

## 12.10 Tannins

Tannins are important PSMs which have been historically used in the leather production industry for tanning of animal hides. They are water-soluble phenol derivatives naturally produced and hoarded by higher plants as PSMs. Chemically, they are polyphenols with molecular weights between 500 and 3000 Da, while in the form of complexes formed with saccharides, alkaloids, and proteins, their MW can increase even up to 20,000 Da. Tannins are the only phenol which has the ability to precipitate in protein synthesis. The use of these compounds to convert the hides of raw animal into leather has been in practice from decades. Many substances by virtue of their biosynthetic origin and structure are considered to be tannins. Tannins are majorly divided into two types: condensed tannins and hydrolyzable tannins. Several molecules of phenolic acids combined to form hydrolyzable tannins, which include hexahydroxydiphenic acids, and esters unite these compounds with glucose molecule in the center. Two forms of principal hydrolyzable tannins are ellagitannins and gallotannins, composed of ellagic and gallic acid units. However, ellagitannins are found in many of the medicinal plants which elucidate the geraniin structure (Catarino et al. 2017). On the other hand, condensed tannins are also known as proanthocyanidins; these are the compounds structurally based on precursors of oligomeric flavonoid and typically differ on the basis of linkages present between the units: carbon stereochemistry, hydroxylation patterns, and the additional substitution present in the compound. *Camellia sinensis* (tea) and *Hamamelis virginiana* leaves are some drugs which contain both hydrolyzable and condensed tannins

(Puneet et al. 2013). Antidiarrheal activity of tannin containing drugs adds in their properties along with this they have been used as an antidote against heavy metals and alkaloid poisoning.

## 12.11 Coumarins

Coumarins include the compounds derived from benzo-α-pyrone, which is a known lactone of *o*-hydroxycinnamic acid. Almost thousands of coumarins are isolated from natural sources. Melilot or sweet clover, *Galium odoratum* (sweet woodruff), and *Dipteryx odorata* (tonka bean) are the richest sources of coumarin (Hoffmann 2003). Umbelliferone, aesculetin, and scopoletin are the plant coumarins present in free state or in the glycoside form. Coumarin-rich plants include *Datura stramonium*, *Jatropha belladonna*, *Ruta graveolens*, *Aesculus hippocastanum*, and Rosaceae (Evans 2009). Coumarins report certain important biological activities which include anti-cancer, anti-inflammatory, anti-Alzheimer properties (Xu et al. 2015).

## 12.12 Flavonoids

Most of the naturally occurring phenols are flavonoids. The structure of flavonoids consists of a chroman ring along with an aromatic ring at positions 2, 3, or 4. Depending upon the central ring's oxidation level (ring C), flavonoids can be divided into various classes including flavones, anthocyanins, and flavonols which are the most common. Flavones mostly occur in yellow color, are mostly found in young tissues and higher plants, and are particularly present in cell sap. Compositae, Umbelliferae, Rutaceae, Leguminosae, and Polygonaceae families are significantly rich in flavonoid content. Recent studies have laid significant grounds for using various herbals as essential food ingredients owing to their immense medicinal properties. Many functional foods and drugs have been derived from herbals like *Chamaemelum nobile* (Roman chamomile), *Glycyrrhiza glabra* (liquorice root), and *Ginkgo biloba* (gingko), which are very rich in flavonoids. The group of flavonoids is known for its anti-allergic and anti-inflammatory effects and for its vasoprotective and antithrombotic properties and is very well recognized for its protective property against tumor and gastric mucosa (Montanher et al. 2007; Serafini et al. 2010).

## 12.13 Xanthones and Chromones

Xanthones fall among the biggest classes of naturally occurring compounds in natural chemistry. A large number of xanthones have been isolated from various natural sources including higher plants, fungi, ferns, and lichens. They have gradually risen to a great importance because of their medicinal properties. These are the structural derivatives of benzo-γ-pyrone, but the pharmaceutical significance of many of these compounds is not so noteworthy except for eugenin and khellin found in cloves and mustard seeds, respectively (Evans 2009). Furanochrome has the most complex structure, which is an active constituent in the fruits of *Ammi visnaga.* Guttiferae and Gentianaceae, which mainly contain xanthones, are otherwise sporadically scattered in the plant kingdom (Polygalaceae and Moraceae). Malawi highlanders and bordering countries use *Polygala nyikensis* against fungal originated skin problems and disorders. In recent studies, this plant has shown to exert antifungal activity due to the presence of xanthones in its roots (Johann et al. 2011). So far they have been explored for their potential hepatoprotective, anticarcinogenic, antileprosy, antimalarial, antioxidant, anticholinergic, mutagenic, radioprotective, immunomodulatory, antiparasitic, antihistaminic, larvicidal, and ovicidal activities.

## 12.14 Stilbenes

They are small-sized compounds and are widely distributed secondary metabolites among plant genera mostly as heartwood constituents in heterogeneous plant assembly. The most widespread stilbene is resveratrol which is a *para-hydroxylated* compound possessing estrogen-like activity present in Fabaceae, Pinaceae, Myrtaceae, and Vitaceae families. It is well reported to act like antioxidants, protecting the body against oxidative damage which is the potential cause of putting the body at higher risk for cancer and several other heart diseases. It is naturally present in very high concentration in the skin of red grapes, while significant volumes can also be found in peanuts and berries.

## 12.15 Alkaloids

They are organic compounds with heterocyclic ring in which at least one nitrogen atom is present. They are not the homogeneous group of compounds considering any point of view, chemical, biochemical, or physiological; except for the presence of nitrogen making them nitrogen-containing compounds, no other general definitions suits for all the alkaloids. Chemical structure of alkaloid is further divided into different types. Aromatics, acridones, ephedras, carbolines, imidazoles, ergots, indoles, indolizidines, bisindoles, manzamines, quinolones, oxindoles, quinazolines, phenethylamines, and phenylisoquinolines are the basic types of alkaloids. Alkaloids present in plants have been traditionally used by man as medicines for over 3000 years. Presence of alkaloids is not definite in lower plants: sulfur-containing alkaloids and lysergic acid derivatives. Gymnosperms and pteridophyte alkaloids are known for their medicinal uses which include ephedra, lycopodium, and texus alkaloids. Angiosperms have uneven distribution of alkaloids. Alkaloids show diverse pharmacological actions which include local anesthesia, analgesia, cardiac stimulation, relaxation, respiratory stimulation, muscle relaxation, vasoconstriction and toxicity and hypertension, and antineoplastic and hypotensive properties. Alkaloids' activity against vertebrate's toxicity, cytotoxicity, mutagenic or carcinogenic activity, and antifungal, antibacterial, allelopathic, and antiviral properties has been reported in literature. The toxicity of certain alkaloids against animals when eaten to death has also been reported. Several alkaloids are also used as insecticides (anabasine and nicotine) (Hoffmann 2003). Examples of some alkaloids are as follows:

### 12.15.1 Nicotine

*Nicotiana tabacum* (tobacco plant) and other species of *Nicotiana* contain nicotine which has tranquilizing properties and is the main addictive component present in tobacco. Nicotine is extremely toxic in nature, sometimes causing respiratory paralysis when consumed at high doses. Nicotine is a ganglion cholinergic-receptor against with complex pharma complex pharmacological actions, the neuromuscular junction, the adrenal medulla, and the brain (Benowitz 2009).

### 12.15.2 Caffeine

Botanically unrelated species constitute caffeine, which include tea (*Camellia sinensis*), coffee (*Coffea* spp.), mate (*Ilex paraguariensis*), and kola (*Cola acuminate*). In case of coffee beans, caffeine is bound to chlorogenic acid. The process of roasting liberates the other compounds along with caffeine which further contribute to coffee's aroma. Caffeine produces stimulation effects on respiratory, central nervous, and cardiovascular systems.

### 12.15.3 Vinblastine

*Catharanthus roseus* contains a significant amount of vinblastine which is further isolated to treat high blood pressure and diabetes and can also be used as disinfectant. Studies show vinblastine as an efficient fighter against cancer (Moudi et al. 2013).

## 12.16 Lipids

Plants have the ability to synthesize low-molecular-weight compounds at least a million of them (Delgoda and Murray 2017) which don't have a direct function in the internal cellular system of the plants but provide them survival advantages by integrating environmental interaction like protection against pests, herbivores, pathogens, UV rays, etc. (Weng 2013). Most of the secondary metabolite pathways are situated in plastids, and synthesis of these metabolites is organ specific like roots and leaves synthesize compounds to ward off pathogens, herbivores, etc. and flowers synthesize compounds to attract pollinators, etc. (Böttger et al. 2018). The mostly known secondary metabolites are phenolic compounds, carotenoids, lignans, essential oils, etc. but the untapped lipid-based secondary metabolites are explored under this section.

## 12.17 Lipid-Based Secondary Metabolites

Plants produce a variety of lipid-based secondary metabolites like steroids, phytoestrogens, waxes, polyacetylenes, etc.

### 12.17.1 Polyacetylenes

Plants like red ginseng, carrots, and devil's club produce polyacetylenes like falcarinol and falcarindiol which tend to have anti-inflammatory (Christensen 2011), anti-cancer (Cheung et al. 2019), antitumor (Kobæk-Larsen et al. 2005), antiplatelet aggregatory, etc. activities. But their major function in plants is to protect roots from fungal diseases.

### 12.17.2 Phytosterols

Plant sterols, stanols, are present in the form of hydroxycinnamic acid esters, fatty acid esters, and also glycosides (Moreau et al. 2018), and few dominant compounds in our diet are sitosterol, sistostanol, campestanol, campesterol, and stigmasterol (Klingberg et al. 2008). These sterols and stanols have health beneficial effects in humans like decreasing intestinal cholesterol absorption (Davis Jr. et al. 2004), immune modulation (paper), lowering CVD risk (Gylling et al. 2014), and reducing liver inflammation (Plat et al. 2014; De Smet et al. 2015).

### 12.17.3 Carbohydrates

#### 12.17.3.1 Glycosides

Glycosides include sugar moiety bound to a functional group via glycosidic bond. Plants store chemicals in inactivate glycoside form and perform various significant beneficial roles, e.g., arbutin is a glycoside found in bearberries, and various researches have shown that it possesses antitussive, antibacterial, anti-cancer, and anti-inflammatory activities (Mustapha et al. 2016). Loganin (a major glycoside of iridoid class) obtained from *Cornus officinalis* has been proven to have antidiabetic, anti-inflammatory, anti-osteoporosis, and neuroprotective effects in certain neurodegenerative diseases like Parkinson's disease (Tseng et al. 2019). Sinigrin, a thioglycoside, is usually found in mustard seeds, brussels sprouts, etc. They have diverse pharmacological effects like antifungal, anti-inflammatory, antioxidant, antibacterial, anti-cancer, and wound healing properties (Mazumder et al. 2016). Hesperidin is a flavanone glycoside found in citrus peel that has therapeutic effects like anti-inflammatory, antioxidant, antidepressive, and neuroprotective effects, eases memory impairments, etc. (Kim et al. 2019). An animal study with coumarin glycoside extracted from *Hydrangea paniculata* is reported to have kidney protective properties and also eases the escalation of diabetic nephropathy (Sen et al. 2019).

#### 12.17.3.2 Terpenoids

The term isoprenoids is commonly used as an alternative term for terpenoids and is the most diverse on structural basis. Terpenes are mostly known for their essential oil components which can be obtained from steam distillation, hydro-distillation, or dry distillable fractions having characteristic smell or odor (Ludwiczuk et al. 2017). Terpenoids have positive effects on health like anxiolytic, analgesia, anti-inflammatory, cancer chemopreventive, antimicrobial, anti-hyperglycemic, and anti-depressant effects and provide protection to the skin. Most of these effects were reported by preclinical studies with animal or in vitro methods (Baron 2018). The terpenes are classified into eight subclasses based on the number of carbon atoms and isoprene unit. Another important class includes the hemiterpenes which possess five carbon atoms and one isoprene unit and are thus considered as the simplest terpenoids. Hemiterpenes are easily emitted from the plant parts due to low boiling point of isoprene (34 °C). Common examples of hemiterpenes are tiglic, angelic, senecioic, and isovaleric acids and isoamyl alcohol. Monoterpenoids are another type of terpenoids which have two isoprene units and ten carbon atoms. The monoterpene class is further classified into three subgroups, i.e., acyclic, monocyclic, and bicyclic. In each subgroup, these monoterpenoids have unsaturated hydrocarbons and may have alcohols, ketones, and aldehydes as functional groups. Common examples of this class of aliphatic terpenoids are myrcene, geraniol, citral, and linalool. Monocyclic monoterpenoids are limonene, α-terpineol, thymol, eucalyptol, carvone, α-pinene, and cineol. Limonene is the oil extracted and the chief constituent of peels of oranges and other citrus fruits. Recently, another important class of terpenoids has emerged with great therapeutic potential referred to as the sesquiterpenoids which comprise of 3 isoprene units and 15 carbon atoms in different frameworks in linear, mono-, bi-, and tricyclic forms. Sesquiterpene lactones are different from other sesquiterpenoids. These terpenes are further classified into two subclasses based on their lactone ring present on 6,12-olides and 8,12-olides. Examples of the first class include santonin, costunolide, artabsin, parthenolide, and matricin, whereas the second subclass 8,12-olides include inunolide, thapsigargin, alantolactone, and helenalin. The major lactones present in sesquiterpene are eudesmanolides, germacranolides, and guaianolide. Diterpenoids have 20 carbon atoms as a backbone with 4 isoprene units. This class is also further classified into linear, bi-, tri-, tetra-, penta-, or macrocyclic diterpenes based on their core structure. They are present in polyoxygenated form having keto and OH-groups which are esterified by aromatic acids. Examples of this class are sclareol, carnosic acid, phytol, abietic acid, and gibberellin. Sesterterpenoids are terpenoids having 25 carbons and 5 isoprene units. As with other terpenes, these also further classified as linear, mono-, di-, tri-, tetra-, and macrocyclic frameworks. Nitiol belongs to this class. Triterpenes comprise of 30 carbon atoms and 6 isoprene units. Triterpenes also have alcohol, carboxylic acids, or aldehyde with complex cyclic structures. Plant sterols are called phytosterols which are triterpenes having cyclopentane perhydro-phenanthrene ring. Examples of plant sterols are stigmasterol, sitosterol, camperterol, and betulinic acid. Tetraterpenoids on the other hand are also commonly known as carotenoids having 40 carbon atoms and 8 isoprene units. They are fat-soluble pigments. The 8 isoprene units are connected head-to-tail in lycopene which is responsible for producing color. Polyterpenoids are terpenoids with more than 8 isoprene units and 40 carbon atoms. Natural rubber molecules are included in this class of terpene. In some plants, polyisoprene has trans-double bond configuration, while in natural rubber, isoprene units exist in cis-configuration.

#### 12.17.3.3 Saponins

Saponins are sterol glycosides which are amphiphalic compounds consisting of sugar unit (pentose, hexose, or uronic acid) and nonpolar compounds called sapogenin (sterol or triterpene) which originated from plants like soybeans, peanuts, chickpeas, and spinach. Due to the amphiphilic nature of saponins, they behave like emulsifying agents and form stable foams. Structurally, sugar units (glucose, glucuronic acid, galactose, xylose) of saponins are linked to aglycone unit which is steroid in nature. Subclass of saponins is triterpenoid glycosides and steroid glycosides. The subclass triterpenoid glycosides is majorly present in soybeans, beans, peas, sugar beet, spinach, tea, and sunflower, whereas the subclass steroid glycosides is derived from furostan and spirostan skeletons which have five to six double bonds. Major sources of steroid saponins are capsicum peppers, tomato seeds, fenugreek, oats, and asparagus. The dietary saponins are reported to have a wide range of health benefits which include reducing plasma cholesterol in animal models by inhibiting fat absorption in the gut lining. Other related health benefits are protection against cancer, reduction of blood lipids, and blood glucose response. Dietary saponins are used to treat hypercalciuria and also showed relationship (inversely) with renal stone incidences (Shi et al. 2004). Major classes of dietary PSMs along with their plant sources and potential therapeutic role in health management are discussed in Table 12.1.

## 12.18 Extraction of Plant Secondary Metabolites from Plants

Plants are the natural machineries' major sources of bioactive compounds where they are produced in large quantity. As scientifically proven, these compounds have effect on human health and they can prevent many serious health conditions. Thus, they can act as diet supplements to fulfill the need of antioxidants which are otherwise supplied as pharmaceutical products in the form of pills. The major concern about commercial applications of the PSMs as bioactive components is their low concentration in various plant sources, which however can be overcome by

**Table 12.1** Major plant secondary metabolites and their sources and therapeutic applications

| Compound | Plant source(s) | Therapeutic application | Reference |
|---|---|---|---|
| **Phenols (flavonoids)** | | | |
| Apigenin (4′,5,7-trihydroxyflavone) | *Chamaemelum nobile, Glycyrrhiza glabra, Achillea millefolium, Matricaria recutita, Anthemis nobilis, Apium graveolens,, Camellia sinensis, Coriandrum sativum, Chamaemelum nobile, Gingko biloba, Marrubium vulgare, Artemisia dracunculus* | Anti-infection, antiviral, anticarcinogenic activity | Duke and Beckstrom-Sternberg (2000) |
| Diosmin (3′,5,7-trihydroxy-4′-methoxyflavone 7-rutinoside) | *Hyssopus officinalis, Citrus aurantium, Vicia ervilia* | For the treatment of chronic venous insufficiency, varicose veins, and lymphedema | Sandhar et al. (2011) |
| Luteolin (2-(3,4-dihydroxyphenyl)-5,7-dihydroxychromen-4-one) | *Halenia corniculata, Pyrola rotundifolia, Gentiana tenella, Reseda luteola, Achillea millefolium, Thymus vulgaris, Limonium sinuatum, Vitex rotundifolia, Erigeron canadensis, Sophora angustifolia* | Antioxidant, cardiovascular activity, anticarcinogenic, antidiabetic activity, anti-inflammatory activity, anti-allergic activity and anti-microorganism activity | Ko et al. (2005), Lee et al. (2006), Gutiérrez-Venegas et al. (2006), Odontuya et al. (2005) |
| Quercetin (2-(3,4-dihydroxyphenyl)-3,5,7-trihydroxy-4 H-chromen4-one) | *Podophyllum peltatum, Cephalotaxus harringtonia, Dysoxylum malabaricum, Maytenus serrata, Thapsia garganica* | Antihypertensive, antimutagenic, antioxidant anticarcinogenic, and anticoagulant | Cragg and Newman (2005), Davis et al. (2009), Larson et al. (2010) |
| Kaempferol (3,5,7-trihydroxy-2-(4-hydroxyphenyl)-4H-chromen-4-one) | *Foeniculum vulgare, Kaempferia pulchra, Zingiber zerumbet, Alpinia conchigera, Pedilanthus tithymaloide, Jatropha podagrica, Tibouchina semidecandra* | Strong anticarcinogenic, anti-inflammatory, and antioxidant activity | Hämäläinen et al. (2007), Chew et al. (2009) |
| | | | Kaur et al. (2010) |

| | | | |
|---|---|---|---|
| Galangin (3,5,7-trihydroxy-2-phenylchromen-4-one) | *Alpinia officinarum, Helichrysum aureonitens, Rubia cordifolia, Glycyrrhiza glabra* | Antioxidant, anticarcinogenic, anti-inflammatory, anti-microorganism activity | |
| Fisetin (2-(3,4-hihydroxyphenyl)-3,7- dihydroxychromen-4-one) | *Fragaria spp., Rhus verniciflua* | Antioxidant, anticarcinogenic, anti-microorganism, anti-inflammatory activity | Gideon (2003), Lee et al. (2002) |
| Myricetin (3,5,7-trihydroxy-2-(3,4,5- trihydroxyphenyl)-4-chromenone) | *Cephalotaxus harringtonia, Bleekeria vitensis, Dysoxylum binectariferum, Euphorbia semiperfoliata* | Anticarcinogenic, antioxidant, antidiabetic, antimicrobial | Ong and Khoo (2000) |
| **Phenols (non flavonoids)** | | | |
| Tannins (gallotannins and ellagitannins) | Immature fruits (unripe hazelnuts) | Antioxidant, radical scavenging, antimicrobial, antiviral, antimutagenic, and antinutrient | Sieniawska (2015) |
| Hydroxycinnamates (p-coumaric and caffeic acid, chlorogenic acid) (*E*)-3-(3-hydroxyphenyl) prop-2-enoic acid) | Leaves of green mate (*Ilex paraguariensis*), processed seeds of *Coffea canephora* | Antiviral, antibacterial, antimutagenic, anti-inflammatory, and antioxidant effects | Shahidi and Chandrasekara (2010) |
| Stilbenes (resveratrol) (5-[(*E*)-2-(4-hydroxyphenyl) ethenyl] benzene-1,3-diol) | Grapes, wine, soya and peanuts, itadori plant (*Polygonum cuspidatum*) | Antidiabetic, anti-inflammatory, and antioxidant effects | Akinwumi et al. (2018) |
| **Alkaloid** | | | |
| Ephedrine, pseudo-ephedrine, norpseudoephedrine (cathine), norephedrine, methylephedrine, methylpseudoephedrine | *Ephedra sinica* | Cardiac system and central nervous; semi-adrenaline, stimulation of α- and ß-receptors; antihypersensitivity, anti-inflammatory, anti-asthma activity, anti-allergic | Parsaeimehr et al. (2010), Furu et al. (2007) |
| Caffeine (1,3,7-trimethylpurine-2,6-dione) | *Camellia sinensis, Theobroma cacao, Coffea arabica* | Decreased sleep, increased arousal and cautiousness by influence on CNS | Lane et al. (1990, 2002) |

(continued)

**Table 12.1** (continued)

| Compound | Plant source(s) | Therapeutic application | Reference |
|---|---|---|---|
| Physostigmine (eserine) [(3*aR*,8*bS*)-3,4,8*b*-trimethyl-2,3*a*-dihydro-1*H*-pyrrolo[2,3-b]indol-7-yl] *N*-methylcarbamate | *Physostigma venenosum* | Inhibitors of neurotransmitter – Actions on the neuromuscular junction, acetylcholinesterase, degrading enzyme inhibitor | Lane et al. (1990, 2002) |
| Berberine, coptisine, norlaudanosoline | *Bocconia frutescens, Papaver rhoeas, Berberis aquifolium, Hydrastis canadensis, Phellodendron amurense, Coptis chinensis, Eschscholzia californica* | Anti-inflammatory activity, anti-cancer, anti-microorganism, positive inhibitory effects on HIV antidiabetic, antidepressant, and LDL reducer | Lin et al. (2004) |
| Coralyne, chaconine, demissine, solamargine, solanine, solanidine | *Solanum lycopersicum, S. tuberosum, S. melongena* | Anti-cancer, anti-inflammatory agent, antioxidant, anti-microorganism, DNA-binding agent | Lin et al. (2004) |
| Harmaline, harmine | *Peganum harmala, Banisteriopsis caapi* | Anti-cancer, antibacterial; central nervous stimulator | Zaker et al. (2007) |
| Rauwolscine, corynanthine | *Rauwolfia canescens* | Central nervous stimulator | Zaker et al. (2007) |
| **Terpenes** | | | |
| Cardenolides | Apocynaceae, Asclepiadaceae, Cruciferae, Liliaceae, Moraceae, Ranunculaceae, and Scrophulariaceae | Influence Na+ /K+ ATPase of heart muscle and it is used to treat heart disease | Liu et al. (2000) |
| Saponins | Agavaceae, Alliaceae, Asparagaceae, Dioscoreaceae, Liliaceae, Amaryllidaceae, Bromeliaceae, Palmae, Scrophulariaceae | Anti-inflammatory, anti-cancer, antiviral; antifungal, or even anti-HIV-1 | Liu et al. (2000) |
| Limonene 1-methyl-4-(1-methylethenyl)-cyclohexene | Grapefruit, lemon, lime | Dietary anticarcinogen | Dell'Eva et al. (2004), Nakase et al. (2008) |
| Artemisinin | Sweet wormwood plant | Used as anti-cancer, antimalarial | Dell'Eva et al. (2004), Nakase et al. (2008) |

| | | | |
|---|---|---|---|
| Taxol | *Taxus brevifolia* Nutt | Antitumor, antileukemic (Pandi et al. 2011) | |
| **Fatty acids** | | | |
| Linoleic acid (cis, cis-9,12-octadecadienoic acid | Sunflower, safflower, soybean, corn, and canola oils | Decrease in the ratio of unsaturated fatty acid (USFA) and increased saturated fatty acid (SFA) in the yolk of the egg | Hur et al. (2007) |
| α-Linolenic acid ((9Z,12Z,15Z)-octadeca-9,12,15-trienoic acid) | Flaxseed (linseed oil), rapeseed (canola), and soybeans | Potential nutraceutical to protect the brain from stroke, characterized by its pleiotropic effects in neuroprotection, vasodilation of brain arteries, and neuroplasticity | Blondeau et al. (2015) |
| Stearic acid (octadecanoic acid) | *Stemodia foliosa* | Antibacterial | da Silva et al. (2002) |

processing the raw material into concentrated extracts. The extracts are highly purified products having few or negligible toxins rendering their ample application in pharmaceutical and food processing industry. Therefore, to enhance the consumption of these bioactive compounds, various extraction processes are suggested by various scientific communities. Extraction process of biologically active compounds from plant material depends on various factors starting from plant sample selection and sample preparation to the extraction protocol of target compound. The extraction methods are influenced by chemical properties of plant compounds, size of plant material to be extracted, and other interfering compounds. Other critical factors which directly affect the yield of extracts are extraction time, temperature, sample-solvent ratio, repetition of extraction, and type of solvent. Extraction time and temperature influence the solubility, solvent viscosity, and mass transfer. To remove interfering compounds such as chlorophylls or waxes, added techniques are applied. Other aspects which accelerate the extraction technique of plant secondary metabolites are their bioavailability and bio-accessibility. Many of these health-promoting compounds are greatly affected by various food processing and unit operations.

Several extraction methods are used at domestic levels since decades. The historical application of maceration begins during the wine preparation. Maceration in wine preparation helps in diffusion of various compounds of grapes all together. The major purpose of maceration was to improve the aroma and color of final product. These parameters are important with respect to the wine quality and consumer acceptance. Maceration during wine preparation is done as both pre- and post-fermentive steps. Maceration is also reported to increase the amount of tannins extracted from grapes as well as phenolic compound and color pigment. The maceration time influences the total yield of bioactive compounds from plant matrices (Ghanem et al. 2019). Other traditional extraction techniques like Soxhlet extraction methods are nowadays assisted with other techniques to reduce the losses. One such technique is high-pressure Soxhlet extraction, used to isolate pesticides from food products like potato, carrot, and olive oils. Automated Soxhlet extraction is used to analyze the total fat content. Soxhlet extraction with ultrasound has many advantages in extraction of fat from sunflower, soybean, and rape oil (De Castro and Priego-Capote 2010). The most common traditional extraction technique, hydro-distillation, is used for extracting aroma and essential oils from various plants. A study by Lucchesi et al. (2004) showed the comparison between hydro-distillation and solvent free microwave extraction for extraction efficiency of essential oils. The hydro-distillation showed high extraction yield of linalool. The monoterpene, limonene results in better hydro-distillation extraction when compared with microwave extraction. The traditional extraction techniques for bioactive compounds are maceration, Soxhlet, and hydro-distillation which possess serious drawbacks like being potentially hazardous to the environment due their high requirement of petroleum-based organic solvents and their low extraction efficiency. Also, these convectional processes work at high temperature and are time-consuming procedures which promote the need for novel techniques for extraction of bioactive compounds. To recover the processing losses and improve extraction yield, various novel extraction techniques are introduced. Extraction techniques are recently focused on green extraction to reduce the environmental stress and energy expenditure caused by convectional extraction techniques.

The improved techniques have mostly focused on improving the yield and conserving loss occurrence during the extraction protocols. Ultrasound-assisted extraction (UAE) has the potential in restricting the losses of heat-sensitive bioactive compounds during extraction. Hence, it has been used in various pharmaceutical industries for extraction of phyto-nutraceutical compounds. Several reports suggested the impressive yield by application of ultrasounds in extraction process. The biologically active compounds (BAC), isoflavones, from ground soybeans showed 15% increased extraction by UAE (Vilkhu et al. 2008). Also, UAE is a promising extraction technique for carotenoid extraction from different fruit parts (skin, pulp, or seed). UAE showed approximately 143% improved extraction yield without any losses of carotenoids in comparison to traditional methods (Chemat et al. 2017b). In terms of time and energy saving during extraction, the first technique which comes in mind is microwave-assisted extraction (MAE). Several reports show the effective extraction of bioactive compounds using MAE compared to traditional method (27.62–90.23% increased yield). Anthocyanin, phenols, and flavonoids are successfully extracted using MAE. Other bioactive compounds such as glucosinolates (cabbage), pectin (pomelo peel), and transresveratrol (peony seed) are studied for their MAE (Ekezie et al. 2017). The SCF (supercritical fluid) extraction was first used in 1978 for decaffeination of green coffee beans. Later, it was used for extraction of flavor by using $CO_2$ (liquid). These events accelerated the commercial and industrial application of SCF extraction with numerous modifications. The major applications of supercritical fluid extraction are decaffeination process for tea, flavor extraction from a variety of herbs and plants, and extraction of oils and fats. Also, it is extensively used for dealcoholization of alcohols (Raventós et al. 2002). Among few techniques, pulsed electric field (PEF) extraction are used at both lab and pilot scales. PEF extraction has permeabilization ability and wide applicability. Many bioactive compounds like polyphenols and colorants such as carotenoids and betalains have been investigated for their improved extraction efficiency by PEF. The application of PEF before fermentation in wine production showed high polyphenol extraction and varied color attributes. The enhancement of anthocyanin by application of PEF was also highlighted. The PEF application doesn't induce the off odors and potential in reducing turbidity in oils (Chemat et al. 2017a, b). Enzymes are well-known for their specificity since decades. Therefore, enzyme-assisted extraction (EAE) is widely used and results in extraction of highly specific compounds. The EAE technique is widely used to get high yields of polysaccharides, oils, flavors, colorants, and BAC. These involve a variety of enzymes and show faster extraction with high recovery in comparison with nonenzymatic methods. Extraction of polyphenols from grapes involves enzymes that are used in industrial scales. As reported, 2 h of treatment with enzyme results in 65.8% improved anthocyanin yield from pomace. Also, enzymes are widely used to clarify the wins as well. Enzyme extraction is appropriate for extraction of catechins from milk tea and extraction of oligosaccharides from rice bran. These enzymes break the intermolecular bonds which result in the release of bioactive compound. Similarly, lignans are successfully extracted from flax hull by using cellulase enzymes (Puri et al. 2012).

Pressurized liquid extraction (PLE) is one of the potential green techniques which is widely used for nutraceutical extraction from herbs and food materials. As reported, polyphenols showed better stability and yield by PLE. Also, isoflavones (soybeans), carotenoids (herbal tea and vegetables), and capsaicinoids (peppers) were extracted from PLE. The essential oil extraction from coriander seeds was also reported (Mustafa and Turner 2011). Recently, many essential steps are applied to protect the environment from the harmful effects of petroleum-based solvents. Recently, many studies come up with production of ionic liquids (IL) or deep eutectic solvents (DES) to replace organic solvents. These DESs are successfully used for extraction of catechin, terpenoids, flavonoids, saponins, and phenolic acids, whereas ILs are used in extraction of phenolic acids like protocatechuic acid, ferulic acid, and caffeic acid (García et al. 2016; Tang et al. 2012). The traditional extraction techniques are now replaced with promising techniques such as ultrasound-assisted extraction (UAE), microwave-assisted extraction (MAE), supercritical fluid extraction (SFE), pulsed electric field (PEF) extraction, pressurized solvent extraction, enzyme-assisted extraction (EAE), deep eutectic solvent (DES) extraction, and ionic liquid extraction (Brglez Mojzer et al. 2016). As mentioned, the recent trends on extraction protocol enable good recovery of biologically active compounds and improve the extract quality. In the line of novel techniques, UAE has been utilized widely due to its important advantages such as simplicity, being inexpensive, low solvent requirement, and being eco-friendly. Ultrasound-microwave-assisted extraction was also reported by various researchers which enhances the extraction efficiency. Therefore, there is a possibility of combining two or more novel techniques which may improve recovery of bioactive compounds. Infrared-assisted extraction is recently approached by various researchers for extraction of various bio-resources due its high efficiency. Another important technique accelerated solvent extraction, i.e., pressurized liquid extraction (PLE), is studied for extraction of natural products (Gullón et al. 2017). The exploration of DESs and ILs for their utilization in potential target-based studies for various food and nonfood applications has shown to be very proficient in the extraction of bioactive compounds from numerous plant matrices. Various applications, advantages, and drawbacks of convectional and non-convectional techniques used for the extraction of bioactive compounds are discussed in Table 12.2.

## 12.19 Food Processing and Bioavailability of Plant Secondary Metabolites

Bioavailability of the PSMs is the most important aspect about their potential food and pharmaceutical application. Food applications and bioavailability of the PSMS is a multistep process and involves liberation of the bioactive compounds from the food matrix, its successive absorption, distribution, metabolism, and finally

**Table 12.2** Extraction techniques used for various bioactive compounds

| | Type of extraction method used | Commercial application | Advantages | Disadvantages |
|---|---|---|---|---|
| Conventional techniques | Maceration | – Extraction and pre-fermentation of grapes for wine production (extraction of anthocyanins) | – Simpler process of extraction<br>– Cold maceration results in similar odor to that of original plant<br>– No degradation of heat-sensitive compounds | – Consist of several steps<br>– 3–4 days required for extraction |
| | Soxhlet extraction | – Oil extraction from soybeans<br>– Extraction of vanillin from vanilla pods | – Samples are continuously brought into contact with fresh extraction solvent<br>– No need of filtration<br>– Low cost of basic equipment | – High solvent requirement, large amount of extractant wasted<br>– Thermal decomposition of bioactive compounds<br>– Time required between 3 and18 h |
| | Hydro-distillation | – Essential oil extraction from plants (spices) | – Organic solvents are not involved<br>– Used for extraction of volatile compounds | – Time-consuming process<br>– High degradation of heat-sensitive bioactive compounds<br>– Low extraction selectivity |
| | Organic solvent extraction | – Extraction of phlorotannins (phenolics)-(ethanol or methanol)<br>– Omega-3 fatty acids (chloroform)<br>– Carotene extraction (hexane) | – Extraction of both nonpolar and polar compounds | – Potential cyto- and environmental toxicity<br>– Uneconomical<br>– Consumption of time, energy, and polluting solvents<br>– Low yield compared to solvent used<br>– Volatile and petrochemical origin |
| Non-conventional techniques | Supercritical extraction | – Extraction of essential oil from spices (cumin) and ginger<br>– Extraction of | – Short extraction time<br>– High yield<br>– Repeated reflux can be done for | – Temperature, pressure, extraction time, particle size of plant, moisture |

(continued)

**Table 12.2** (continued)

| | Type of extraction method used | Commercial application | Advantages | Disadvantages |
|---|---|---|---|---|
| | | carotenoids<br>– Extraction of caffeine<br>– Extraction of polyphenols, lycopene, and phytosterols | complete extraction<br>– Easy separation of solute and solvent<br>– Small amount of sample (in mg) for laboratory can be extracted to tonnes of extraction for industries | content of feed, and flow rate of $CO_2$ affect extraction efficiency |
| | Microwave-assisted extraction | – Extraction of flavonoids from onion and tea polyphenols<br>– Alkaloid extraction using ILs was reported | – Improved yield in less time<br>– Low solvent requirement<br>– Faster and selective heating<br>– High efficiency of extraction<br>– Most economical<br>– Generation of residue is avoided | – Only for small-scale system<br>– High investment cost of the instrument<br>– High cost of electricity than fuels |
| | Ionic liquid solvent extraction | – Extraction of amino acids and phenolic compounds<br>– Extraction of tannins and essential oils | – Do not crystallize at room temperature<br>– Immiscibility in oil<br>– Higher extraction yields and purification<br>– High selectivity | – High price<br>– Sometimes instability |
| | Ultrasonic-assisted extraction | – Extraction of protein, bioactive compounds like polyphenols, flavones, soy isoflavones, caffeine, and herbal oils | – Simple, faster extraction rate of thermosensitive components at mild/low temperature range<br>– Low cost; it is a viable option for green extraction of bioactive compound<br>– Low lab wastes; solvent penetration is | – High wave ultrasound results in the undesirable changes in extraction due to formation of free radicals<br>– Extraction dilution is the main drawback when dynamic ultrasound-assisted extraction is used. |

(continued)

**Table 12.2** (continued)

| | Type of extraction method used | Commercial application | Advantages | Disadvantages |
|---|---|---|---|---|
| | | | higher in sample matrices | – Production of free radicals |
| | Deep eutectic solvent extraction | – Extraction of phenols, protein, and flavonoid is reported | – Inexpensive and simple preparation<br>– Doesn't generate waste, sustainable, different combinations can be used<br>– Wide polarity and excellent solubility to various compounds<br>– Compatible with food, pharmaceuticals, and cosmetic<br>– Negligible vapor pressure, high conductivities, thermal stability, surface tensions, and polarity | – Lower extraction efficiencies due to high viscosity, requires high energy<br>– Recycling and recovery of targeted compounds could be difficult due to zero vapor<br>– These extraction techniques are possible in industrial scale when these extracts are applied directly without involving the purification step |
| | Pulsed electric field extraction | – Oil and phytosterol—Maize germ (higher oil yield and increased phytosterols)<br>– Polyphenolics—Grapes (increase of total polyphenolic content in fresh pressed grape juice) | – Increase mass transfer during extraction<br>– Minimize degradation of heat-sensitive compounds<br>– Low energy consumption<br>– Improved stability of bioactive compounds | – Effectiveness of PEF extraction depends on energy provided, strength of pulsed field, extraction temperature, and number of pulse<br>– Extraction yield varies with properties of treated material |
| | Enzyme-assisted extraction | – Used for extraction of polysaccharides, oils, natural pigments, flavors, and medicinal compounds (terpenoids, | – Improved release of bioactive compounds<br>– High recovery<br>– Eco-friendly technology<br>– Nontoxic | – Extraction depends on factors like enzyme concentration, particle size of plant material, time for |

(continued)

Table 12.2 (continued)

| | Type of extraction method used | Commercial application | Advantages | Disadvantages |
|---|---|---|---|---|
| | | flavonoids, alkaloids, etc.) | – Noninflammable properties | hydrolysis, and solid to water ratio |
| | Pressurized solvent extraction | – Extraction of carotenoids, lipids, isoflavones, polyphenols, essential oils, xanthones, etc. | – Requires small amount of solvent<br>– Faster extraction process<br>– Higher analyte solubility<br>– Enhanced mass transfer during extraction<br>– Extraction of heat-stable compounds | – Depends on pressure, temperature, size of plant matrix, volume of flush, solid-solvent ratio, and static time |

elimination from the body. This key process ensures the bio-efficacy of bioactive compounds present in food material. To provide beneficial effect on health, bioactive compounds need to be bioavailable. Bioactivity can be achieved by better understanding of digestion of bioactive compounds and their distribution in various body systems. Few important factors which affect the bioavailability include bio-accessibility, type of food matrix, transporter elements, molecular structure of bioactive compounds, and the enzymes involve in metabolic process. The terms bio-accessibility and bioavailability of bioactive compounds are significant for understanding and establishing the connection between food and nutrition. The abortion rate of bioactive compounds and its extent vary from individual to individual. Type of food processing affects the bioavailability of bioactive compounds in terms of either enhancement or losses. The plant produces secondary metabolites as a defense mechanism under biotic and abiotic conditions. Various postharvesting techniques for stimulating the bioactive compounds accumulation. However, food processing often results in degradation of bioactive compounds which led to their reduced amount in final processed products. Therefore, various physical and chemical modifications are introduced in food processing operations to provide safe and nutritious food to consumers with highest amount of photoactive compounds present.

In one of the studies conducted by Rein et al. (2013) on bioavailability of (-)-epicatechin from cocoa and flavan-3-ols from black and green tea, it was reported that bio-accessibility of epicatechin is similar in both cocoa-based beverage and chocolate bar. Some researchers claimed that there is no effect of milk incorporation on the bioavailability of cocoa polyphenols. In contrast, other scientific articles reported the effect of milk proteins on (-)-epicatechin bioavailability. Further, the author stated that addition of milk to tea possesses negligible effect on bio-accessibility and bioavailability of flavan-3-ols. Another secondary plant metabolite is lycopene which is responsible for antioxidant protection against lipids, protein, and DNA. These compounds are reported with improved bioavailability by food processing as it is released from plant cell during thermal and mechanical processing. Also, heat is responsible for their trans- to cis-isomerization (Agarwal et al. 2001). Another example stated that the glucosinolates from *Brassica* plants reduced by 75% over 6 h when finely shredded, whereas thermal cooking-steam cooking, microwave cooking, and stir-frying didn't induce significant changes in glucosinolates' content. Boiling reduces 90% of glucosinolates by leaching, and hence, it is suggested that boiling should be avoided to improve the bioavailability of isothiocyanates in vegetables (Barba et al. 2016). Sonication extraction technique was reported to improve total phenolic bio-accessibility by 400% in dried cashew apple bagasse. This technique induces changes in microstructure by ultrasounds and is used as a unit operation before drying to enhance the mass transfer and drying efficiency. Ultrasound technique has the potential implication for extraction of bioactive compounds from plant matrix (Ribas-Agustí et al. 2018). The bio-accessibility of plant secondary metabolites is influenced by other compounds such as lipids, proteins, carbohydrates, etc. In support to this, a study conducted by Colle et al. (2013) revealed that by adding 5% of lipids before processing of tomato enhances the lycopene bio-accessibility, whereas the type of lipid added was given minor attention. Also, the type of thermal processing conditions adopted is highly dependent on the commodity and its inherent qualities which vary with size and shape of the food, so do with the variations of processing on PSMs and their bioavailability. Effect of different methods of food processing on plant secondary metabolites has been depicted in Fig. 12.2.

## 12.20 Improvement in Delivery of Bioactive Compounds

Recently, the advances in food processing technology paved the way for novel processing methods to aid in traditional process which improves the nutritional aspects of food and product safety. To enhance the bioavailability and bio-accessibility of bioactive compounds, the variables and their intensities can be manipulated. Initially, traditional techniques majorly focused on production of stabilized and long-lasting food products which nowadays has shifted to improve and increase the bioavailability of bioactive compounds by modulating physical and chemical controls during food processing. Now the interest of the food processing sector has shifted the interest toward effective delivery of the bioactive compounds to the target sites. One such technique used for better delivery of bioactive health-promoting compounds is encapsulation. In the past few years, several methods have been developed for the successful encapsulation of such compounds by various techniques like spray drying, fluidized bed coating, extrusion, spray chilling and cooling, liposome entrapment, inclusion complexation, and rational suspension separation which also provide their controlled release. Among various drug delivery systems, self-emulsifying drug delivery system (SEDDS) is the most popular and

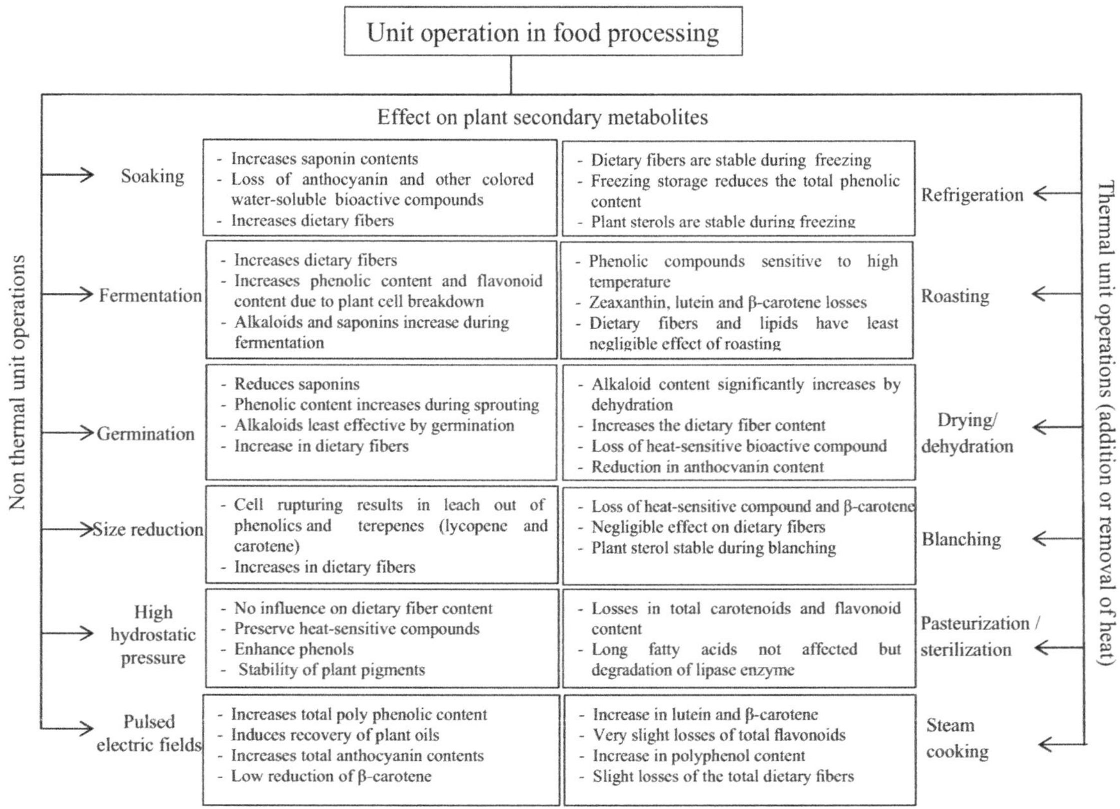


**Fig. 12.2** Effect of method of food processing on various plant secondary metabolites

most viable approach to solve low oral bioavailability problem which is a lipid-based formulation. With drug mixture, they form very fine emulsion having the size between 100 nm and <50 nm. A study by Kumar and Bhopal (2012) that designed the SEDDS with rutin in edible oil reported that rutin containing edible oil added with Tween 80 and Ca-8 (1:1:8) results in droplets of 30 nm and releases maximum drug after 30 min, i.e., 93%. Similarly, cyclodextrin can be used for encapsulation of biologically active compounds to improve their delivery to target sites. Microencapsulation can be used by various food processors to safeguard the sensitive food bioactive compounds, to protect compounds from nutritional losses, to optimize the time-release action into the formulation, and to preserve or mask the flavors and aroma and for easy handling.

The flexibility in this technique offers food researchers great opportunities to develop foods with enhanced nutritious and flavorful product to achieve today's consumer demand. Moreover, properties of microcapsule can be modified for specific application by changing its composition, release mechanism, size of particle, physical structure of final product, and its cost. In another study, the unprotected and encapsulated riboflavin, niacin, and thiamine were compared with each other as they are generally degraded or destroyed during processing in spaghetti. The study revealed that riboflavin, niacin, and thiamine were highest in cooked pasta where they were encapsulated as compared to unprotected vitamins in spaghetti pasta. Nano-capsule technique is most recently developed to mark the link between nanotechnology with food and agriculture for ensured food security and improved delivery of bioactive compounds, flavor and nutrition, ability to absorb nutrients, functionality of foods, and cost-effectiveness of distribution and storage. Many delivery systems like biopolymer matrices, emulsions, simple solutions, and colloidal association were introduced to maintain the bioactive compounds at suitable consumption level for a long term. These nano-carriers may reduce the toxicity and enhance the distribution of these bioactive compounds. The nanoparticle system possesses better encapsulation properties than the traditional encapsulation system such as release efficiency (Rashidi and Khosravi-Darani 2011). A broad outline of the nanoparticle and nanoemulsion formation by various techniques and encapsulation of various bioactive compounds and their delivery have been depicted in Fig. 12.3.

## 12.21 Therapeutic Role of PSMs in Management of Chronic Health Diseases

Adoption of healthy lifestyle is very important as it has been estimated that 30–50% of cancer and cardiovascular diseases (CVDs) can be prevented by behavioral risk factors like tobacco use, unhealthy food diet, low physical activity, and extensive alcohol consumption. Changing these behavioral attributes may result in the improvement of intermediate risk factors like blood glucose, blood pressure, blood lipids, obesity, and overweight which ultimately may also result in complex diseases. Preventive approach by changing the lifestyle is less expensive and

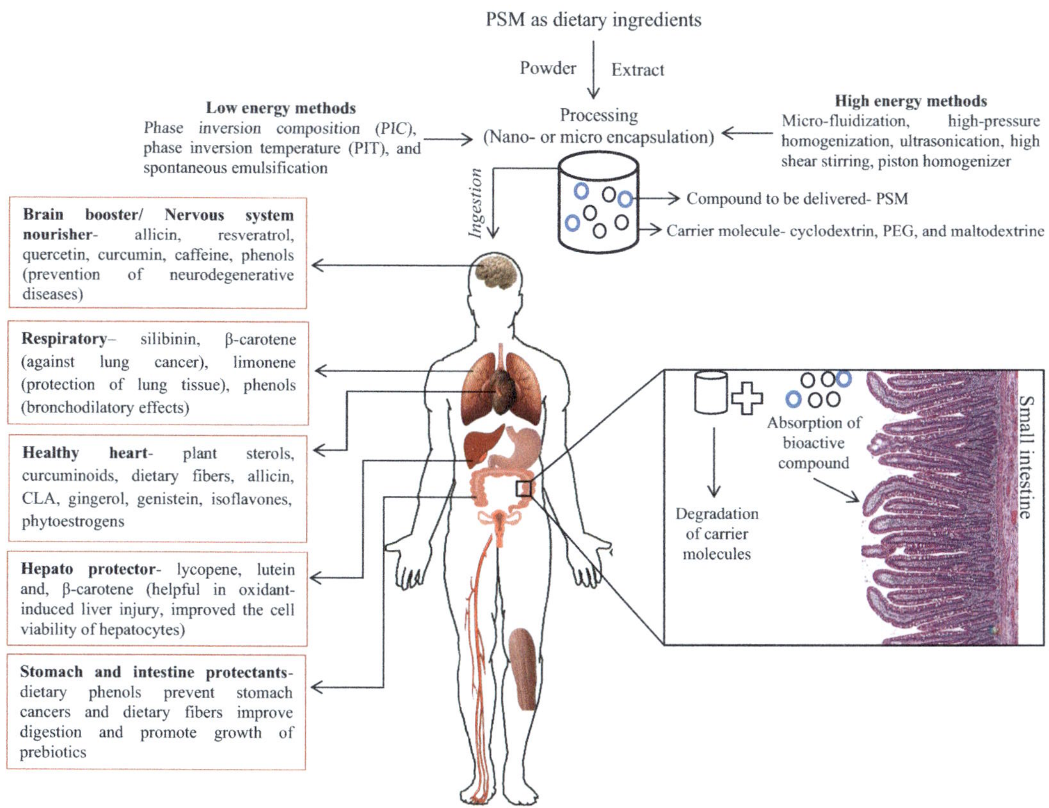


**Fig. 12.3** Delivery of plant secondary metabolites by encapsulation technique and their impact on human health

economic than their treatment after disease condition. Therefore, consuming a healthy food diet can be helpful in prevention of noncommunicable diseases like cancer, diabetes, cardiovascular disease, etc., even though healthy diet depends on individual features like age and gender and also the availability of food in an area and the cultural context. Effect of various PSMs on various body organs is depicted in Fig. 12.3.

In 2016, Lin et al. stated that the phenolic antioxidants present in berries have excellent potential in management of type 2 diabetes. The phenolic compounds control hyperglycemia, and therefore when high phenolic antioxidant food is added to human diet, it will exhibit beneficial effects. Yi et al. (2005) showed that food diets that are rich in phenolic compounds are associated with low risk of severe health condition like cancer. The anti-cancer activity was demonstrated against two colon cancer cell lines, HT-29 and caco-2, by blueberry phenolic compounds. The phenolic compounds have antiproliferative effect and inhibitory action to apoptosis induction. Many simple phenolic compounds like cinnamic acid, *p*-coumaric acid, caffeic acid, ferulic acid, chlorogenic acid, and rosmarinic acid possess potent anti-inflammatory and antioxidant properties. Also, these compounds are valuable in management and treatment of diabetes and hyperlipidemia. Some recent evidence suggests that PSMs are helpful in treatment of obesity-related health condition. In the adipose tissues, the derivative of hydroxycinnamic acid restricts macrophage infiltration and activation of nuclear factor-κB (NF-KB) in obese animals. They further suppress the expression of tumor necrosis factor-α (TNF-α) which increase secretion of anti-inflammatory agents from adipocytes (Alam et al. 2016). According to Jiang and Dusting (2003), diet rich in plant phenolic compounds may reduce the risk associated with coronary heart disease. The hypolipidemic activity and antioxidant properties of these bioactive phenolic compounds possess a critical role in preventing oxidation of lipoproteins and atherosclerotic lesion development. Alkaloids, like berberine, are present in various plant parts like roots, stem, or bark in Berberidaceae family and are reported to have beneficial therapeutic effect. Debnath et al. (2018) reviewed the therapeutic effects of berberine for its potent antidiabetic properties. These bioactive compounds improve the sensitivity of cells toward insulin with synergistic effects like anti-inflammatory, antihypertensive, antioxidant, hepatoprotective, and antidepressant. Another alkaloid, cynarin, from *Cynara cardunculus* plant possesses high antioxidant properties along with antiproliferative activity. Promising chemotherapeutic and chemoprotective effect of solanine has been reported by Jayakumar and Murugan (2016), while apoptosis, restriction of anti-proliferation, migration, invasion, and angiogenesis in different cancer cell lines were proposed to be possible mechanisms for the induction. PSMs like tomatidine when administered orally in mice show lower levels of LDL cholesterol, serum cholesterol, and atherosclerotic lesions. This indicates inhibition of atherogenesis by tomatidine. Saponins possess several health benefits and positively contribute to treat diabetes, cardiovascular problems like high blood pressure, blot clots, and high cholesterol and to cure physical and mental stress (Astuti et al. 2011). Saponins from various pulses are reported to have anti-cancerous activity and potential to reduce lipids (Rochfort et al. 2011). Saponins at appropriate concentrations are responsible for lowering blood glucose level by increasing plasma insulin. It encourages insulin secretion from the pancreas and inhibits glucose in the bloodstream (El Barky et al. 2016). Terpenoids are the PSMs which have strong antioxidative effect, while monoterpenoids possess chemopreventive action by different mechanisms. Primarily, they act by suppressing phase I and phase II enzyme induction and interaction with rat sarcoma (RAS) signal transduction pathway. Monoterpene and limonene are effective in suppressing tumors. Other terpenes, like carotene, are effective in controlling risk of many coronary heart diseases. It has been reported that they effectively lead the process of oxidation of low-density lipoprotein, a critical factor in development of atherosclerosis. Carotenoids play an important role in decreasing the risk of coronary heart disease by inhibiting peroxidation in LDL. Carotenoids recently have been reported to neutralize carcinogens and ROS (reactive oxygen species) (Wagner and Elmadfa 2003; Styrczewska et al. 2013). Conjugated linoleic acids (CLA) belong to lipid class of plant secondary metabolites. As mentioned earlier, many epidemiological studies on CLA showed positive impact on animal and human health. The CLA plays a significant role in preventing cancer in all the stages of carcinogenesis, i.e., initiation, promotion, and metastasis and neovascularization or angiogenesis. The evidence obtained from the animal-based studies shows that CLA isomers have potential to restrict the tumor development of colon and mammary glands. Also, 0.5% CLA with either low-fat or high-fat diet supplement to mice showed reduction in 40–80% fat. In a study on humans, healthy exercising human and overweight and obese human given with CLA diet of 1.8 g and 1.7–6.8 g/day, respectively, for 12 weeks reported to have a fat mass reduction of up to 4% without any changes in their body weight. CLA is reported to have restoring properties like insulin sensitivity and to normalize glucose tolerance in fatty rats. Food diet along with CLA showed resolution of atherosclerosis at very low level (0.1% of diet). Different studies on rabbits, hamsters, and mice reported reduction of atherosclerotic lesions and local inflammatory and plasma lipoproteins by mixed or single isomers of CLA (Yang et al. 2015). Carbohydrates also characterize a broad and very important class of PSMs present in different plant species in the form of fibers. The plant-based dietary fibers when consumed up to a proportion of 10 g/day showed reduction in elevated serum C-reactive protein (CRP) by 11% without kidney disease and 38% reduction in those who have kidney problems. The total dietary fibers are associated with mortality in kidney disease (Krishnamurthy et al. 2012). In the many studies and also health organizations like European Prospective Investigation into Cancer and Nutrition (EPIC) supports that by quintile intake of dietary fiber significant reduction in risk of colorectal cancer can be achieved. Other clinical evidences showed effective role of fiber-rich diet in lowering GI and improving glycated protein levels which act as a marker for glycemic control. High fibers in food showed positive effect on type 2 diabetes and its management (Kendall et al. 2010). According to Bazzano et al. (2003), soluble fibers are effective in reducing the concentration of serum cholesterol (LDL) and have a protective effect against risk of coronary disease. In many clinical studies, the role of dietary fibers to control the glycemic index and weight gain of the subjects has been well established. The role of

PSMs in management and treatment of various diseases with their specific mode of action has been discussed hereunder:

### 12.21.1 Cardiovascular Disease

Presently, cardiovascular diseases are the most prominent chronic diseases causing mortality of approximately one third of the world's population. As reported, many factors which influence cardio-related diseases are modifiable, unlike genetic factors that are non-modifiable. Modifiable factors are majorly concerned with the lifestyle and diet of the subjects. Diet modulates the response of various body organs if incorporated with enough quantity of PSMs and thus may reduce the degree and severity of cardiovascular diseases like ischemic strokes, atherosclerosis, and neurodegenerative diseases (Malaguti et al. 2015). Nutraceutical medicine has successfully proved the role of bioactive compounds in protecting cardiovascular health. The PSMs like lycopene and glucan play an active role in cardiovascular diseases. Quercetin has antiarrhythmic properties and significantly lowers the effect of lipids of fish oils. Phytosterols have also been reported to pose LDL cholesterol lowering properties. Other PSMs like phytoestrogens inhibit the proliferation of myocardial cells, hence reducing the risk of chronic degenerative diseases. B-carotene, glutathione, and lipoic acid are reported to show an active role in preventing CVD. PUFA in corn oil, soybean oil, safflower oil, mustard oil, and flex oil plays a defensive role in cardiovascular diseases. The dietary fibers also have an inhibitory action on myocardial cells. Recently, soy isoflavones and genistein have been studied for their potential cardioprotective effects (Sharma and Singh 2010).

### 12.21.2 Cancer

Cancer is another top chronic disease with very high mortality rate worldwide. Majority of cancers occur due to the uncontrolled cell division (DNA damage) which ultimately leads to the formation of tumors. PSMs cover an array of bioactive compounds which have potential to control DNA damage, a limiting step in formation of tumors. Cancers have adverse effects on the entire body. Many types of treatment for cancers have been suggested like chemotherapy, radiation, surgery, and phytotherapy. Among all the treatments, the natural nutritional compounds have advantages over other treatments in terms of side effects. Essential oils like DHA (docosahexaenoic acid), EPA (eicosapentaenoic acid), and omega-3 fatty acids are reported with anti-cancerous properties. Carotenoids have active free radical scavenging activity (which are potent in damaging DNA) and are potential remedy to treat various types of cancers. The dietary fibers have been reported to reduce the possibility of colon cancer occurrence. The plant-based polysaccharides play an active role in colon lesions. The PSMs of green and black tea like flavones, amines, and saponins play an active role in alteration of metabolic enzymes which induce the signals for cell arrest and apoptosis. In contrast, some studies suggest that the involvement of high black tea in diet reduces risk of gastrointestinal-assisted cancer. Curcumins from turmeric are clinically used as anti-cancerous compound. They affect the expression of protein in colon cancer and are well-known for their chemoprotective effects. Eugenol, the bioactive compound in honey, cinnamon, clove oil, and citrus, is reported to stimulate the cell division and apoptosis in G1 phase. It supports the apoptosis in colon cancer cells and is effective in controlling the growth of tumor cells (Kuppusamy et al. 2014).

### 12.21.3 Hyperlipidemia

Lipid homeostasis is believed to lead to hyperlipidemic body condition which is associated with development of cardiovascular diseases. Reducing serum lipids is beneficial in other disease complications like CVD, obesity, and diabetes. Control over serum lipids can be achieved by cutting off total cholesterol, triglyceride, and LDL cholesterol but increasing HDL cholesterol intake in diet and also minimizing trans- and saturated diet and elevating hypolipidemic-based nutraceuticals such as dietary PUFA (DHA, EPA, and ALA (α-linolenic acid). These healthy compounds have potential hypotriglycemic, antithrombotic, antiarrhythmic, and anti-inflammatory activities. Several mechanisms are suggested like the activity of PUFA in up-regulating expression of gene involved in fatty acid oxidation. Phytosterols have poor absorption in the intestine compared with cholesterol and thereby inhibit cholesterol absorption. Another mechanism suggests that phytosterols may inhibit the synthesis of cholesterol. The soluble dietary fibers like β-glucan and pectin are effective in decreasing serum low-density lipoprotein cholesterol. The mechanism and activity of dietary fibers suggest the inhibition of reabsorption of bile salts and result in maximum secretion of fecal bile salt. This endorses hepatic conversion of cholesterol into bile acids and up-regulates LDL-R to accept more cholesterol and reduces low-density lipoprotein cholesterol in plasma. The tea polyphenol, catechins, exerts many health-promoting effects like antiproliferative, anti-inflammation, and hypolipidemic effects. The mechanism suggested for hyperlipidemic activity is inference of catechin in emulsification, micellar solubilization, and hydrolysis of lipids, thereby hindering their absorption in the intestine. Also, catechin present in tea limits pancreatic lipase activity by reducing its absorption in a dose-dependent manner. The epidemiological study on rat showed that tea catechin and black tea polyphenol restrict the in vitro micellar solubility of cholesterol (Chen et al. 2014).

### 12.21.4 Obesity

Obesity is one of the fast-growing chronic diseases which is directly or indirectly linked with complications like diabetes, CVD, hypertension, mental health diseases, few forms of cancer, and strokes. Obesity is a complex condition involving excessive amount of body fat which is the result of imbalance ratio of energy consumption and expenditure. The high-calorie, high-fat diet and insufficient activity are considered as external factors of obesity. Genetic predisposition is another important factor involved in obesity. Among all treatment options, alteration of or changing of lifestyle is the best approach to cure obesity. Many PSMs play an active role in curing obesity. Resveratrol is a plant-based nutrient found in a variety of food like cranberries, strawberry, and peanuts. This compound acts on various lipids and reduces their levels in plasma serum (Sergent et al. 2012). The phytonutrient curcumin possesses direct action on adipocytes. Curcumin acts on leptin by blocking its signal from the liver which further elevates adiponectin secretion and may reduce its deposition. Adiponectin secretion plays a critical role in improving obesity-related inflammations. Catechins are also reported to have anti-adipogenic activity. Various phenolic compounds like hydroxycinnamic acid and ferulic acid showed activity against adipocytes. The mechanism of action of phenolic compounds is the restriction of adipocyte proliferation or restriction of lipid and carbohydrate metabolic enzymes, thereby overall improving the metabolism of adipocytes. The PUFA like EPA and DHA are reported with their beneficial activity on adipose tissues. Both fatty acids restrict adipocyte proliferation and their up-regulation (Alves et al. 2012).

### 12.21.5 Diabetes

Diabetes is the most prominent metabolic health condition which is majorly influenced by high glucose intake. Symptoms of the disease include frequent hunger or thirst or urination. The well-known reason may be either reduced beta insulin cells or reduced response toward insulin. Diabetes carries several risks of other chronic diseases like obesity, aging, and mutation of functional insulin cells. Many natural herbs having bioactive compounds have potential in controlling diabetes like alkaloids, phenolics, anthocyanins, terpenes, curcuminoids, etc. possessing antidiabetic effects. They exert inhibitory effects on α-glycosidase enzyme. These natural sources are cheap and more effective and less harmful compared to their chemical counterparts. Consumption of *p*-coumaric acid and gallic acid showed altered glucose and insulin levels in rats, while allicin possesses direct activity on blood glucose levels by inhibiting α- and β-amylase activity. CLA reduces the risk of type 2 diabetes by its antioxidant property which provides defense mechanism against cellular oxidative damage. As reported, flavonoids have antihyperglycemic effects through their antioxidant potential. Other compounds like tannins, alkaloids, terpenoids, saponins, and phenolic compounds are reported to have antidiabetic properties (IiD et al. 2020).

## 12.22 Renal Disorder

Chronic kidney disease is the result of oxidative stress and inflammation in the body. Mainly, oxidative stress has direct linkage with kidney disease. In general, diet rich in phytochemicals and PSMs from fruits/vegetables or spices possesses protection against such predominant chronic disease. Phytosterols have been reported to offer both antioxidant and anti-inflammatory activities (Al-Okbi et al. 2014). Curcumins have been reported to have potential action on nephrons by increasing blood pressure, plasma urea, albumin, and creatinine ratio in urine. They provide protection against inflammatory markers (TNF-α, IL-1β, and IL-6) and morphological destruction occurring during chronic kidney disease (de Almeida Alvarenga et al. 2018). Other plant nutrients, phytoestrogens, isoflavones (soybeans), and lignans (flaxseeds), possess estrogen-like properties and have evidence in their beneficial role in renal-associated diseases (Ranich et al. 2001).

## 12.23 Alzheimer's Disease

Recently, Alzheimer's disease (AD) has shown global concerns due to the increase in the total number of cases worldwide. AD is the most common disease related with neurological disorder which is due to degeneration or dysfunction of neurons. The factor which influences the occurrence of AD are either genetic or the environmental. Low consumption or being deficient of linoleic acid and elevated concentration of arachidonic acid in diet are common in patients suffering from AD. Therefore, linoleic acid has been reported to have a protective effect on neurons. CLA have been reported with restricting effects on Aβ oligomerization/fibrillation. Along with CLA, other essential PUFA showed neuroprotective and antioxidant effects when consumed in appropriate volumes. Phytosterol also plays a significant role in Aβ formation and directly affects the expression and availability and activity of secretases. Also, phytosterol exerts protection against decreased learning abilities in pregnant mice and their offspring and reduces the LDL cholesterol level in serum and reported with alteration of brain's mRNA and NMDAR1 (N-methyl-D-aspartate receptor1). Similar effect was reported for gallic acids which when administered in microglial cells showed inhibitory action on NF-κB acetylation. The study also revealed the reduced production of cytokine and protective effects on neurons from neurotoxicity. Another bioactive compound anthocyanin also reported protective action of Aβ-induced neuron toxicity in AD mouse model. Also, anthocyanins have defensive effect against memory by their interaction with GABA receptors (Gorji et al. 2018).

## 12.24 Diarrhea

Digestion is commonly known for its delivery purpose of nutrients and bioactive compounds to the entire body parts. All the food compounds and digestive enzymes together interact and reflect human health. Diarrhea is the digestive-related disorder occurring during improper digestion. To cure this disease, deep understanding between food and the changes happening during the entire digestion process is required. Certain bioactive compounds have potential to solve digestive system-related issues. Polyphenols may have both positive and negative impact on food digestion. The positive effect is their inhibitory activity on energy-rich food which in turn may result in weight gain. However, poor absorption of polyphenols may have a negative impact on heath as it restricts the bioavailability of the bioactive compounds to exert its effect. Also, antioxidant activity of phenolics changes the pH and may interfere the other compounds (Cirkovic Velickovic and Stanic-Vucinic 2018; Schulz et al. 2017). This refers to the importance of selecting appropriate proportions of various bioactive compounds for exploring specific health benefits.

## 12.25 Conclusion and Future Thrust Area

Large quantity of agricultural crops, herbals, spices, condiments, and medicinal plants are grown, transported, and consumed all over the world. The agricultural commodities specifically are the incorporation of our daily meal. These commodities are loaded with abundance of PSMs or the bioactive compounds which have many health-promoting properties. Studies on these PSMs as potential diet ingredients for countering several health-related issues served as the basis for development of many new drugs in medical industry. The well-established role of these bioactive compounds or PSMs in management of overall health has led to their increased acceptability as potential substitutes to chemical medicines with several concerns and side effects on the body. Being natural in origin with no or almost negligible reported side effects on human health has served as an additional factor for their increased demand and acceptability among mass population. In the twentieth century, the concept of food as a mere source of energy to sustain life has transformed drastically, and nowadays nutraceutical and functional foods have started to lead the market with a positive growth. There is vast information already reported and published in the literature regarding therapeutic applications of PSM; however, their minimum effective concentration, bioavailability, effect on specific organs, metabolic pathway affected, and effects of processing on the form and bioavailability of plant secondary metabolites are the areas which need exploration. Further, the extraction protocols for these PSMs from the plant cells need standardization to minimize the quantitative and qualitative losses, improve the recovery, and retain highest bioactivity. Many efforts and advancement have been made in the past few decades on standardization of the extraction technique to address these issues. These PSMs have no magic bullets to manage or treat each disease; however, their unrestricted long-term and regular consumption can be correlated with reduced susceptibility to different diseases. In the future, the research work should be focused on identification of specific PSMs to treat various diseases, protocols for their extraction, stabilization, effective delivery, bioavailability, etc. Their specific health benefits should be well documented and supported with both in vitro and in vivo studies to increase their acceptability and recommendations by the medical practitioners.

## References

Agarwal A, Shen H, Agarwal S, Rao AV (2001) Lycopene content of tomato products: its stability, bioavailability and in vivo antioxidant properties. J Med Food 4(1):9–15

Akinwumi BC, Bordun KAM, Anderson HD (2018) Biological activities of stilbenoids. Int J Mol Sci 19(3):792

Alam MA, Subhan N, Hossain H, Hossain M, Reza HM, Rahman MM, Ullah MO (2016) Hydroxycinnamic acid derivatives: a potential class of natural compounds for the management of lipid metabolism and obesity. Nutr Metab 13(1):27

Al-Okbi SY, Mohamed DA, Hamed TE, Esmail RS, Donya SM (2014) Plant food extracts as a source of bioactive compounds for prevention of cisplatin-induced kidney dysfunction in rats. Polish J Food Nutr Sci 64(1):49–57

Alves N, Valdes S, Silveira C, Duarte-Martino H, Milagro-Yoldi FI, Moreno-Aliaga MJ, Ribeiro S (2012) Studies on mechanistic role of natural bioactive compounds in the management of obesity an overview

Astuti SM, Sakinah MA, Andayani RB, Risch A (2011) Determination of saponin compound from Anredera cordifolia (Ten) Steenis plant (binahong) to potential treatment for several diseases. J Agric Sci 3(4):224

Auger C, Al-Awwadi N, Bornet A, Rouanet JM, Gasc F, Cros G, Teissedre PL (2004) Catechins and procyanidins in Mediterranean diets. Food Res Int 37(3):233–245

Balentine DA, Wiseman SA, Bouwens LC (1997) The chemistry of tea flavonoids. Crit Rev Food Sci Nutr 37(8):693–704

Barba FJ, Nikmaram N, Roohinejad S, Khelfa A, Zhu Z, Koubaa M (2016) Bioavailability of glucosinolates and their breakdown products: impact of processing. Front Nutr 3:24

Baron EP (2018) Medicinal properties of cannabinoids, terpenes, and flavonoids in cannabis, and benefits in migraine, headache, and pain: an update on current evidence and cannabis science. Headache 58(7):1139–1186

Bazzano LA, He J, Ogden LG, Loria CM, Whelton PK (2003) Dietary fiber intake and reduced risk of coronary heart disease in US men and women: the National Health and Nutrition Examination Survey I Epidemiologic Follow-up Study. Arch Intern Med 163(16):1897–1190

Benowitz NL (2009) Pharmacology of nicotine: addiction, smoking-induced disease, and therapeutics. Annu Rev Pharmacol Toxicol 49:57–71. https://doi.org/10.1146/annurev.pharmtox.48.113006.094742

Besancon E, Guos Lok J, Tymianski M, Lo EH (2008) Beyond NMDA and AMPA glutamate receptors: emerging mechanisms for ionic imbalance and death in stroke. Trends Pharm Sci 29: 268–267

Blondeau N, Lipsky RH, Bourourou M, Duncan MW, Gorelick PB, Marini AM (2015) Alpha-linolenic acid: an omega-3 fatty acid with neuroprotective properties—ready for use in the stroke clinic? Biomed Res Int 2015. https://doi.org/10.1155/2015/519830

Böttger A, Vothknecht U, Bolle C, Wolf A (2018) Secondary metabolites in plants: general introduction. In: Lessons on caffeine, Cannabis & Co. Springer, Cham, pp 143–152

Brglez Mojzer E, Knez Hrnčič M, Škerget M, Knez Ž, Bren U (2016) Polyphenols: extraction methods, antioxidative action, bioavailability and anticarcinogenic effects. Molecules 21(7):901

Burnham BS, Gupton JT, Krumpe K, Webb T, Shuford J, Bowers B, Warren AE, Barnes C, Hall IH (1998) Cytotoxicity of substituted Alkyl-3, 4-bis (4-methoxyphenyl) pyrrole-2-carboxylates in L1210 Lymphoid Leukemia Cells. Archiv der Pharmazie: Int J Pharm Med Chem 331(11):337–341

Catarino MD, Silva AMS, Cruz MT, Cardoso SM (2017) Antioxidant and anti-inflammatory

Chapman KD, Moore TS Jr (1993) Catalytic properties of a newly discovered acyltransferase that synthesizes N-acylphosphatidylethanolamine in cottonseed (Gossypium hirsutum L.) microsomes. Plant Physiol 102(3):761–769

Chemat F, Rombaut N, Meullemiestre A, Turk M, Perino S, Fabiano-Tixier AS, Abert-Vian M (2017a) Review of green food processing techniques. Preservation, transformation, and extraction. Innovative Food Sci Emerg Technol 41:357–377

Chemat F, Rombaut N, Sicaire AG, Meullemiestre A, Fabiano-Tixier AS, Abert-Vian M (2017b) Ultrasound assisted extraction of food and natural products. Mechanisms, techniques, combinations, protocols and applications. A review. Ultrason Sonochem 34:540–560

Chen G, Wang H, Zhang X, Yang ST (2014) Nutraceuticals and functional foods in the management of hyperlipidemia. Crit Rev Food Sci Nutr 54(9):1180–1201

Cheung SS, Hasman D, Khelifi D, Tai J, Smith RW, Warnock GL (2019) Devil's Club falcarinol-type polyacetylenes inhibit pancreatic cancer cell proliferation. Nutr Cancer 71(2):301–311

Chew YL, Goh JK, Lim YY (2009) Assessment of in vitro antioxidant capacity and polyphenolic composition of selected medicinal herbs from Leguminosae family in Peninsular Malaysia. Food Chem 116:13–18

Christensen LP (2011) Aliphatic C17-polyacetylenes of the falcarinol type as potential health promoting compounds in food plants of the Apiaceae family. Recent Pat Food Nutr Agric 3: 64–77

Cirkovic Velickovic TD, Stanic-Vucinic DJ (2018) The role of dietary phenolic compounds in protein digestion and processing technologies to improve their antinutritive properties. Compr Rev Food Sci Food Saf 17(1):82–103

Colle IJ, Lemmens L, Van Buggenhout S, Met K, Van Loey AM, Hendrickx ME (2013) Processing tomato pulp in the presence of lipids: the impact on lycopene bioaccessibility. Food Res Int 51(1):32–38

Cragg GM, Newman DJ (2005) Plants as a source of anti-cancer agents. J Ethnopharmacol 100:72–79

Croteau R, Kutchan TM, Lewis NG (2000) Natural products (secondary metabolites). In: Buchanan BB, Gruissem W, Jones RL (eds) Biochemistry & molecular biology of plants. Courier Companies, Inc, pp 1250–1318

da Silva LLD, Nascimento M, Silva DHS, Furlan M, da Silva Bolzani V (2002) Antibacterial activity of a stearic acid derivative from Stemodia foliosa. Planta Med 68(12):1137–1139

Davis HR Jr, Zhu LJ, Hoos LM, Tetzloff G, Maguire M, Liu J et al (2004) NiemannPick C1 Like 1 (NPC1L1) is the intestinal phytosterol and cholesterol transporter and a key modulator of whole-body cholesterol homeostasis. J Biol Chem 279:33586–33592

Davis JM, Murphy EA, Carmichael MD, Davis B (2009) Quercetin increases brain and muscle mitochondrial biogenesis and exercise tolerance. AJP-Regu Physiol 296:1071–1077

de Almeida Alvarenga L, de Oliveira Leal V, Borges NA, de Aguiar AS, Faxén-Irving G, Stenvinkel P, Lindholm B, Mafra D (2018) Curcumin-A promising nutritional strategy for chronic kidney disease patients. J Funct Foods 40:715–721

De Castro ML, Priego-Capote F (2010) Soxhlet extraction: past and present panacea. J Chromatogr A 1217(16):2383–2389

De Smet E, Mensink RP, Boekschoten MV, de Ridder R, Germeraad WT, Wolfs TG et al (2015) An acute intake of plant stanol esters alters immune-related pathways in the jejunum of healthy volunteers. Br J Nutr 113:794–802

Debnath B, Singh WS, Das M, Goswami S, Singh MK, Maiti D, Manna K (2018) Role of plant alkaloids on human health: a review of biological activities. Mater Today Chem 9:56–72

Del Rio D, Stewart AJ, Mullen W et al (2004) HPLC-MSn analysis of phenolic compounds and purine alkaloids in green and black tea. J Agric Food Chem 52:2807–2815
Delgoda R, Murray JE (2017) Evolutionary perspectives on the role of plant secondary metabolites. In: Pharmacognosy. Academic Press, pp 93–100
Dell'Eva R, Pfeffer U, Vené R, Anfosso L, Forlani A, Albini A, Efferth T (2004) Inhibition of angiogenesis in vivo and growth of Kaposi's sarcoma xenograft tumors by the anti-malarial artesunate. Biochem Pharmacol 68:2359–2366
Duke JA, Beckstrom-Sternberg SM (2000) Apigenin. Dr. Duke's phytochemical and ethnobotanical databases. http://www.ars-grin.gov/cig-bin/duke
Ekezie FGC, Sun DW, Han Z, Cheng JH (2017) Microwave-assisted food processing technologies for enhancing product quality and process efficiency: a review of recent developments. Trends Food Sci Technol 67:58–69
El Barky AR, Hussein SA, Alm-Eldeen AA, Hafez YA, Mohamed TM (2016) Anti-diabetic activity of Holothuria thomasi saponin. Biomed Pharmacother 84:1472–1487
Eric Y (2002) Botanical medicines for the urinary tract. World J Urol 20(5):285–293
Evans WC (2009) Pharmacognosy, 16th edn, Edinburgh
Furu K, Skurtveit S, Langhammer A, Nafstad P (2007) Use of anti-asthmatic medications as a proxy for prevalence of asthma in children and adolescents in Norway: a nationwide prescription database analysis. Eur J Clin Pharmacol 63:7693–7698
García A, Rodríguez-Juan E, Rodríguez-Gutiérrez G, Rios JJ, Fernández-Bolaños J (2016) Extraction of phenolic compounds from virgin olive oil by deep eutectic solvents (DESs). Food Chem 197:554–561
Ghanem C, Taillandier P, Rizk Z, Nehme N, Souchard JP, El Rayess Y (2019) Evolution of polyphenols during Syrah grapes maceration: time versus temperature effect. Molecules 24(15):2845
Gideon MP (2003) Protein and non-protein protease inhibitors from plants. Stud Nat Prod Chem 29:567–641
Goławska S, Sprawka I, Łukasik I, Goławski A (2014) Are naringenin and quercetin useful chemicals in pest-management strategies? J Pest Sci 87(1):173–180
Gorji N, Moeini R, Memariani Z (2018) Almond, hazelnut and walnut, three nuts for neuroprotection in Alzheimer's disease: a neuropharmacological review of their bioactive constituents. Pharmacol Res 129:115–127
Gu L, House SE, Wu X, Ou B, Prior RL (2006) Procyanidin and catechin contents and antioxidant capacity of cocoa and chocolate products. J Agric Food Chem 54(11):4057–4061
Gullón B, Lú-Chau TA, Moreira MT, Lema JM, Eibes G (2017) Rutin: a review on extraction, identification and purification methods, biological activities and approaches to enhance its bioavailability. Trends Food Sci Technol 67:220–235
Gutiérrez-Venegas G, Kawasaki-Cárdenas P, Arroyo-Cruz SR, Maldonado-Frías S (2006) Luteolin inhibits lipopolysaccharide actions on human gingival fibroblasts. Eur J Pharmacol 541:95–105
Gylling H, Plat J, Turley S, Ginsberg HN, Ellegard L, Jessup W et al (2014) Plant sterols and plant stanols in the management of dyslipidaemia and prevention of cardiovascular disease. Atherosclerosis 232:346–360
Hämäläinen M, Nieminen R, Vuorela P, Heinonen M, Moilanen E (2007) Antiinflammatory effects of flavonoids: genistein, kaempferol, quercetin, and daidzein inhibit STAT-1 and NF-kappaB activations, whereas flavone, isorhamnetin, naringenin, and pelargonidin inhibit only NF-kappaB activation along with their inhibitory effect on iNOS expression and NO production in activated macrophages. Mediat Inflamm 2007:45673
Hoffmann D (2003) Medical herbalism: the science and practice of herbal medicine. Healing Arts Press One Park Street, Rochester, Varmont. ISBN: 978-089281749-8
Hur SJ, Park GB, Joo ST (2007) Biological activities of conjugated linoleic acid (CLA) and effects of CLA on animal products. Livest Sci 110(3):221–229
IiD II, Kumar S, Shukla S, Kumar V, Sharma R (2020) Putative antidiabetic herbal food ingredients: nutra/functional properties, bioavailability and effect on metabolic pathways. Trends Food Sci Technol. https://doi.org/10.1016/j.tifs.2020.01.017

Jayakumar K, Murugan K (2016) Solanum alkaloids and their pharmaceutical roles: a review. J Anal Pharm Res 3(6):00075

Jiang F, Dusting GJ (2003) Natural phenolic compounds as cardiovascular therapeutics: potential role of their antiinflammatory effects. Curr Vasc Pharmacol 1(2):135–156

Johann S, Mendes BG, Missau FC, Resende MAD, Pizzolatti MG (2011) Antifungal activity of five species of Polygala. Braz J Microbiol 42:1065–1075

Kaur P, Kaur S, Kumar S, Singh P (2010) Rubia cordifolia L. and Glycyrrhiza glabra L. medicinal plants as potential source of COX-2 inhibitors. Am J Biomed Sci 2:108–120

Kendall CWC, Esfahani A, Jenkins DJA (2010) The link between dietary fibre and human health. Food Hydrocoll 24(1):42–48

Kim J, Wie MB, Ahn M, Tanaka A, Matsuda H, Shin T (2019) Benefits of hesperidin in central nervous system disorders: a review. Anat Cell Biol 52(4):369–377

Klingberg S, Andersson H, Mulligan A, Bhaniani A, Welch A, Bingham S et al (2008) Food sources of plant sterols in the EPIC Norfolk population. Eur J Clin Nutr 62:695–703

Ko WC, Shih CM, Leu IJ, Chen TT, Chang JP (2005) Mechanisms of relaxant action of luteolin in isolated guinea pig trachea. Planta Med 71:406–411

Kobæk-Larsen M, Christensen LP, Vach W, Ritskes-Hoitinga J, Brandt K (2005) Inhibitory effects of feeding with carrots or (−)-falcarinol on development of azoxymethane-induced preneoplastic lesions in the rat colon. J Agric Food Chem 53:1823–1827

Krishnamurthy VMR, Wei G, Baird BC, Murtaugh M, Chonchol MB, Raphael KL et al (2012) High dietary fiber intake is associated with decreased inflammation and all-cause mortality in patients with chronic kidney disease. Kidney Int 81(3):300–306

Kumar PV, Bhopal AKP (2012) Formulation design and evaluation of rutin loaded self-emulsifying drug delivery system (SEDDs) using edible oil. Asian J Pharm Clin Res 5(1):76–78

Kuppusamy P, Yusoff MM, Maniam GP, Ichwan SJA, Soundharrajan I, Govindan N (2014) Nutraceuticals as potential therapeutic agents for colon cancer: a review. Acta Pharm Sin B 4(3):173–181

Lane JD, Adcock RA, Williams RB, Kuhn CM (1990) Caffeine effects on cardiovascular and neuroendocrine responses to acute psychosocial stress and their relationship to level of habitual caffeine consumption. Psychosom Med 52:320–336

Lane JD, Pieper CF, Phillips-Bute BG, Bryant JE, Kuhn CM (2002) Caffeine affects cardiovascular and neuroendocrine activation at work and home. Psychosom Med 64(4):595–603

Larson AJ, Symons JD, Thunder J (2010) Quercetin: a treatment for hypertension? A review of efficacy and mechanisms. Pharmaceuticals 3:237–250

Lee JC, Lim KT, Jang YS (2002) Identification of Rhus verniciflua Stokes compounds that exhibit free radical scavenging and anti-apoptotic properties. Biochim Biophys Acta 1570:181–191

Lee WJ, Wu LF, Chenb WK, Wang CJ, Tseng TH (2006) Inhibitory effect of luteolin on hepatocyte growth factor/scatter factor-induced HepG2 cell invasion involving both MAPK/ERKs and PI3K–Akt pathways. Chem Biol Interact 160:123–133

Lin S, Tsai SC, Lee CC, Wang BW, Liou JW, Shyu KG (2004) Berberine inhibits HIF-1α expression via enhanced proteolysis. Mol Pharmacol 66(3):612–619

Liu WK, Xu SX, Che CT (2000) Anti-proliferative effect of ginseng saponins on human prostate cancer cell line. Life Sci 67:1297–1306

Lucchesi ME, Chemat F, Smadja J (2004) Solvent-free microwave extraction of essential oil from aromatic herbs: comparison with conventional hydro-distillation. J Chromatogr A 1043(2):323–327

Ludwiczuk A, Skalicka-Woźniak K, Georgiev MI (2017) Terpenoids. In: Pharmacognosy. Academic Press, pp 233–266

Malaguti M, Angeloni C, Hrelia S (2015) Nutraceutical bioactive compounds promote healthspan counteracting cardiovascular diseases. J Am Coll Nutr 34(sup1):22–27

Mazumder A, Dwivedi A, Du Plessis J (2016) Sinigrin and its therapeutic benefits. Molecules 21(4):416

Montanher AB, Zucolotto SM, Schenkel EP, Fröde TS (2007) Evidence of anti-inflammatory effects of Passiflora edulis in an inflammation model. J Ethnopharmacol 109(2):281–288

Moreau RA, Nystrom L, Whitaker BD, Winkler-Moser JK, Baer DJ, Gebauer SK et al (2018) Phytosterols and their derivatives: structural diversity, distribution, metabolism, analysis, and health-promoting uses. Prog Lipid Res 70:35–61
Moudi M, Go R, Yong Seok Yien C, Nazre M (2013) Vinca alkaloids. Int J Prev Med 4(11):1231–1235
Mustafa A, Turner C (2011) Pressurized liquid extraction as a green approach in food and herbal plants extraction: a review. Anal Chim Acta 703(1):8–18
Mustapha N, Mokdad-Bzeouich I, Maatouk M, Ghedira K, Hennebelle T, Chekir-Ghedira L (2016) Antitumoral, antioxidant, and antimelanogenesis potencies of Hawthorn, a potential natural agent in the treatment of melanoma. Melanoma Res 26:211–222
Nakase I, Lai H, Narendra PS, Sasaki T (2008) Anticancer properties of artemisinin derivatives and their targeted delivery by transferrin conjugation. Int J Pharm 354:28–33
Odontuya G, Hoult JRS, Houghton PJ (2005) Structure-activity relationship for antiinflammatory effect of luteolin and its derived glycosides. Phytother Res 19:782–786
Ong KC, Khoo HE (2000) Effects of myricetin on glycemia and glycogen metabolism in diabetic rats. Life Sci 67:1695–1705
Parsaeimehr A, Sargsyan E, Javidnia K (2010) A comparative study of the antibacterial, antifungal and antioxidant activity and total content of phenolic compounds of cell cultures and wild plants of three endemic species of Ephedra. Molecules 15:1668–1678
Pierpoint WS (2000) Why do plants make medicines. Biochemist 22:37–40
Plat J, Hendrikx T, Bieghs V, Jeurissen ML, Walenbergh SM, van Gorp PJ et al (2014) Protective role of plant sterol and stanol esters in liver inflammation: insights from mice and humans. PLoS One 9:e110758
Puneet G, Aditi S, Vinayak K, Anant GN, Samir D, Nikhil M (2013) Camellia sinensis (tea): implications and role in preventing dental decay. Pharmacogn Rev 7(14):152–156. https://doi.org/10.4103/0973-7847.120515
Puri M, Sharma D, Barrow CJ (2012) Enzyme-assisted extraction of bioactives from plants. Trends Biotechnol 30(1):37–44
Ranich T, Bhathena SJ, Velasquez MT (2001) Protective effects of dietary phytoestrogens in chronic renal disease. J Ren Nutr 11(4):183–193
Rashidi L, Khosravi-Darani K (2011) The applications of nanotechnology in food industry. Crit Rev Food Sci Nutr 51(8):723–730
Raventós M, Duarte S, Alarcón R (2002) Application and possibilities of supercritical CO2 extraction in food processing industry: an overview. Food Sci Technol Int 8(5):269–284
Rein MJ, Renouf M, Cruz-Hernandez C, Actis-Goretta L, Thakkar SK, da Silva Pinto M (2013) Bioavailability of bioactive food compounds: a challenging journey to bioefficacy. Br J Clin Pharmacol 75(3):588–602
Ribas-Agustí A, Martín-Belloso O, Soliva-Fortuny R, Elez-Martínez P (2018) Food processing strategies to enhance phenolic compounds bioaccessibility and bioavailability in plant-based foods. Crit Rev Food Sci Nutr 58(15):2531–2548
Rochfort S, Ezernieks V, Neumann N, Panozzo J (2011) Pulses for human health: changes in isoflavone and saponin content with preparation and cooking. Aust J Chem 64(6):790–797
Sandhar HK, Prasher S, Tiwari P, Salhan M, Sharma P (2011) A review of phytochemistry and pharmacology of flavonoids. Internationale Pharmaceutica Sciencia 1:25–41
Schulz M, Biluca FC, Gonzaga LV, Borges GDSC, Vitali L, Micke GA et al (2017) Bioaccessibility of bioactive compounds and antioxidant potential of juçara fruits (Euterpe edulis Martius) subjected to in vitro gastrointestinal digestion. Food Chem 228:447–454
Sen Z, Weida W, Jie M, Li S, Dongming Z, Xiaoguang C (2019) Coumarin glycosides from Hydrangea paniculata slow down the progression of diabetic nephropathy by targeting Nrf2 anti-oxidation and smad2/3-mediated profibrosis. Phytomedicine 57:385–395
Serafini M, Peluso I, Raguzzini A (2010) Flavonoids as anti-inflammatory agents. Proc Nutr Soc 69(3):273–278. https://doi.org/10.1017/S002966511000162X)
Sergent T, Vanderstraeten J, Winand J, Beguin P, Schneider YJ (2012) Phenolic compounds and plant extracts as potential natural anti-obesity substances. Food Chem 135(1):68–73

Shahidi F, Chandrasekara A (2010) Hydroxycinnamates and their in vitro and in vivo antioxidant activities. Phytochem Rev 9(1):147–170

Sharma R, Singh RB (2010) Bioactive foods and nutraceutical supplementation criteria in cardiovascular protection. Open Nutraceuticals J 3(1)

Shi J, Arunasalam K, Yeung D, Kakuda Y, Mittal G, Jiang Y (2004) Saponins from edible legumes: chemistry, processing, and health benefits. J Med Food 7(1):67–78

Sieniawska E (2015) Activities of tannins–from in vitro studies to clinical trials. Nat Prod Commun 10(11):1934578X1501001118

Spiller F, Alves MK, Vieira SM, Carvalho TA, Leite CE, Lunardelli A, Poloni JA, Cunha FQ, de Oliveira JR (2008) Anti-inflammatory effects of red pepper (Capsicum baccatum) on carrageenan- and antigen-induced inflammation. J Pharm Pharmacol 60(4):473–478. https://doi.org/10.1211/jpp.60.4.0010

Styrczewska M, Kulma A, Kostyn K, Hasiewicz-Derkacz K, Szopa J (2013) Flax terpenoid pathway as a source of health promoting compounds. Mini Rev Med Chem 13(3):353–364

Tang B, Bi W, Tian M, Row KH (2012) Application of ionic liquid for extraction and separation of bioactive compounds from plants. J Chromatogr B 904:1–21

Tseng YT, Lin WJ, Chang WH, Lo YC (2019) The novel protective effects of loganin against 1-methyl-4-phenylpyridinium-induced neurotoxicity: enhancement of neurotrophic signaling, activation of IGF-1R/GLP-1R, and inhibition of RhoA/ROCK pathway. Phytother Res 33(3):690–701

Unno T, Sugimoto A, Kakuda T (2000) Scavenging effect of tea catechins and their epimers on superoxide anion radicals generated by hypoxanthine and xanthine oxidase. J Sci Food Agric 80:601–606

US Department of Agriculture, Agricultural Research Service (2002) USDA-Iowa State University database on the isoflavone content of foods. Nutrient data laboratory web site. http://www.nal.usda.gov/fnic/foodcomp/Data/isoflav/isoflav.html

Verpoorte R, Alfermann AW (2000) Metabolic engineering of plant secondary metabolism. Kluwer Academic Publishers, Dordrecht

Vilkhu K, Mawson R, Simons L, Bates D (2008) Applications and opportunities for ultrasound assisted extraction in the food industry—a review. Innovative Food Sci Emerg Technol 9(2):161–169

Wagner KH, Elmadfa I (2003) Biological relevance of terpenoids. Ann Nutr Metab 47(3–4):95–106

Weng J (2013) The evolutionary paths towards complexity: a metabolic perspective. New Phytol 201:1141–1149

Winkel-Shirley B (2001) Flavonoid biosynthesis. A colourful model for genetics, biochemistry, cell biology and biotechnology. Plant Physiol 126:485–493

Xu L, Wu Y, Zhao X, Zhang W (2015) The study on biological and pharmacological activity of coumarins. In: Asia-Pacific Energy Equipment Engineering Research Conference; (AP3ER 2015). pp 135–138

Yang B, Chen H, Stanton C, Ross RP, Zhang H, Chen YQ, Chen W (2015) Review of the roles of conjugated linoleic acid in health and disease. J Funct Foods 15:314–325

Yi W, Fischer J, Krewer G, Akoh CC (2005) Phenolic compounds from blueberries can inhibit colon cancer cell proliferation and induce apoptosis. J Agric Food Chem 53(18):7320–7329

Zaker F, Oody A, Arjman A (2007) A study on the antitumoral and differentiation effects of Peganum harmala derivatives in combination with ATRA on leukaemic cell. Arch Pharm Res 30:844–849

Zbigniew S, Beata Ż, Kamil J, Roman F, Barbara K, Andrzej D (2014) Antimicrobial and antiradical activity of extracts obtained from leaves of three species of the genus Pyrus. Microb Drug Resist 20(4):337–343

# Role of Plant Secondary Metabolites as Modulators of Multidrug Resistance in Cancer Therapy

# 13

Mayank Chaudhary

**Abstract**

Cancer is one of the leading causes of global mortality and morbidity. It is caused by dysregulation of key cellular processes. Treatment of cancer focuses on re-regulation of these functions, and different approaches of cancer treatment include surgery, chemotherapy, radiotherapy and immunotherapy. Immunotherapy approach is highly specific but is expensive with limited target range. Radiotherapy and chemotherapy can have severe toxic effects even on normal cells. In addition to this, problem of relapse and acquired resistance is common in certain types of cancer. Another probable hindrance in the efficacy of chemotherapy is the phenomenon of multidrug resistance (MDR). It occurs mainly due to efflux of anticancer drugs by the action of ATP-binding cassette (ABC) transporters. Various sources are routinely screened for identification of novel molecule with greater efficacy and reduced side effects. Identification of naturally occurring phytochemicals has shown great potential in targeting multidrug resistance with reduced cytotoxic effect. As a result, various plant-based secondary metabolites are routinely screened for their MDR reversal effect. Plant secondary metabolites (PSMs) belonging to three major groups (terpenoids, phenolics and alkaloids) have shown great potential in reducing efflux along with accumulation of anticancer drug within cancerous cells. Thus, PSMs hold immense value in modulating multidrug resistance and increasing efficacy of chemotherapy by targeting ABC transporters.

M. Chaudhary (✉)
Department of Biotechnology, Maharishi Markandeshwar (Deemed to be) University, Mullana (Ambala), Haryana, India





A. K. Sharma, A. Sharma (eds.), *Plant Secondary Metabolites*,
https://doi.org/10.1007/978-981-16-4779-6_13

**Keywords**

Plant secondary metabolites · ABC transporters · Multidrug resistance · Terpenoids · Phenolics · Alkaloids

## 13.1 Introduction

Plants synthesize a variety of metabolites using enzyme-catalysed metabolic pathways. Primary metabolites are those compounds that are essential for proper growth, development and reproduction of plants. Thus, they are crucial metabolites required for survival of plants. Compounds considered as primary metabolites include energy-rich molecules, structural components, genetic information carriers and pigments. Some of the primary metabolites are also used in industries for commercial purpose (Rani et al. 2018). Contrary to this, secondary metabolites are not essential for growth and development of plants and are basically derived from primary metabolites (Fig. 13.1).

## 13.2 Plant Secondary Metabolites (PSMs)

Plants are fixed living organisms that cannot escape from herbivorous attackers by running away nor do they have immune system to fight against invading parasites. So either they have developed physical constraints or evolved a variety of defence chemicals to cope such situation (Wink et al. 2012). These chemicals are termed as secondary metabolites that can show toxicity to susceptible host as preventive measure or can possess beneficial properties having therapeutic potential (Birudu and Naik 2014). In addition to defensive role, some secondary metabolites (SM) also function as signalling molecules and abiotic stress protectors. Main targets of SM within microbes include nucleic acids, proteins and membranes. A variety of SM possesses hydrophilic properties because of which they can cross the cell membrane via simple diffusion. These are synthesized in small quantities and are unique to different plant species. Plant secondary metabolites (PSMs) can broadly be classified into alkaloids, phenolics and terpenoids. Each of these broader categories contains several individual compounds having antioxidant, anti-inflammatory, anti-mutagenic and anti-carcinogenic properties (Ferguson 2001).

## 13.3 Multidrug Resistance (MDR)

Multidrug resistance has become a matter of global concern challenging the efficacy of current cancer chemotherapy and antibiotic treatment of bacterial infections (Chambers et al. 2020). Mechanism of MDR can be broadly divided into pump resistance and non-pump resistance. Non-pump resistance mechanisms include degradation of drugs by lysozymes and anti-apoptotic and antioxidant defences.

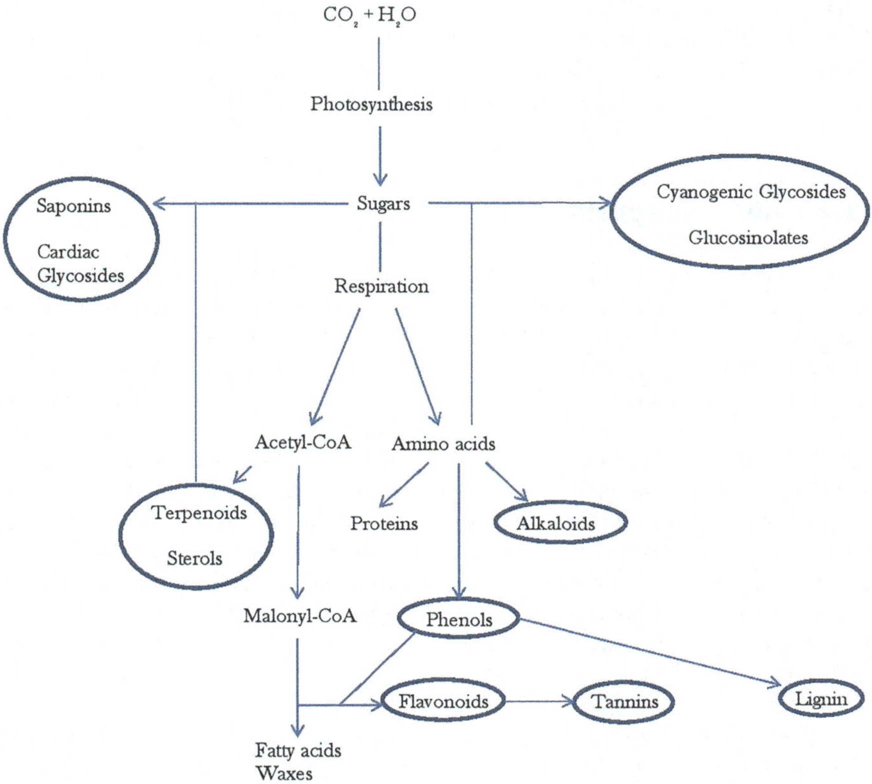


**Fig. 13.1** Synthesis of secondary metabolites (Rani et al. 2018)

Pump resistance depends upon efflux of anticancer drugs by membrane-bound efflux pumps (Kumar and Jaitak, 2019). Multidrug resistance transporters include ATP-binding cassette (ABC) superfamily, resistance nodulation division (RND) superfamily, major facilitator superfamily (MFS), small multidrug resistance (SMR) family and multidrug and toxic compound extrusion (MATE) family (Mousa and Bruner 2016). Reasons for multidrug resistance (MDR) development include activation of ATP-binding cassette (ABC) transporters that causes efflux of lipophilic compounds, activation of cytochrome p450 oxidases (CYP) that causes oxidation of lipophilic agents and activation of glutathione transferase (GST) that conjugates xenobiotics with glutathione (Wink et al. 2012). Another reason for drug resistance can be decreased drug uptake by solute carrier (SLC) transporters expressed in cancer cells. Most of the SLC transporters are secondary active transporters that use energy within concentration gradients for transport of molecules (Joyce et al. 2015). SLC transporters are responsible for uptake of several anticancer drugs, and downregulation/mutation of this transport protein within cancer cells can result in development of drug resistance.

However, raised activity of efflux pumps, mainly ABC transporters, has been found as a leading cause of all the mechanisms involved in MDR of cancer cells. Overexpression of these membrane-bound drug transporters in cancer cells is responsible for fall of cancer chemotherapy (Kumar and Jaitak 2019).

## 13.4 ABC Transporters

Cancer cells have developed resistance against defence chemicals through overexpression of ABC transporters. These are mostly active in the intestine, kidney and liver (Nooter and Stoter 1996; Steinbach et al. 2002). These cause efflux of drug from the cell upon hydrolysis of ATP, i.e. drug is removed from the interior of the cell using active transport mechanism (Fig. 13.2). This drug efflux is mainly mediated by transmembrane transporters belonging to ATP-binding cassette (ABC) protein superfamily. Typical composition of ABC transporters involves two transmembrane domains (TMDs) and two nucleotide (ATP) binding domains (NBDs) on cytosolic surface. In comparison to NBD, TMD represents highly variable structure responsible for substrate specificity of ABC transporters. Binding of ATP induces conformational change in TMD resulting in substrate efflux upon hydrolysis of ATP (Kumar and Jaitak 2019). Out of wide ABC superfamily, those most frequently implied with MDR phenotype include (Eckford and Sharom 2009) the following:

1. P-glycoprotein: P-gp was the first ABC transporter that was cloned consisting of transmembrane and ATP binding domain. It is also known as MDR1 protein (multiple drug resistant protein) directed to the gut lumen (Wink et al. 2012). P-gp removes a wide range of lipophilic chemotherapeutics that enters tumour cells through simple diffusion (Loo and Clarke 2005). Upon ATP hydrolysis, this

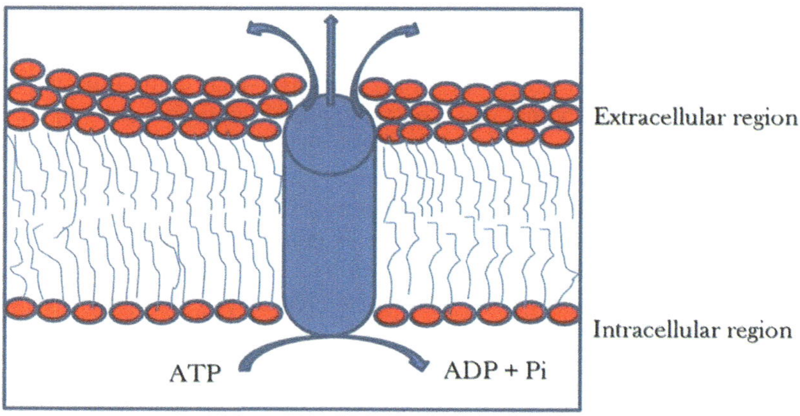


**Fig. 13.2** Mechanism of efflux by ABC transporters upon ATP hydrolysis (Chambers et al. 2020)

efflux pump exports molecules to extracellular region that are bound to its transmembrane domain. In addition to tumour cells, P-gp is also present in normal tissues like the kidney, liver, intestine, brain and adrenal glands. This indicates role of P-gp in the excretion of xenobiotics and metabolites in urine and bile into the intestinal lumen (Hohmann et al. 2002).
2. Multiple resistance-associated protein 1 (MRP1): This pump is encoded by *ABCC1* gene. It was firstly discovered in pulmonary carcinoma cells and then in other polarized cells. This pump is able to transport both hydrophilic and hydrophobic xenobiotics (Chambers et al. 2020). It utilizes glutathione either in free form or bound form for drug transport (Van der Kolk et al. 1999). MRP1 is also involved in export of endobiotics (pro-inflammatory molecules, hormones and antioxidants) (Johnson and Chen 2017). Similarity in biochemical and pharmacological properties is found between MRP1 and P-gp (Wijnholds et al. 2000). However, only 23% sequence similarity is found between both proteins where MRP1 has one additional transmembrane domain ($TMD_0$) and single functional site for ATP hydrolysis (Johnson and Chen 2017).
3. Breast cancer resistance protein (BCRP): It was discovered in human placenta and then found in other tissues. This protein acts as primary barrier in different barrier systems (Chambers et al. 2020). This protein pump shows wide substrate specificity including anticancer drugs (doxorubicin, camptothecin and imatinib) (Krishnamurthy and Schuetz 2006) and positively and negatively charged molecules along with other hydrophilic molecules (Kumar and Jaitak 2019). This protein is encoded by *ABCG2* gene and is considered as half transporter comprising of single TMD and single NBD. Thus, it becomes functionally active only after dimerization which can be either homo- or hetero-dimerization (Kumar and Jaitak 2019).

BCRP and P-gp efflux pumps are highly expressed at apical membrane of the blood-brain barrier (BBB), liver, intestine and placenta (Schinkel and Jonker 2003). These ATP-driven pumps can cause efflux of lipophilic compounds to the gut lumen or blood system from the cell, thereby reducing the concentration of toxic compounds within the cell. These transporters are critical at BBB as they cause rapid removal of lipophilic xenobiotics from the central nervous system (CNS) (Mahringer and Fricker 2010). The concept of multidrug resistance (MDR) was discovered in cancer patients undergoing chemotherapy where resistance was developed against the cytotoxic drug. Later on, it was found that tumour cells were able to pump out lipophilic alkaloids at the speed similar to entry of drug. In addition, these ATP transporters are overexpressed in tumour cells upon application of anticancer drugs (Schinkel and Jonker 2003). As a result, MDR is one of the key reasons creating hindrance in the success of chemotherapy (Gottesman 2002). These ABC transporters are also responsible for microbial MDR (Steffens et al. 1996).

The agents that are used to reverse the effect of MDR are known as chemosensitizers or modulators. They overcome MDR by enhancing the cytotoxicity or by inhibiting the efflux activity of ABC transporters (Robert and Jarry 2003). Simultaneous expression of different ABC transporters is seen in multiresistant cancer cells (Annereau et al. 2004). As a result co-expression of these transporters can result in more diverse resultant profiles in comparison to expression of any single member alone. This demands the need of broad-spectrum reversal agents (Maliepaard et al. 2001; Brooks et al. 2003). Both natural and synthetic compounds have been used to resensitize resistant tumour cells (He and Liu 2002). But contradictory results were seen in different clinical trials (Solary et al. 2000; Dantzig et al. 2001) with some agents showing severe side effects. These modulators either act as competitive inhibitors of ABC transporters or alter the expression of genes coding for these transporters (Li et al. 2011). These can also be administered as co-drugs in chemosensitization affecting transport of drugs by causing physiological changes in the lipid bilayer (Malik et al. 2017). The first generation of modulators including verapamil, reserpine and tamoxifen showed great toxicity, while the second generation which mostly covered analogues of the first generation including dexverapamil, valspodar and biricodar were tolerable with low transporter selectivity. The third generation of modulators including elacridar, laniquidar and mitotane were developed using QSAR approach to overcome the limitations of first- and second-generation modulators, but most of these failed in clinical trials exhibiting undesirable side effects (Mohana et al. 2016; Kumar and Jaitak 2019). The common reasons for failure of most of these modulators belonging to different generations include toxicity, non-specificity towards transporters and interference with pharmacokinetics of chemotherapeutic drugs. This shifted the focus towards natural products that are included in the fourth generation having lower toxicity with higher selectivity. Plant kingdom therefore provides various natural compounds in the form of secondary metabolites that are either less toxic or non-toxic and biologically active (Cheynier et al. 2013). For discovery of suitable agents, different phytochemicals are being isolated and tested for their effectiveness against MDR (Fig. 13.3).

Plant secondary metabolites (PSMs) that are currently used as therapeutics are terpenoids, phenolics or alkaloids (Table 13.1). Most of the compounds belonging to these broader categories function as competitive inhibitors competing for the active site of ABC transporters resulting in reduced efflux of anticancer drug from cancerous cells.

The following section covers detailed information about different plant compounds belonging to these broad categories that are studied/screened for their activity against MDR.

## 13.5 Terpenoids

Terpenoids represent diverse group of natural compounds found in plants. Most of the terpenoids are biologically active possessing vast therapeutic potential (Ludwiczuk et al. 2017). These are composed of isoprene as the basic unit and subcategorized on the basis of the number of isoprene units and subsequent carbon atoms (Gill et al. 2016). Synthesis of terpenoids occurs via two major pathways: mevalonic acid pathway and methylerythritol phosphate (MEP) pathway. Terpenoids are used therapeutically as antimicrobial, anti-inflammatory, antioxidant

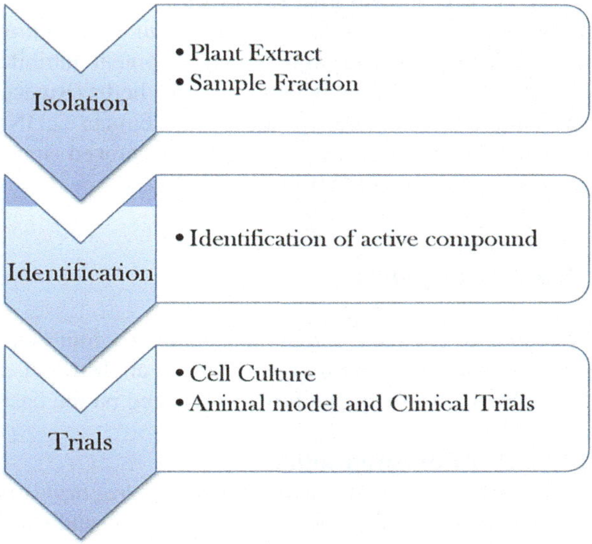


**Fig. 13.3** Workflow in identification of novel bioactive agent (Budisan et al. 2017)

**Table 13.1** Plant secondary metabolites (PSMs) as multidrug resistance modulators (Wink et al. 2012)

| Broader category | Subcategory |
|---|---|
| Terpenoids | Monoterpenes |
| | Diterpenes |
| | Triterpenes |
| | Steroids |
| | Tetraterpenes |
| Phenolics | Phenylpropanoids |
| | Flavonoids, chalcones, xanthones, anthocyanins and related polyphenols |
| | Quinones, anthraquinones |
| | Lignans |
| | Coumarins |
| Alkaloids | True, proto- and pseudoalkaloids |

and immunomodulatory agent along with use in aromatherapy (Brahmkshatriya and Brahmkshatriya 2013).

### 13.5.1 Monoterpenes

These are composed of two isoprene units with ten carbon atoms having molecular formula of $C_{10}H_{16}$.

#### 13.5.1.1 Citronellal and Citronellol

These monoterpenoids from essential oil of *Zanthoxyli* Fructus showed strong inhibition of P-gp and were found as potent inhibitors. For isolation of potent inhibitor, fraction from six types of citrus herbs (Rutaceae) was prepared and tested on cultured cells transfected with human cDNA encoding P-gp where monoterpenoids from *Zanthoxyli* Fructus showed significant inhibitory effect on P-gp-mediated transport (Yoshida et al. 2005).

### 13.5.2 Diterpenes

This class of chemical compounds comprises four isoprene units having molecular formula of $C_{20}H_{32}$ and derived from geranylgeranyl diphosphate (Toyomasu and Sassa 2010). These can be further classified on the basis of their skeletal core.

#### 13.5.2.1 Andrographolide

This active diterpenoid is isolated from *Andrographis paniculata*. The effect of this compound on transport activity of P-gp was studied by measuring the amount of inorganic phosphate that is liberated upon ATP hydrolysis. This assay method distinguishes substrate from inhibitor (Ishikawa et al. 2004). The effect of this compound was studied along with other compounds that were tested on crude intestinal membrane fractions prepared from jejunal mucosa for ATPase activity (Yumoto et al. 2001). Biphasic effect of andrographolide was observed where low concentration stimulated and higher concentration inhibited P-gp ATPase activity (Najar et al. 2010).

#### 13.5.2.2 Jatrophane Diterpene Polyesters

The effect of 15 Jatrophane diterpene polyesters isolated from Euphorbiaceae family on reversal of MDR in mouse lymphoma cells was examined. Natural products in Jatrophane diterpene group isolated from *Euphorbia esula*, *E. peplus*, *E. salicifolia* and *E. serrulata* were found as effective compounds for MDR reversal. The MDR-reversing activity of extracts isolated from *E. peplus*, *E. salicifolia* and *E. serrulata* decreased at higher concentrations which can be due to toxicity leading to cell shrinkage and granulation (Hohmann et al. 2002). Similarly, different Jatrophane diterpenoids isolated from *Euphorbia sororia* were studied for their effectiveness against MDR. Out of the various compounds studied, Compound 1 (Euphosorophane A) was found to be highly potent with high therapeutic index and lower cell toxicity. Moreover, it reversed P-gp-mediated resistance to doxorubicin (DOX) by acting as competitive inhibitor to DOX in binding to P-gp resulting in significant increase in cellular concentrations of DOX (Hu et al. 2018). Structure-activity relationship (SAR) studies of Jatrophane diterpenes isolated from *Euphorbia helioscopia* was carried out to study MDR reversal activity and cytotoxicity (Li et al. 2018). Previous studies have highlighted the importance of acylation of free hydroxyl group at C-14 in increasing the activity (Zhu et al. 2016). Similarly, SAR study conducted by Li et al. (2018) on mini compound library of Jatrophane by chemical modification revealed that introduction of alkyl acyl group or aryl acyl group at C-14 is favourable for MDR reversal activity of Jatrophane diterpenoids.

#### 13.5.2.3 Lathyrane Diterpenes

Effect of Lathyrane diterpenes along with other known Jatrophane diterpenes isolated from *Euphorbia lagascae* on MDR reversal was studied by Duarte et al. (2006). The effect of isolated Lathyrane diterpenes on inhibition of rhodamine 123 efflux from MDR1-transfected mouse lymphoma cells was compared. Out of the three Lathyrane diterpenes studied, latilagascene B displayed highest inhibition in comparison to latilagascene A and B.

### 13.5.3 Triterpenes

These comprise three terpene units having molecular formula of $C_{30}H_{48}$. Triterpenes are produced by animals, plants and fungi.

#### 13.5.3.1 Betulinic Acid, Pomolic Acid and Oleanolic Acid

These active compounds are derived from Chrysobalanaceae family. Betulinic acid is derived from $CH_2Cl_2$ fraction of leaves of *Licania tomentosa*, whereas pomolic acid is isolated from $CH_2Cl_2$ fraction of leaves of *Chrysobalanus icaco*. In addition to this, AcOEt fraction of *L. tomentosa* fruits yielded oleanolic acid. Activity of these compounds was tested on human leukaemic cell line, K562 and Lucena 1 (resistant cell line with MDR characteristics) (Fernandes et al. 2003). The results obtained showed that all these triterpenes possessed anti-MDR properties.

#### 13.5.3.2 Limonoids

These secondary metabolites possess insecticidal, antimicrobial, anti-cancerous and other pharmacological activities (Roy and Saraf 2006). Three limonoids were extracted from MeOH extract of the bark of *Phellodendron amurense*, and their cytotoxicity and MDR reversal potential were studied (Min et al. 2007). Compounds obacunon, limonin and 12α-hydroxylimonin showed less cytotoxicity against studied human cancer cell lines and further inhibited MDR activity in P-gp expressing multidrug resistant cell line. Similarly, limonoids isolated from *Citrus species* were tested against multidrug resistance in cancerous cell lines having higher expression of P-gp (El-Readi et al. 2010). Limonin and deacetylnomilin were isolated from *C. pyriformis* using petroleum ether as solvent. These compounds caused reduction in P-gp-mediated efflux in drug resistant leukaemia cell line (ARD5000). These compounds not only reversed doxorubicin resistance, rather they restored its cytotoxicity in doxorubicin-treated Caco-2 cells. *Citrus* limonoids such as nomilin and obacunon inhibited proliferation of many cancerous cells and also enhanced cytotoxic activities of certain compounds in multidrug resistant BK-V1 cells (Poulose et al. 2006). Use of limonoids is advantageous as they lack toxicity even at higher doses and induce specific carcinogen-metabolizing enzymes (Roy and Saraf 2006). Certain studies have showed the criticality of rings in limonoid skeleton for antineoplastic activity. Molecular changes in ring 'A' can decrease or remove anticancer activity as deacetylnomilin showed lower inhibition of P-gp in comparison to limonin (El-Readi et al. 2010). Additionally, presence of furan ring and epoxide groups in limonoid skeleton is essential for formation of covalent bonds with active site of the proteins.

#### 13.5.3.3 Cumingianol A–F

These triterpenes were isolated from MeOH extract of the leaves of *Dysoxylum cumingianum*. Cytotoxicity assays and MDR-reversing effects of isolated compounds were studied in cancer cell lines which also included multidrug resistant cell line (KB-C2). Some of the compounds from cumingianol group exhibited enhanced cytotoxicity against KB-C2 cell line in the presence of colchicine suggesting possible role as MDR-reversing agents (Kurimoto et al. 2011).

#### 13.5.3.4 Tormentic Acid

Cytotoxicity of triterpenoids extracted from *Cecropia lyratiloba* was studied on K562 (leukaemic cell line) and Lucena-1 (K562 derivative overexpressing P-gp) by Rocha et al. (2007). Cytotoxic potential of isolated triterpenoids (tormentic acid, 2α-acetyl tormentic acid, 3β-acetyl tormentic acid) was found. Additionally, acetylation of tormentic acid at C2 increased cytotoxicity, whereas C3 acetylation showed no effect on cytotoxic potential.

#### 13.5.3.5 Glycyrrhizin

It is basically isolated from *Glycyrrhiza glabra*. Biphasic effect of this compound was found by Najar et al. (2010) where lower concentration stimulated and higher concentration inhibited ATPase activity.

#### 13.5.3.6 Uvaol

Earlier studies showed that methanolic extract of *Carpobrotus edulis* is active against MDR in different bacterial species. But study by Ordway et al. (2003) demonstrated that same extract is also suitable for reversing MDR in cancer cells. This work was further extended where constituents of extract were identified and studied for multidrug resistance potential (Martins et al. 2010). Anti-proliferative activity of isolated compounds was studied on mouse lymphoma cells and human *MDR1*-transfected mouse lymphoma cells. Uvaol was found to be the most effective compound against mouse lymphoma cells. Additionally, uvaol inhibited extrusion of rhodamine 123 and ethidium bromide by MDR1-transfected cell line proving its potential as an adjuvant in chemotherapy (Martins et al. 2010).

### 13.5.4 Steroids

#### 13.5.4.1 Cardenolides

Bioactive cardenolides were isolated from *Nerium oleander*, and cytotoxicity along with MDR reversal activity of these compounds was studied (Zhao et al. 2007).

These compounds were isolated from methanol extract of *Nerium oleander*. MDR reversal activity was studied through assaying of cellular accumulation of calcein dye in MDR A2780AD cells. Compound cardenolide (N-9) showed least cytotoxicity with significant accumulation of calcein suggesting role as MDR reversal agent. On the contrary, compound cardenolide (N-4) showed significant to moderate toxicity in studied cell lines.

#### 13.5.4.2 Cycloartanes

Potential of isolated cycloartanes from *Euphorbia* species was studied as multidrug resistance reversal agent and apoptosis inducer (Madureira et al. 2004). These compounds were isolated from acetone extract of *Euphorbia segetalis* and *E. portlandica*. Some of the studied compounds inhibited efflux pump activity, while others showed cytotoxicity. Out of the studied compounds, cycloartane-3β,24,25-triol and cycloartane-23-ene-3β,25-diol caused significant accumulation of rhodamine 123 within cells and exhibited highest MDR reversal activity.

#### 13.5.4.3 Methylprototribestin

Steroidal saponins isolated from fractions of *Tribulus terrestris* were evaluated for MDR modulation (Ivanova et al. 2009). Methylprototribestin showed best result in accumulation of rhodamine 123 in *MDR1*-transfected mouse lymphoma cells. In addition to this, combinatorial study of methylprototribestin and anticancer drug doxorubicin exhibited synergistic and additive effects.

#### 13.5.4.4 Protopanaxatriol (Ginsenoside)

Role of total saponins, protopanaxadiol ginsenosides (PDG), protopanaxatriol ginsenosides (PTG) and other major components of PDG and PTG extracted from *Panax ginseng* as MDR modulators was studied (Choi et al. 2003). Cytotoxicity and MDR reversal activities of these drugs were tested on AML-2/WT, AML-2/D100 (P-gp overexpressing) and AML-2/DX100 (MRP overexpressing) cell lines, respectively. Only PTG showed cytotoxic and chemosensitizing activity among all the studied compounds. PTG showed cytotoxicity in all the targeted cell lines, but chemosensitization was observed in only AML-2/D100, which is P-gp overexpressing cell line. Application of PTG led to accumulation of anticancer drug (daunorubicin) in AML-2/100 cell line by preventing binding of anticancer drug to drug binding sites on P-gp. This was validated by inhibition of photoaffinity labelling agent ([$^3$H] azidopine) for P-gp by PTG. Results also suggested that saponins are specific in their action as PTG caused inhibition of P-gp-mediated drug efflux and not MRP pump-mediated drug efflux.

#### 13.5.4.5 Stigmasterol and β-Sitosterol-O-Glucoside (Sterols)

These compounds reduced P-gp-mediated drug efflux in resistant cell line (CEM/ADR5000) at non-toxic concentration. The isolated compounds extracted from fractions of *Citrus jambhiri* restored cytotoxicity of doxorubicin in previously resistant cell line (El-Readi et al. 2010).

#### 13.5.4.6 Withaferin A

This compound is isolated from *Withania somnifera*. It is believed that overexpression of drug efflux protein (P-glycoprotein) can be due to hyperactivation of NF-kB (nuclear factor kappa-light-chain-enhancer of activated B-cells) (Suttana et al. 2010). In extension of such studies, NF-kB inhibitor (withaferin A) is considered as suitable drug for cancer chemotherapy. In addition to drug efflux, P-gp overexpression also inhibits protease activation. However, Suttana et al. (2010) found that withaferin A removed attenuation of caspase activation and apoptosis in drug resistant cell line (K562/adr). As a result, this compound was considered as a potential phytochemical to overcome drug resistance and induce death in chemoresistant cells.

### 13.5.5 Tetraterpenes

These comprise eight isoprene units with molecular formula of $C_{40}H_{64}$.

#### 13.5.5.1 Carotenoids

Isolated carotenoids from different vegetables and other sources were studied for their MDR reversal activity and apoptosis induction capability (Molnar et al. 2004). On the basis of MDR reversal potential, carotenoids were classified as inactive, moderately active and very active compounds. MDR reversal potential was seen on the basis of accumulation of rhodamine 123 in human *MDR1* gene-transfected L1210 mouse lymphoma cells. No effect on MDR reversal was observed for alpha- and beta-carotene (from *Daucus carota*), whereas capsanthin and capsorubin (from *Capsicum annuum*) and lycophyll (from *Solanum dulcamara*) showed highest potential. In addition to these carotenoids, violaxanthin and antheraxanthin (from yellow flowers of *Viola tricolor*), lycopene (from *Lycopersicon esculentum*), lutein (from *Caltha palustris*), and zeaxanthin (from *Lycium halimifolium*) exhibited moderate effect. Similarly, differential response was observed for studied carotenoids on induction of apoptosis. It was observed that structural differences were responsible for differential MDR reversal potential of studied carotenoids. Moderate effect on MDR reversal was observed for carotenoids having hydroxylation on the right six-membered ring. Contrary to this, carotenoids having hydroxylation on the left ring or on the right five-membered ring reversed multidrug resistance to highest potential in the studied cell line (Molnar et al. 2004). Similar study on the role of carotenoids as MDR reversal agents was done by Gyemant et al. (2006). An important observation of the study was that the configuration of (9Z) compounds showed greater binding potential for P-gp than linear forms (all-E) of zeaxanthin, neoxanthin and violaxanthin. Additionally, structural features (like size, shape and polarity) of carotenoid compounds can affect their interaction with lipid membrane, thus affecting overall reversal mechanism (Gyemant et al. 2006).

## 13.6 Phenolics

The term phenol in general defines phenyl ring carrying one or more hydroxyl groups. Plant phenolics and polyphenols are secondary metabolites synthesized either by shikimate/phenylpropanoid pathway or by polyketide acetate/malonate pathway. Phenolic compounds are among the most widely distributed secondary metabolites having vast applications (Lattanzio 2013).

### 13.6.1 Phenylpropanoids

#### 13.6.1.1 Chlorogenic Acid

This compound is isolated from *Coffea arabica*. Effect of this compound along with other compounds on P-gp ATPase activity was studied by Najar et al. (2010). Chlorogenic acid showed inhibitory effect on P-gp activity in concentration-dependant manner. Increased inhibition was found with an increase in the concentration of studied compound. Release of inorganic phosphate was measured for determining P-gp-dependent ATPase activity. Thus, chlorogenic acid worked as P-gp inhibitor and can be a potent MDR reversal agent.

#### 13.6.1.2 Curcumin

This phytochemical compound is isolated from rhizomes of *Curcuma longa* L. (turmeric) and widely used as herbal medicine. This compound has vast pharmaceutical potential with lesser toxicity (Rodrigues et al. 2016). Curcumin exhibits antioxidant, anti-inflammatory and anticancer effects. This compound downregulates AKT kinase signalling to show antitumor effect. It is also capable of altering expression of P-gp to represent MDR reversal phenotype. Three major curcuminoids from curcumin mixture were studied for MDR modulation activity on KB-V1 (multidrug resistant cervical carcinoma cell line) (Chearwae et al. 2004). Studied curcuminoids (I, II and III) stimulated ATPase activity of P-gp at lower concentration but worked as potent inhibitors with increase in concentration where curcumin I was the most effective. It was also found that these compounds showed inhibition by binding to substrate binding site of P-gp pump. Out of individual compounds studied, curcumin I was the most effective, but its effect was slightly less than curcumin mixture. In an extension to this work, same group targeted active metabolite form of curcuminoids where effect of tetrahydrocurcumin (THC) was studied on ABC transporter proteins (Limtrakul et al. 2007). THC inhibited the efflux action of all the studied ABC transporters and acted as potential MDR reversal agent. Similarly, effect of curcuma drugs and curcumin on activity of ABC transporters was studied in Caco-2 cells (Hou et al. 2008). Both curcuma drugs and curcumin showed opposing effects where curcuma drugs increased expression and activity of P-gp and MDR1 contrary to curcumin which acted as inhibitor of both efflux pumps.

### 13.6.2 Flavonoids and Related Polyphenols

Flavonoids exhibits broad spectrum of pharmacological activities (Ravishankar et al. 2013). These are polyphenolic compounds possessing phenyl benzopyrone structure categorized into flavones, isoflavones, flavanones, flavanonols and chalcones on the basis of saturation level, substitution pattern of C-ring and opening of central pyran ring (Middleton et al. 2000). These are considered as promising anticancer agents because of their ability to block cell cycle and induce apoptosis, mitotic spindle disruption and angiogenesis inhibition (Ravishankar et al. 2013). In addition to modulation of protein kinases, flavonoids show chemopreventive effect by inhibiting cytochrome P450. Flavonoids are also linked with multidrug resistance modulation by inhibition of *MDR1* overexpression, by direct binding to nucleotide binding domain in cytosolic region of efflux pump or through inhibition of ATPase activity (Ren et al. 2003).

#### 13.6.2.1 Chrysin, Amorphigenin and Epigallocatechin

Chrysin, amorphigenin and epigallocatechin are isolated from several plant species, Fabaceae family and *Carpobrotus edulis*, respectively. Different flavonoids were tested for their potential as MDR reversal agents in human *MDR-1* gene-transfected mouse lymphoma cells and *MRP* expressing human breast cancer cell line (MDA-MB-231) (Gyemant et al. 2005). Results showed that chrysin and amorphigenin acted as major modifier of P-gp-mediated efflux pump, and epigallocatechin acted as *MRP*-mediated efflux pump modifier. In addition to this, chrysin and amorphigenin showed anti-proliferative activity individually which enhanced in synergistic association with epirubicin in mouse lymphoma cell line.

#### 13.6.2.2 Genistein and Derivatives

These are isolated from several species of Fabaceae family. Effect of these flavonoids on efflux activity of P-gp was studied by Taur and Rodriguez-Proteau (2008) in Caco-2 and LLC-PK1 cells. Genistein and its derivatives inhibited P-gp-mediated efflux of cimetidine in both studied cell lines. However, differential expression/activity of P-gp in both cell lines was responsible for difference in efflux levels of cimetidine. Additionally, cimetidine showed no effect on rhodamine 123 accumulation. This was related to different binding sites of both compounds on P-gp resulting in no interaction. Their results proved that genistein and its derivatives altered P-gp-mediated disposition profile of cimetidine indicating their role as suitable MDR reversal agents.

#### 13.6.2.3 Hesperidin and Neohesperidin

These compounds were isolated from ethyl acetate fractions of *Citrus jambhiri*. In addition to sterols and limonoids, flavonoids were also studied for MRD reversal potential (El-Readi et al. 2010). P-gp-mediated efflux was significantly reduced by the studied flavonoids at non-toxic concentration targeting drug resistant human leukaemia cells (CEM/ARD5000). These compounds further reversed doxorubicin resistance to restore its toxicity suggesting their role as potent MDR reversal agents.

#### 13.6.2.4 Mangiferin and Norathyriol

Mangiferin is isolated from *Mangifera indica*. Effect of phytoconstituents present in several medicinal herbs on activity of P-gp was studied by Najar et al. (2010). Along with other studied compounds, mangiferin showed biphasic effect where at lower concentration it stimulated ATPase activity and at higher concentration inhibited ATPase activity of P-glycoprotein. Similarly effect of compounds isolated from *Mangifera indica* on efflux activity was studied in P-gp expressing cell lines by Chieli et al. (2009). Mangiferin and its aglycone, norathyriol, significantly inhibited activity of P-gp multidrug transporter which was confirmed through accumulation of rhodamine 123 in studied cell lines.

#### 13.6.2.5 Tricin

This compound is obtained from ethyl acetate extract of *Sasa borealis*. Role of tricin as effective P-gp modulator was studied by Jeong et al. (2007). In addition to enhancement in accumulation of daunomycin in P-gp overexpressing cell line (MCF-7/ADR), this compound also increased the cytotoxic potential of daunomycin. Inhibitory effect shown by tricin on P-gp-mediated drug efflux is due to the structural presence of 3,5-dimethoxy-4-hydroxylphenyl group.

### 13.6.3 Quinones, Anthraquinones and Naphthoquinones

#### 13.6.3.1 Emodin, Aloe Emodin and Rhein

Comparative study between cytotoxic activities of hydroxyanthraquinone and hydroxynaphthoquinone derivatives was done by Cui et al. (2008). Hydroxyanthraquinone derivatives were isolated from *Rheum palmatum*, whereas hydroxynaphthoquinone derivatives were isolated from *Lithospermum erythrorhizon*. Different fractions of the extract yielded a variety of compounds. Hydroxynaphthoquinones were found to be more cytotoxic as compared to hydroxyanthraquinone derivatives. Hydroxyanthraquinone derivatives showed greater toxicity against HCT 116 cells than HepG2 (P-gp overexpressing cell line). On the contrary, hydroxynaphthoquinone derivatives showed similar cytotoxic activities for both cell lines. This indicated that hydroxynaphthoquinone derivatives might not be affected by P-gp-mediated efflux, so they can be used as potential MDR reversal agent.

### 13.6.4 Lignans

#### 13.6.4.1 Syringaresinol

This compound is isolated from *Sasa borealis*. MDR reversal potential of this compound was studied along with another flavone compound by Jeong et al. (2007). This compound increased accumulation and cytotoxicity of daunomycin with reduction in efflux from P-gp overexpressing cell line (MCF-7/ADR). Thus, this compound showed inhibitory effect on P-gp activity.

### 13.6.5 Coumarins and Furanocoumarins

#### 13.6.5.1 Bergamottin

It is isolated from grapefruit juice (GFJ), basically *Citrus* hybrids (Rutaceae). In vivo study of this grapefruit juice component on P-gp activity was done in Sprague-Dawley rats (de Castro et al. 2008). Bergamottin along with 6',7'--dihydroxybergamottin and 6',7'-epoxybergamottin has been reported as a contributor to GFJ-drug interaction (Bailey et al. 2000). Effect of bergamottin (BG) was studied using talinolol as a substrate for P-gp (de Castro et al. 2008). The studied compound showed greatest effect on disposition of talinolol through inhibition of P-gp-mediated efflux.

## 13.7 Alkaloids

These include large family of heterocyclic nitrogenous compounds that are physiologically active and are more common in dicots. These are more commonly synthesized by lysine, tyrosine, ornithine or tryptophan amino acids (Seigler, 1998). Alkaloids are basically classified on the basis of their chemical structure. They can also be divided into true, proto- and pseudoalkaloids True alkaloids contain heterocyclic nitrogen and originate from amino acids. Protoalkaloids also originate from amino acids and contain nitrogen but not in heterocyclic ring. Pseudoalkaloids are alkaloid-like compounds with non-amino acid origin.

### 13.7.1 Acridone Alkaloids

#### 13.7.1.1 Arboririne, Evoxanthine, Isogravacridone Chloride, Rutacridone and Others

These acridone alkaloids isolated from *Ruta graveolens* were tested for MDR reversal activity in human *MDR1*-transfected L5178 mouse lymphoma cells (Rethy et al. 2008). Effect of isolated acridone alkaloids on P-gp activity and accumulation of rhodamine 123 was studied. Isogravacridone chloride and rutacridone compounds efficiently inhibited P-gp activity. Gravacridonediol even at lower concentration caused increased accumulation of rhodamine 123 to a great extent. Isogravacridone chloride showed greatest anti-proliferative/cytotoxic potency, whereas gravacridonetriol and gravacridonediol monomethyl ether significantly reduced P-gp expression.

### 13.7.2 Indole Alkaloids

#### 13.7.2.1 Reserpine, Catharanthine, Vindoline and Others

Reserpine is isolated from *Rauwolfia serpentina*. Beck et al. (1988) studied role of different indole alkaloids as modulators of MDR. Studied alkaloids enhanced vinblastine cytotoxicity in multidrug resistant human leukaemic cell line ($CEM/VLB_{1K}$) where reserpine came out to be the most potent modulator. Additionally, some and not all of the studied indole alkaloids that acted as MDR modulators competed with the labelling compound ([$^{125}$I] NASV) to bind to P-gp. Possible explanation for such observation was different binding site of some of the indole compounds and labelling agent ([$^{125}$I] NASV) to P-gp.

### 13.7.3 Bisbenzylisoquinoline Alkaloids

#### 13.7.3.1 Tetrandrine and Fangchinoline

These alkaloids are isolated from *Stephania tetrandra.* Study related to MDR reversal capability of bisbenzylisoquinoline alkaloids, tetrandrine and fangchinoline, highlighted significant results (Choi et al. 1998). Studied alkaloids enhanced the cytotoxicity of drugs in HCT15 (P-gp positive) cell line, whereas no effect on cytotoxicity of same drugs was observed in SK-OV-3 (P-gp negative) cell line. Similarly, targeted alkaloids increased accumulation of rhodamine 123 in HCT-15 cells but not in SK-OV-3 cells. Thus, the study showed that tetrandrine and fangchinoline increased the cytotoxicity of anticancer drugs through modulation of P-glycoprotein efflux pump.

## 13.8 Conclusion

Selected plant secondary metabolites (PSMs) can result in modulation of multidrug resistance in cancer cells. Terpenoid and alkaloid metabolites act as competitive inhibitors of ABC transporters (P-gp, MRP1 or BCRP) as they are supposed to be substrate of these efflux pumps. On the other hand, phenolic compounds show their activity by altering protein conformation upon binding to the transporter. Thus, combinatorial approach including cytotoxic agent/anticancer drug along with natural chemosensitizer/modulator can prove effective in improving efficacy of chemotherapy by overcoming multidrug resistance in cancer patients.

## References

Annereau JP, Szakacs G, Tucker CJ, Arciello A, Cardarelli C, Collins J et al (2004) Analysis of ATP-binding cassette transporter expression in drug-selected cell lines by a microarray dedicated to multidrug resistance. Mol Pharmacol 66(6):1397–1405

Bailey DG, Dresser GK, Kreeft JH, Munoz C, Freeman DJ, Bend JR (2000) Grapefruit-felodipine interaction: effect of unprocessed fruit and probable active ingredients. Clin Pharmacol Ther 68(5):468–477

Beck WT, Cirtain MC, Glover CJ, Felsted RL, Safa AR (1988) Effects of indole alkaloids on multidrug resistance and labelling of P-glycoprotein by a photoaffinity analog of vinblastine. Biochem Biophys Res Commun 153(3):959–966

Birudu RB, Naik MJ (2014) Anticancer properties of secondary metabolites of medicinal plants in carcinoma. Br Biomed Bull 2(4):662–668

Brahmkshatriya PP, Brahmkshatriya PS (2013) Terpenes: chemistry, biological role, and therapeutic applications. In: Ramawat K, Mérillon JM (eds) Natural products. Springer, Berlin, Heidelberg, pp 2665–2691

Brooks TA, Minderman H, O'Loughlin KL, Pera P, Ojima I, Baer MR et al (2003) Taxane-based reversal agents modulate drug resistance mediated by P-glycoprotein, multidrug resistance protein, and breast cancer resistance protein. Mol Cancer Ther 2(11):1195–1205

Budisan L, Gulei D, Zanoaga OM, Irimie AI, Sergiu C, Braicu C et al (2017) Dietary intervention by phytochemicals and their role in modulating coding and non-coding genes in cancer. Int J Mol Sci 18(6):1178

Chambers CS, Viktorova J, Rehorova K, Biedermann D, Turkova L, Macek T et al (2020) Defying multidrug resistance! Modulation of related transporters by flavonoids and Flavonolignans. J Agric Food Chem 68(7):1763–1779

Chearwae W, Anuchapreeda S, Nandigama K, Ambudkar SV, Limtrakul P (2004) Biochemical mechanism of modulation of human P-glycoprotein (ABCB1) by curcumin I, II and III purified from turmeric powder. Biochem Pharmacol 68:2043–2052

Cheynier V, Comte G, Davies KM, Lattanzio V, Martens S (2013) Plant phenolics: recent advances on their biosynthesis, genetics and ecophysiology. Plant Physiol Biochem 72:1–20

Chieli E, Romiti N, Rodeiro I, Garrido G (2009) In vitro effects of Mangifera indica and polyphenols derived on ABCB1/P-glycoprotein activity. Food Chem Toxicol 47:2703–2710

Choi SU, Park SH, Kim KH, Choi EJ, Kim S, Park WK et al (1998) The bisbenzylisoquinoline alkaloids, tetrandine and fangchinoline enhance the cytotoxicity of multidrug resistance related-drugs via modulation of P-glycoprotein. Anticancer Res 9:255–261

Choi CH, Kang G, Min YD (2003) Reversal of P-glycoprotein mediated multidrug resistance by protopanaxatriol ginsenosides from Korean red ginseng. Planta Med 69(3):235–240

Cui XR, Tsukada M, Suzuki N, Shimamura T, Gao L, Koyanagi J et al (2008) Comparison of the cytotoxic activities of naturally occurring hydroxyanthraquinones and hydroxynaphthoquinones. Eur J Med Chem 43(6):1206–1215

Dantzig AH, Law KL, Cao J, Starling JJ (2001) Reversal of multidrug resistance by the P-glycoprotein modulator, LY335979, from the bench to the clinic. Curr Med Chem 8(1):39–50

de Castro WC, Talcott SM, Derendorf H, Butterweck V (2008) Effect of grapefruit juice, naringin, naringenin and bergamottin on the intestinal carrier-mediated transport of talinolol in rats. J Agric Food Chem 56:4840–4845

Duarte N, Gyemant N, Abreu PM, Molnar J, Ferreira MJU (2006) New macrocyclic Lathyrane diterpenes from *Euphorbia lagascae* as inhibitors of multidrug resistance of tumour cells. Planta Med 72(2):162–168

Eckford PDW, Sharom FJ (2009) ABC efflux pump-based resistance to chemotherapy drugs. Chem Rev 109(7):2989–3011

El-Readi MZ, Hamdan D, Farrag N, El-Shazly A, Wink M (2010) Inhibition of P-glycoprotein activity by limonin and other secondary metabolites from *citrus* species in human colon and leukemia cell lines. Eur J Pharmacol 626:139–145

Ferguson LR (2001) Role of plant polyphenols in genomic stability. Mutat Res 475(1-2):89–111

Fernandes J, Castilho RO, da Costa MR, Wagner-Souza K, Kaplan MAC, Gattass CR (2003) Pentacyclic triterpenes from Chrysobalanaceae species: cytotoxicity on multidrug resistant and sensitive leukemia cell lines. Cancer Lett 190(2):165–169

Gill BS, Kumar S, Navgeet G (2016) Triterpenes in cancer: significance and their influence. Mol Biol Rep 43:881–896

Gottesman MM (2002) Mechanisms of cancer drug resistance. Annu Rev Med 53:615–627

Gyemant N, Tanaka M, Antus S, Hohmann J, Csuka O, Mandoky L et al (2005) In vitro search for synergy between flavonoids and epirubicin on multidrug resistant cancer cells. In Vivo 19:367–374

Gyemant N, Tanaka M, Molnar P, Deli J, Mandoky L, Molnar J (2006) Reversal of multidrug resistance of cancer cells in vitro: modification of drug resistance by selected carotenoids. Anticancer Res 26:367–374

He L, Liu GQ (2002) Effects of various principles from Chinese herbal medicine on rhodamine 123 accumulation in brain capillary endothelial cells. Acta Pharmacol Sin 23(7):591–596

Hohmann J, Molnar J, Redei D, Evanics F, Forgo P, Kalman A et al (2002) Discovery and biological evaluation of a new family of potent modulators of multidrug resistance: reversal of multidrug resistance of mouse lymphoma cells by new natural Jatrophane diterpenoids isolated from euphorbia species. J Med Chem 45(12):2425–2431

Hou XL, Takahashi K, Tanaka K, Tougou K, Qiu F, Komatsu K et al (2008) Curcuma drugs and curcumin regulate the expression and function of P-gp in Caco-2 cells in completely opposite ways. Int J Pharm 358:224–229

Hu R, Gao J, Rozimamat R, Aisa HA (2018) Jatrophane diterpenoids from *Euphorbia sororia* as potent modulators against p-glycoprotein based multidrug resistance. Eur J Med Chem 146: 157–170

Ishikawa T, Hirano H, Onishi Y, Sakurai A, Tarui S (2004) Functional evaluation of ABCB1 (P-glycoprotein) polymorphisms: high speed screening and structure-activity relationship analyses. Drug Metab Pharmacokinet 19(1):1–14

Ivanova A, Serly J, Dinchev D, Oscovszki I, Kostova I, Molnar J (2009) Screening of some saponins and phenolic components of *Tribulus terrestris* and *Smilax excelsa* as MDR modulators. In Vivo 23:545–550

Jeong YH, Chung SY, Han AR, Sung MK, Jang DS, Lee J et al (2007) P-glycoprotein inhibitory activity of two phenolic compounds, (-)-Syringaresinol and Tricin from Sasa Borealis. Chem Biodivers 4(1):12–16

Johnson ZL, Chen J (2017) Structural basis of substrate recognition by the multidrug resistance protein MRP1. Cell 168:1075–1085

Joyce H, McCann A, Clynes M, Larkin A (2015) Influence of multidrug resistance and drug transport proteins on chemotherapy drug metabolism. Expert Opin Drug Metab Toxicol 11(5):795–809

Krishnamurthy P, Schuetz JD (2006) Role of ABCG2/BCRP in biology and medicine. Annu Rev Pharmacol Toxicol 46:381–410

Kumar A, Jaitak V (2019) Natural products as multidrug resistance modulators in cancer. Eur J Med Chem 176:268–291

Kurimoto S, Kashiwada Y, Lee KH, Takaishi Y (2011) Triterpenes and a triterpene glucoside from *Dysoxylum cumingianum*. Phytochemistry 72:2205–2211

Lattanzio V (2013) Phenolic compounds: introduction. In: Ramawat K, Mérillon JM (eds) Natural products. Springer, Berlin, Heidelberg, pp 1544–1573

Li S, Lei Y, Jia Y, Li N, Wink M, Ma Y (2011) Piperine, a piperidine alkaloid from piper nigrum re-sensitizes P-gp, MRP1 and BCRP dependant multidrug resistant cancer cells. Phytomedicine 19(1):83–87

Li J, Wang WQ, Tang S, Song WB, Xuan LJ (2018) Jatrophanes as promising multidrug resistance modulators: advances of structure-activity relationships. Fitoterapia 127:138–145

Limtrakul P, Chearwae W, Shukla S, Phisalphong C, Ambudkar SV (2007) Modulation of function of three ABC drug transporters, P-glycoprotein (ABCB1), mitoxantrone resistance protein (ABCG2) and multidrug resistance protein (ABCC1) by tetrahydrocurcumin, a major metabolite of curcumin. Mol Cell Biochem 296:85–95

Loo TW, Clarke DM (2005) Recent progress in understanding the mechanism of P-glycoprotein mediated drug efflux. J Membr Biol 206(3):173–185

Ludwiczuk A, Skalicka-Wozniak K, Georgiev MI (2017) Terpenoids. In: Delgoda R (ed) Badal S. Academic Press, Pharmacognosy, pp 233–266

Madureira AM, Spengler G, Molnar A, Varga A, Molnar J, Abreu PM et al (2004) Effect of cycloartanes on reversal of multidrug resistance and apoptosis induction on mouse lymphoma cells. Anticancer Res 24(2B):859–864

Mahringer A, Fricker G (2010) BCRP at the blood brain barrier: genomic regulation by 17-β-estradiol. Mol Pharm 7(5):1835–1847

Maliepaard M, Van Gastelen MA, Tohgo A, Hausheer FH, Van Waardenburg RC, de Jong LA (2001) Circumvention of breast cancer resistance protein (BCRP)-mediated resistance to Camptothecins in vitro using non-substrate drugs or the BCRP inhibitor GF120918. Clin Cancer Res 7(4):935–941

Malik A, Sharma B, Jain P (2017) Introduction to P-glycoprotein/ABCB1/MDR1 and their modulator extracted from plant. Int J Pharm Sci Rev Res 44(1):112–119

Martins A, Vasas A, Schelz Z, Viveiros M, Molnar J (2010) Hohmann, et al. constituents of Carpobrotus edulis inhibit P-glycoprotein of MDR1-transfected mouse lymphoma cells. Anticancer Res 30(3):829–835

Middleton E Jr, Kandaswami C, Theoharides TC (2000) The effects of plant flavonoids on mammalian cells: implications for inflammation, heart disease and cancer. Pharmacol Rev 52(4):673–751

Min YD, Kwon HC, Yang MC, Lee KH, Choi SU, Lee KR (2007) Isolation of limonoids and alkaloids from *Phellodendron amurense* and their multidrug resistance (MDR) reversal activity. Arch Pharm Res 30(1):58–63

Mohana S, Ganesan M, Agilan B, Karthikeyan R, Srithar R, Mary RB et al (2016) Screening dietary flavonoids for the reversal of P-glycoprotein-mediated multidrug resistance in cancer. Mol BioSyst 12(8):2458–2470

Molnar J, Gyemant N, Mucsi I, Molnar A, Szabo M, Kortvelyesi T et al (2004) Modulation of multidrug resistance and apoptosis of cancer cells by selected carotenoids. In Vivo 18:237–244

Mousa JJ, Bruner SD (2016) Structural and mechanistic diversity of multidrug transporters. Nat Prod Rep 33(11):1255–1267

Najar IA, Sachin BS, Sharma SC, Satti NK, Suri KA, Johri RK (2010) Modulation of P-glycoprotein ATPase activity by some phytoconstituents. Physiother Res 24:454–458

Nooter K, Stoter G (1996) Molecular mechanisms of multidrug resistance in cancer chemotherapy. Pathol Res Pract 192(7):768–780

Ordway D, Hohmann J, Viveiros M, Viveiros A, Molnar J, Leandro C et al (2003) Carpobrotus edulis methanol extract inhibits the MDR efflux pumps, enhances killing of phagocytosed S. aureus and promotes immune modulation. Phytother Res 17(5):512–519

Poulose SM, Harris ED, Patil BS (2006) Antiproliferative effects of citrus limonoids against human neuroblastoma and colonic adenocarcinoma cells. Nutr Cancer 56:103–112

Rani AS, Neelima G, Mukhopadhyay R, Jyothi KSN, Sulakshana G (2018) In silico characterization of plant secondary metabolites. In: Choudhary DK, Kumar M, Prasad R, Kumar V (eds) In silico approach for sustainable agriculture. Springer, Singapore, pp 251–273

Ravishankar D, Rajora AK, Greco F, Osborn HMI (2013) Flavonoids as prospective compounds for anti-cancer therapy. Int J Biochem Cell Biol 45:2821–2831

Ren W, Qiao Z, Wang H, Zhu L, Zhang L (2003) Flavonoids: promising anticancer agents. Med Res Rev 23(4):519–534

Rethy B, Hohmann J, Minorics R, Varga A, Ocsovszki I, Molnar J et al (2008) Antitumor properties of acridone alkaloids on a murine lymphoma cell line. Anticancer Res 28:2737–2744

Robert J, Jarry C (2003) Multidrug resistance reversal agents. J Med Chem 46(23):4805–4817

Rocha G, Simoes M, Lucio KA, Oliveira RR, Kaplan MAC, Gattass CR (2007) Natural triterpenoids from Cecropia lyratiloba are cytotoxic to both sensitive and multidrug resistant leukemia cell lines. Bioorg Med Chem 15(23):7355–7360

Rodrigues VL, Sousa E, Vasconcelos MH (2016) Curcumin as a modulator of P-glycoprotein in cancer: challenges and perspectives. Pharmaceut 9:71

Roy A, Saraf S (2006) Limonoids: overview of significant bioactive triterpenes distributed in plant kingdom. Biol Pharm Bull 29(2):191–201

Schinkel AH, Jonker JW (2003) Mammalian drug efflux transporters of the ATP binding cassette (ABC) family: an overview. Adv Drug Deliv Rev 55(1):3–29

Seigler DS (1998) Introduction to alkaloids. In: Seigler DS (ed) Plant secondary metabolism. Springer, Boston, MA, pp 506–512

Solary E, Mannone L, Moreau D, Caillot D, Casasnovas RO, Guy H et al (2000) Phase I study of Cinchonine, a multidrug resistance reversing agent, combined with the CHVP regimen in relapsed and refractory lymphoproliferative syndromes. Leukemia 14(12):2085–2094

Steffens JJ, Pell EJ, Tien M (1996) Mechanisms of fungicide resistance in phytopathogenic fungi. Curr Opin Biotechnol 7(3):348–355

Steinbach D, Sell W, Voigt A, Hermann J, Zintl F, Sauerbrey A (2002) BRCP gene expression is associated with a poor response to remission induction therapy in childhood acute myeloid leukemia. Leukemia 16(8):1443–1447

Suttana W, Mankhetkorn S, Poompimon W, Palagani A, Zhokhov S, Gerlo S et al (2010) Differential chemosensitization of P-glycoprotein overexpressing K562/adr cells by withaferin a and siamois polyphenols. Mol Cancer 9:99

Taur JS, Rodriguez-Proteau R (2008) Effects of dietary flavonoids on the transport of cimetidine via P-glycoprotein and cationic transporters in Caco-2 and LLC-PK1 cell models. Xenobiotica 38: 1536–1550

Toyomasu T, Sassa T (2010) Diterpenes. In: Sassa T, Toyomasu T (eds) Comprehensive Natural Products II. Elsevier, pp 643–672

Van der Kolk DM, Vellenga E, Muller M, de Vries EG (1999) Multidrug resistance protein MRP1, glutathione and related enzymes. Their importance in acute myeloid leukemia. Adv Exp Med Biol 457:187–198

Wijnholds J, deLange EC, Scheffer GL, Van den berg DJ, Mol CA, Van der Valk M et al (2000) Multidrug resistance protein 1 protects the choroid plexus epithelium and contributes to the blood-cerebrospinal fluid barrier. J Clin Invest 105(3):279–285

Wink M, Ashour ML, El-Readi MZ (2012) Secondary metabolites from plants inhibiting ABC transporters and reversing resistance of cancer cells and microbes to cytotoxic and antimicrobial agents. Front Microbiol 3:130

Yoshida N, Takagi A, Kitazawa H, Kawakami J, Adachi I (2005) Inhibition of P-glycoprotein mediated transport by extracts of and monoterpenoids contained in *Zanthoxyli* Fructus. Toxicol Appl Pharmacol 209:167–173

Yumoto R, Murakami T, Sanemasa M, Nasu R, Nagai J, Takano M (2001) Pharmacokinetic interaction of cytochrome p450 3A related compounds with rhodamine 123, a P-glycoprotein substrate in rats pre-treated with dexamethasone. Drug Metab Dispos 29(2):145–151

Zhao M, Bai L, Wang L, Toki A, Hasegawa T, Kikuchi M et al (2007) Bioactive cardenolides from the stems and twigs Nerium oleander. J Nat Prod 70(7):1098–1103

Zhu J, Wang R, Lou L, Li W, Tang G, Bu X et al (2016) Jatrophane diterpenoids as modulators of P-glycoprotein-dependent multidrug resistance (MDR): advances of structure-activity relationships and discovery of promising MDR reversal agents. J Med Chem 59(13):6353–6369

# Plant Secondary Metabolites: Functions in Plants and Pharmacological Importance

# 14

Priya Katyal

**Abstract**

Plant secondary metabolites are the organic compounds that are not necessary for growth or reproduction of plants, but which can be demonstrated genetically, physiologically, or biochemically. Of the estimated 400,000–500,000 plant species, only a small percentage has been investigated for their ethanopharmacological value, and even a lower fraction has been exploited to get the active pharmaceutical ingredients. Studies have revealed that the ability of plants to synthesize secondary metabolites is directly related to their course of evolution, and these metabolites fulfill physiological needs of plants such as attracting pollinators, protecting from pests and pathogens, enabling the plant to tolerate environmental stress, etc. The evolution of secondary metabolism is primarily from the primary metabolism by alterations in certain enzymes and pathways. This chapter provides an overview of the diversity of plant secondary metabolites in terms of their structure, function, and biosynthesis. Further research is needed to understand the chemical interactions between different plant species, and between plants and other organisms so that the potential secondary metabolites can be traced for novel uses by mankind.

**Keywords**

Terpenes · Alkaloids · Cyanogenic glycosides · Phenolics · Biosynthesis

P. Katyal (✉)
Department of Microbiology, Punjab Agricultural University, Ludhiana, India



A. K. Sharma, A. Sharma (eds.), *Plant Secondary Metabolites*,
https://doi.org/10.1007/978-981-16-4779-6_14

## 14.1 Why Plants Produce Secondary Metabolites?

Organic compounds, produced by bacteria, fungi, and plants that are not essential for their growth, photosynthesis, development, or reproduction or other vital functions, are called as secondary metabolites. Their absence is not lethal to the organism but can decrease their ability to survive, compete, or attract. These secondary metabolites are extremely diverse in their structure and function and have been identified in almost all families of plants. A complex mixture of different secondary metabolites is produced by different families of plants and hence is an important taxonomic feature for their classification. In contrast to primary metabolites that are indispensable for growth, secondary metabolites play an important role in signaling and regulating most of the primary metabolic pathways. Plant hormones that are regulating the development and metabolism of plants are also important secondary metabolites.

Plant secondary metabolites are being exploited as medicines, as recreational drugs, and as spices or flavors for culinary applications. In the plant itself, these chemicals are playing important function as signals and antioxidants, so the term "secondary" seems to be a misnomer. Many plants have been known to produce chemical constituents like alkaloids (nicotine, cocaine) and terpenes (curcumin) that are having pharmaceutical importance. So the study of such plants for pharmaceutical applications is called "ethnopharmacology." Some of the plant-based chemicals are "psychoactive," while others give characteristic aroma and flavor to the food that are responsible for varying food preferences. For example, flavors and aroma of family Cruciferae are primarily due to nitrogen- and sulfur-containing chemicals, and their ability to resist many predators is due to the glucosinolates present in them. Similarly, tannins are responsible for the astringency of wine and chocolate (Bidlack et al. 2000). The use of spices and other seasonings is developed from their combined uses as preservatives (since they are antibiotic).

Plants produce a variety of secondary metabolites that are produced for three main reasons: specific physiological functions, intraspecific interactions, and interspecific interactions. Among the interspecific interaction, there are plant-plant interactions (allelopathy), plant-microbe interactions, and plant-animals interactions. Some of the secondary metabolites play important ecological function by attracting pollinators (insects or animals) or by acting as natural pesticides, and these can account for 10% of the dry matter (Hans-Walter 2005).

Some of the plants produce toxic proteins like amylase and proteinase inhibitor and lectins that can impair the digestion of consumers. For example, in response to infestation of caterpillars, maize plants release a protease to disrupt the intestine of caterpillars. Similarly, hydrolysis of starch is inhibited in digestive tract of insects when they feed on beans that contain amylase inhibitors. The seeds of some legumes and other plants also contain proteinase inhibitors, which block the digestion of proteins by inhibiting proteinases in the animal digestive tract. Lectins present in some legumes bind to free sugars, glycolipids, or glycoproteins in the intestine, thereby inhibiting the food absorption. Ricin is an extremely toxic protein found in castor beans that can kill a human consuming even a few milligrams.

Defense substances produced by plants against herbivores are usually constitutive, but in some instances, browsing damage can induce their synthesis. For example, *Acacias* produce more tannins after feeding damage, thus making the leaves inedible. Similarly, plants that are damaged by caterpillars produce certain aromatic secondary metabolites to attract parasitic wasps. These wasps lay eggs in the caterpillars leading to their death.

Defense substances are synthesized mostly in response to an infection by microorganisms, especially fungi. Within hours of infection, plants produce these defense substances called as phytoalexins (alekein means "to defend"). These defense substances include isoprenoids, flavonoids, stilbenes, etc. Plants also produce enzymes, such as β-glucanases, chitinases, and proteinases and aggressive oxygen compounds (superoxide radical, hydrogen peroxide, nitrogen monoxide) that damage the cell walls of bacterial and fungal pathogens. Synthesis of the abovementioned defense substances is induced by "elicitors."

Elicitors are either cell wall degrading enzymes produced by the pathogens to attack host plant cell wall, polysaccharide segments of the host plant's cell wall, or cell wall fragments of the pathogen itself. The binding of the elicitor on specific receptors on plasma membrane of plant cells initiates a signaling pathway, in which protein kinases and signaling agents like salicylic acid and jasmonic acid induce the synthesis of phytoalexins, reactive oxygen compounds, and defense enzymes.

Some of the secondary metabolites are not only toxic to animals but also toxic to humans. Therefore, several breeding experiments were initiated to eliminate or reduce the toxic secondary metabolites in cultivated varieties. These cultivated varieties of plants are more sensitive to pests than their wild relatives. Attempts were also being made to produce more resistant plants by crossing them with wild plants. However, it can sometimes lead to reduce acceptability, e.g., insect-resistant potato was unacceptable, because of the high toxic solanine content. Similarly, insect-resistant celery caused severe skin damage due to high content of psoralens. A number of plant secondary metabolites that are harmful to humans (lectins, enzyme inhibitors, and glucosinolates) are destroyed by cooking, while others are carcinogenic. It has been estimated that in industrialized countries more than 99% of all carcinogenic substances that humans normally consume with their diet are plant secondary metabolites that are natural constituents of the food.

This chapter mainly focuses on the variety of plant secondary metabolites in terms of their structure, function, biosynthesis, and uses. Secondary metabolites produced by the plants protect the plants from attack of bacterial, fungal, and viral pathogens (Dicke and Baldwin 2010; Hopkins et al. 2009; Kessler and Halitschke 2007; Wittstock and Gershenzon 2002). Some of the secondary metabolites (phenylpropanoid) protect the plants from UV damage. Similarly, some have medicinal value, e.g., artemisinin, a widely used component of Chinese medicines, is produced from the herb *Artemisia annua*. In addition to medicines, secondary metabolites are also used in flavorings and pigments.

## 14.2 Selection of Plant Source and Isolation of Secondary Metabolites

Strategies for selection of particular plant or plant part can be either based on ethnobotany, chemical ecology, or anatomical clues. The ethnobotanical studies were based on crude extract containing many inert compounds along with the active compound(s). But the advent of modern organic chemistry led to the concept of an active principal ingredient that can have applications in medicines and pest control. Chemical ecology is based on understanding the chemical interactions between different plant species, and between plants and other organisms, and has led to the discovery of bioactive compounds with diverse applications. For example, sorgoleone—a very potent natural herbicide—was discovered from *Sorghum* species by exploiting chemical ecology approach. Different types of secondary metabolites have been found in specialized cell layers of plants or in their epidermal cell secretions. Detailed studies on the anatomical specialization of plants at the subcellular, cellular, tissue, or organ level provide clues related to the synthesis, storage, and activity of these SMs (Bidlack et al. 2000).

Once the selection of a particular plant is made, attempts are made to isolate the active principal ingredient. This process involves the selection of suitable bioassays, use of diverse analytical approaches, and study of bioinformatics. After discovery of active compounds, their synthetic analogues can be generated by using bioinformatics or use of computational chemistry-based quantitative structure-activity relationship (QSAR) analysis and their activity can be standardized.

## 14.3 Secondary Metabolites and Their Functions

*Albrecht Kossel*, a German biochemist, coined the term secondary metabolites, and these were described as end product of nitrogen metabolism by a botanist, *Friedrich Johann Franz Czapek*. Usually, secondary metabolites (SMs) are specific to an individual species. These metabolites can affect a variety of species in varying ways, and certain species exploit these secondary metabolites for their own protection. For example, monarch butterflies are able to feed on milkweed (*Asclepias*) in spite of the toxic metabolites present in it. These secondary metabolites present in the butterfly and caterpillars protect them from predators.

Other functions of SMs include attracting animals for pollination and seed dispersal. SMs with bright colors and fragrance attract pollinators and reward them with nectar. These are bioactive substances, interfering with multiple targets in cells, tissues, or organs of animal or microbe. Several SMs due to their structural similarity with neurotransmitters can act as agonists or antagonists of signal transduction and are either neuroactive, psychoactive, or habit forming. Food preferences of insects for a single or a few phylogenetically related plants are often related to difference in the composition of plant's SMs that serve as olfactory cues for insects (Brown and Trigo 1995).

Some insects can feed on plants that are toxic as they can tolerate toxins and store it in their body to protect themselves from predators. Moths of family Arctiidae and a few other arthropods actively sequester pyrrolizidine alkaloids (Pas) and use it not only as defense but also as a precursor of male pheromones (Wink 2018). Plants cannot run away from predators and lack adaptive immune system as present in vertebrates so they had to cope with herbivores and pathogens present around them (Wink 2003). Therefore, plants produce a wide variety of chemical constituents as secondary metabolites that enable plants to ward off herbivores and pathogens.

Not only secondary metabolites of plants but also other strategies such as presence of an impermeable bark, cuticle, thorns, and spikes and their capacity to renew parts that are damaged by herbivores are playing protective role. Structural diversity of SMs and their mode of action are presented in Table 14.1.

Usually, plants produce a complex mixture of different SMs which vary w.r.t. concentration and composition in different organs at different stages of growth and differ within a population. This strategy is adopted to cope up with the adaptation development by insects and other herbivores. It is also useful to prevent resistance development by pathogens such as bacteria and fungi (Mason and Singer 2015). Some of the SMs mimics the structure of important neurotransmitters (serotonin, dopamine, adrenaline, noradrenaline, histamine, acetylcholine, GABA, etc.) and act as CNS stimulant, psychoactive, and hallucinogens.

## 14.4 General Mode of Action of SMs

1. Among alkaloids, several of them are either agonists or antagonists of membrane spanning ion channels, neurotransmitter receptors, and neurotransmitter inactivating enzymes and transporters.
2. Some of the secondary metabolites also form covalent bonds with $-NH_2$, $-SH$, or $-OH$ groups of amino acid residues of protein. Some of lipophilic terpenes can assemble in the inner hydrophobic core of globular proteins and change their 3D structure (Wink and Schimmer 2010).
3. Lectins, emetine, and lycorine interfere with protein synthesis in ribosomes.
4. Lipophilic SMs (mono-, sesqui-, di-, and triterpenes, steroids, phenylpropanoids) can be trapped inside the biomembranes of living cells and change its fluidity and permeability.
5. Saponins that contain lipophilic or triterpene moiety steroid and hydrophilic sugar chain can lyse biomembranes by complexing membrane cholesterol.
6. SMs can interfere with metabolic pathways involved in the synthesis and degradation of nucleic acids. Different SMs can intercalate between stacks of DNA base pairs preventing activity of helicase and RNA polymerase bringing about frame shift mutations, hence are genotoxic and cytotoxic.
7. Some of the plants' SMs are estrogenic in nature called as "phytoestrogens." Phytoestrogens can mimic the hormone estrogen at low doses, while at high doses, they function as natural estrogen blockers (Barbour and Devine 2010). These estrogenic SMs have the ability to induce tumor growth of the estrogen

**Table 14.1** Important classes of plant secondary metabolites and their mode of action

| Class | Major compounds (plant) | Mode of action |
|---|---|---|
| *Containing nitrogen* | | |
| Alkaloids | Morphine (poppy), nicotine (tobacco), atropine (*Atropa*), colchicines (*Colchicum*), quinine (*Cinchona*), ephedrine (*Ephedra*), cocaine (coca plant leaves), caffeine (*Coffea arabica*), theobromine (cocoa beans) | Agonists and antagonists of neurotransmitters, enzyme inhibitors |
| *Containing nitrogen and sulfur* | | |
| Glucosinolates | Glucotropaeolin, gluconasturtiin, glucoraphanin, goitrin, sinigrin (all obtained from cruciferous vegetables) | Antifeedants, natural pesticides, goitrogens, and antithyroid agents |
| *Terpenes and terpenoids* | | |
| Monoterpenes | Menthol (mint), linalool (many plants), camphor (*Cinnamomum camphora*), and eucalyptol (*Eucalyptus globulus*) | Interfere with neurotransmission, block ion transport, anesthetic, insecticide, repellent, flavoring substance |
| Sesquiterpenes | Zingiberene (ginger), humulones (*Humulus lupulus*), parthenolide (*Parthenium* and relatives of family Asteraceae ) | Flavoring, $GABA_A$ receptor activity and antibacterial, contact dermatitis |
| Diterpenes | Taxanes like paclitaxel (*Taxus* or yews) | Chemotherapeutic agent |
| | Salvinorin (*Salvia divinorum*) | Psychotropic drug |
| | Gossypol (cotton) | Phosphorylation inhibitor |
| Triterpenes, cardiac glycosides | Squalene (vegetable oils) | Key component for synthesis of cell membrane |
| | Ginsenoside (*Ginseng*) | Neuroprotective, agonists of steroid hormone receptors |
| | Digitogenin (*Digitalis* or foxglove) | Stimulate heart muscle and alter ion transport |
| Tetraterpenoids | Curcumin (*Curcuma longa)* | Antioxidant and anticarcinogenic |
| | Carotene (orange/yellow vegetables) | Antioxidant and coloring compound |
| Terpene polymers | Rubber (*Hevea* (rubber trees) and *dandelion*) | Gum up insects; airplane tires |
| npo | Gutta-percha (*Palaquium gutta*) | Endodontic fillings |
| Sterols | Spinasterol (spinach) | Antiproliferative action |
| *Phenolics* | | |
| Phenolic acids | Cinnamic acid, gallic acid, ferulic acid caffeic acid, chlorogenic acid (produced by many plants) | Responsible for plant–microbe interactions, oxidative damage, browning in cut fruits, affect color, flavor, and aroma of wine |
| Coumarins | Esculetin (*Cortex fraxini*) | Antioxidant and ultraviolet filter |
| | Dicoumarol (Sweet clover hay) | Anticoagulant and vitamin K depletion activities |

(continued)

**Table 14.1** (continued)

| Class | Major compounds (plant) | Mode of action |
|---|---|---|
| | Umbelliferone (members of family Umbelliferae) | Anti-inflammatory, anti-hyperglycemic, molluscicidal, and antitumor activities, control cell cycle, apoptosis, and DNA fragmentation |
| Lignans | Secoisolariciresinol (flax and sesame seeds) | Antioxidant |
| | Hinokinin (many plants) | Anti-inflammatory, antioxidant |
| | Podophyllin urushiol (mayapple or ground lemon) | Emetic, cathartic, anti-helminthic |
| Flavonoids | Luteolin (*Reseda luteola*) | Antioxidant |
| | Apigenin (many plants) | Used to dye wool |
| | Anthocyanin and catechins (all plants) | Enzyme inhibitor, antioxidant and prooxidants, estrogenic |
| Tannins | Gallotannin, condensed tannin (oak, hemlock trees, birdsfoot trefoil, legumes) | Block digestion by inhibiting enzymes, antioxidants |

receptor cells and also act as cardio-protectant (Peter 1998). It has been reported that some herbs as forage grasses such as clover or alfalfa can express estrogenic properties and interact with fertility of animals (Hussein and El-Anssary 2018).

8. Alkylating agents (aristolochic acid, cycasin, safrole, and phenylpropanoid) intercalate between nucleotide bases leading to mutation and genotoxicity.
9. Secondary metabolites like alkaloids follow different molecular routes to block, suppress, and prevent the metastasis of cancerous cells. Alkaloids are plant-derived secondary metabolites that contain heterocyclic nitrogen and are well exploited as immunomodulators to prevent mutagenesis and inflammation along with antibacterial and anticancerous agents (Katyal and Sharma 2019).

Herbivorous insects have developed varied strategies to inactivate or detoxify different plant SMs. These insects have well-developed monooxygenases that can either inactivate SMs via cytochrome P-mediated pathways or insects can eliminate the toxic SMs via ATP-binding cassette (ABC) transporters. Some of the insects harbor symbiotic intestinal microorganisms that produce enzymes to degrade or inactivate SMs (Pennisi, 2017). For example, plant *Senecio jacobaea* is eaten by *Tyria jacobaeae* (moth). So the population of this plant will be seriously affected by moth population though the moth cannot completely eliminate its host (Wink and Legal 2001).

## 14.5 Classification, Structure, and Distinctive Role of Secondary Metabolites

Depending upon their structure and biosynthetic pathways, plant secondary metabolites can be classified into four main groups: terpenes (made from mevalonic acid, composed almost entirely of carbon and hydrogen), phenolics (made from simple sugars, containing benzene rings, hydrogen, and oxygen), glycosides (N-containing SMs, may contain sulfur, e.g., glucosinolates), and alkaloids (N-containing SMs).

**Terpenes** Terpenes constitute a large group of secondary metabolites that are made up of isoprene units. These are mainly containing carbon and hydrogen, but those containing oxygenated hydrocarbons are called as terpenoids. Their chemical formula is $(C_5H_8)_n$. Here $n$ represents the number of linked isoprenoid units. So depending upon the value of $n$, these can be classified as hemiterpene ($n = 1$, $C_5$), monoterpene ($n = 2$, $C_{10}$), sesquiterpene ($n = 3$, $C_{15}$), diterpene ($n = 4$, $C_{20}$), sesterterpene ($n = 5$, $C_{25}$), triterpene ($n = 6$, $C_{30}$), sesquarterterpene ($n = 7$, $C_{35}$), tetraterpene ($n = 8$, $C_{40}$), and polyterpene ($n > 8$, $C_{>40}$).

Primary function of isoprenoid unit is as important constituents of membrane, photosynthetic pigments, and plant hormones controlling growth and reproduction and as carriers of electrons and glucosyl carriers involved in glucosylation. As secondary metabolites, the isoprenoids play ecological role as component of latex, resins, oils, and waxes making the plant parts either toxic or indigestible to protect them from herbivores. They also act as intrinsic antibiotics protecting the plants from attack of pathogenic bacteria and fungi and inhibit the germination and development of competing plant. Isoprenoids are important constituents of pigments as well as scents of flowers or fruits, thereby playing a role in pollination and seed dispersal. Commercially important products of isoprenoids are vitamins (A, D, E), aroma substances (used in food, beverages, and cosmetics), rubber, natural insecticides (e.g., pyrethrin), and solvents (e.g., turpentine) and as pharmaceutical components.

Turpentine oil is a source of many cyclic compounds containing 10, 15, 20, or more C atoms. Such cyclic compounds were collectively named as terpenes. Figure 14.1 shows some representatives of terpenes. Limonene is a monoterpene containing 10 C-atoms and is an important aroma component of lemon oil used as flavoring agent. Carotene is a tetraterpene containing 40 C atoms, accumulated in most plant parts as red, yellow, and orange pigment for attracting pollinators. Rubber is obtained from the latex of the rubber tree *Hevea brasiliensis* and is a polyterpene (about 1500 C atoms) present in the liquid state when freshly synthesized and is translocated to different plant parts as oil-water emulsion.

Isoprenoids was recognized as the basic constituent of terpenes by Otto Wallach (Bonn, Göttingen). Isoprenoid units are synthesized by higher plants and some groups of algae through different pathways by enzymes present not only in the cytosol but also in the plastids. As shown in Fig. 14.2, acetyl-CoA is the precursor for the synthesis of isoprenoids in the cytosol leading to the synthesis of sesquiterpenes and triterpenes. While in plastids, pyruvate and glyceraldehyde

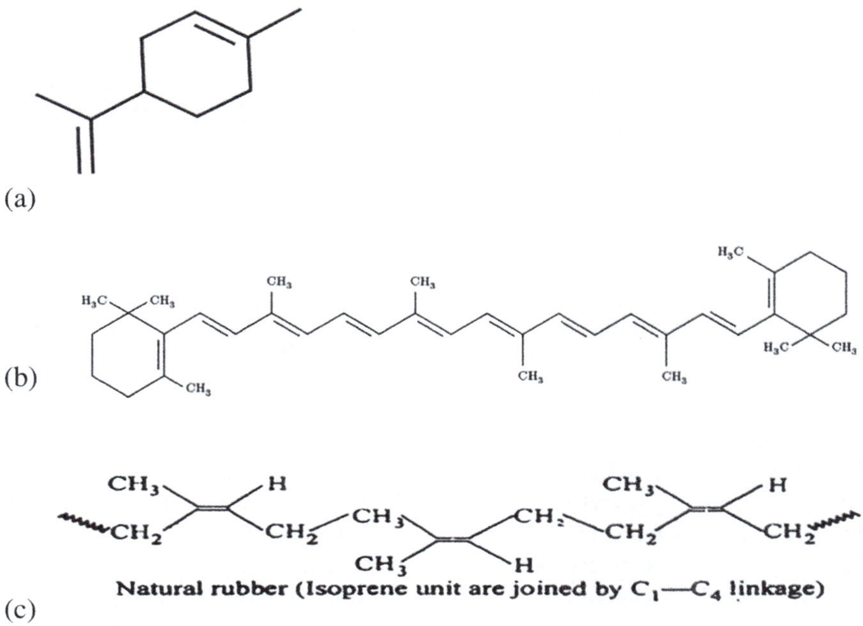


**Fig. 14.1** Structure of important terpenes: (**a**) limonene, (**b**) beta-carotene, and (**c**) natural rubber

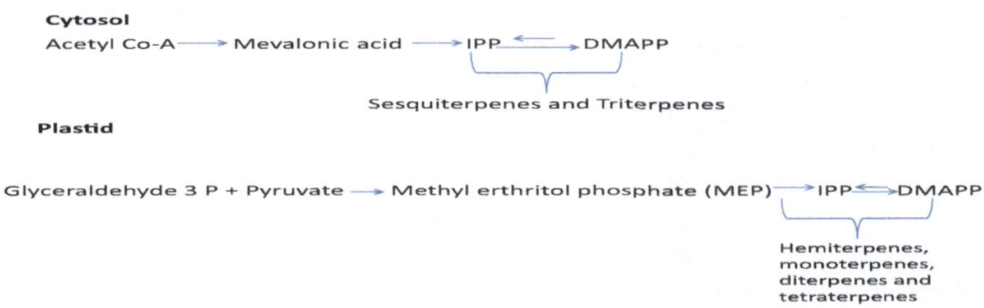


**Fig. 14.2** Precursors for synthesis of terpenes are present in cytosol and plastid

3-phosphate are the precursors for the synthesis of isopentyl pyrophosphate forming hemi-, mono-, di-, and tetraterpenes.

Sterols and saponins are the terpenoids with a particular ring structure, e.g., cycloartenol. Among the plant sterols, cycloartenol is an important triterpenoid and is the precursor for the synthesis of almost all plant steroids (Hubert 2003) in comparison to lanosterol which is the precursor for the synthesis of steroids of fungi and animals. Examples of terpenes are azadirachtin (*Azadirachta indica*), artemisinin (*Artemisia annua*), and tetrahydrocannabinol (*Cannabis sativa*).

**Fig. 14.3** Structure of important phenolics

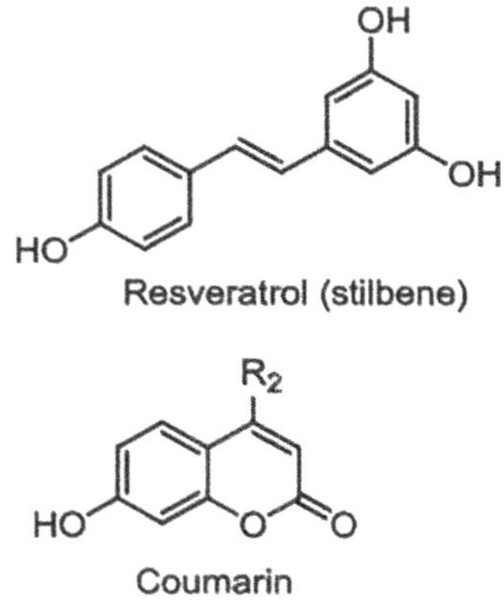


**Phenolics** These are the organic compounds containing an aromatic ring structure with one or more hydroxyl groups. Some phenolics interfere with digestion, have slow growth, block enzyme activity and cell division, or just taste awful. These are the most abundant secondary metabolites of plants and can be as simple as phenolic acid or as complex as tannins (Fig. 14.3). Tannins are described as polyphenols that can be either condensed tannins, made up of flavan-3-ols polymer subunits linked via 4–6 and 4–8 interflavan bonds, or can be hydrolysable tannins that are esters of gallic acid with a central polyol (Ossipov et al. 2003). Though they contain phenol ring (resveratrol), their basic skeleton is quite variable—C6-C1 (phenolic acids), C6-C2 (acetophenone), C6-C3 (phenylpropanoid, hydroxycinnamate, and coumarins), C6-C4 (naphthoquinone), C6-C1-C6 (xanthone), C6-C2-C6 (stilbene, anthraquinone), C6-C3-C6 (flavonoids, isoflavonoids), $(C6\text{-}C3)_2$ lignins, and $(C6\text{-}C3\text{-}C6)_2$ biflavonoids.

Flavonoids are the most important class of plant and fungal secondary metabolites that derive their name from the Latin word flavus (meaning yellow) and governs several important organoleptic characteristics (aroma, color, nutritional composition, etc.) of the plant (Pott et al. 2019). These secondary metabolites are known for their antioxidant and anticancerous properties, and their high dietary intake is known to protect from cardiovascular and cancer diseases (Rodriguez et al. 2014; Giampieri et al. 2015).

**Glycoside** These are the organic compounds in which carbohydrate is attached via glycosidic bond to a noncarbohydrate moiety containing a hydroxyl group (Fig. 14.4). Glucose is the most commonly found sugar in the glycosides. Important representatives are cardiac glycosides, cyanogenic glycosides, and saponins. Examples of glycosides are nojirimycin, glucosinolates, and cardenolides.

Cardiac glycosides of plant origin are made up of a steroid backbone containing a lactone ring at the 17-βposition and a sugar moiety at the 3-βposition (Kytidou et al. 2020). Positive effect of cardenolides on cardiac arrhythmia, congestive heart failure, and atrial fibrillation has been reported by a number of workers (Nesher

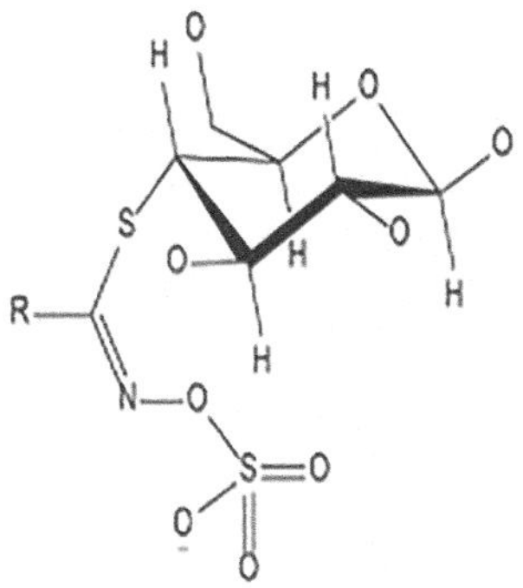

Glucosinolates

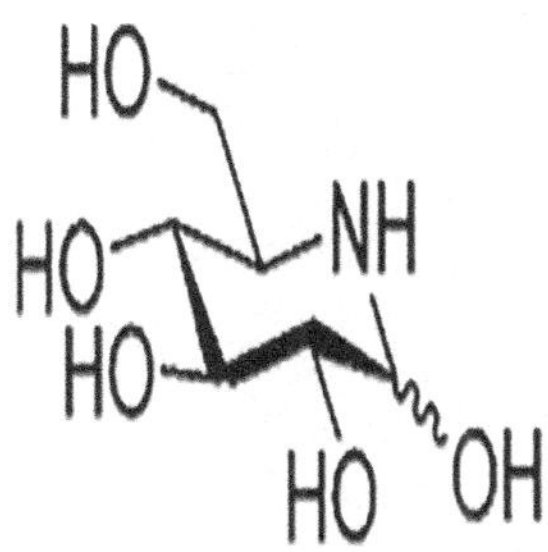

Nojirimycin

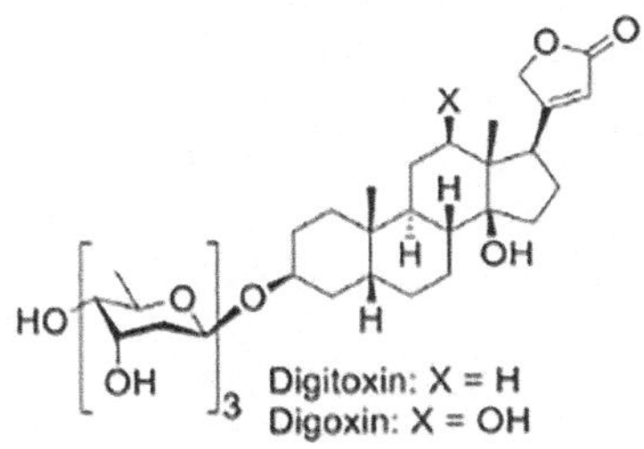


Cardenolides : Cardiac Glycoside

**Fig. 14.4** Structures of important glycosides

et al. 2007). Cardenolides have the ability to inhibit the $Na^{+}/K^{+}$ ATPase ion pump and thereby act as antiarrhythmic agents by increasing intracellular potassium concentrations (Kepp et al. 2012; Patel 2016). High intracellular potassium in turn increases the intracellular calcium and thereby improves myocardial contraction and cardiac pump activity (Newman et al. 2008).

Glucosinolates (mustard oil glycosides) are the secondary metabolites that are protective in function against grazing animals. These secondary metabolites have been reported in radish, cabbage, and mustard plants (Fahey et al. 2001). Glucobrassicin is an important glycoside formed from tryptophan and is present in cabbage. The enzyme thioglucosidase is responsible for the hydrolysis of the glycoside followed by the liberation and rearrangement of the sulfate residue. This spontaneously results in the formation of an isothiocyanate or mustard oil that is toxic at higher concentrations (Halkier and Gershenzon 2006). Cyanogenic glycosides, glucosinolates, and the enzyme thioglucosidase are located in separate compartments of the plant tissues and come into contact only after physical damage to the plant or grazing. High glucosinolate content in some of the early varieties of

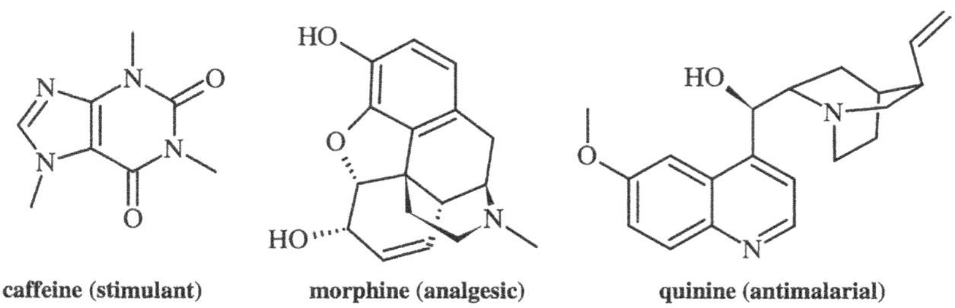


**Fig. 14.5** Structure of important alkaloids

rapeseed made them unsuitable for fodder, and several breeding experiments have resulted in rapeseed varieties that lack glucosinolate in the seeds.

Plants contain prussic acid (HCN) in the bound form as cyanogenic glycoside to protect its own mitochondrial respiratory chain. HCN is a respiratory inhibitor that inhibits cytochrome oxidase, i.e., the final step of the respiration. Plants use this poison to protect themselves from grazing animals. Similarly, the consumption of peach kernels or bitter almonds can be toxic for humans. The amygdalin in the kernels and roots of peaches is an example of cyanogenic glycoside and is stored as stable compound in the vacuole. After the hydrolysis by glucosidase, the remaining cyanohydrin decomposes spontaneously to prussic acid and an aldehyde by a hydroxynitrile lyase enzyme. The aldehydes formed from cyanogenic glycosides are often very toxic for a feeding animal. Therefore, the cyanogenic glycosides are an effective defense strategy utilized by a number of plants (Hopkins et al. 2009; Li et al. 2000; Müller et al. 2010).

**Alkaloids** Alkaloids are heterogenous, basic compounds containing one or more nitrogen atoms that are typically derived from plant sources (Fig. 14.5). These are synthesized from amino acids and contain heterocyclic ring containing one to several N atoms. Being basic in nature, alkaloids are stored in the acidic vacuole in the protonated form. Plant alkaloids are being used as poisons, stimulants, narcotics, and medicine. Their mode of action is quite variable – as blockers of ion channels and inhibitors of enzyme and neurotransmission – leading to hallucinations, loss of coordination, convulsions, vomiting, and death. On the basis of chemical structures, they may be classified into two broad categories: heterocyclic (quinine, caffeine, and nicotine) and nonheterocyclic (hordenine, colchicine, taxol) alkaloids.

A well-known alkaloid, atropine, is found in the plant *Atropa belladonna* (deadly nightshade). It is used to cause dilation of pupils of the eye for eye examination and to treat bradycardia. Before surgery, it is administered to reduce salivation and bronchial secretions. It is also used as an antidote against mushroom poisoning. Another tropane derivative is cocaine present in *Erythroxylum coca* the coca plant and is a well-known narcotic agent. An isoquinoline alkaloid, morphine, has medicinal application as a painkiller and is also a precursor for the synthesis of heroin. A piperidine alkaloid, coniine, is derived from plant poison hemlock and is a respiratory poison with toxicity to animals and humans. In literature, this is the poison used to kill Socrates. Nicotine is formed in the roots of tobacco plants and is carried along with the xylem sap into the stems and leaves. Its sulfate derivative is used as a very potent insecticide and is a byproduct of tobacco industry. A quinoline alkaloid, quinine, is obtained from *Cinchona* bark and is used as a potent antimalarial drug. Caffeine contains a purine as the heterocycle and is the stimulant present in coffee. The alkaloid taxol, isolated from the yew tree (*Taxus brevifolia*), is a potent anticarcinogenic agent. Various derivatives of the alkaloid camptothecin obtained from *Camptotheca acuminate* (happy tree) and taxol are used as cancer therapeutics.

Other examples of alkaloids include hyoscyamine obtained from *Datura stramonium*, scopolamine from the Solanaceae (nightshade) plant family, and codeine and morphine from *Papaver somniferum* (opium poppy). Similarly, vincristine and vinblastine from *Catharanthus roseus* are strong mitotic inhibitors.

In addition to those secondary metabolites that are toxic to herbivores, plants synthesize certain unusual amino acids (nonprotein amino acids) having structure very similar to that of normal amino acids, e.g., canavanine from jack bean (*Canavalia ensiformis*) is a nonprotein amino acid recognized by the arginine tRNAs of animals. Grazing animals feeding on these plants take up canavanine with their food and incorporate canavanine instead of arginine into their proteins resulting in altered three-dimensional structure of proteins. These altered proteins lose their biological activity; hence, canavanine is toxic for herbivores. On the other hand, plants synthesizing canavanine have the arginine tRNA that does not react with canavanine. This same protective mechanism is used by some insects, which are specialized in eating leaves containing canavanine.

## 14.6 Pathway of Synthesis of Secondary Metabolites

Intermediates from glycolysis (EMP pathway), Krebs (TCA) cycle, or the shikimate pathway often serve as precursors for the synthesis of different secondary metabolites (Kroymann 2011). Studies have revealed the origin of secondary metabolism by modification or rerouting of primary metabolic pathways (de Kraker and Gershenzon 2011; Kroymann 2011). Gene duplication is the suggested mechanism by which secondary metabolism is expanding at a very fast pace and with so much diversity (Carrington et al. 2018). Most of the gene families involved in secondary metabolism have evolved from primary metabolism protein folds (Weng et al. 2012). Most of the secondary metabolites are produced from rerouting of Krebs cycle, EMP, amino acids, pentose phosphate, and shikimate pathways (Herrmann 1995). The aromatic amino acids phenylalanine and tyrosine are the final products of the shikimate pathway. These amino acids are the phenylpropanoid precursors and are acted upon by amino acid lyases for the synthesis of both volatile and nonvolatile phenylpropanoids. Phenylpropanoids can be as simple as phenolic acids, or can be as complex as stilbenes, lignans, or flavonoids. Phenylalanine ammonia lyase (PAL) is the key enzyme for directing carbon flow from primary to secondary metabolism. This enzyme deaminates phenylalanine to cinnamic acid to start the phenylpropanoid pathway. Along with PAL, several other enzymes such as

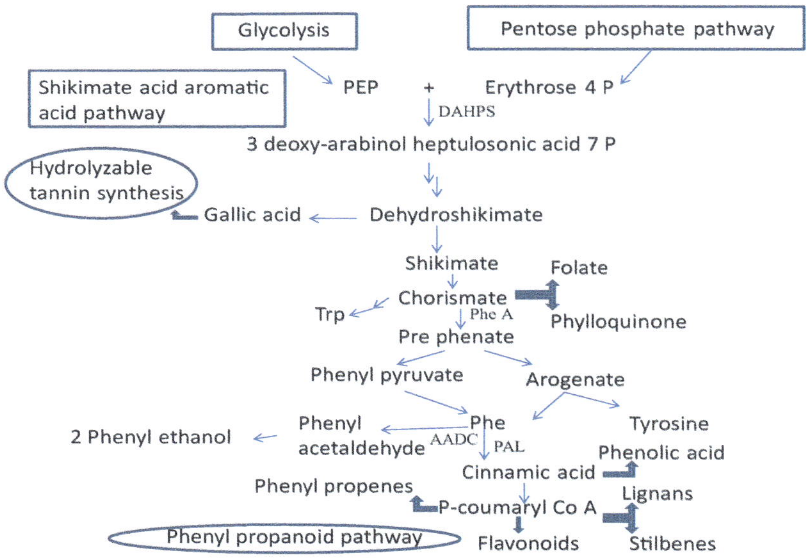


**Fig. 14.6** Shikimate pathway for the synthesis of phenolic acid compounds (*DAHPS* 3-deoxy-d-arabino-heptulosonate 7-phosphate synthase, *PheA* chorismate mutase/prephenate dehydratase, *AADC* aromatic amino acid decarboxylase, *PAL* phenylalanine ammonia lyase)

oxygenases, ligases, oxidoreductases, and transferases are involved in the synthesis of diverse phenylpropanoids. Formation of 4-coumaroyl-coenzyme A is the main branch point within the pathway (Vogt 2010). Condensation of coumaroyl-CoA and malonyl-CoA molecules leads to the formation of flavanones and initiates the flavonoid pathway. Flavanones then synthesize dihydroflavonols that can also be converted to anthocyanidins. Anthocyanins are formed by oxidation and glycosylation of anthocyanidins and confer different colors to different fruits (Petrussa et al. 2013). Shikimate pathway starts with the key enzyme (Fig. 14.6) 3-deoxy-D-arabino-heptulosonate 7-phosphate synthase (DAHPS) that controls the amount of carbon entering into the secondary metabolism. This enzyme combines phosphoenolpyruvate and erythrose 4-phosphate to form 3-deoxy-D-arabino-heptulosonic acid 7-phosphate (Herrmann 1995). This pathway is not only essential for the synthesis of phenylpropanoids but is also involved in the synthesis of essential micronutrients such as tetrahydrofolate (vitamin B9) (Wolak et al. 2017) and phylloquinone (vitamin K1) (Basset et al. 2017). These essential micronutrients are synthesized via chorismate. Chorismate is therefore an important branch point within the shikimate pathway.

Isopentenyl diphosphate (IPP) and dimethylallyl diphosphate (DMAPP) are the building blocks for synthesis of terpenoids and are synthesized through two different biosynthetic pathways. The pathway operative in the plastids provides the substrates for the production of mono-, di-, and tetraterpenoids. Pathway operative in cytosol is the mevalonate (MVA) pathway, and the IPP and DMAPP derived from this

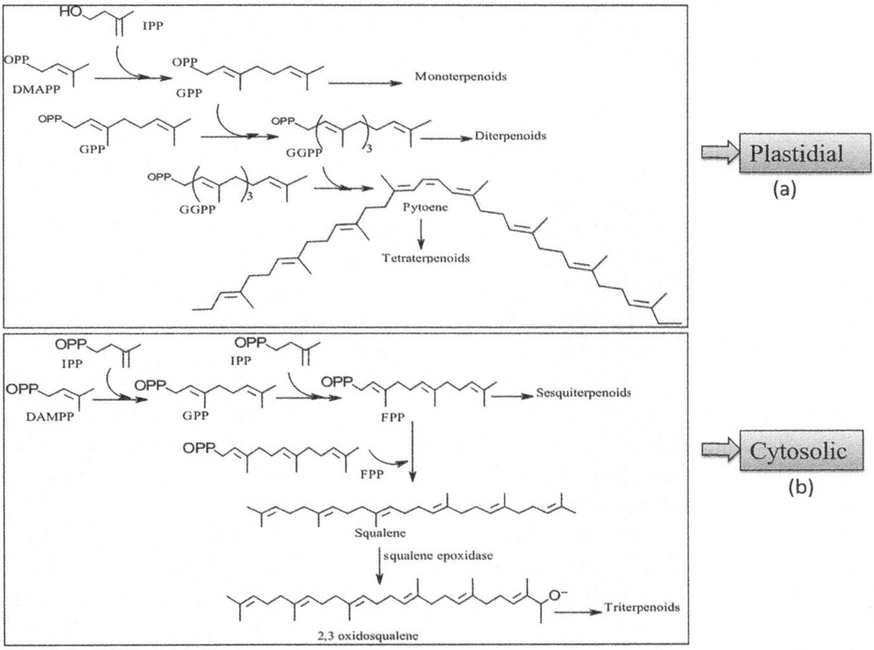


**Fig. 14.7** Biosynthesis of mono-, di-, and tetraterpenoids in the plastids (**a**) and biosynthesis of sesqui- and triterpenoids in the cytosol (**b**)

pathway are used in the production of sesquiterpenoids and triterpenoids (Fig. 14.7). Condensation of IPP and DMAPP catalyzed by the enzyme geranyl diphosphate synthase (GPS) leads to the synthesis of C10 (geranyl diphosphate (GPP)). GPP undergoes condensation with IPP by enzyme farnesyl diphosphate synthase (FPS) to form farnesyl diphosphate (FPP). GPP and FPP act as the precursors for synthesis of monoterpenoids and sesquiterpenoids, respectively. Condensation of FPP with IPP results in the formation of the C20 precursor of the diterpenes. On the other hand, two FPP molecules dimerize with simultaneous removal of the diphosphate groups to form squalene (C30) by the activity of squalene synthase (SQS) (Blagg et al. 2002). Sterols and steroids in plants are synthesized by the addition of oxygen to squalene with the help of enzyme squalene monooxygenase or epoxidase. This enzyme is responsible for synthesis of 2,3-oxidosqualene, the precursor for synthesis of triterpenoids, sterols, and steroids in plants. The precursor of the tetraterpenoids or carotenoids, i.e., phytoene, is synthesized by the dimerization of two GGPP molecules followed by elimination of the two diphosphate groups by the activity of enzyme phytoene synthase (PS) (Kashkooli et al. 2018).

Novel alkaloid biosynthetic enzymes are being characterized in terms of their structural biochemistry, molecular biology, and biotechnological applications (Cordell and Choi 2011; Ziegler and Facchini 2008). Most of the alkaloids are synthesized from amino acids such as aspartic acid, histidine, lysine, ornithine,

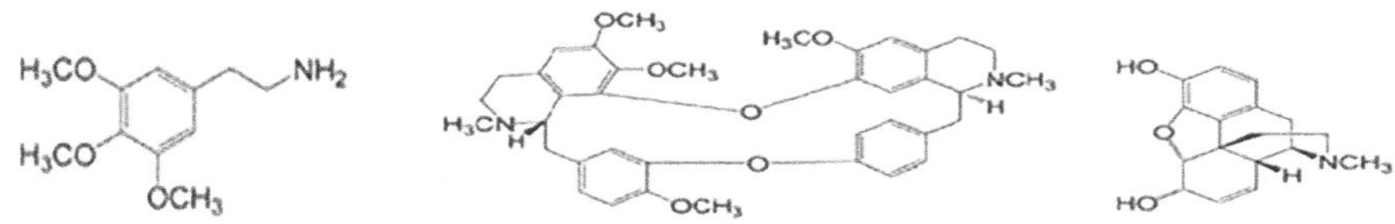


Mescaline Tetrandrine Morphine

**Fig. 14.8** Phenylalanine- and tyrosine-derived alkaloids

phenylalanine, tryptophan, and tyrosine via diverse biosynthetic pathways. Alkaloids derived from phenylalanine/tyrosine are shown in Fig. 14.8. Phenylalanine is the key precursor for synthesis of a large number of alkaloids having a wide range of structural diversity. Another amino acid, tyrosine, is produced from phenylalanine by an oxidation reaction catalyzed by a monooxygenase (phenylalanine hydroxylase) that causes addition of hydroxyl group to the aromatic ring. Phenylalanine and tyrosine are then involved in the synthesis of simple alkaloid derivatives like mescaline, morphine, and the complex alkaloids like tetrandrine (Fig. 14.8).

Biosynthetic pathway for synthesis of cyanogenic glycosides (CGs) has been studied in different plants like sorghum, cassava, and barley by different workers (Jones et al. 1999; Andersen et al. 2000; Nielsen et al. 2002; Forslund et al. 2004). These pathways involve the synthesis of CGs from six different amino acids like valine, isoleucine, leucine, phenylalanine, tyrosine, and cyclopentenyl-glycine (Vetter 2000). Dhurrin is the most well-studied cyanogenic glycoside being synthesized in sorghum plant. In general, three major steps are involved in biosynthesis of cyanogenic glycosides, and these enzymes are arranged in the form of a single metabolon (Fig. 14.9). The first step involves the conversion of L-amino acid precursor to aldoxime by two successive N-hydroxylations by cytochrome P450 family of enzyme. Aldoxime is converted to cyanohydrin by another cytochrome P450 enzyme. The last step involves glycosylation of cyanohydrin by the UDP-glucosyltransferase enzyme. Two different cytochrome enzymes – CYP79A1 and CYP71E1 – are responsible for the first and second step, while UDP-glucosyltransferase is involved in the final step of dhurrin biosynthesis (Ganjewala et al. 2010).

## 14.7 Medicinal Applications of Some Important Secondary Metabolites

*Artemisinin*: It is a widely used secondary metabolite in most of the Chinese medicine and was also reported as a powerful antimalarial drug by a Chinese scientist (Tu Youyou). Several studies have revealed that malarial parasite (*Plasmodium falciparum*) has become resistant to this drug, but it is still being used as a combination therapy along with other antimalarial drugs.

*Taxol*: Paclitaxel is the active component of widely used anticancer drug – Taxol. It is being used in chemotherapy for a variety of cancers including ovarian, breast,

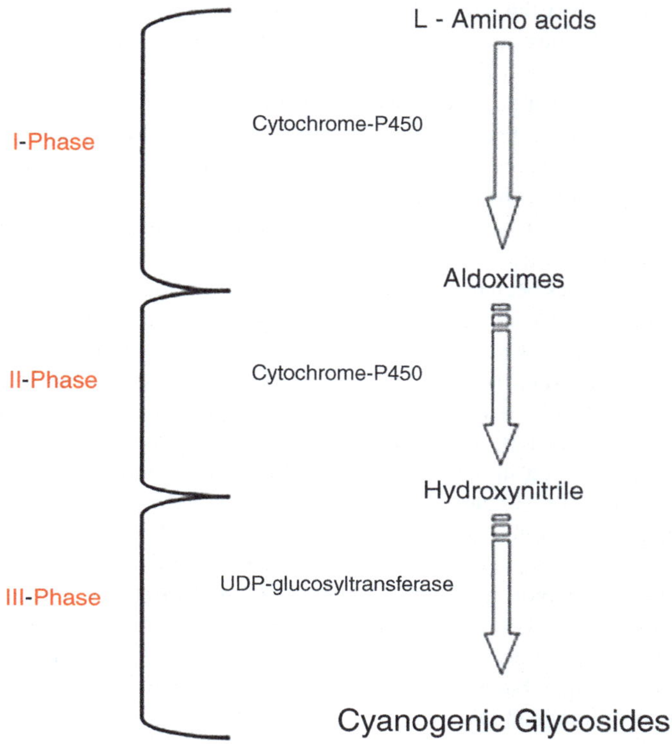


**Fig. 14.9** Overview of steps involved in the synthesis of cyanogenic glycosides. (Source: Ganjewala et al. 2010)

hepatic, lung, cervical, and pancreatic cancers and AIDS-related Kaposi sarcoma. It was first isolated in 1973 from a bark of a coniferous tree (Pacific yew).

*Morphine*: This is the first active alkaloid obtained from opium poppies and was first discovered by Friedrich Sertürner in the year 1804. It is used as an analgesic, for treatment of shortness of breath, and for de-addiction of strong opiates (heroin). Side effects of its use include addiction, hormonal imbalance, and constipation.

*Codeine*: It is also an opium poppy-derived alkaloid, isolated by *Pierre Jean Robiquet* in 1832. Primarily, codeine is an analgesic that is used to treat pain, coughing, diarrhea, and irritable bowel syndrome. Codeine is much milder than morphine, so has been reported to be much safer to use. The process of extraction of codeine from opium poppy is not economical, so it is prepared by the chemical process of methylation of morphine.

*Atropine*: It is an alkaloid found in plants belonging to family Solanaceae such as *Atropa belladonna*, *Datura* sp., and *Hyoscyamus*. It is injected intravenously or intramuscularly to treat bradycardia (slow heartbeat), and as an antidote to organophosphates. Its side effects include impaired vision, nausea, lack of sweating, dry mouth, urinary retention, constipation, and tachycardia (fast heartbeat).

*Resveratrol*: It is a flavonoid that is abundant in grapes, blueberries, bilberries, cranberries, raspberries, cocoa, and peanuts. Flavonoids are important dietary supplements that are used to reduce the risk of cancer and heart diseases. Certain studies have shown that flavonoids like quercetin have antibiotic activity.

*Digoxin*: It is a cardiac glycoside derived from foxglove plant by William Withering in 1785. It is used to treat different heart ailments. Side effects of its use include nausea, bradycardia, diarrhea, and arrhythmia.

Biotechnological approaches have been used to decrease the concentration of unwanted secondary metabolites or to increase the concentration of desired ones. Selective breeding has been used to decrease the concentration of naringin which is a secondary metabolite causing bitterness in grapefruit. Similarly, to increase the concentration of desired metabolites, different plant tissue culture techniques that allow growth under controlled conditions, overcome seasonal variability, and protect from parasites and pathogenic microbes are being exploited. To increase the synthesis of secondary metabolites in the plants, certain elicitors (jasmonates, UV-B, ozone) can be used that induce stress on the plant and thereby increase SMs level (Hartmann 2007). To synthesize these secondary metabolites in living factories (microbes), genetic engineering techniques have been used. Evolva used a recombinant *S. cerevisiae* strains to produce a secondary metabolite—vanillin—that is widely used as a flavoring agent. Currently, *Evolva* produces a wide array of chemicals such as stevia, resveratrol, or nootkatone. With the development of recombinant technologies, the Nagoya Protocol on Access to Genetic Resources and the Fair and Equitable Sharing of Benefits Arising from their Utilization to the Convention on Biological Diversity was signed in 2010.

## 14.8 Conclusion

Secondary metabolites of plants with known biological activities can be considered as important substitute to other synthetic drugs and chemicals that are used for culinary or pharmaceutical values. With continuous increase in development of resistance to existing chemicals by pathogens and pests, there is tremendous scope of exploiting new sources of novel secondary metabolites. These novel SMs can play an important role in finding alternate pharmaceutical agents to fight cancers, autoimmune disorders, and infectious diseases, gastrointestinal and cardiac ailments. These bioactive agents can also be crucial for controlling newly emerging pests and pathogens of plants especially in changing climatic conditions. Due to their role in enabling plants to deal with environmental stress, metabolic engineering of plants to either increase or decrease the production of SMs is rapidly gaining interest. Scientific community has to put endless efforts to cope up with ever-increasing resistant species of pathogens and pests by tracing new sources of secondary metabolites with novel activities.

## References

Andersen MD, Busk PK, Svendsen I, Moller BL (2000) Cytochromes P-450 from cassava (*Manihot esculenta* Crantz) catalyzing the first steps in the biosynthesis of the cyanogenic glucosides linamarin and lotaustralin cloning, functional expression in *Pichia pastoris*. J Biol Chem 275: 1966–1975

Barbour WS, Devine C (2010) Phytoestrogens and breast cancer. Cornell University, 31/03/2010 Web 1 Apr 2011

Basset GJ, Latimer S, Fatihi A, Soubeyrand E, Block A (2017) Phylloquinone (vitamin K1): occurrence, biosynthesis and function. Mini Rev Med Chem 17:1028–1038. https://doi.org/10.2174/1389557516666160623082714

Bidlack WR, Omaye ST, Meskin MS, Topham DKW (2000) Phytochemicals as bioactive agents. CRC Press LLC, Florida

Blagg BS, Jarstfer MB, Rogers DH, Poulter CD (2002) Recombinant squalene synthase. A mechanism for the rearrangement of presqualene diphosphate to squalene. J Am Chem Soc 124:8846–8853

Brown KS, Trigo JR (1995) The ecological activity of alkaloids. In: Cordell GA (ed) The alkaloids, vol 47. Academic Press, New York, NY, pp 227–356

Carrington Y, Guo J, Fillo A, Kwon J, Tran LT, Ehlting J (2018) Evolution of a secondary metabolic pathway from primary metabolism: shikimate and quinate biosynthesis in plants. Plant J 95:823–883. https://doi.org/10.1111/tpj13990

Cordell GA, Choi T (2011) Alkaloids and their biosynthesis. In: Phytochemistry and pharmacognosy section. Chemical sciences, engineering and technology resources, Encyclopedia of Life Support Systems (EOLSS) developed under the auspices of the UNESCO, EOLSS Publishers, Oxford. http://wwweolssnet

de Kraker JW, Gershenzon J (2011) From amino acid to glucosinolate biosynthesis: protein sequence changes in the evolution of methylthioalkylmalate synthase in *Arabidopsis*. Plant Cell 23:38–53. https://doi.org/10.1105/tpc110079269

Dicke M, Baldwin IT (2010) The evolutionary context for herbivore-induced plant volatiles: beyond the 'cry for help'. Trends Plant Sci 15:167–175

Fahey JW, Zalcmann AT, Talalay P (2001) The chemical diversity and distribution of glucosinolates and isothiocyanates among plants. Phytochemistry 56:5–51

Forslund K, Morant M, Jorgensen B, Olsen CE, Asamizu E, Sato S, Tabata S, Bak S (2004) Biosynthesis of the nitrile glucosides rhodiocyanoside A and D and the cyanogenic glucosides lotaustralin and linamarin in *Lotus japonicus*. Plant Physiol 135:71–84

Ganjewala D, Kumar S, Devi A, Ambika K (2010) Advances in cyanogenic glycosides biosynthesis and analyses in plants: a review. Acta Biol Szegediensis 54(1):1–14. http://wwwsciu-szegedhu/ABS

Giampieri F, Forbes-Hernandez TY, Gasparrini M, Alvarez-Suarez JM, Afrin S, Bompadre S et al (2015) Strawberry as a health promoter: an evidence based review. Food Funct 6:1386–1398. https://doi.org/10.1039/C5FO00147A

Halkier BA, Gershenzon J (2006) Biology and biochemistry of glucosinolates. Annu Rev Plant Biol 57:303–333

Hans-Walter H (2005) Plant biochemistry. Elsevier Academic Press, USA

Hartmann T (2007) From waste products to ecochemicals: fifty years research of plant secondary metabolism. Phytochemistry 68:2831–2846

Herrmann KM (1995) The shikimate pathway as an entry to aromatic secondary metabolism. Plant Physiol 107:7–12. https://doi.org/10.1104/pp10717

Hopkins RJ, Van Dam NM, Van Loon JJA (2009) Role of glucosinolates in insect–plant relationships and multitrophic interactions. Annu Rev Entomol 54:57–83

Hubert S (2003) The role of sterols in plant growth and development. Prog Lipid Res 42(3):163–175. https://doi.org/10.1016/S0163-7827(02)00047-4

Hussein RA, El-Anssary AA (2018) Plants secondary metabolites: the key drivers of the pharmacological actions of medicinal plants. IntechOpen. https://doi.org/10.5772/intechopen76139
Jones PR, Moller BL, Hoj PB (1999) The UDP-glucose: p-hydroxymande-lonitrile-O-glucosyltransferase that catalyzes the last step in synthesis of the cyanogenic glucoside dhurrin in *Sorghum bicolor*. J Biol Chem 274:35483–35491
Kashkooli AB, van der Krol A, Bouwmeester HJ (2018) Terpenoid biosynthesis in plants. In: Siegmund B, Leitner E (eds) Flavour science. Verlag der Technischen Universität Graz. doi:10.3217/978-3-85125-593-5-1, CC BY-NC-ND 40
Katyal P, Sharma S (2019) Emerging alkaloids against cancer: a peep into factors, regulation and molecular mechanisms. In: Sharma AK (ed) Bioactive natural products for the management of cancer from bench to bed side. Springer, Singapore, pp 37–60
Kepp O, Menger L, Vacchelli E, Adjemian S, Martins I, Ma Y et al (2012) Anticancer activity of cardiac glycosides. Oncoimmunology 1(9):1640–1642. https://doi.org/10.4161/onci21684
Kessler A, Halitschke R (2007) Specificity and complexity: the impact of herbivore-induced plant responses on arthropod community structure. Curr Opin Plant Biol 10:409–414
Kroymann J (2011) Natural diversity and adaptation in plant secondary metabolism. Curr Opin Plant Biol 14:246–251. https://doi.org/10.1016/jpbi201103021
Kytidou K, Artola M, Overkleeft HS, Aerts JMFG (2020) Plant glycosides and glycosidases: a treasure-trove for therapeutics. Front Plant Sci 11:357. https://doi.org/10.3389/fpls202000357
Li Q, Eigenbrode SD, Stringam GR, Thiagarajah MR (2000) Feeding and growth of *Plutella xylostella* and *Spodoptera eridania* on *Brassica juncea* with varying glucosinolate concentrations and myrosinase activities. J Chem Ecol 26:2401–2419
Mason PA, Singer MS (2015) Defensive mixology: combining acquired chemicals towards defence. Funct Ecol 29:441–450. https://doi.org/10.1111/1365-243512380
Müller R, de Vos M, Sun JY, Sønderby IE, Halkier BA, Wittstock U, Jander G (2010) Differential effects of indole and aliphatic glucosinolates on lepidopteran herbivores. J Chem Ecol 36:905–913
Nesher M, Shpolansky U, Rosen H, Lichtstein D (2007) The digitalis-like steroid hormones: new mechanisms of action and biological significance. Life Sci 80:2093–2107. https://doi.org/10.1016/jlfs200703013
Newman RA, Yang P, Pawlus AD, Block KI (2008) Cardiac glycosides as novel cancer therapeutic agents. Mol Interv 8:36–49. https://doi.org/10.1124/mi818
Nielsen KA, Olsen CE, Pontoppidan K, Moller BL (2002) Leucine derived cyano-glucosides in barley. Plant Physiol 129:1066–1075
Ossipov V, Salminen JP, Ossipova S, Haukioja E, Pihlaja K (2003) Gallic acid and hydrolysable tannins are formed in birch leaves from an intermediate compound of the shikimate pathway. Biochem Syst Ecol 31:3–16. https://doi.org/10.1016/S0305-1978(02)00081-9
Patel S (2016) Plant-derived cardiac glycosides: role in heart ailments and cancer management. Biomed Pharmacother 84:1036–1041. https://doi.org/10.1016/jbiopha201610030
Pennisi E (2017) How do gut microbes help herbivores counting the ways. Science 355:236. https://doi.org/10.1126/science3556322236
Peter K (1998) Resveratrol, a phytoestrogen found in red wine. A possible explanation for the conundrum of the 'French paradox'? Eur J Endocrinol 138:619–620
Petrussa E, Braidot E, Zancani M, Peresson C, Bertolini A, Patui S et al (2013) Plant flavonoids-biosynthesis, transport and involvement in stress responses. Int J Mol Sci 14:14950–14973. https://doi.org/10.3390/ijms140714950
Pott DM, Osorio S, Vallarino JG (2019) From central to specialized metabolism: an overview of some secondary compounds derived from the primary metabolism for their role in conferring nutritional and organoleptic characteristics to fruit. Front Plant Sci 10:835. https://doi.org/10.3389/fpls2019
Rodriguez MA, Heiss C, Borges G, Crozier A (2014) Berry (poly)phenols and cardiovascular health. J Agric Food Chem 62:3842–3851. https://doi.org/10.1021/jf403757g
Vetter J (2000) Plant cyanogenic glycosides. Toxicon 38:11–36

Vogt T (2010) Phenylpropanoid biosynthesis. Mol Plant 3:2–20. https://doi.org/10.1093/mp/ssp106

Weng JK, Philippe RN, Noel JP (2012) The rise of chemodiversity in plants. Science 336:1667–1670. https://doi.org/10.1126/science1217411

Wink M (2003) Evolution of secondary metabolites from an ecological and molecular phylogenetic perspective. Phytochemistry 64:3–19. https://doi.org/10.1016/S0031-9422(03)00300-5

Wink M (2018) Plant secondary metabolites modulate insect behavior – steps toward addiction? Front Physiol 9:364

Wink M, Legal L (2001) Evidence for two genetically and chemically defined host races of *Tyria jacobaeae* (Arctiidae, Lepidoptera). Chemoecology 11:199–207. https://doi.org/10.1007/PL00001852

Wink M, Schimmer O (2010) Molecular modes of action of defensive secondary metabolites. In: Annual plant reviews volume 39: functions and biotechnology of plant secondary metabolites, 2nd edn, pp 21–161. https://doi.org/10.1002/9781444318876ch2

Wittstock U, Gershenzon J (2002) Constitutive plant toxins and their role in plant defense. Curr Opin Plant Biol 5:300–307

Wolak N, Zawrotniak M, Gogol M, Kozik A, Rapala-Kozik M (2017) Vitamins B1, B2, B3 and B9 – occurrence, biosynthesis pathways and functions in human nutrition. Mini Rev Med Chem 17:1075–1111. https://doi.org/10.2174/1389557516666160725095729

Ziegler J, Facchini PJ (2008) Alkaloid biosynthesis: metabolism and trafficking. Annu Rev Plant Biol 59:735–769

# Plant Secondary Metabolites in Antiviral Applications

# 15

Neeru Dhamija and AnitaGarg Mangla

**Abstract**

Plants possess a unique defense system by virtue of secreting low-molecular-weight secondary metabolites that protects them from harmful microorganisms to herbivores. These compounds are primarily found in essential oils of the plants and can act as antibacterial/ antifungal/antiviral agents in combating various human diseases. To add to the list is the recent Covid-19 virus which is responsible for the current ongoing pandemic and is caused by coronavirus. Ancient civilization in India, China, as well as other western countries used plants' secreted compounds as medicines. With the advent of modern era and progress in scientific endeavor, there was enormous growth of allopathic medicine in the nineteenth century and that in turn caused downfall in conservative alternate plant-based traditional medicine system. Now again in the twenty-first century, we are going back to rediscover plant-based medicines to combat deadly human diseases as plant-based antiviral compounds are one of the best alternatives to various antiviral drugs/inhibitors/medicines since they offer safety and less toxicity (Lillehoj et al., Vet Res 49(1):76, 2018). This chapter focuses on the role of plant-based secondary metabolites in antiviral applications.

**Keywords**

Plant defense system · Antiviral drugs · Inhibitors · Plant-based medicines · Protection · Diseases

N. Dhamija · A. Mangla (✉)
Department of Biochemistry, Daulat Ram College, University of Delhi, Delhi, India
e-mail: anitamangla@dr.du.ac.in





A. K. Sharma, A. Sharma (eds.), *Plant Secondary Metabolites*,
https://doi.org/10.1007/978-981-16-4779-6_15

## 15.1 Introduction

Plants can synthesize a wide spectrum of secondary metabolites and have been used as remedies and treatment for a variety of viral diseases. Based on various studies and evidences, a number of alkaloids, flavonoids, saponins, terpenoids, carotenoids, glucosides, and polyphenols which are produced by medicinal plants have been widely studied, tested, and used for their potential antiviral activities and have been shown to inhibit/block viral progression by affecting adherence, penetration, replication, maturation, and/or packaging of viral particles. Viruses have been responsible for causing various pandemic events. In the twenty-first century, the first pandemic event was caused by influenza virus H1N1, and the current ongoing coronavirus pandemic has caused global threat and affected the world's population adversely. These distressing events call for better strategy and efficient remedies to handle the outburst of various viral infections every now and then, since viruses pose problems associated with the ability of their genomes to mutate as well as become resistant to drugs (Irwin et al. 2016). Plant-based antiviral natural compounds have been evaluated for emerging viral diseases because of their minimal side effects and numerous health benefits. Plant-based drugs have been evaluated for the past many years for having antioxidant, antibacterial, antiviral, and anticancerous effects (Biswas et al. 2020). Plant secondary metabolites with antiviral properties can be combined with preexisting therapies and delivery mechanisms to enhance antiviral effects along with good bioavailability (Kapoor et al. 2017).

Viruses depend on the host metabolism and environment for their replication and hence survival. Virus genome consists of single- or double-stranded DNA/RNA genome. For their survival, they exploit host cellular machinery to divide and disseminate throughout the body (Helenius 2018). Secreted plant secondary metabolites have a wide variety of functions and can be targeted for developing antivirals by effectively studying their mechanisms of action to inhibit virus entry, division, packaging, etc., and also to handle problems related to drug resistance, viral latency, and conflicting efficacy in recurrent infection in immune-compromised patients with the currently available antivirals (Irwin et al. 2016; Sumithira et al. 2012). Therefore, viral spread can be inhibited at various stages including adhesion and entry, replication within host, transcription, translation and further modifications, and finally packaging and dissemination in the host cells (Subudhi et al. 2018).

## 15.2 Alkaloids

Alkaloids are present in nature as a class of basic, naturally occurring organic compounds in plants, fungi, and bacteria that contain at least one nitrogen atom (Harborne 1998). Structurally, alkaloids are made up of carbon, hydrogen, and nitrogen and may also contain oxygen, sulfur, and rarely other elements such as phosphorus, bromine, and chlorine. They are important active ingredients in Chinese herbal medicine because of their significant biological activities. Alkaloids are

**Plant Secondary Metabolites**

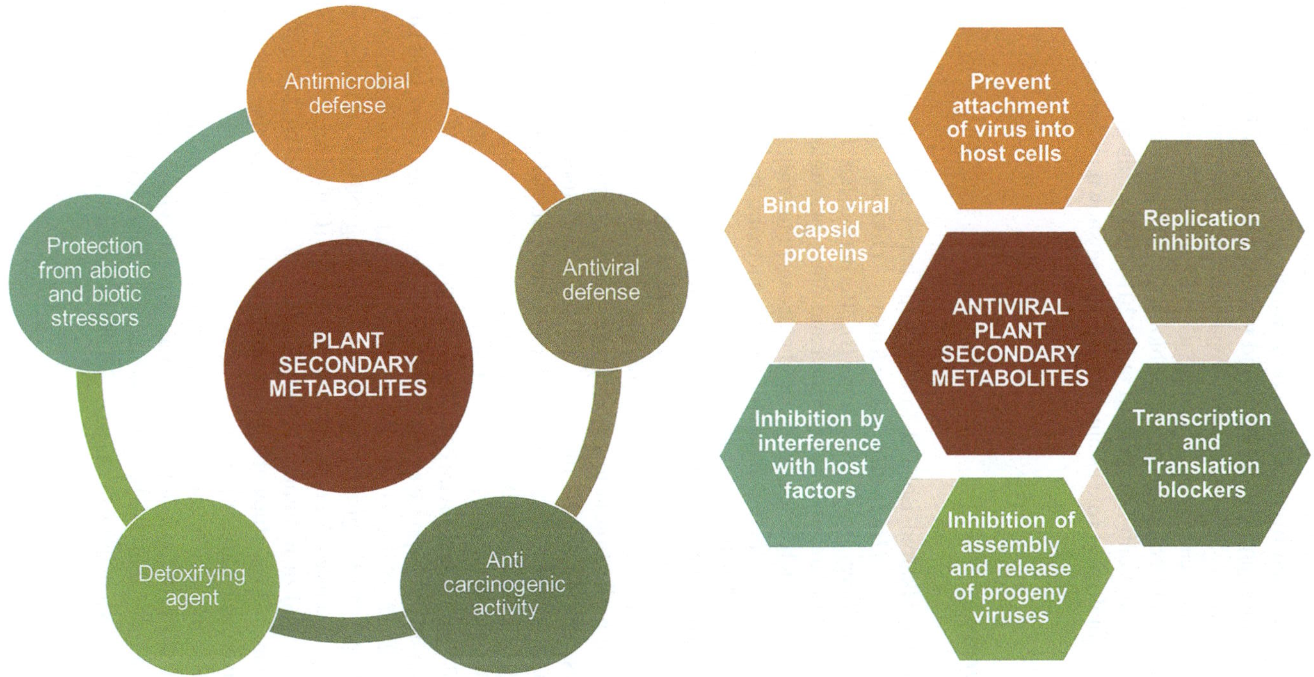


Various functions

Antiviral mode of action

Amongst these compounds one of the most effective natural component are Alkaloids (Moradi et al. 2018).

majorly found in higher plants especially in dicots and few in lower plants (Dang et al. 2012) and are the largest category of plant-based secondary metabolites. Alkaloids can be classified by the source from which they are obtained combined with chemical structures and are mainly found in higher plants such as those from families Ranunculaceae, Leguminosae, Papaveraceae, Menispermaceae, and Loganiaceae and are categorized using a biosynthetic-based classification (Lu et al. 2012a, b) and comprise of many biological activities like antibacterial, antiviral, anticancer, and anti-asthma (Shi et al. 2014). Many alkaloids have been evaluated for their biological role as antivirals and have shown promising results.

### 15.2.1 Berberine

It is an isoquinoline alkaloid isolated from natural herbs such as *Coptis (Coptis chinensis), goldenseal (Hydrastis canadensis), Oregon grape (Mahonia aquifolium)*, and *barberry (Berberis vulgaris)*. Berberine has been used as an antibacterial drug for the treatment of diarrhea and other gastrointestinal infections. It has diverse functions such as lowering blood glucose, regulating blood triglyceride and cholesterol, and influencing the function of pancreatic beta cell (Shen et al. 2010). Other biological activities also include anticancer, anti-inflammatory, and antiviral effects. Berberine has been shown to effectively inhibit the growth of cytomegalovirus (CMV) and herpes simplex virus (Chin et al. 2010; Hayashi et al. 2007). The natural compound berberine hydrochloride (BBR) has a potent inhibitory activity on HCMV replication at low micromolar concentrations (Mercorelli et al. 2016). In Chinese and Ayurvedic medicine, BBR has been used since ages for its antimicrobial and anti-parasitic effects (Imenshahidi and Hosseinzadeh 2016; Kumar et al. 2015). BBR has been shown to have inhibitory activity against a range of viruses including influenza virus (Cecil et al. 2011), respiratory syncytial virus (Shin et al. 2015), enterovirus (Wang et al. 2017a, b), flavivirus (Robinson et al. 2018), alphavirus (Varghese et al. 2016a, b), herpes simplex virus, and HCMV (Hayashi et al. 2007; Song et al. 2014). BBR affects cellular processes by modulating multiple host signaling pathways including NF-κB (Pandey et al. 2008) and mitogen-activated protein kinase (MAPK) (Cui et al. 2009). These pathways are activated in a number of viral infections such as HCMV infection and may be responsible for efficient expression of viral genes and thus progression of viral replicative cycle (Johnson et al. 2000; De Meritt et al. 2004; Caposio et al. 2007, 2010). BBR may hinder the full activation of these transduction pathways in HCMV-infected cells, thus resulting in impairment of viral replication. It has recently been shown that berberine inhibits cytomegalovirus replication by interfering with the viral immediate-early 2 (IE2) protein transactivating activity (Anna et al. 2019).

### 15.2.2 Hirsutine

It is an indole alkaloid isolated from *Uncaria rhynchophylla* and possesses anti-arrhythmic, antihypertensive, cardioprotective, and anti-metastatic properties via virtue of its effects on the inhibition of $Ca^{2+}$ influx and the release of intracellular $Ca^{2+}$ (Horie et al. 1992; Wu et al. 2011; Lou et al. 2014). Calcium homeostasis is associated with the disease severity of dengue (Shivanthan and Rajapakse 2014). Hirsutine has been shown to inhibit the viral particle assembly, budding, and release, however did not show affect replication and translation machinery of DENV life cycle (Takayuki et al. 2017). It was identified as a potent anti-DENV compound, exhibiting low cytotoxicity and high efficacy. In the past, hirsutine has been reported as exhibiting inhibitory effects against influenza A virus (subtype H3N2) and was found to inhibit the replication of influenza A (subtype H3N2) (Takayama et al. 1997). Therefore, hirsutine is effective against a wide variety of viruses by exhibiting inhibitory effects at various stages of the virus life cycle.

### 15.2.3 Thalimonine

It is an isoquinoline alkaloid belonging to genus *Thalictrum* (Ranunculaceae) (Umarov et al. 1970) and isolated from the Mongolian plant Thalictrum simplex L in 1991 (Velcheva et al. 1992). In cell cultures, thalimonine has been shown to irreversibly inhibit the replication of herpes simplex virus type I and improved the T cell mitogen-induced proliferation and suppressed the action of B cell mitogens in vitro. Also it was found to compromise the process of viral morphogenesis and the budding of viral particles (Varadinova et al. 1996). Thalimonine has been shown to markedly inhibit the reproduction of influenza virus A/Germany/27, str. Weybridge (H7N7) and A/Germany/34, str. Rostock (H7N1) in cell cultures of chicken embryo fibroblasts, and also results indicated that production of hemagglutinin (HA) and the virus-induced cytopathic effects (CPE) as measures of viral reproduction were markedly reduced (Serkedjieva and Velcheva 2003). Recently, in silico analysis has shown that thalimonine has potential activity to inhibit the main protease (Mpro) of SARS-CoV-2 (Saksham and Arpit 2020).

Quinolizidine and isoquinoline groups of alkaloids have been shown to have an effect on synthesis of viral proteins and influence various stages of viral replication. Homonojirimycin (HNJ), an alkaloid extracted from *Commelina communis*, has been shown to exhibit strong inhibitory activity against influenza viral infection (Zhang et al. 2013). Plant root of *Dendrobium nobile* contains the alkaloid dendrobine which reduced the titer of influenza virus during its replication (Li et al. 2017). *Enantia chlorantha* (Oliver) is a traditional medicinal plant that has been used in Nigeria for curing malaria fever, caused by *Plasmodium* species. This plant has been shown to have antiviral activity against hepatitis A, B, C, and D.

## 15.3 Phenolics

Plant secondary metabolites possessing one or more phenol rings are termed as phenolics. These compounds are widely distributed in plants. Phenolics are found in fruits, vegetables, chocolates, legumes, tea, coffee, etc.

### 15.3.1 Flavonoids

Flavonoids are a group of low-molecular-weight phytonutrients that are polyphenolic in nature and found in greater amounts in many fruits, vegetables, roots, stems, grains, and seeds (Carletti et al. 2014; Zakaryan et al. 2017). These secondary metabolites have diverse activities associated with them like anticancerous, antiviral, and antimicrobial. All the flavonoids contain three components: (1) a flavan nucleus, (2) two benzene rings, and (3) heterocyclic pyrene ring connecting the two benzene rings. There are different subgroups of around 6000 flavonoids found in nature. They include flavanones, flavanonols, flavanols/catechins, flavonols, flavones, anthocyanins, chalcones, etc., differing from one another in the extent of oxidation and substituent groups in the benzene ring. In isoflavonoids, the link between the two benzene rings is different than in the flavonoids.

Flavonoids have diverse functions, to name a few:

1. They protect plants from abiotic and biotic stressors.
2. They act as detoxifying agents.
3. They have anticarcinogenic activities.
4. They have antimicrobial and antiviral defensive properties (Panche et al. 2016; Ghildiyal et al. 2020).

#### 15.3.1.1 Flavonoids as Antiviral Agents

Flavonoids are long being explored for their antiviral activities (Kaul et al. 1985; Jacob and Thomas 2019). In vitro studies were employed to look for antiviral effects of flavonoids like quercetin, hesperetin, catechin, and naringin on infectivity and replication of herpes simplex virus type 1, parainfluenza virus type 3, and respiratory syncytial virus (Kaul et al. 1985). They have been shown to work on a variety of RNA and DNA viruses like Coxsackie B virus, poliomyelitis virus, cytomegalovirus, coronavirus, rhinovirus, rotavirus, and rabies virus (Evers et al. 2005; Chávez et al. 2006; Nowakowska 2007). Recently, it has been tested in another non-enveloped RNA virus, enterovirus A71, one of the leading causes of hand, foot, and mouth disease (HFMD) (Lalani and Poh 2020).

Flavonoids are known to work via various mechanisms:

1. bind to specific viral capsid proteins
2. prevent entry of virus into host cells
3. replication inhibitors and transcription and translation inhibitors

4. packaging and release inhibitors
5. interference with host factors

#### 15.3.1.2 Flavonols

Quercetin, belonging to subclass flavonol, is found in red grape wines, leaves of radish, and fennel and pepper seeds. It was initially demonstrated to inhibit rabies virus (Cutting et al. 1949) and vesicular stomatitis virus (Veckenstedt and Pusztai 1981). Tetra-O-methyl derivative inhibited influenza virus in vitro (Roschek et al. 2009), while rhamnoside derivative of quercetin popularly known as quercitrin is an early replication inhibitor of influenza virus. It is also a known inhibitor of hepatitis C virus and herpes simplex virus (HSV-1) (Choi et al. 2009; Bachmetov et al. 2012; Lu et al. 2012a, b). Yet another derivative of quercetin, 3-β-O-D-glucoside, was very effective against Ebola virus (Qiu et al. 2016). Quercetin also inhibits rhinovirus infection by acting as transcription and translation inhibitor. It's a known inhibitor of encephalomyocarditis virus (EMCV), while its methyl derivative is a poliovirus inhibitor and trimethyl/dihydro derivative inhibits Coxsackie B4 virus (Veckenstedt and Pusztai 1981; Gonzalez. et al. 1990; Vrijsen et al. 1987; Ishitsuka et al. 1982; Galochkina et al. 2016). Another flavonol, rutin, is found in tartary buckwheat, rhubarb, orange, and lemon. Rutin in its sulfated form, sodium rutin sulfate (SRS), shows antihuman immunodeficiency virus (HIV-1) and anti-HSV activity in vitro potentially acting as an entry and fusion inhibitor (Tao et al. 2007). Fisetin, a flavonol found in the acacia leaves and strawberries, is an inhibitor of chikungunya virus (Lani et al. 2016) and dengue virus 2 (Zandi et al. 2011). Yet another flavonol, kaempferol, predominantly found in raspberry, capers, grapes, and black bean, possesses anti-influenza A activity by non-competitively inhibiting neuraminidase activity (Jeong et al. 2009). It is also known for its antiviral activity against coronavirus inhibiting the release of the progeny viruses (Schwarz et al. 2014; Table 15.1)

#### 15.3.1.3 Flavone

Apigenin, belonging to class flavonoids and subclass flavones, is found in leaves of parsley, celery, and spinach. It employs host factor modulation as its antiviral strategy toward HCV. It effectively reduces production of host microRNA, namely, miRNA122, involved in HCV infection (Shibata et al. 2014). Baicalein, another flavone, is majorly found in roots of baical skullcap and blue skullcap. It has anti-HIV activity. It blocks the interactions of HIV-1 envelope protein gp120 with host immune cells (Li et al. 2000). Baicalein also possesses anti-dengue activity in cell cultures by not letting DENV-2 virus attach to the host cells (Moghaddam et al. 2014). It also possesses antiviral activities against Japanese encephalitis virus (JEV), acting as replication inhibitor and influenza virus (Nayak et al. 2014; Johari et al. 2012; Hassandarvish et al., 2016). Luteolin, found in leaves of basil, parsley, and spinach and pepper seeds, is an entry inhibitor of influenza A virus and severe acute respiratory syndrome coronavirus (*SARS-CoV*) (Yi et al. 2004; Yan et al. 2019)

**Table 15.1** Flavonoids: antiviral activity of class flavanol

| Class | | Plant source | Antiviral activity against | |
|---|---|---|---|---|
| Flavonol | Quercetin 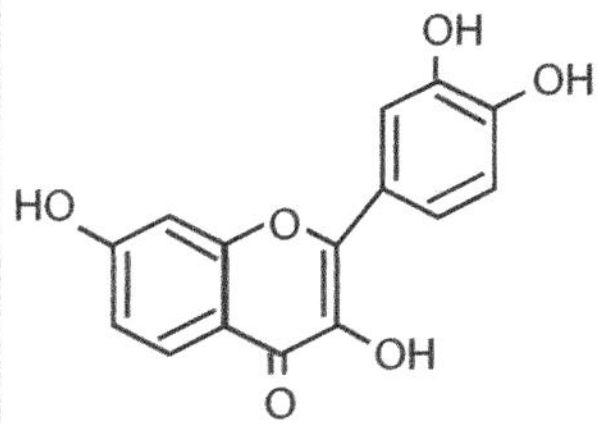 | Red grape wines, leaves of radish, fennel, pepper seeds | Rabies virus | Cutting et al. (1949) |
| | | | Vesicular stomatitis virus | Veckenstedt and Pusztai (1981) |
| | | | Tetra-O-methyl derivative—influenza virus | Roschek Jr et al. (2009) |
| | | | Rhamnoside derivative—influenza virus, hepatitis C, HSV | Choi et al. (2009), Bachmetov et al. (2012), Lee et al. (2017) |
| | | | 3-β-O-D-glucoside derivative—Ebola virus | Qiu et al. (2016) |
| | | | Rhinovirus | Veckenstedt and Pusztai (1981) |
| | | | EMCV | Gonzalez et al. (1990) |
| | | | Methyl derivative—poliovirus | Vrijsen et al. (1987) |
| | | | Trimethyl/dihydro derivative—coxsackievirus B4 | Ishitsuka et al. (1982), Galochkina et al. (2016) |
| | Rutin 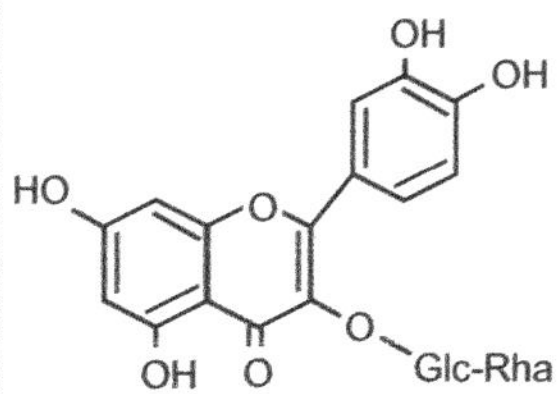 | Tartary buckwheat, rhubarb, orange, and lemon | Sulfated form of Rutin—HCV, HIV | Tao et al. (2007) |

| | | | | |
|---|---|---|---|---|
| | Fisetin 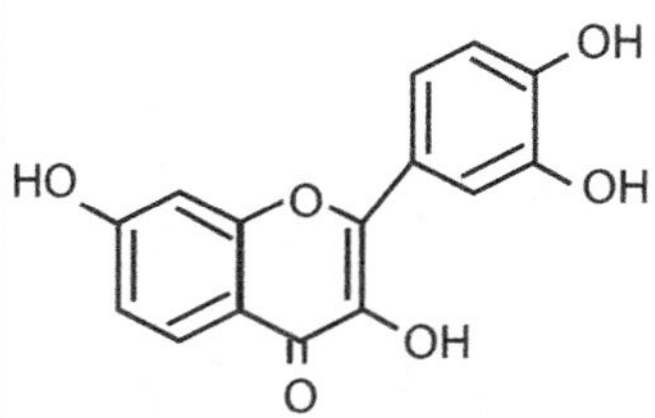 | Acacia leaves and strawberries | Chikungunya virus | Lani et al. (2016) |
| | | | Dengue virus 2 | Zandi et al. (2011) |
| | 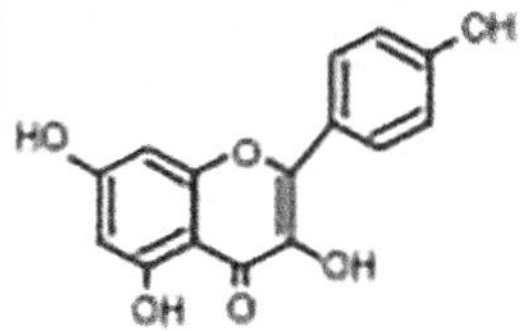 | Raspberry, capers, grapes, and black bean | Influenza virus | Jeong et al. (2009) |
| | | | Coronavirus | Schwarz et al. (2014) |

#### 15.3.1.4 Isoflavone

Isoflavones are found in *Vicia faba* (seeds of fava beans) and soybeans. Genistein, glycitein, and ononin belong to this subclass of flavonoids. Genistein is inhibitory to HIV-1 inhibition potentially working at the assembly and release of the progeny viruses from the host cells. It inhibits HIV-1 accessory protein, Vpu, to mediate this antiviral effect (Sauter et al. 2014).

#### 15.3.1.5 Flavan

Catechins and epigallocatechin gallate (EGCG) are two flavans. Catechins are found in cocoa beans, argan oil, and tea leaves. Catechins inhibit influenza A virus infection by binding to hemagglutinin and preventing virus entry (Song et al. 2005), while EGCG has anti-HIV-1 activity. EGCG binds to $CD4^+$ T cells of the host and does not let gp120 viral envelope protein to bind to CD4 receptor blocking the entry of the virus (Kawai et al. 2003). EGCG inhibits chikungunya (Weber et al. 2015), HSV (Isaacs et al. 2008), HBV (He et al. 2011a, b), EBV (Chang et al. 2003), and Zika virus (Carneiro et al. 2016; Table 15.2).

### 15.3.2 Coumarins

Coumarins belong to phenolics class of plant secondary metabolites. They are also found in bacteria and fungi. They are found predominantly in various families of plant kingdom, namely, Clusiaceae and Rutaceae (Stefanachi et al. 2018). Its structure comprises of a benzene ring fused to α-pyrene ring. Coumarins are low-molecular-weight compounds with no known side effects. They are promising drug candidates for a plethora of viral diseases (Mishra et al. 2020). Potential of coumarins as an anti-HIV agent has been explored by a number of researchers (Zhu and Jiang 2018; Laila et al. 2019; Srikrishna et al. 2018; Kudo et al. 2013). These studies open avenues for antiviral activities of coumarins against other viruses too. Coumarins were seen to possess anti-HCV activity. Coumarin derivative 6,8-diallyl-5,7-dihydroxycoumarin inhibited the HCV polymerase (Nichols et al. 2013). Coumarins and its derivatives are also potential drug candidates for HBV (Su et al. 2009), dengue (Yusufzai et al. 2018), chikungunya (Gómez-Calderón et al. 2017), and influenza virus (Wang et al. 2017a, b). Influenza virus like any other retrovirus has high mutation rate due to absence of proofreading activity in reverse transcriptase. For such viruses, redox-sensitive pathways prove to be better choice as drug target (Escuret et al. 2008). Oxidized coumarins possess anti-influenza activity and target these redox pathway proteins (Bizzarri et al. 2017). The current pandemic by retrovirus SARS-CoV-2 is also posing hindrance in vaccine development because of the same reason. Thus, we should explore more pathways which are independent of strain.

**Table 15.2** Flavonoids: antiviral activity of classes flavone, isoflavone, and flavans

| Class | | Plant source | Antiviral activity against | |
|---|---|---|---|---|
| Flavone | Apigenin 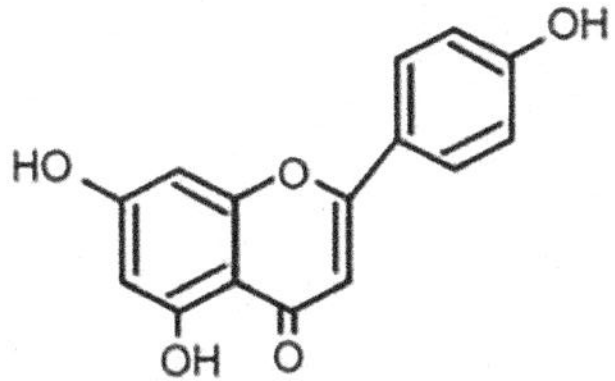 | Leaves of parsley, celery, and spinach | HCV | Shibata et al. (2014) |
| | Baicalein 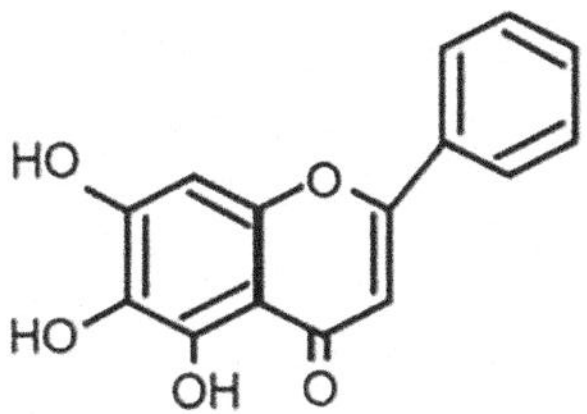 | Roots of baical skullcap and blue skullcap | HIV-1 | Li et al. (2000) |
| | | | Dengue virus | Moghaddam et al. (2014) |
| | | | Japanese encephalitis virus | Nayak et al. (2014), Johari et al. (2012), Hassandarvish (2016) |
| | Luteolin 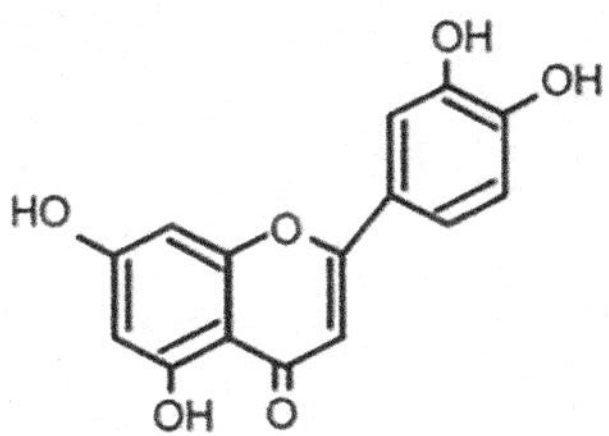 | Leaves of basil, parsley, and spinach and pepper seeds | Influenza A virus and severe acute respiratory syndrome coronavirus | Yi et al. (2004), Yan et al. (2019) |

(continued)

**Table 15.2** (continued)

| Class | | Plant source | Antiviral activity against | |
|---|---|---|---|---|
| Isoflavone | Genistein 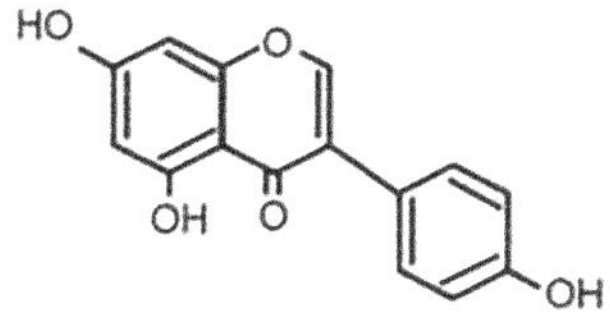  | *Vicia faba* (seeds of fava beans) and soybeans | HIV-1 | Sauter et al. (2014) |
| Flavan | Catechins 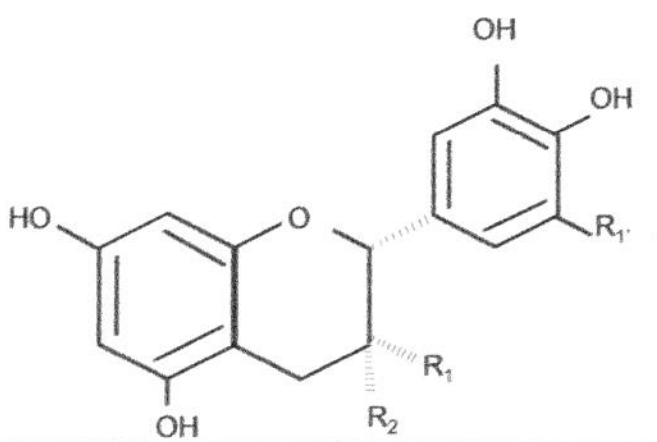  | Cocoa beans, argan oil, and tea leaves | Influenza A | Song et al. (2005) |
| | | | HIV-1 | Kawai et al. (2003) |

### 15.3.3 Tannins

Tannins are extensively used in traditional medicine since ages. They are predominantly found in all parts of the plant – its stem, root, fruits, and seeds. Many herbs contain tannins. Tannins are used in treatment of burns, gastritis, irritable bowel syndrome (IBS), etc. (Cheng et al. 2002; Ashok and Upadhyaya 2012; Jaiswal et al. 2018). Tannins have blood pressure lowering effects in hypertensive patients by inhibiting angiotensin I-converting enzyme (ACE) (Liu et al. 2003). They also have anticancer properties. They exhibit antitumor activity against prostate, breast, lung, stomach, and cervical cancers (Ascacio-Valdés et al. 2011; Yıldırım and Kutlu 2015). They have antimicrobial properties by virtue of inhibition of microbial enzymes such as peroxidase, pectinase, and xylanase. This confers protection to plants from various microbes that degrade plant cell wall (Heldt and Piechulla 2011; Ribeiro et al. 2018).

Like other plant phenolic compounds, tannins confer protective mechanism against oxidative stress created by a variety of viral infections, namely, hepatitis B and C viruses, HIV, HSV, EBV, etc. (De Marco 2013; Vilhelmova-Ilieva et al. 2019). Antioxidant activity harbored by many plant polyphenolics like flavonoids, tannins, and phenolic acids shows promising therapeutic potential against aforementioned viral infections. Tannins obtained from apple are found to be very effective against influenza virus (He et al. 2011b). Out of the four hydrolyzable tannins (HTs) (types I to IV), ellagitannins (ETs) are HTs that belong to HT types II–IV. ET geraniin has anti-HSV activity, while HT casuarinin possesses anti-HSV-2 activity (Vilhelmova-Ilieva et al. 2019). Comprehensive study of antiviral role of seven tannins was evaluated on 12 different viruses including a list of six enveloped and six non-enveloped viruses. Persimmon extracts (PE) prepared from immature green persimmon fruit showed inhibition of almost all viruses tested, namely, H3N2, H5N3, HSV, vesicular stomatitis virus, Sendai virus, Newcastle disease virus, poliovirus, coxsackie virus, adenovirus type 5, rotavirus, feline calicivirus, and mouse norovirus (Ueda et al. 2013). Such plant metabolites offer hope to rescue the world from the current coronavirus pandemic and should be tested for their plausible anti-SARS-CoV-2 potential (Fig. 15.1).

### 15.3.4 Stilbenoids

Among plant secondary metabolites, stilbenoids represent a diverse class of non-flavonoid polyphenolic compounds with varied therapeutic benefits. They are produced by plants as a protective mechanism against pathogens or stress factors (Akinwumi et al. 2018). Stilbenoids are present in plants both as constitutive and inducible metabolites suggesting their important role in providing resistance against diseases (Niesen et al. 2013). In plants, they are present as both monomers and oligomers. Resveratrol, pterostilbene, piceatannol, and oxyresveratrol are monomers and are characterized by two aromatic rings (can bear prenyl, geranyl, or farnesyl chains) linked by olefin, and can be present as glycones or glycosylated forms

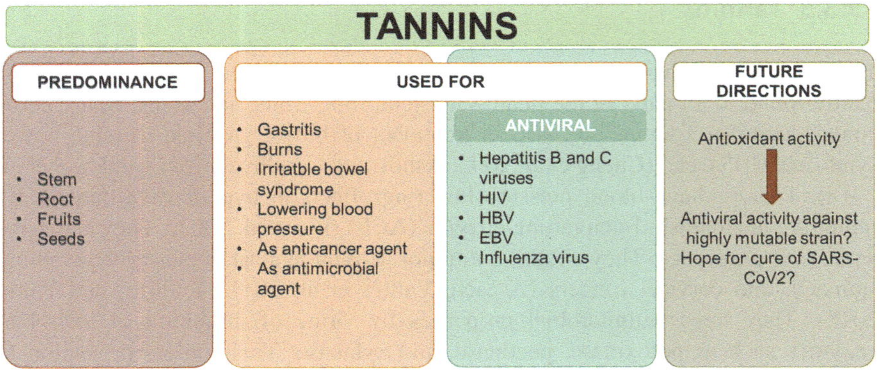


**Fig 15.1** Tannins: their predominance, usage, and future directions

(Riviere et al. 2012). Sources of stilbenoids are grapes and wine from Vitaceae family (*Vitis vinifera* L.) majorly and are also present in cocoa, blueberry, bilberry, cowberry, red berry, strawberry, and red currant (Niesen et al. 2013). Resveratrol has been shown to have antioxidant, anti-inflammatory, anticancer, antidiabetic, antimicrobial, and antiviral properties (Weiskirchen and Weiskirchen 2016; Abba et al. 2015; Yang et al. 2015). Hopeaphenol and shoreaketone are stilbenoid compounds isolated from some species of Dipterocarpaceaeous plants and shown to exhibit antiviral activity against influenza A virus (IAV, A/NWS/33 strain, H1NI subtype) (Ito et al. 2018). Z-3,5,4$^{0}$-trimethoxystilbene (Z-TMS) is a natural compound isolated from the bark of *Virola elongata* (Donald Macre and Neil Towers 1984) and has been shown to have anti-HCV activities. Z-TMS was found to be 100 times more potent than resveratrol in downregulating the levels of HCV polymerase NS5B in a dose-dependent manner (Nguyen et al. 2016). Derivatives of resveratrol have shown dose-dependent inhibition of dengue virus (DENV2) (Han et al. 2017). These compounds were shown to inhibit RNA viral synthesis in DENV2 replication and thus were suggested to effect post-fusion events of the virus.

## 15.4 Terpenes and Terpenoids

Terpenes are unsaturated hydrocarbons of more than 30,000 natural compounds with the formula $(C_5H_8)_n$ and are predominantly produced by plants, namely, conifers (Davis and Croteau 2000). Terpenes belong to the largest category of secondary metabolites and consist of five carbon isoprene units with simple hydrocarbons. Terpenoids are natural compounds formed by addition of various functional groups onto terpenes. Based on the number of carbons, terpenes are further classified into monoterpenes ($C_{10}$), sesquiterpenes ($C_{15}$), diterpenes ($C_{20}$), etc. Terpenoids have a significant role in therapeutics, pharmaceuticals, and herbal treatments and in the area of medicine (Jaeger and Cuny 2016; Perveen and AI-Taweel 2018). Medicinal plant *Marrubium vulgare* grows in the wild and is commonly known as "horehound" belonging to the Lamiaceae family (Masoodi et al. 2008). Methanolic crude extract and hexane fraction of *M. vulgare* showed presence of terpenes with anti-viral activity HSV-1 (Fayyad et al. 2014). They found that terpenes could disrupt the attachment of HSV to the cell membrane and interfere with the viral replication, thus decreasing the viral yield. Dammarenolic acid, a secodammarane triterpenoid found in *Aglaia* sp., showed potent antiviral activity against HIV-1 (Esimone et al. 2010). Triterpenoid was shown to inhibit cell cycle stages between S and G2/M phase, thereby inhibiting the viral replication and dissemination. Also dammarenolic acid potently inhibited in vitro replication of other retroviruses such as murine leukemic virus and simian immunodeficiency virus in vector-based antiviral screening studies. Loliolide, a monoterpene found in extracts of *Phyllanthus urinaria* plant, showed antiviral activity against HCV (Chung et al. 2016). The presence of loliolide inhibited the viral attachment to the host membrane and impeded viral entry/fusion, with minimal effect on replication/translation and particle production. Monoterpenes beta-pinene and limonene were shown to reduce viral infectivity of HSV by 100% (Astani and Schnitzler 2014). These terpenes were shown to inhibit HSV before adsorption and not after entry of the virus into the cell. Another monoterpene, isoborneol, also has shown antiviral activity against HSV-1 and inhibited glycosylation of the viral proteins (Armaka et al. 1999). Diterpenes obtained from buds of *Wikstroemia chamaedaphne* showed antiviral activity against hepatitis B virus (Shi-Fei et al. 2018). Triterpenoid betulinic acid and dihydroquinopimaric acid derivatives showed antiviral effect against HSV-I and HIV-I (Kazakova et al. 2018). Structural modifications at C3 and C28 of betulinic acid enhanced the antiviral activity against HSV-I and HIV-1. Acylated forms of betulinic acid showed antiviral activity against influenza virus. Also modifications of C4 position of dihydroquinopimaric acid enhanced the antiviral activity against H7N1. Anti-HIV activity was also seen with the acylated forms of betulinic acid and shown to effect later stages of HIV replicative cycle, based on presence of substituted functional group at C3 (Flekhter et al. 2002; Table 15.3).

## 15.5 Carotenoids

This class can be classified into two groups: (1) those compounds having single long carbon chain known as carotenes and (2) those compounds having oxygen atoms in its structure known as xanthophylls. All carotenoids are produced from 8 isoprene molecules and contain 40 carbon atoms, i.e., are tetraterpenoids. Carotenoids can contain substituted beta-ionone rings that have vitamin A activity and the ones with non-provitamin A activity. Carotenoids have been found to have antioxidant, anti-inflammatory, antibacterial, antiviral, and anticancer activities (Takshma and Kirtan 2020). Recent discovery has shown that non-provitamin carotenoids, namely, phytoene and phytofluene, are highly effective in reducing the number of virus particles of influenza virus (Kalo et al. 2017). Also both carotenoids have been shown to reduce the expression of genes essential for the entry of virus into host cells as well as genes encoding motor proteins that facilitate the virus movement within

**Table 15.3** Terpenes and terpenoids: general information, classification, their role, and antiviral activity

| Terpenes and terpenoids | | | |
|---|---|---|---|
| **General information** | | | |
| Unsaturated hydrocarbon $(C_5H_8)_n$ | Terpenes consist of five carbon isoprene units with simple hydrocarbons. Terpenoids are natural compounds formed by addition of various functional groups into terpenes | | Found in conifers |
| **Classification** | | | |
| Monoterpene (C10) | Sesquiterpenes (C15) | | Diterpenes (C20) |
| **Play role in** | | | |
| Therapeutics | Herbal treatments | | |
| **Plant species/ terpene/terpenoid** | **Antiviral activity for** | **Mechanism of action** | **Reference** |
| *Marrubium vulgare* | HSV-1 | Disrupt attachment of HSV-1 to host cell membrane | Fayyad et al. (2014) |
| Algae sp. Dammarenolic acid | HIV-1 | Inhibition of cell cycle stages between S and G2/M phase inhibiting viral replication | Esimone et al. (2010) |
| *Phyllanthus urinaria* Loliolide | HCV | Inhibited the viral attachments to the host membrane, impeded viral entry/fusion | Chung et al. (2016) |
| Monoterpenes beta-pinene and limonene | HSV | Viral adsorption | Astani and Schnitzler (2014) |
| Monoterpene isoborneol | HSV-1 | Inhibited glycosylation of the viral proteins | Armaka et al. (1999) |

the cell, reducing the synthesis of viral RNA, as well as translation and thus delaying viral replication. Crocin and crocetin are two natural carotenoids of saffron, a spice derived from the flower of *Crocus sativus* (Azam et al. 2013). Iranian saffron extract and its major ingredients crocin and picrocrocin showed cytotoxicity against HSV-I and HIV-I in vitro. Carotenoid crocin was shown to inhibit replication of HSV after entry into cells as well as suppressed HSV penetration in the target cells. Virus entry as well as replication was also shown to be inhibited by picrocrocin (Sepehr et al. 2018).

## 15.6 Conclusion and Future Perspectives

This chapter gives an insight into the role of plant secondary metabolites in combating various viral infections. Therapeutic activity of BBR, a member of alkaloid group, is very promising as a candidate that can be successfully applied for antiviral pharmacological strategies, and plausibly against the current SARS-CoV-2 viral infection affecting globally. These phytochemicals, in addition to having an effect on different stages of viral replication, also improve the disease with inhibition of various pro-inflammatory cytokines. Alkaloid berberine has been shown to inhibit the production of TNF-α and prostaglandin E2 (PGE2) in cells infected with influenza virus (H1N1) (Cecil et al. 2011). Structural modifications of a number of plant secondary metabolites like terpenoids and carotenoids have shown to improve antiviral properties against viruses such as influenza virus, HSV-I, and HPV-I. Combination of plant secondary metabolites with different compounds and drugs may provide potential benefits and better antiviral approach. Further evaluation will need to be done so as to minimize side effects of pharmaceutical drugs to the mankind. Mankind today is posed with some life-threatening viral diseases such as HIV and chikungunya virus, for which we still do not have the curable antiviral therapies in place. Consuming diets rich in antioxidants act as a defensive mechanism rendering protection from various diseases like atherosclerosis, cancer, etc. This is the nutraceutical potential that phenolics hold. Phenolics, namely, flavonoids and tannins, are potent antioxidants and may also show tremendous antioxidant activity in rescuing the virus-infected cells from oxidative stress inflicted by the virus. This may weaken the effect of the virus rescuing the host cell. This pathway is essentially important for highly mutable viruses that change one strain to another quickly. Influenza virus and SARS-CoV-2 belong to this class of viruses. Exploration of the antioxidant activity of phenolic group of the plant secondary metabolites is still in its infancy. It appears to hold tremendous antiviral potential especially for highly mutable strain. Does this hold the cure of the current Covid-19 pandemic?

This chapter has revealed a rich source of medicinal plants and their mechanism of action to inhibit viral progression. Further extensive research is required to explore biodiverse regions to isolate still unexplored plant metabolites with potent antiviral potential.

## References

Abba Y, Hassim H, Hamzah H, Noordin MM (2015) Antiviral activity of resvera-trol against human and animal viruses. Adv Virol 2015:184241. https://doi.org/10.1155/2015/184241

Akinwumi BC, Bordun KAM, Anderson HD (2018) Biological activities of stilbenoids. Int J Mol Sci 19:1–25. https://doi.org/10.3390/ijms19030792

Anna L, Beatrice M, Lorenzo M, Giorgio P, Giorgio G, Arianna L (2019) The isoquinoline alkaloid berberine inhibits human cytomegalovirus replication by interfering with the viral immediate early-2 (IE2) protein transactivating activity. Antiviral Res 64:52–60

Armaka M, Papanikolaou E, Sivropoulou A, Arsenakis M (1999) Antiviral properties of isoborneol, a potent inhibitor of herpes simplex virus type 1. Antiviral Res 43:79–92

Ascacio-Valdés JA, Buenrostro-Figueroa JJ, Aguilera-Carbo A, Prado-Barragán A, Rodríguez-Herrera R, Aguilar CN (2011) Ellagitannins: biosynthesis, biodegradation and biological properties. J Med Plants Res 5(19):4696–4703

Ashok PK, Upadhyaya K (2012) Tannins are astringent. J Pharmacognosy Phytochem 1(3):45–50

Astani A, Schnitzler P (2014) Antiviral activity of monoterpenes beta-pinene and limonene against herpes simplex virus in vitro. Iran J Microbiol 6(3):149–155

Azam B, Afshin K, Zahra Bathaie S (2013) Saffron and natural carotenoids: biochemical activities and anti-tumor effects. Biochim Biophys Acta 1845:20–30

Bachmetov L, Gal-Tanamy M, Shapira A, Vorobeychik M, Giterman-Galam T, Sathiyamoorthy P, Zemel R (2012) Suppression of hepatitis C virus by the flavonoid quercetin is mediated by inhibition of NS3 protease activity. J Viral Hepatitis 19(2):e81–e88

Biswas D, Nandy S, Mukherjee A, Pandey DK, Dey A (2020) Moringa oleifera Lam and derived phytochemicals as promising antiviral agents: a review. South Afr J Bot 129:272–282. https://doi.org/10.1016/j.sajb.2019.07.049

Bizzarri BM, Botta L, Capecchi E, Celestino I, Checconi P, Palamara AT, Saladino R (2017) Regioselective IBX-mediated synthesis of coumarin derivatives with antioxidant and anti-influenza activities. J Nat Prod 80(12):3247–3254

Caposio P, Luganini A, Hahn G, Landolfo S, Gribaudo G (2007) Activation of the virus-induced IKK/NF-kappaB signalling axis is critical for the replication of human cytomegalovirus in quiescent cells. Cell Microbiol 9:2040–2054

Caposio P, Luganini A, Bronzini M, Landolfo S, Gribaudo G (2010) The Elk-1 and serum response factor binding sites in the major immediate-early promoter of human cytomegalovirus are required for efficient viral replication in quiescent cells and compensate for inactivation of the NF-κB sites inproliferating cells. J Virol 84:4481–4493

Carletti G, Nervo G, Cattivelli L (2014) Flavonoids and melanins: a common strategy across two kingdoms. Int J Biol Sci 10(10):1159

Carneiro BM, Batista MN, Braga ACS, Nogueira ML, Rahal P (2016) The green tea molecule EGCG inhibits Zika virus entry. Virology 496:215–218

Cecil CE, Davis JM, Cech NB, Laster SM (2011) Inhibition of H1N1 influenza A virus growth and induction of inflammatory mediators by the isoquinoline alkaloid berberine and extracts of goldenseal (Hydrastis canadensis). Int Immunopharmacol 11:1706–1714

Chang LK, Wei TT, Chiu YF, Tung CP, Chuang JY, Hung SK, Liu ST (2003) Inhibition of Epstein–Barr virus lytic cycle by (−)-epigallocatechin gallate. Biochem Biophys Res Commun 301(4):1062–1068

Chávez JH, Leal PC, Yunes RA, Nunes RJ, Barardi CR, Pinto AR, Simões CM, Zanetti CR (2006) Evaluation of antiviral activity of phenolic compounds and derivatives against rabies virus. Vet Microbiol 116:53–59

Cheng HY, Lin CC, Lin TC (2002) Antiherpes simplex virus type 2 activity of casuarinin from the bark of Terminalia arjuna Linn. Antiviral Res 55(3):447–455

Chin LW, Cheng YW, Lin SS et al (2010) Anti-herpes simplex virus effects of berberine from Coptidis rhizoma, a major component of a Chinese herbal medicine, Ching-Wei-San. Arch Virol 155:1933–1941

Choi HJ, Song JH, Park KS, Kwon DH (2009) Inhibitory effects of quercetin 3-rhamnoside on influenza A virus replication. Eur J Pharmaceut Sci 37(3–4):329–333

Chung CY, Liu CH, Burnouf T, Wang GH, Chang SP, Jassey A, Tai CJ, Tai CJ, Huang CJ, Richardson CD, Yen MH (2016) Activity-based and fraction-guided analysis of Phyllanthus urinaria identifies loliolide as a potent inhibitor of hepatitis C virus entry. Antiviral Res 130:58–68

Cui G, Qin X, Zhang Y, Gong Z, Ge B, Zang YQ (2009) Berberine differentially modulates the activities of ERK, p38 MAPK, and JNK to suppress Th17 and Th1 T cell differentiation in type 1 diabetic mice. J Biol Chem 284:28420–28429

Cutting WC, Dreisbach RH, Neff BJ (1949) Antiviral chemotherapy; flavones and related compounds. Stanford Med Bull 7(3):137

Dang TT, Onoyovwi A, Farrow SC et al (2012) Biochemical genomics for gene discovery in benzylisoquinoline alkaloid biosynthesis in opium poppy and related species. Methods Enzymol 515:231–266

Davis EM, Croteau R (2000) Cyclization enzymes in the biosynthesis of monoterpenes, sesquiterpenes, and diterpenes. Biosynthesis. Topics Curr Chem 209:53–95

De Marco F (2013) Oxidative stress and HPV carcinogenesis. Viruses 5(2):708–731

De Meritt IB, Milford LE, Yurochko AD (2004) Activation of the NF-kappaB pathway in human cytomegalovirus-infected cells is necessary for efficient transactivation of the major immediate-early promoter. J Virol 78:4498–4507

Donald Macre W, Neil Towers GH (1984) An ethnopharmacological examination of Virola elongata bark: a South American arrow poison. J Ethnopharmacol 12:75–92. https://doi.org/10.1016/0378-8741(84)90087-4

Escuret V, Frobert E, Bouscambert-Duchamp M, Sabatier M, Grog I, Valette M, Ferraris O (2008) Detection of human influenza A (H1N1) and B strains with reduced sensitivity to neuraminidase inhibitors. J Clin Virol 41(1):25–28

Esimone CO, Eck G, Nworu CS, Hoffmann D, Überla K, Proksch P (2010) Dammarenolic acid, a secodammarane triterpenoid from Aglaia sp. shows potent anti-retroviral activity in vitro. Phytomedicine 17(7):540–547

Evers DL, Chao CF, Wang X, Zhang Z, Huong SM, Huang ES (2005) Human cytomegalovirus-inhibitory flavonoids: studies on antiviral activity and mechanism of action. Antiviral Res 68: 124–134

Fayyad AG, Ibrahim N, Yaakob WA (2014) Phytochemical screening and antiviral activity of Marrubium vulgare. Malays J Microbiol 10(2):106–111

Flekhter OB, Karachurina LT, Nigmatullina LR, Sapozhnikova TA, Baltina LA, Zarudii FS, Galin FZ, Spirikhin LV, Tolstikov GA, Plyasunova OA, Pokrovskii AG (2002) Synthesis and pharmacological activity of betulin dinicotinate. Russ J Bioorg Chem 28:494–500

Galochkina AV, Anikin VB, Babkin VA, Ostrouhova LA, Zarubaev VV (2016) Virus-inhibiting activity of dihydroquercetin, a flavonoid from Larix sibirica, against coxsackievirus B4 in a model of viral pancreatitis. Arch Virol 161(4):929–938

Ghildiyal R, Prakash V, Chaudhary VK, Gupta V, Gabrani R (2020) Phytochemicals as antiviral agents: recent updates. In: Plant-derived bioactives. Springer, Singapore, pp 279–295

Gómez-Calderón C, Mesa-Castro C, Robledo S, Gómez S, Bolivar-Avila S, Diaz-Castillo F, Martínez-Gutierrez M (2017) Antiviral effect of compounds derived from the seeds of Mammea americana and Tabernaemontana cymosa on Dengue and Chikungunya virus infections. BMC Complem Altern Med 17(1):57

Gonzalez ME, Martínez-Abarca F, Carrasco L (1990) Flavonoids: potent inhibitors of poliovirus RNA synthesis. Antiviral Chem Chemother 1(3):203–209

Han YS, Penthala NR, Oliveira M, de Mesple T, Xu H, Quan Y, Crooks PA, Wainberg MA (2017) Identification of resveratrol analogs as potent anti-dengue agents using a cell-based assay. J Med Virgol 89:397–407. https://doi.org/10.1002/jmv.24660

Harborne JB (1998) Phytochemical methods a guide to modern techniques of plant analysis, 3rd edn. Springer Science & Business Media, New York

Hassandarvish P, Rothan HA, Rezaei S, Yusof R, Abubakar S, Zandi K (2016) In silico study on baicalein and baicalin as inhibitors of dengue virus replication. RSC Adv 6(37):31235–31247

Hayashi K, Minoda K, Nagaoka Y et al (2007) Antiviral activity of berberine and related compounds against human cytomegalovirus. Bioorg Med Chem Lett 17:1562–1564

He RR, Wang M, Wang CZ, Chen BT, Lu CN, Yao XS, Kurihara H (2011a) Protective effect of apple polyphenols against stress-provoked influenza viral infection in restraint mice. J Agric Food Chem 59(8):3730–3737

He W, Li LX, Liao QJ, Liu CL, Chen XL (2011b) Epigallocatechin gallate inhibits HBV DNA synthesis in a viral replication-inducible cell line. World J Gastroenterol 17(11):1507

Heldt HW, Piechulla B (2011) Phenylpropanoids comprise a multitude of plant secondary metabolites and cell wall components. Plant Biochem 4:446–447

Helenius A (2018) Virus entry: looking back and moving forward. J Mol Biol 430(13):1853–1862

Horie S, Yano S, Aimi N, Sakai S, Watanabe K (1992) Effects of hirsutine, an antihypertensive indole alkaloid from *Uncaria rhynchophylla*, on intracellular calcium in rat thoracic aorta. Life Sci 50:491–498. https://doi.org/10.1016/0024-3205(92)90388-6

Imenshahidi M, Hosseinzadeh H (2016) Berberis vulgaris and berberine: an update review. Phytother Res 30:1745–1764

Irwin KK, Renzette N, Kowalik TF, Jensen JD (2016) Antiviral drug resistance as an adaptive process. Virus Evol 2(1):vew014

Isaacs CE, Wen GY, Xu W, Jia JH, Rohan L, Corbo C, Hillier S (2008) Epigallocatechin gallate inactivates clinical isolates of herpes simplex virus. Antimicrob Agents Chemother 52(3):962–970

Ishitsuka H, Ohsawa C, Ohiwa T, Umeda I, Suhara Y (1982) Antipicornavirus flavone Ro 09-0179. Antimicrob Agents Chemother 22(4):611–616

Ito T, Hayashi K, Nishiguchi H, Iinuma M (2018) Resveratrol oligomer C-glucosides and anti-viral resveratrol tetramers isolated from the stem bark of Shorea uliginosa. Phytochem Lett 28:1–7. https://doi.org/10.1016/j.phytol.2018.07.026

Jacob A, Thomas J (2019) Therapeutic potential of dietary flavonoids against viral-borne infections: a review. Drug Invent Today 11(2):469–479

Jaeger R, Cuny E (2016) Terpenoids with special pharmacological significance: a review. Nat Prod Commun 11(9):1373–1390. https://doi.org/10.1177/1934578X1601100946

Jaiswal H, Singh OJ, Chauhan A, Sahu MK, Dv SP (2018) A review on tannins. Biosci J 6:16–17

Jeong HJ, Ryu YB, Park SJ, Kim JH, Kwon HJ, Kim JH, Lee WS (2009) Neuraminidase inhibitory activities of flavonols isolated from Rhodiola rosea roots and their in vitro anti-influenza viral activities. Bioorganic Med Chem 17(19):6816–6823

Johari J, Kianmehr A, Mustafa MR, Abubakar S, Zandi K (2012) Antiviral activity of baicalein and quercetin against the Japanese encephalitis virus. Int J Mol Sci 13(12):16785–16795

Johnson RA, Huong SM, Huang ES (2000) Activation of the mitogen-activated protein kinase p38 by human cytomegalovirus infection through two distinct path- ways: a novel mechanism for activation of p38. J Virol 74:1158–1167

Kalo E, Havas F, Von Oppen-Bezalel L, Ben-Chitrit O (2017) Carotenoid compositions having antiviral activities and uses thereof International Publication Number WO 2017/029674 Al 23 February 2017

Kapoor R, Sharma B, Kanwar SS (2017) Antiviral phytochemicals: an overview. Biochem Physiol 6(2):7

Kaul TN, Middleton E Jr, Ogra PL (1985) Antiviral effect of flavonoids on human viruses. J Med Virol 15(1):71–79

Kawai K, Tsuno NH, Kitayama J, Okaji Y, Yazawa K, Asakage M, Nagawa H (2003) Epigallocatechin gallate, the main component of tea polyphenol, binds to CD4 and interferes with gp120 binding. J Allergy Clin Immunol 112(5):951–957

Kazakova OB, Smirnova IE, Baltina LA, Boreko EI, Savinova OV, Pokrovskii AG (2018) Antiviral activity of acyl derivatives of betulin and betulinic and dihydroquinopimaric acids. Russ J Bioorgan Chem 44(6):740–744

Kudo E, Taura M, Matsuda K, Shimamoto M, Kariya R, Goto H, Okada S (2013) Inhibition of HIV-1 replication by a tricyclic coumarin GUT-70 in acutely and chronically infected cells. Bioorganic Med Chem Lett 23(3):606–609

Kumar A, Ekavali, Chopra K, Mukherjee M, Pottabathini R, Dhull DK (2015) Current knowledge and pharmacological profile of berberine: an update. Eur J Pharmacol 761:288–297

Laila U, Akram M, Shariati MA, Hashmi AM, Akhtar N, Tahir IM, Shaheen G (2019) Role of medicinal plants in HIV/AIDS therapy. Clin Exp Pharmacol Physiol 46(12):1063–1073

Lalani S, Poh CL (2020) Flavonoids as antiviral agents for Enterovirus A71 (EV-A71). Viruses 12(2):184

Lani R, Hassandarvish P, Shu MH, Phoon WH, Chu JJH, Higgs S, Zandi K (2016) Antiviral activity of selected flavonoids against Chikungunya virus. Antiviral Res 133:50–61

Lee S, Lee HH, Shin YS, Kang H, Cho H (2017) The anti-HSV-1 effect of quercetin is dependent on the suppression of TLR-3 in Raw 264.7 cells. Arch Pharm Res 40(5):623–630

Li BQ, Fu T, Dongyan Y, Mikovits JA, Ruscetti FW, Wang JM (2000) Flavonoid baicalin inhibits HIV-1 infection at the level of viral entry. Biochem Biophys Res Commun 276(2):534–538

Li R, Liu T, Liu M, Chen F, Liu S, Yang J (2017) Anti-influenza A virus activity of dendrobine and its mechanism of action. J Agric Food Chem 65(18):3665–3674

Lillehoj H, Liu Y, Calsamiglia S, Fernandez-Miyakawa ME, Chi F, Cravens RL, Oh S, Gay CG (2018) Phytochemicals as antibiotic alternatives to promote growth and enhance host health. Vet Res 49(1):76

Liu JC, Hsu FL, Tsai JC, Chan P, Liu JYH, Thomas GN, Lin JY (2003) Antihypertensive effects of tannins isolated from traditional Chinese herbs as non-specific inhibitors of angiontensin converting enzyme. Life Sci 73(12):1543–1555

Lou C, Takahashi K, Irimura T, Saiki I, Hayakawa Y (2014) Identification of Hirsutine as an anti-metastatic phytochemical by targeting NF-kappaB activation. Int J Oncol 45:2085–2091. https://doi.org/10.3892/ijo.2014.2624

Lu JJ, Bao JL, Chen XP et al (2012a) Alkaloids isolated from natural herbs as the anticancer agents. Evid Based Complement Alternat Med 2012:485042

Lu N, Khachatoorian R, French SW (2012b) Quercetin: bioflavonoids as part of interferon-free hepatitis C therapy? Expert Rev Anti-Infect Ther 10(6):619–621

Masoodi M, Ahmed B, Zargar I, Khan S, Khan S, Singh P (2008) Antibacterial activity of whole plant extract of Marrubium vulgare. Afr J Biotechnol 7:86–87

Mercorelli B, Luganini A, Nannetti G, Tabarrini O, Palù G, Gribaudo G, Loregian A (2016) Drug repurposing approach identifies inhibitors of the prototypic viral transcription factor IE2 that block human cytomegalovirus replication. Cell Chem Biol 23:340–351

Mishra S, Pandey A, Manvati S (2020) Coumarin: an emerging antiviral agent. Heliyon 6(1): e03217

Moghaddam E, Teoh BT, Sam SS, Lani R, Hassandarvish P, Chik Z, Zandi K (2014) Baicalin, a metabolite of baicalein with antiviral activity against dengue virus. Scientific Rep 4:5452

Moradi MT, Karimi A, Lorigooini Z (2018) Alkaloids as the natural anti-influenza virus agents: a systematic review. Toxin Rev 37(1):11–18

Nayak MK, Agrawal AS, Bose S, Naskar S, Bhowmick R, Chakrabarti S, Chawla-Sarkar M (2014) Antiviral activity of baicalin against influenza virus H1N1-pdm09 is due to modulation of NS1-mediated cellular innate immune responses. J Antimicrob Chemother 69(5):1298–1310

Nguyen CB, Kotturi H, Waris G, Mohammed A, Chandrakesan P, May R, Sureban S, Weygant N, Qu D, Rao CV, Dhanasekaran DN, Bronze MS, Houchen CW, Ali N (2016) (Z)-3,5,4'-‐Trimethoxystilbene limits hepatitis C and cancer pathophysiology by blocking microtubule dynamics and cell-cycle progression. Cancer Res 76:4887–4896. https://doi.org/10.1158/0008-5472.CAN-15-2722

Nichols DB, Leão RA, Basu A, Chudayeu M, de Moraes PDF, Talele TT, Kaushik-Basu N (2013) Evaluation of coumarin and neoflavone derivatives as HCV NS 5 B polymerase inhibitors. Chem Biol Drug Des 81(5):607–614

Niesen DB, Hessler C, Seeram NP (2013) Beyond resveratrol: a review of natural stilbenoids identified from 2009-2013. J Berry Res 3:181–196. https://doi.org/10.3233/JBR-130062

Nowakowska Z (2007) A review of anti-infective and anti-inflammatory chalcones. Eur J Med Chem 42:125–137

Panche AN, Diwan AD, Chandra SR (2016) Flavonoids: an overview. J Nutr Sci 5:e47

Pandey MK, Sung B, Kunnumakkara AB, Sethi G, Chaturvedi MM, Aggarwal BB (2008) Berberine modifies cysteine 179 of IkappaBalpha kinase, suppresses nuclear factor-kappaB-regulated antiapoptotic gene products, and potentiates apoptosis. Cancer Res 68:5370–5379

Perveen S, Al-Taweel A (2018) Introductory chapter: terpenes and terpenoids. In: Perveen S, Al-Taweel A (eds) Terpenes and terpenoids. IntechOpen, London

Qiu X, Kroeker A, He S, Kozak R, Audet J, Mbikay M, Chrétien M (2016) Prophylactic efficacy of quercetin 3-β-Od-glucoside against Ebola virus infection. Antimicrob Agents Chemother 60(9):5182–5188

Ribeiro M, Simões LC, Simões M (2018) Biocides. In: Cotter P (ed) Reference module in life sciences. Elsevier. https://doi.org/10.1016/B978-0-12-809633-8.12118-1

Riviere C, Pawlus AD, Merillon JM (2012) Natural stilbenoids: distribution in the plant kingdom and chemotaxonomic interest in Vitaceae. Nat Prod Rep 29:1317–1333. https://doi.org/10.1039/c2np20049j

Robinson CL, Chong ACN, Ashbrook AW, Jeng G, Jin J, Chen H, Tang EI, Martin LA, Kim RS, Kenyon RM, Do E, Luna JM, Saeed M, Zeltser L, Ralph H, Dudley VL, Goldstein M, Rice CM, Cheng CY, Seandel M, Chen S (2018) Male germ cells support long-term propagation of Zika virus. Nat Commun 9:2090

Roschek B Jr, Fink RC, McMichael MD, Li D, Alberte RS (2009) Elderberry flavonoids bind to and prevent H1N1 infection in vitro. Phytochemistry 70(10):1255–1261

Saksham G, Arpit R (2020) In silico analysis of selected alkaloids against main protease ($M^{pro}$) of SARS-CoV-2. Chem-Biol Interact 332:109309

Sauter D, Schwarz S, Wang K, Zhang R, Sun B, Schwarz W (2014) Genistein as antiviral drug against HIV ion channel. Planta Med 80(08/09):682–687

Schwarz S, Sauter D, Wang K, Zhang R, Sun B, Karioti A, Schwarz W (2014) Kaempferol derivatives as antiviral drugs against the 3a channel protein of coronavirus. Planta Med 80(02–03):177

Sepehr S, Rezvan Z, Sepideh S, Azam B (2018) Antiviral effects of saffron and its major ingredients. Curr Drug Deliv 15(5):698–704

Serkedjieva J, Velcheva M (2003) In vitro anti-influenza virus activity of the pavine alkaloid (-)-thalimonine isolated from Thalictrum simplex L. Antivir Chem Chemother 14:75–80

Shen N, Li CN, Huan Y et al (2010) Advances of the mechanism study on berberine in the control of blood glucose and lipid as well as metabolism disorders. Acta Pharm Sin 45(6):699–704

Shi QI, Hui SU, Zhang AH, Hong-Ying XU, Guang-Li YA, Ying HA, Xi-Jun WA (2014) Natural alkaloids: basic aspects, biological roles, and future perspectives. Chin J Nat Med 12(6):401–406

Shibata C, Ohno M, Otsuka M, Kishikawa T, Goto K, Muroyama R, Koike K (2014) The flavonoid apigenin inhibits hepatitis C virus replication by decreasing mature microRNA122 levels. Virology 462:42–48

Shi-Fei L, Ying-Ying J, Zhi-Qiang Z, Jian-Bin C, Jie J, Xun-Long S, Li-Wei Z (2018) Diterpenes from buds of *Wikstroemia chamaedaphne* showing antihepatitis B virus activities. Phytochemistry 151:17–25. https://doi.org/10.1016/j.phytochem.2018.01.021

Shin H-B, Choi M-S, Yi C-M, Lee J, Kim N-J, Inn K-S (2015) Inhibition of re spiratory syncytial virus replication and virus-induced p38 kinase activity by ber berine. Int Immunopharmacol 27: 65–68

Shivanthan MC, Rajapakse S (2014) Dengue and calcium. Int J Crit Illn Inj Sci 4:314–316. https://doi.org/10.4103/2229-5151.147538

Song JM, Lee KH, Seong BL (2005) Antiviral effect of catechins in green tea on influenza virus. Antiviral Res 68(2):66–74

Song S, Qiu M, Chu Y, Chen D, Wang X, Su A, Wu Z (2014) Downregulation of cellular c-Jun N-terminal protein kinase and NF-kappaB activation by berberine may result in inhibition of herpes simplex virus replication. Antimicrob Agents Chemother 58:5068–5078

Srikrishna D, Godugu C, Dubey PK (2018) A review on pharmacological properties of coumarins. Mini Rev Med Chem 18(2):113–141

Stefanachi A, Leonetti F, Pisani L, Catto M, Carotti A (2018) Coumarin: a natural, privileged and versatile scaffold for bioactive compounds. Molecules 23(2):250

Su CR, Yeh SF, Liu CM, Damu AG, Kuo TH, Chiang PC, Wu TS (2009) Anti-HBV and cytotoxic activities of pyranocoumarin derivatives. Bioorganic Med Chem 17(16):6137–6143

Subudhi B, Chattopadhyay S, Mishra P, Kumar A (2018) Current strategies for inhibition of chikungunya infection. Viruses 10(5):235

Sumithira P, Mangala SD, Sophie AM, Latha CP (2012) Antiviral and antioxidant activities of two medicinal plants. Int J Curr Sci:256–261

Takayama H, Iimura Y, Kitajima M, Aimi N, Konno K, Inoue H et al (1997) Discovery of anti-influenza a virus activity of a corynanthe-type indole alkaloid, hirsutine, in vitro and the structure-activity relationship of natural and synthetic analogs. Bioorg Med Chem Lett 7: 3145–3148. https://doi.org/10.1016/S0960-894X(97)10154-8

Takayuki H, Fumihiro K, Shigeru T, Kazufumi T, Masahito U, Tomohiko T, Tomoyuki M (2017) Hirsutine, an indole alkaloid of uncaria rhynchophylla, inhibits late step in dengue virus lifecycle. Front Microbiol 8:1674

Takshma B, Kirtan P (2020) Carotenoids: potent to prevent diseases review. Nat Prod Bioprospect 10:109–117

Tao J, Hu Q, Yang J, Li R, Li X, Lu C, Ben K (2007) In vitro anti-HIV and-HSV activity and safety of sodium rutin sulfate as a microbicide candidate. Antiviral Res 75(3):227–233

Ueda K, Kawabata R, Irie T, Nakai Y, Tohya Y, Sakaguchi T (2013) Inactivation of pathogenic viruses by plant-derived tannins: strong effects of extracts from persimmon (Diospyros kaki) on a broad range of viruses. PLoS One 8(1):e55343

Umarov KS, Telezhenetskaya MV, Ismailov ZF et al (1970) Alkaloids of Thalictrum simplex. Chem Nat Compounds 6:219–220

Varadinova TL, Shishkov SA, Ivanovska ND et al (1996) Antiviral and immunological activity of a new pavine alkaloid (-)-thalimonine isolated from Thalictrum simplex. Phytother Res 10:414–417

Varghese FS, Kaukinen P, Glasker S, Bespalov M, Hanski L, Wennerberg K, Kummerer BM, Ahola T (2016a) Discovery of berberine, abamectin and ivermectin as antivirals against chikungunya and other alphaviruses. Antivir Res 126:117–124

Varghese FS, Thaa B, Amrun SN, Simarmata D, Rausalu K, Nyman TA, Merits A, McInerney GM, Ng LFP, Ahola T (2016b) The antiviral alkaloid berberine reduces chikungunya virus-induced mitogen-activated protein kinase signaling. J Virol 90:9743–9757

Veckenstedt A, Pusztai R (1981) Mechanism of antiviral action of quercetin against cardiovirus infection in mice. Antiviral Res 1(4):249–261

Velcheva MP, Petrova RR, Danghaaghiin S et al (1992) The structure of (-)-thalimonine. J Nat Prod 55:679–680

Vilhelmova-Ilieva N, Galabov AS, Mileva M (2019) Tannins as antiviral agents. In: Tannins-structural properties, biological properties and current knowledge. IntechOpen

Vrijsen R, Everaert L, Van Hoof LM, Vlietinck AJ, Berghe DV, Boeye A (1987) The poliovirus-induced shut-off of cellular protein synthesis persists in the presence of 3-methylquercetin, a flavonoid which blocks viral protein and RNA synthesis. Antiviral Res 7(1):35–42

Wang H, Li K, Ma L, Wu S, Hu J, Yan H, Jiang J, Li Y (2017a) Berberine inhibits enterovirus 71 replication by downregulating the MEK/ERK signaling pathway and autophagy. Virol J 14:2

Wang Y, Yan W, Chen Q, Huang W, Yang Z, Li X, Wang X (2017b) Inhibition viral RNP and anti-inflammatory activity of coumarins against influenza virus. Biomed Pharmacother 87:583–588

Weber C, Sliva K, von Rhein C, Kümmerer BM, Schnierle BS (2015) The green tea catechin, epigallocatechin gallate inhibits chikungunya virus infection. Antiviral Res 113:1–3

Weiskirchen S, Weiskirchen R (2016) Resveratrol: how much wine do you have to drink to stay healthy? Adv Nutr An Int Rev J 7:706–718. https://doi.org/10.3945/an.115.011627

Wu LX, Gu XF, Zhu YC, Zhu YZ (2011) Protective effects of novel single compound, hirsutine on hypoxic neonatal rat cardiomyocytes. Eur J Pharmacol 650:290–297. https://doi.org/10.1016/j.ejphar.2010.09.057

Yan H, Ma L, Wang H, Wu S, Huang H, Gu Z, Li Y (2019) Luteolin decreases the yield of influenza A virus in vitro by interfering with the coat protein I complex expression. J Nat Med 73(3):487–496

Yang T, Shungang L, Zhang X, Pang X, Lin Q, Cao J (2015) Resveratrol, sirtuins, and viruses. Rev Med Virol 25:431–445. https://doi.org/10.1002/rmv.1858

Yi L, Li Z, Yuan K, Qu X, Chen J, Wang G, Chen L (2004) Small molecules blocking the entry of severe acute respiratory syndrome coronavirus into host cells. J Virol 78(20):11334–11339

Yıldırım I, Kutlu T (2015) Anticancer agents: saponin and tannin. Int J Biol Chem 9:332–340

Yusufzai SK, Osman H, Khan MS, Abd Razik BM, Ezzat MO, Mohamad S, Parumasivam T (2018) 4-Thiazolidinone coumarin derivatives as two-component NS2B/NS3 DENV flavivirus serine protease inhibitors: synthesis, molecular docking, biological evaluation and structure–activity relationship studies. Chem Central J 12(1):69

Zakaryan H, Arabyan E, Oo A, Zandi K (2017) Flavonoids: promising natural compounds against viral infections. Arch Virol 62(9):2539–2551

Zandi K, Teoh BT, Sam SS, Wong PF, Mustafa MR, AbuBakar S (2011) Antiviral activity of four types of bioflavonoid against dengue virus type-2. Virol J 8(1):1–11

Zhang G, Zhang B, Zhang X, Bing F (2013) Homonojirimycin, an alkaloid from dayflower inhibits the growth of influenza A virus in vitro. Acta Virol 57(1):85–86

Zhu JJ, Jiang JG (2018) Pharmacological and nutritional effects of natural coumarins and their structure–activity relationships. Mol Nutr Food Res 62(14):1701073